MODERN PHYSICAL CHEMISTRY

Engineering Models, Materials, and Methods with Applications

Innovations in Physical Chemistry: Monograph Series

MODERN PHYSICAL CHEMISTRY

Engineering Models, Materials, and Methods with Applications

Edited by

Reza Haghi, PhD

Emili Besalú, PhD

Maciej Jaroszewski, PhD

Sabu Thomas, PhD

Praveen K. M.

AAP | APPLE ACADEMIC PRESS

Apple Academic Press Inc.
3333 Mistwell Crescent
Oakville, ON L6L 0A2 Canada

Apple Academic Press Inc.
9 Spinnaker Way
Waretown, NJ 08758 USA

International Standard Book Number-13: 978-1-77188-643-7 (Hardcover)
International Standard Book Number-13: 978-1-315-14311-8 (eBook)

Library and Archives Canada Cataloguing in Publication

Modern physical chemistry : engineering models, materials, and methods with applications / edited by Reza Haghi, PhD, Emili Besalú, PhD, Maciej Jaroszewski, PhD, Sabu Thomas, PhD, Praveen K.M.

(Innovations in physical chemistry : monograph series)
Includes bibliographical references and index.
Issued in print and electronic formats.
ISBN 978-1-77188-643-7 (hardcover).--ISBN 978-1-315-14311-8 (PDF)

1. Chemistry, Physical and theoretical. 2. Chemical engineering.
I. Haghi, Reza K., editor II. Series: Innovations in physical chemistry.
Monograph series

QD453.3.M63 2018 541 C2018-902794-0 C2018-902795-9

Library of Congress Cataloging-in-Publication Data

Names: Haghi, Reza K., editor. | Besalú, Emili, editor. | Jaroszewski, Maciej, editor. | Thomas, Sabu, editor. | M., Praveen K., editor. | Apple Academic Press.
Title: Modern physical chemistry : engineering models, materials, and methods with applications / edited by Reza Haghi, PhD, Emili Besalu, PhD, Maciej Jaroszewski, PhD, Sabu Thomas, PhD, Praveen K.M.
Other titles: Modern physical chemistry (Apple Academic Press)
Description: First edition. | Toronto ; Waretown, NJ, USA : Apple Academic Press, 2018. | Series: Innovations in physical chemistry | Includes bibliographical references and index.
Identifiers: LCCN 2018022214 (print) | LCCN 2018022973 (ebook) | ISBN 9781315143118 (ebook) | ISBN 9781771886437 (hardcover : alk. paper)
Subjects: LCSH: Chemical engineering. | Chemistry, Physical and theoretical.
Classification: LCC TP145 (ebook) | LCC TP145 .M536 2018 (print) | DDC 660--dc23
LC record available at https://lccn.loc.gov/2018022214

Apple Academic Press also publishes its books in a variety of electronic formats. Some content that appears in print may not be available in electronic format. For information about Apple Academic Press products, visit our website at **www.appleacademicpress.com** and the CRC Press website at **www.crcpress.com**

ABOUT THE EDITORS

Reza Haghi, PhD

Reza Haghi, PhD, is a research assistant at the Institute of Petroleum Engineering at Heriot-Watt University, Edinburgh, Scotland, United Kingdom. Dr. Haghi has published several papers in international peer-reviewed scientific journals and has published several papers in conference proceedings, technical reports, and lecture notes. He is expert in the development and application of spectroscopy techniques for monitoring hydrate and corrosion risks and developed techniques for early detection of gas hydrate risks. He conducted integrated experimental modeling in his studies and extended his research to monitoring system to pH and risk of corrosion. During his PhD work at Heriot-Watt University, he developed various novel flow assurance techniques based on spectroscopy, as well as designed and operated test equipment. He received his MSc in advanced control systems from the University of Salford, Manchester, England, United Kingdom. E-mail: RKHaghi@gmail.com

Emili Besalú, PhD

Emili Besalú, PhD is a lecturer in physical chemistry at the University of Girona, Spain. He has contributed more than 120 international papers and book chapters on theoretical chemistry, mainly devoted to methodologies in SAR and QSAR fields. He is the referee for various journals. His preliminary interests were related to molecular quantum similarity, perturbation methods, and multilinear regression. His interests today are focused on the treatment and ranking of congeneric molecular database families and especially the interplay between statistically based and computational procedures. E-mail: emili.besalu@udg.edu

Maciej Jaroszewski, PhD

Maciej Jaroszewski, PhD, is an assistant professor and head of the High Voltage Laboratory at Wroclaw University of Technology in Wroclaw, Poland. He received his MS and PhD degrees in high-voltage (HV) engineering from the same university in 1993 and 1999, respectively. Dr. Jaroszewski was a contractor/prime contractor of several grants and a head of a grant project on "Degradation processes and diagnosis methods for high-voltage ZnO arresters for distribution systems" and is currently a contractor of a key project cofinanced by the foundations of the European Regional Development Foundation within the framework of the Operational Programme Innovative Economy. His current research interests include HV techniques, HV equipment diagnostics, HV test techniques, degradation of ZnO varistors, and dielectric spectroscopy. E-mail: maciej.jaroszewski@pwr.edu.pl

Sabu Thomas, PhD

Sabu Thomas, PhD, is the Pro-Vice Chancellor of Mahatma Gandhi University and Founding Director of the International and Inter University Center for Nanoscience and Nanotechnology, Mahatma Gandhi University, Kottayam, Kerala, India. He is also a full professor of polymer science and engineering at the School of Chemical Sciences of the same university. He is a fellow of many professional bodies. Professor Thomas has (co-)authored many papers in international peer-reviewed journals in the area of polymer science and nanotechnology. He has organized several international conferences. Professor Thomas's research group has specialized in many areas of polymers, which includes polymer blends, fiber-filled polymer composites, particulate-filled polymer composites and their morphological characterization, ageing and degradation, pervaporation phenomena, sorption and diffusion, interpenetrating polymer systems, recyclability and reuse of waste plastics and rubbers, elastomeric crosslinking, dual porous nanocomposite scaffolds for tissue engineering, etc. Professor Thomas's research group has extensive exchange programs with different industries and research and academic institutions all over the world and is performing world-class collaborative research in various fields. Professor Thomas's Center is equipped with various sophisticated instruments and has established state-of-the-art experimental facilities, which cater to the needs of researchers within the country and abroad.

Professor Thomas has published over 750 peer-reviewed research papers, reviews, and book chapters and has a citation count of 31,574. The H index of Prof. Thomas is 81, and he has six patents to his credit. He has delivered over 300 plenary, inaugural, and invited lectures at national/international meetings over 30 countries. He is a reviewer for many international journals. He has received MRSI, CRSI, nanotech medals for his outstanding work in nanotechnology. Recently Prof. Thomas has been conferred an Honoris Causa (DSc) by the University of South Brittany, France, and University Lorraine, Nancy, France.

Praveen K. M.

Praveen K. M. is an Assistant Professor of Mechanical Engineering at SAINTGITS College of Engineering, India. He is currently pursuing a PhD in Engineering Sciences at the University of South Brittany (Université de Bretagne Sud) – Laboratory IRDL PTR1, Research Center "Christiaan Huygens," in Lorient, France, in the area of coir-based polypropylene micro composites and nanocomposites. He has published an international article in *Applied Surface Science* (Elsevier) and has also presented poster and conference papers at national and international conferences. He also has worked with the Jozef Stefan Institute, Ljubljana, Slovenia; Mahatma Gandhi University, India; and the Technical University in Liberec, Czech Republic. His current research interests include plasma modification of polymers, poly

INNOVATIONS IN PHYSICAL CHEMISTRY: MONOGRAPH SERIES

This new book series, Innovations in Physical Chemistry: Monograph Series, offers a comprehensive collection of books on physical principles and mathematical techniques for majors, non-majors, and chemical engineers. Because there are many exciting new areas of research involving computational chemistry, nanomaterials, smart materials, high-performance materials, and applications of the recently discovered graphene, there can be no doubt that physical chemistry is a vitally important field. Physical chemistry is considered a daunting branch of chemistry— it is grounded in physics and mathematics and draws on quantum mechanics, thermodynamics, and statistical thermodynamics.

Innovations in Physical Chemistry has been carefully developed to help readers increase their confidence when using physics and mathematics to answer fundamental questions about the structure of molecules, how chemical reactions take place, and why materials behave the way they do. Modern research is featured throughout also, along with new developments in the field.

Editors-in-Chief

A. K. Haghi, PhD
Editor-in-Chief, International Journal of Chemoinformatics and Chemical Engineering and Polymers Research Journal; Member, Canadian Research and Development Center of Sciences and Cultures (CRDCSC), Montreal, Quebec, Canada
Email: AKHaghi@Yahoo.com

Lionello Pogliani, PhD
University of Valencia-Burjassot, Spain
Email: lionello.pogliani@uv.es

Ana Cristina Faria Ribeiro, PhD
Researcher, Department of Chemistry, University of Coimbra, Portugal
Email: anacfrib@ci.uc.pt

BOOKS IN THE SERIES

- High-Performance Materials and Engineered Chemistry
- Applied Physical Chemistry with Multidisciplinary Approaches
- Methodologies and Applications for Analytical and Physical Chemistry
- Physical Chemistry for Engineering and Applied Sciences: Theoretical and Methodological Implication
- Theoretical Models and Experimental Approaches in Physical Chemistry: Research Methodology and Practical Methods
- Engineering Technology and Industrial Chemistry with Applications
- Modern Physical Chemistry: Engineering Models, Materials, and Methods with Applications
- Engineering Technologies for Renewable and Recyclable Materials: Physical-Chemical Properties and Functional Aspects
- Physical Chemistry for Chemists and Chemical Engineers: Multidisciplinary Research Perspectives
- Chemical Technology and Informatics in Chemistry with Applications

CONTENTS

LIST OF CONTRIBUTORS

Ann Rose Abraham
School of Pure and Applied Physics, Mahatma Gandhi University, Kottayam, Kerala, India

Premlata Ambre
Department of Pharmaceutical Chemistry, Bombay College of Pharmacy, Mumbai 400098, India

Teuku Beuna Bardant
Department of Computer Science and Information Management, Tunghai University, Taichung, Taiwan

Andreea Irina Barzic
"Petru Poni" Institute of Macromolecular Chemistry, 41A Grigore Ghica Voda Alley, 700487 Iasi, Romania

Kottur Anver Basha
P.G. and Research Department of Chemistry, C. Abdul Hakeem College, Melvisharam, Vellore, Tamil Nadu 632509, India, Tel.: +914172266187, Fax: +914172269487
E-mail: kanverbasha@gmail.com

Sukhwinder Kaur Bhullar
Department of Mechanical Engineering, Bursa Technical University, Osmangazi Campus, Gaziakdemir Mah., Mudanya Cad. No. 4/10, 16190 Osmangazi, Bursa, Turkey

Luminita Ioana Buruiana
"Petru Poni" Institute of Macromolecular Chemistry, 41A Grigore Ghica Voda Alley, 700487 Iasi, Romania

Gloria Castellano
Departamento de Ciencias Experimentales y Matemáticas, Facultad de Veterinaria y Ciencias Experimentales, Universidad Católica de Valencia San Vicente Màrtir, Guillem de Castro-94, E-46001 València, Spain

Natarajan Chandrasekaran
Centre for Nanobiotechnology, VIT University, Vellore, Tamil Nadu, India

Chin Kang Chen
Computational Science, Research Center for Chemistry, The Indonesian Institute of Sciences, Serpong, Indonesia

Mehmet Copur
Chemical Engineering Department, Bursa Technical University, 16310 Bursa, Turkey

Evans Coutinho
Department of Pharmaceutical Chemistry, Bombay College of Pharmacy, Mumbai 400098, India

Theresa O. Egbuchunam
Department of Chemistry, Federal University of Petroleum Resources, Effurun, Delta State, Nigeria
E-mail: egbuchunam.theresa@fupre.edu.ng

Chris Ellingford
International Institute for Nanocomposites Manufacturing (IINM), WMG, University of Warwick, CV4 7AL, United Kingdom

Leu Fang-Yie
Department of Computer Science and Information Management, Tunghai University, Taichung, Taiwan

Lukasz Figiel
International Institute for Nanocomposites Manufacturing (IINM), WMG, University of Warwick, CV4 7AL, United Kingdom

Dinesh Goyal
Department of Biotechnology, Thapar University, Patiala, Punjab 147004, India

Camelia Hulubei
"Petru Poni" Institute of Macromolecular Chemistry, 41A Grigore Ghica Voda Alley, 700487 Iasi, Romania

Jissy Jacob
School of Chemical Sciences, Mahatma Gandhi University, Kottayam, Kerala, India

Nandakumar Kalarikkal
School of Pure and Applied Physics, International and Inter University Centre for Nanoscience and Nanotechnology, Mahatma Gandhi University, Kottayam, Kerala 686560, India
E-mail: nkkalarikkal@mgu.ac.in

T. Kannan
LAMP, Department of Physics, National Institute of Technology, Calicut, Kerala, India

Rajpreet Kaur
Department of Chemistry, B.B.K. D.A.V. College for Women, Amritsar, Punjab 143005, India

Poonam Khullar
Department of Chemistry, B.B.K. D.A.V. College for Women, Amritsar, Punjab 143005, India

Sravanthi Loganathan
School of Chemical Sciences, Mahatma Gandhi University, Kottayam, Kerala, India

Divya Mandial
Department of Chemistry, B.B.K. D.A.V. College for Women, Amritsar, Punjab 143005, India

Tony McNally
International Institute for Nanocomposites Manufacturing (IINM), WMG, University of Warwick, CV4 7AL, United Kingdom

Raghvendra Kumar Mishra
International and Inter University Centre for Nanoscience and Nanotechnology, Mahatma Gandhi University, Kottayam, Kerala, India. E-mail: raghvendramishra4489@gmail.com

Amitava Mukherjee
Centre for Nanobiotechnology, VIT University, Vellore, Tamil Nadu, India

Grace Obi
Department of Chemistry, Federal University of Petroleum Resources, Effurun, Delta State, Nigeria
E-mail: obi.grace@fupre.edu.ng

Felix E. Okieimen
Centre for Biomaterials Research, University of Benin, Benin City, Edo State, Nigeria
E-mail: felix.okieimen@uniben.edu

Mehtap Ozekmekci
Chemical Engineering Department, Bursa Technical University, 16310 Bursa, Turkey

Sukanchan Palit
Department of Chemical Engineering, University of Petroleum and Energy Studies, Post-Office Bidholi via Premnagar, Dehradun 248007, India E-mail: sukanchan68@gmail.com, sukanchan92@gmail.com

Joginder Singh Paneysar
Department of Pharmaceutical Chemistry, Bombay College of Pharmacy, Mumbai 400098, India

Prerna
Department of Biotechnology, Thapar University, Patiala, Punjab 147004, India

Ilangovan Pugazhenthi
P.G. and Research Department of Chemistry, C. Abdul Hakeem College, Melvisharam, Vellore, Tamil Nadu 632509, India

Andrianopsyah Mas Jaya Putra
Department of Computer Science and Information Management, Tunghai University, Taichung, Taiwan

P. Predeep
LAMP, Department of Physics, National Institute of Technology, Calicut, Kerala, India

Sakvai Mohammed Safiullah
P.G. and Research Department of Chemistry, C. Abdul Hakeem College, Melvisharam, Vellore, Tamil Nadu 632509, India

Prosanjit Saha
Dr. M. N. Dastur School of Materials Science and Engineering, Indian Institute of Engineering Science and Technology, Shibpur, Howrah 711103, India

Heru Susanto
Department of Computer Science and Information Management, Tunghai University, Taichung, Taiwan; Computational Science, Research Center for Chemistry, The Indonesian Institute of Sciences, Serpong, Indonesia E-mail: heru.susanto@lipi.go.id, susanto.net@gmail.com

Lavnaya Tandon
Department of Chemistry, B.B.K. D.A.V. College for Women, Amritsar, Punjab 143005, India

S. Thanigaivel
Centre for Nanobiotechnology, VIT University, Vellore, Tamil Nadu, India

John Thomas
Centre for Nanobiotechnology, VIT University, Vellore, Tamil Nadu, India, Tel.: +914162202876, Fax: +91 416 2243092 E-mail: john.thomas@vit.ac.in, th_john28@yahoo.co.in

Sabu Thomas
International and Inter University Centre for Nanoscience and Nanotechnology, Mahatma Gandhi University, Kottayam, Kerala, India; School of Chemical Sciences, Mahatma Gandhi University, Kottayam, Kerala, India

Francisco Torrens
Institut Universitari de Ciència Molecular, Universitat de València, P. O. Box 22085, E-46071 València, Spain

S. Vijayakumar
Centre for Nanobiotechnology, VIT University, Vellore, Tamil Nadu, India

Chaoying Wan
International Institute for Nanocomposites Manufacturing (IINM), WMG, University of Warwick, CV4 7AL, United Kingdom E-mail: Chaoying.wan@warwick.ac.uk

Senem Yetgin
Department of Food Engineering, Kastamonu University, Turkey E-mail: syetgin@kastamonu.edu.tr

Dr. Yunlong Zi
Georgia Institute of Technology, Atlanta, USA E-mail: yunlongzi@gmail.com

LIST OF ABBREVIATIONS

ABC	amphiphilic block copolymer
AMIA	American Medical Informatics Association
ATRP	atom transfer radical polymerization
BM	Brownian motion
BNC	bacterial nanocellulose
BOD	biochemical oxygen demand
BSA	bovine serum albumin
CB	carbon black
CCRD	central composite rotatable design
CD	circular dichroism
CNMs	carbon nanomaterials
CNT	carbon nanotubes
COD	chemical oxygen demand
CSOM	chronic suppurative otitis media
CuPc	copper phthalocyanine
CVD	chemical vapor deposition
DE	dielectric elastomers
DEA	dielectric elastomer actuator
DMPA	depot medroxyprogesterone acetate
DNA	deoxyribonucleic acid
DSC	differential scanning calorimetry
EAPs	electroactive polymers
EBI	European Bioinformatics Institute
EDA	ethylenediamine
EDTA	polyethylene diamine tetraacetic acid
EFB	empty fruit bunch
ENMs	electrospun nanofibrous membranes
EPD	electrophoretic deposition
EPDM	ethylene-propylene-diene monomer
FAS	fetal alcohol syndrome
FDA	Food and Drug Administration
FE	ferroelectricity
FESEM	field-emission scanning electron microscopy
FM	ferromagnetism

FTIR	Fourier-transform infrared
GC-MS	gas chromatography–mass spectrometry
GNPs	graphite nanoplatelets
GO	graphene oxide
GST	glutathione S-transferase
HAO	hafnium–aluminum oxide
HGP	Human Genome Project
HLB	hydrophile–lipophile balance
HRTEM	high-resolution transmission electron microscopy
ICT	information and communication technology
IET	industrial effluent treatment
JCPDS	Joint Committee on Powder Diffraction Standards
LCE	liquid crystalline elastomers
LCST	lower critical solution temperature
LED	light-emitting diode
LEP	liquid extrusion porosimetry
LEPw	liquid entry pressure of water
LING	lateral integrated nanogenerator
LM	light microscopy
MA	maleic anhydride
MALDI	matrix-assisted laser desorption/ionization
ME	magnetoelectric
MEF	micellar-enhanced filtration
MF	microfiltration
MIM	metal–insulator–metal
MWCO	molecular weight cutoff
NCBI	National Center for Biotechnology Information
NEC	necrotizing enterocolitis
NF	nanofiltration
NIR	near-infrared
NMR	nuclear magnetic resonance
NWFETs	nanowire field-effect transistors
PA	polyaniline
PAE	poly(arylene ether)
PAEK	poly(aryl ether ketones
PCR	polymerase chain reaction
PDB	Protein Data Bank
PDI	polydispersity index
PDMS	polydimethylsiloxane
PENG	piezoelectric nanogenerator

PET	polyethylene terephthalate
PI	polyimides
PLA	polylactic acid
PMMA	poly(methyl methacrylate)
PPG	poly(propylene glycol)
PS	polystyrene
PTFE	poly(tetrafluoroethylene)
PU	polyurethane
PVA	polyvinyl alcohol
PVDF	polyvinylidene fluoride
PVMS	polyvinylmethylsiloxane
PVP	polyvinylpyrolidone
RFLPs	restriction fragment length polymorphisms
RGO	reduced graphene oxide
RO	reverse osmosis
RSM	response surface methodology
SBS	styrene–butadiene–styrene
SEBS	styrene–ethylene–butadiene–styrene
SEM	scanning electron microscopy
SERS	surface-enhanced Raman scattering
SIM	selected ion monitoring
SNP	single nucleotide polymorphism
SOD	superoxide dismutase
SS	suspended solids
SSF	simultaneous saccharification fermentation
STRs	short tandem repeats
SWCNT	single-walled carbon nanotube
TAC	total antioxidant capacity
TC	theory of chaos
TCAA	tetrachloroauric acid
TCBS	thiosulfate-citrate-bile salts-sucrose
TEM	transmission electron microscopy
TNFC	thin-layer nanofiber composite membranes
UF	ultrafiltration
VING	vertical integrated nanogenerator
WZ	wurtzite
XPS	X-ray photoelectron spectroscopy
XRD	X-ray diffraction
ZB	zinc blende
ZFC	zero field cooling

PREFACE

Physical chemistry provides insight into the fundamental reason due to which chemical systems and materials behave the way they do and provides a coherent framework for chemical knowledge, from the molecular to the macroscopic level.

In this new reference book, fundamentals are introduced in a simple manner yet indepth manner, and general principles are induced from key experimental results.

The book features contributions from experts in this field of research and presents a step-by-step guide to the topic with a mix of theory and practice in physical chemistry that is suitable for advanced graduate levels.

Some of the main highlights of this volume are:

- It will serve as a new reference book and as an introduction to many of the more advanced topics of interest to modern researchers.
- It provides up-to-date coverage of the latest research and examines the theoretical and practical aspects of modern physical chemistry.
- It covers key concepts like chemoinformatics and computational chemistry, new nanotechniques, polymer composites, and engineered materials.
- It presents the cutting edge of research in physical chemistry.
- It is an excellent supplement for advanced research students in physical chemistry.
- It highlights some important areas of current interest in polymer products and chemical processes.
- It also focuses on topics with more advanced methods.

PART I
Chemoinformatics and Computational Chemistry

CHAPTER 1

BROWNIAN MOTION, RANDOM TRAJECTORY, DIFFUSION, FRACTALS, THEORY OF CHAOS, AND DIALECTICS

FRANCISCO TORRENS[1,*] and GLORIA CASTELLANO[2]

[1]*Institut Universitari de Ciència Molecular, Universitat de València, P. O. Box 22085, E 46071 València, Spain*

[2]*Departamento de Ciencias Experimentales y Matemáticas, Facultad de Veterinaria y Ciencias Experimentales, Universidad Católica de Valencia San Vicente Màrtir, Guillem de Castro-94, E 46001 València, Spain*

[*]*Corresponding author. E-mail: torrens@uv.es*

CONTENTS

ABSTRACT

Brownian motion (BM) describes the random motions of the microscopic particles that are subjected to the saturation bombing from the invisible molecules of water or gases. The botanist Robert Brown was the first to observe it, while he examined in his microscope study some pollen particles that floated in the water of his slide, but Albert Einstein described it in a mathematical way. BM explains how pollution disperses through air or water and describes many random processes from floods to stock market. Its unpredictable steps are related to fractals. The theory of chaos (TC) states that small changes in circumstances can have further important consequences. If one leaves home 30 s later, besides missing the bus, he perhaps loses a meeting with somebody that goes to redirect him to a new job, changing the course of his life forever. The TC is applied to an overall meteorological weather, where an eddy can burst a hurricane on the other side of the planet (*butterfly effect*). However, chaos is not chaotic in a literal meaning as it causes some patterns. A technique that provides the ability to expand timescales in optics was used to measure ultrafast, intense light pulses directly. Observations from experiments confirmed theoretical predictions made decades ago and could play a role in the prediction of high, sudden, and rare rogue waves on the surface of the oceans or the appearance of other extreme events in nature. Waves similar to rogue waves exist in optics in the form of short and intense light pulses.

1.1 INTRODUCTION

Simonyi reviewed a cultural history of physics.[1] Bensuade-Vincent and Simon revised chemistry as the impure science.[2]

In earlier publications, it was informed by the empirical didactics of molecular shape,[3] the phylogenesis of anthropoid apes,[4] the fractal analysis of the tertiary structure of proteins,[5] fractal hybrid orbitals in biopolymer chains,[6] fractals for hybrid orbitals in protein models,[7] the fractal hybrid orbitals analysis of the tertiary structure of protein molecules,[8] resonance in interacting induced-dipole polarizing force fields, application to force-field derivatives,[9] the modeling of complex multicellular systems, tumor–immune cells competition,[10] molecular diversity classification through information theory,[11] a tool for interrogation of macromolecular structure,[12] a new tool for the study of resonance in chemical education,[13] dialectic walk on science,[14] the work with nanomaterials, and reductionism/positivism philosophical and ethical considerations.[15]

1.2 BROWNIAN MOTION

Brown (19th century) examined pollen grains with the microscope and noticed that they got tangled incessantly.[16] He wondered if they would be living. Whether or not, but they were suffering the collisions of the water molecules covering the glass slide. Pollen particles moved randomly, sometimes not much, some other times much, and, gradually, turned around throughout the slide with unpredictable trajectories. Science community was amazed before his discovery, named Brownian motion (BM).

1.2.1 RANDOM TRAJECTORY

The BM occurs because pollen particles suffer a small shake every time a water molecule collides with them. Water molecules keep moving and constantly collide with each other, in such a way they regularly run into pollen, pushing it through. Although pollen grains size is hundreds of times greater than a water molecule, as pollen is hit in every time by numerous molecules, every one moving in a random direction, an unbalanced force exists making it move, which occurs time and again, so that it follows an irregular trajectory like the run of a drunkard tripping. Its path cannot be predicted in advance because water molecules randomly collide, and pollen flies off in any direction. The BM affects any tiny particle suspended in a fluid. It is noticed in greater-sized particles, for example, smoke particles floating in the air, if they are looked through a magnifying glass. Collision magnitude that the particle receives depends on molecules momentum. When fluid molecules are heavy or move fast, for example, a hot fluid, many more collisions are noticed. Mathematical operations underlying BM were developed at 19th-century end, but Einstein attracted physicists' attention in his 1905 article. He borrowed theory of heat, based on molecular collisions, in order to explain motions observed by Brown. Noticing that BM provided proofs of molecules existence in fluids, physicists accepted the atomic theory, which continued questioned until into 20th century.

1.2.2 DIFFUSION

With time, BM moves particles at a considerable distance but never so far as they advance in a straight line without finding obstacles, because it is more probable that randomness sends a particle backward while moving it forward. If one throws a group of particles at a point of some liquid, it will diffuse out although nobody moves it around or no currents exist inside the liquid. Every particle will move following its own path, and by doing that the concentrated drop spreads out in a diffused cloud, which provides results important for pollution extent from a source, for example, aerosol in the atmosphere. Although no wind exists, chemical substances will become diffused through BM.

1.2.3 FRACTALS

The trajectory followed by a particle going through BM is a fractal example. Every path step is of any size and direction, but a global pattern emerges, which contains inside a structure at all scales from the tiniest to largest outlines, which is the definition of a fractal. Mandelbrot defined fractals in the 1960/1970s as a way to quantify self-similar forms. Fractals are patterns presenting the same appearance on the same scale. If one zooms in on a small pattern fragment, it is impossible to distinguish it from the great-scale pattern, so that one cannot say which rise is simply looking at it. Repetitive patterns without scale appear frequently in nature (e.g., coastlines, trees branches, brackens leaves, and snowflake's sixfold symmetry). Fractal dimensions emerge because their length or dimension depends on the scale to which one look at it. If one measures the distance between two cities by the coastline, he says that 30 km exists between Land's End and Mount's Bay, but if he takes into account all rocks and he measures every one with a rope, he will need a rope of 100 km. If one were further and measured every grain of sand of the coastline, he would need a rope of more than 200 km. Absolute length depends on the scale at which one measures. Fractal dimensions measure the approximate character of something (e.g., cloud, tree, and mountain range). Many fractal shapes, for example, coastline, are produced by a series of phases of a random-like BM. Mathematics of BM is used to generate fractal patterns, which result in great utility in multiple science areas. One creates virtual rustic landscapes with mountains, trees, and clouds for computer games or uses them in codes to draw spatial maps that help robots to conduct themselves by rough lands, modeling crests, and fissures. Physicians find them useful for medical images formation when they need to analyze complex body parts structure, for example, lung, in which branched structures pass from a big to a tiny scale. Ideas on BM are of great utility to predict risks and future incidents, which are the end result of different random events (e.g., floods and stock market fluctuations). The

BM occurs in the configuration of other social processes (e.g., manufacture and decisions making).

1.3 THEORY OF CHAOS

The theory of chaos (TC) states that a butterfly fluttering in Brazil causes a tornado in Texas, recognizing that some systems produce different behaviors although present similar starting points, for example, weather. A minor temperature or pressure change in a place bursts a chain of events that, in turn, erupts a shower in another location. Chaos is not chaotic with the meaning of being completely uncontrolled, unpredictable, or badly structured. Chaos systems are deterministic, that is, if one knows the exact starting point, one is predictable and reproducible. Simple physics describes the series of events that develop, which is the same every time. However, if one pays attention to a final result, it is impossible to go back and determine where it came from, as several paths exist that lead to that result, which is because the differences between the conditions that caused one result and the other were tiny, even impossible to measure. Different results come from slight input changes. Because of divergence, if one is not sure of input values, the variety of subsequent behaviors is enormous. In weather, if eddy temperature differs in only a fraction of degree from that one think, then his predictions result mistaken and outcome results not a violent storm, but a light drizzle or a fierce tornado in the neighboring city. Meteorologists are limited in what advance they foresee weather. Even with the huge amounts of data on atmosphere state, supplied by satellite swarms rotating around the Earth and weather stations spread on the surface, meteorologists foresee weather patterns a few days in advance. Chaos causes enormous uncertainties.

1.3.1 DEVELOPMENT

Lorenz (1960) developed TC through a computer to develop weather models. He noticed that his code generated enormously different output meteorological patterns, because of input numbers rounding. In order to facilitate his calculations, he split the simulations into different fragments and tried to resume them halfway instead of from the first, recopying figures by hand. In his listing, numbers were rounded with three decimals but the computer memory handled six decimal numbers. When shorter form 0.123 substituted 0.123456 in the middle of the simulation, he noticed that outcome weather differed. His models were reproducible and nonrandom but differences were difficult to interpret. Why a tiny change in code produced a wonderful clear weather in one simulation and catastrophic storm in another? In detail, he noticed that resulting meteorological patterns were limited to a certain set, which he named *attractor*. It was not possible to produce any type of weather varying input but a set of meteorological patterns were favored, although it was difficult to foresee which would derive from numerical input, which is a chaos systems key feature: they follow general patterns but one cannot project a specific endpoint back to a particular initial input, because the potential paths are superimposed. Input–output connections are plotted to show the rank of behaviors a particular chaos system presents. The graphic reflects the attractor solutions (*strange attractors*), for example, Lorenz attractor, which looks like a number of overlapped "8" reminding a butterfly wings shape. The TC emerged at the same time in which fractals were discovered, with which it presents a relationship. Attractors' maps of chaotic solutions for many systems appear as fractals, in which fine attractor structure contains another structure in many scales.

1.3.2 FIRST EXAMPLES

Although computers availability made to start TC allowing mathematicians to calculate repeatedly behaviors for different inputs, at 19th-century end, simpler systems were detected showing a chaotic behavior (e.g., billiard balls trajectory and orbits stability). Hadamard studied particle motion mathematics on a curved surface (*Hadamard's billiard*), for example, ball in a golf game. On surfaces, some particles' trajectory turned unstable and fell from the edge. Some others remained on the runner but followed a variable trajectory. Poincaré discovered non-repetitive solutions for three-body orbits under gravity, for example, Earth + two moons, proving orbits instability. The three bodies rotated around each other in loops in continuous change but they did not separate. Mathematicians developed the theory of a many-body-system motion (*ergodic theory*) and applied it to turbulent fluids and electrical oscillations in radio circuits. Since 1950, TC developed at the same time the chaos systems were discovered and digital computer machines were introduced to calculate. Chaotic behavior is common. Chaos occurs in numerous many-body systems, for example, planet orbits. Neptune presents more than 12 moons, which bounce up and down following unstable orbits, which change year after year. Some scientists think the solar system will end in chaos.

1.4 PURPOSE OF DIALECTIC WALK ON SCIENCE

Sanchez-Palencia is a researcher in theoretical mechanics and applied mathematics with more than 45 years of experience. He is Research Master of Centre national de la recherche scientifique and member of the Académie des Sciences of Paris and the Board of Management of the Union Rationaliste in France. His research field is mathematics applied to the mechanics of elastic solids and fluids. He has got scientific recognition at a worldwide level in themes, for example, problems of mechanics depending on small parameters. In the 1970s, he was a pioneer of the homogenization theory.

Sanchez-Palencia has published a monograph *Dialectic Walk on Science*[17,18] astride among science spreading, epistemology, and commented history of science. However, the book shows how it is, above all, the sociology and psychology of research, the methods of production of knowledge, so far from a commonly accepted but little convincing logic, which can provide an sketch of coherence in dialectics, which is not a logic with strict laws but a general framework in which the evolutionary phenomena follow. The work explains how dynamical systems and dialectics, as mathematical and philosophical frameworks, respectively, play an important role in the explanation of the evolutionary phenomena of multiple (sometimes contradictory) and, in general, noninstantaneous causality. However, the monograph goes further than one run as it means an original revision of dialectics itself, on bases extracted from science, a description of its principles in a direct relationship to the most elementary properties of the dynamical systems theory. Precisely, this type of approach has allowed the author an innovating contribution, where he incorporates a new principle into dialectics, taken from (quite general in systems of a certain complexity) deterministic chaos phenomena.

Sanchez-Palencia includes the phenomenon of deterministic chaos in dialectics. He examines the dialectics of complex systems and presents the principle of the erratic behavior on a strange attractor. As in the age in which Engels enunciated the principles of dialectics, deterministic chaos and strange attractors were totally unknown, the author stated a new dialectical principle.

Principle of the erratic behavior of a strange attractor. In complex processes (nonlinear systems described by three or more parameters), the temporal evolution obeys combinations of behaviors described by the principles of dialectics, but it can happen (and the case is normal) that the past behavior of the system can be totally unforeseeable from the (approximate) knowledge of its initial (or present) state. From this past evolution, one knows, in general, only that it consists an erratic movement of the representative point near a certain region (the attractor) of complex structure (unknown in itself, in general); this movement is frequently constituted by several vaguely periodic elements, consisting its erratic character in the tilting, in practically unforeseeable moments, between those elements.

Another original point in Sanchez-Palencia's book is a new treatment of certain questions of game theory (particularly, the *prisoner paradox*), in which new modeling allows a view much closer to the reality and less paradoxical.

We had the pleasure to meet Sanchez-Palencia in the Institute for History of Medicine and Science—French Institute Day *Science and Complexity*,[19] where he presented a conference entitled *Dialectic Walk on Science*.

1.5 TIME MAGNIFICATION USED TO MEASURE CHAOTIC PULSES IN REAL TIME

A technique that provides the ability to expand timescales in optics was used to measure ultrafast, intense light pulses directly. Observations from experiments confirmed theoretical predictions made decades ago and could play a role in the prediction of high, sudden, and rare rogue waves on the surface of the oceans or the appearance of other extreme events in nature. Waves similar to rogue waves exist in optics in the form of short and intense light pulses. *Modulation instability* is a fundamental process of nonlinear science, leading to the unstable breakup of a constant amplitude solution of a physical system. Particular interest existed in studying modulation instability in the cubic nonlinear Schrödinger equation, a generic model for a host of nonlinear systems (e.g., superfluids, fiber optics, plasmas, and Bose–Einstein condensates). Modulation instability is a significant area of study in the context of understanding the emergence of high-amplitude events that satisfy rogue wave statistical criteria. Exploiting advances in ultrafast optical metrology, Dudley group performed real-time measurements in an optical fiber system of the unstable breakup of a continuous wave field, simultaneously characterizing emergent modulation instability breather pulses and their associated statistics.[20] Their results allowed quantitative comparison among experiment, modeling, and theory, and were expected to open perspectives on studies of instability dynamics in physics.

ACKNOWLEDGMENT

The authors acknowledge the support from Generalitat Valenciana (Project No. PROMETEO/2016/094) and Universidad Católica de Valencia San Vicente Mártir (Project No. UCV.PRO.17-18.AIV.03).

KEYWORDS

- **deterministic chaos**
- **dialectics of complex systems**
- **principle of erratic behavior on an attractor**

REFERENCES

1. Simonyi, K. *A Cultural History of Physics*; CRC: Boca Raton, FL, 2012.
2. Bensuade-Vincent, B.; Simon, J. *Chemistry: The Impure Science*; Imperial College Press: London, UK, 2012.
3. Torrens, F.; Sánchez-Pérez, E.; Sánchez-Marín, J. Didáctica empírica de la forma molecular. *Enseñanza de las Ciencias, Número Extra (III Congreso)* **1989,** (1), 267–268.
4. Torrens, F. Filogénesis de los simios antropoides. *Encuentros en la Biología* **2000,** *8*(60), 3–5.
5. Torrens, F. Análisis fractal de la estructura terciaria de las proteínas. *Encuentros en la Biología* **2000,** *8*(64), 4–6.
6. Torrens, F. Fractal Hybrid Orbitals in Biopolymer Chains. *Russ. J. Phys. Chem. (Engl. Transl.)* **2000,** *74*, 115–120.
7. Torrens, F. Fractals for Hybrid Orbitals in Protein Models. *Complexity Int.* **2001,** *8*, 1–13.
8. Torrens, F. Fractal Hybrid Orbitals Analysis of the Tertiary Structure of Protein Molecules. *Molecules* **2002,** *7*, 26–37.
9. Torrens, F.; Castellano, G. Resonance in Interacting Induced-Dipole Polarizing Force Fields: Application to Force-Field Derivatives. *Algorithms* **2009,** *2*, 437–447.
10. Torrens, F.; Castellano, G. Modelling of Complex Multicellular Systems: Tumour-Immune Cells Competition. *Chem. Cent. J.* **2009,** *3*(Suppl I), 75–1–1.
11. Torrens, F; Castellano, G. Molecular Diversity Classification via Information Theory: A Review. *ICST Trans. Complex Syst.* **2012,** *12*(10–12), e4–e1–8.
12. Torrens, F.; Castellano, G. A Tool for Interrogation of Macromolecular Structure. *J. Mater. Sci. Eng. B* **2014,** *4*(2), 55–63.
13. Torrens, F.; Castellano, G. Una nueva herramienta para el estudio de la resonancia en docencia química. *Avances en Ciencias e Ingeniería* **2014,** *5*(1), 81–91.

14. Torrens, F.; Castellano, G. Dialectic Walk on Science. In *Sensors and Molecular Recognition*; Laguarda Miro, N., Masot Peris, R., Brun Sánchez, E., Eds.; Universidad Politécnica de Valencia: València, Spain; Vol. 11, 2017, 271–275.
15. Torrens, F.; Castellano, G. El trabajo con nanomateriales: Historia Cultural, Filosofía Reduccionista/Positivista y ética. In *Tecnología, Ciencia y Sociedad;* Gherab-Martín, K.J., Ed.; Global Knowledge Academics: València, Spain, in press.
16. Baker, J. *50 Physics Ideas You Really Need to Know*; Quercus: London, UK, 2007.
17. Sanchez-Palencia, É. *Promenade Dialéctique dans les Sciences;* Hermann: Paris, France, 2012.
18. Sanchez-Palencia, É. Paseo Dialéctico por las Ciencias; Colección Difunde No. 202: Traducciones No. 8, Universidad de Cantabria: Santander, Spain, 2015.
19. Sanchez-Palencia, É. Book of Abstracts, Jornada sobre Ciencia y Complejidad–Journée Franco-Espagnole sur la Science et la Complexité, València, Spain, October 3, 2016; Institut de Història de la Medicina i de la Ciència López Piñero (Universitat de València)–Institut Français Valencia: València, Spain, 2016; O–2.
20. Närhi, M.; Wetzel, B.; Billet, C.; Toenger, S.; Sylvestre, T.; Merolla, J. M.; Morandotti, R.; Dias, F.; Genty, G.; Dudley, J. M. Real-Time Measurements of Spontaneous Breathers and Rogue Wave Events in Optical Fibre Modulation Instability. *Nat. Commun.* **2016,** *7*, 13675–1361–9.

CHAPTER 2

REVEALING INFORMATICS APPROACH AND ITS IMPACT ON PHYSICAL CHEMISTRY INNOVATION: RESPONSE SURFACE METHODOLOGY (RSM)—CHEMINFORMATICS

HERU SUSANTO[1,2,*], TEUKU BEUNA BARDANT[1], LEU FANG-YIE[1], CHIN KANG CHEN[3], and ANDRIANOPSYAH MAS JAYA PUTRA[1]

[1]*Department of Computer Science and Information Management, Tunghai University, Taichung, Taiwan*

[2]*Computational Science, Research Center for Chemistry, The Indonesian Institute of Sciences, Serpong, Indonesia*

[3]*School of Business and Economics, University of Brunei, Bandar Seri Begawan, Brunei*

**Corresponding author. E-mail: heru.susanto@lipi.go.id, susanto.net@gmail.com*

CONTENTS

ABSTRACT

Informatics research is putting a great emphasis on answering "when," "what if," and "why" questions with the support of information system and technology, and that it could be the key factor in facilitating and attaining an efficient decision-making in research. The main purpose of this study is to explore the application of information and communication technology for physical chemistry area, which technology could bring opportunities as well as challenges in modern science. With the current deluge of data, computational methods have become indispensable to physical chemistry investigations, with the advent of the World Wide Web and fast internet connections, the data contained in these databases and a considerable amount of special-purpose programs can be accessed quickly and efficiently from any location in the world. As a consequence, computer-based tools now play an increasingly significant role in the advancement and development of chemistry research. Hence, this report investigates the relationship between information system and science as well as the consequences and implication of technology in supporting medical research.

2.1 INTRODUCTION

Information system is a part of information technology. It has been well-defined in terms of two perspectives: one relating to its purpose; and the other relating to its structure. From a functional perspective, an information system is a technologically implemented medium for the purpose of recording, storing, and disseminating linguistic expressions as well as for the supporting of inference making. While from a structural perspective, an information system consists of a collection of people, processes, data, models, technology, and partly formalized language, forming a cohesive structure that serves some organizational purpose or function. However, they also can be defined as a set of interconnected components that assemble (or retrieve), process, store, and allocate information in order to support decision-making and control in an organization. In addition to supporting decision-making, coordination, and control, information systems may also aid in helping workers in analyzing problems, visualize complex subjects, and create new products.

2.2 INFORMATICS IN SCIENCE

Owing to the availability of large datasets of digital medical information, the use of informatics to improve healthcare and medical research is made possible where they provide a new trial for investigation and medical discovery. This is because informatics focuses on developing new and effective methods of using technology to process information. In today's society, informatics is being applied at every stage of healthcare from basic research to care delivery and includes many specializations such as bioinformatics, medical informatics, and biomedical informatics.

Furthermore, informatics has also had a huge impact on the field of systems biology as systems biology could use computer modeling and mathematical simulations to predict how complex biological systems would behave. National Institute of Health has claimed that researchers have created models to simulate tumor growths. By applying the computer models in the study, researchers can obtain a better and more comprehensive understanding of how diseases affect an entire biological system in addition to the effects on individual component.

However, information technology is a useful tool in support of healthcare and medical research due to the availability of large datasets of digital medical information. This is due to the development of a new trial for investigation and medical research. Informatics highlights on improving new and effective methods of using technology to process information. In today's society, informatics is being applied at all healthcare phases from elementary study to care delivery including a considerable amount of specializations such as bioinformatics, medical informatics, and biomedical informatics.

Furthermore, the prediction on how complex biological systems will behave also influenced the development of informatics as it had a huge impact on the field of systems biology, and systems biology could use computer modeling and mathematical simulation. National Institute of Health has claimed that researchers have created models to simulate tumor growths. Therefore, researchers can obtain a better and more comprehensive understanding of how diseases may affect an entire biological system in addition to the effects on individual components by applying the computer models in the study.

2.3 THE APPLICATION OF INFORMATICS

2.3.1 RESPONSE SURFACE METHODOLOGY (RSM) APPLICATION IN DEFINING OPTIMUM CONDITION OF EMPTY FRUIT BUNCH (EFB) ENZYMATIC HYDROLYSIS PROCESS

Response surface methodology (RSM) is one of the cheminformatics that explores the relationships between several explanatory variables and one or more response variables. RSM was first introduced by George. E. P Box and K. B. Wilson with the main objective of obtaining optimal response. Box and Wilson suggested a second order polynomial equation as the model template (Vander Wiel et al., 1992; Pena and Fierro, 2001). RSM was already familiar to researchers in bioprocess engineering which includes usually dealing with long-time microbial growth and required to consider many operations' variables. RSM will significantly increase the research efficiency for determining optimum condition and forecasting the results. Bioethanol production from cellulose-based material has been one of the popular bioprocess engineering research fields in Indonesia in the last 5 years. Palm plantation biomass waste became the most concerned potency for bioethanol raw material since Indonesia is the world's leading producer of palm oil. Currently, Indonesia had established ethanol production pilot plant by using palm oil empty fruit bunch (EFB) as a raw material in a fully automatic computerized system. Introducing cheminformatics approach by presenting optimum condition in the mathematical equation will give significant advantages.

The classical method of optimization involves varying one parameter at a time and keeping the other constant. The appearance of surfactant doses as the third variable in determining optimum condition makes the classical method no longer efficient. In two variables system, whether or not there were interrelations among variables, the classical method will give the same optimum condition, regardless which variable was first set as constant. In a system with three variables or more, the classical method becomes inefficient and fails to explain relationships among the variables. RSM was used in several optimizations in enzymatic reaction (Facioli and Barrera-Arellano, 2001). This methodology was recommended due to its ability to consider the multivariable problem. In this study, ester sorbitan (a mixture of 1,4-anhydrosorbitol, 1,5-anhydrosorbitol, and 1,4,3,6-dianhydrosorbitol) or also known as Span 85 was chosen as commercial surfactant based on one-pointtest as shown in Figure 2.1. Span 85 have hydrophile-lipophile balance value only 1.8. These results support the application of Span 85 as a nonionic surfactant in cellulose hydrolysis. The concerned hydrolysis process is the one that was followed by fermentation without any treatment for removing or reducing the effect of surfactant and cellulase prior fermentation. In the hydrolysis system by using 10 FPU/g-substrates, addition 1%v. of Span 85 could increase the cellulose conversion from 38.55% up to 87.30% and clearly more effective than Tween 20. This result is comparable with the hydrolysis system which used 15 FPU/g substrates. Thus, based on one-point test, surfactant additions can save enzymes consumptions by 33%. Since the enzyme cost is still a big part of overall production cost, optimization of surfactant addition in cellulose hydrolysis reaction will be observed by using RSM method with the main objective to save more enzyme.

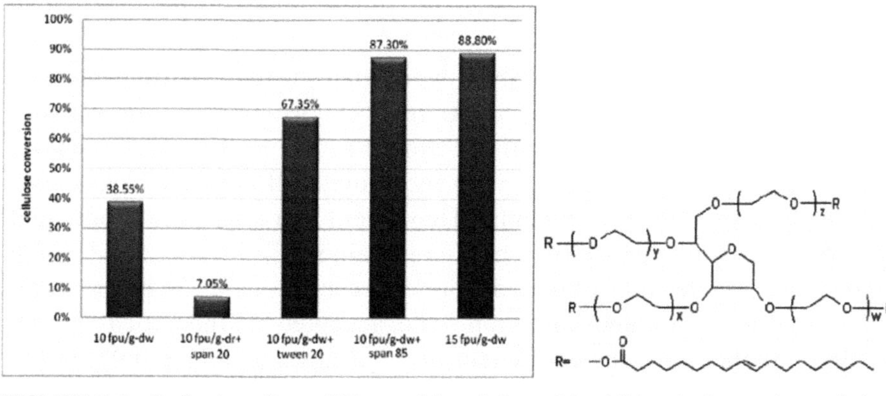

FIGURE 2.1 Left: the effect of Tween 20 and Span 85 addition in increasing cellulose conversion of enzymatic hydrolysis reaction. Right: the molecular structure of Span 85, surfactant that was used in this study.

RSM approach was started by creating a three-level-three-factor central composite rotatable design (CCRD). The variable and their levels selected for the cellulose hydrolysis were Span 85 concentration (0–2%v.), cellulase concentration (10–15 FPU/g substratedw), and substrate loading (20–30%dw). Substrate was palm oil EFB pulp made from EFB from Malimping Indonesia by using kraft pulping with lignin impurities 9.34%wt. Commercial cellulase Novozyme was used and Span 85 was purchased from Merck. Fermentation processes were conducted in the same condition, by adding 1%wt. of powdered yeast, 0.3%wt. of urea, and 0.1%wt. of NPK. Other affecting variables such as operating temperature was kept constant at Indonesian room temperature (25–28°C) and pH solution was set at 4.8 by using 50 mM sodium acetate buffer. All process conducted in a system with total volume 100 ml. Duration of hydrolysis and fermentation were set at 48 and 60 h, respectively. In this study, the observed variables were arranged as tabulated in Table 2.1 by using the least number of required experiments. The data obtained from experiments were resulted reducing sugar, ethanol concentration, and the ethanol conversion. The experimental results, thus called actual value, for respective variables are also shown in the same table. Resulted reducing sugar is the measured reducing sugar concentration that is obtained after the enzymatic hydrolysis process is completed as the feed for the fermentation process. The resulted ethanol concentration is the ethanol concentration in the fermentation broth that is measured after the fermentation process is completed. The ethanol conversion was calculated as % weight, by comparing the resulted ethanol weight to the respective pulp weight as the process feed.

All of these dependent variables were gone to model-fitting process to a second-order polynomial equation by using software for statistic SPSS 14. The basic principle of the model fitting process was creating a mathematical model then use it to recalculate all dependent variables that were tested in experiments. Dependent variables data obtained from recalculations, thus called predicted value, will be compared to the dependent variables data obtained from experiments by using a parameter called coefficient of determination. The accepted mathematical model was the one that gave the coefficient of determination above 0.95, in the range 0–1.

The accepted mathematical models are shown in term of dependent variables, Y_1 for resulted reducing sugar, Y_2 for resulted ethanol concentration, and Y_3 for ethanol conversion. The independent variables are stated as X_1 for substrate loading, X_2 for enzyme loading, and X_3 for the Span 85 doses. Based on their coefficient of determinations, all the values were satisfyingly above 0.95. It can be concluded that this statistical model will give good predictions for ethanol production process by using EFB pulp. By displaying the model prediction in the 3D graph as shown in Figures 2.2–2.4, optimum condition was easier to be determined. These graphs were created by using CAD software and became strong statements for the existing cheminformatics. This is wonderful evidence that information technology gives strong support to chemistry and chemical engineering works.

TABLE 2.1 Central Composite Rotatable Quadratic Polynomial Model, Experimental Data, and Actual Predicted Values for Three-Level-Three-Factor Response Surface Analysis.

Independent variables			**Dependent variables**					
Substrate loading (%dw)	**enzyme (FPU/g substrate)**	**Span 85 concentration (%vol.)**	**Resulted reducing sugar (% div)**		**Resulted ethanol concentration (%v.)**		**Ethanol conversion (%)**	
xl	**x2**	**x3**	**Actual**	**Predicted**	**Actual**	**Predicted**	**Actual**	**Predicted**
20 (−1)	10 (−1)	0 (−1)	11.98	11.94	4.39	4.69	21.95	23.10
25(0)	15 (1)	1 (0)	21.76	20.49	6.97	6.46	27.88	26.18
30(1)	12.5(0)	2 (1)	22.74	25.17	5.6	6.10	18.67	20.05
20 (−1)	15 (1)	2 (1)	10.65	10.65	5.59	5.59	27.95	27.95
30(1)	15 (1)	0 (−1)	23.53	25.93	7.1	7.90	23.67	26.20
25(0)	10 (−1)	2 (1)	24.82	22.39	6.93	6.05	27.72	26.34
25(0)	12.5 (0)	0 (−1)	22.93	20.57	7.16	6.43	28.64	24.96
30(1)	10 (−1)	1 (0)	22.03	23.48	4.89	5.13	16.30	16.68
20 (−1)	12.5(0)	1 (0)	8.92	9.08	6.34	6.67	31.70	33.16
Coefficient of determination (R^2)			0.9906		0.9623		0.9839	

$$Y_1 = -55.3491 + 7.5018X_1 - 5.6530X_2 - 2.0763X_3 - 0.1443X_1^2 + 0.1185X_2^2 + 1.1450X_3^2 + 0.0983X_1X_2 - 0.0519X_1X_3$$

$$Y_2 = 4.1471 + 0.0434X_1 - 0.4240X_2 + 10.8375X_3 - 0.04732X_1^2 - 0.2798X_2^2 - 1.5456X_3^2 + 0.2541X_1X_2 - 0.6038X_1X_3 + 0.7078X_2X_3 \quad (2.1)$$

$$Y_3 = 9.4757 - 0.6878X_1 + 3.2328X_2 + 44.9815X_3 - 0.1651X_1^2 - 1.1526X_2^2 - 6.0872X_3^2 + 0.8819X_1X_2 - 2.2216X_1X_3 + 2.2607X_2X_3$$

However, additional analyses were required for these models to make sure they are aligned with existing biochemical theories. The first analysis was focused on resulted reducing sugar that gave different trends compared to resulted ethanol concentration and ethanol conversion. Analyzing resulted reducing sugar was mainly about the enzyme performance and its kinetics. Based on the basic theory of reaction kinetics, more substrate concentration will give more products. This theory was suitable with the model prediction that gave minimum value and the value increase aligned with substrate loading. Even though there is substrate inhibition theory in enzymatic reactions, but it still not happening in our substrate loading scope, 20–30%dw.

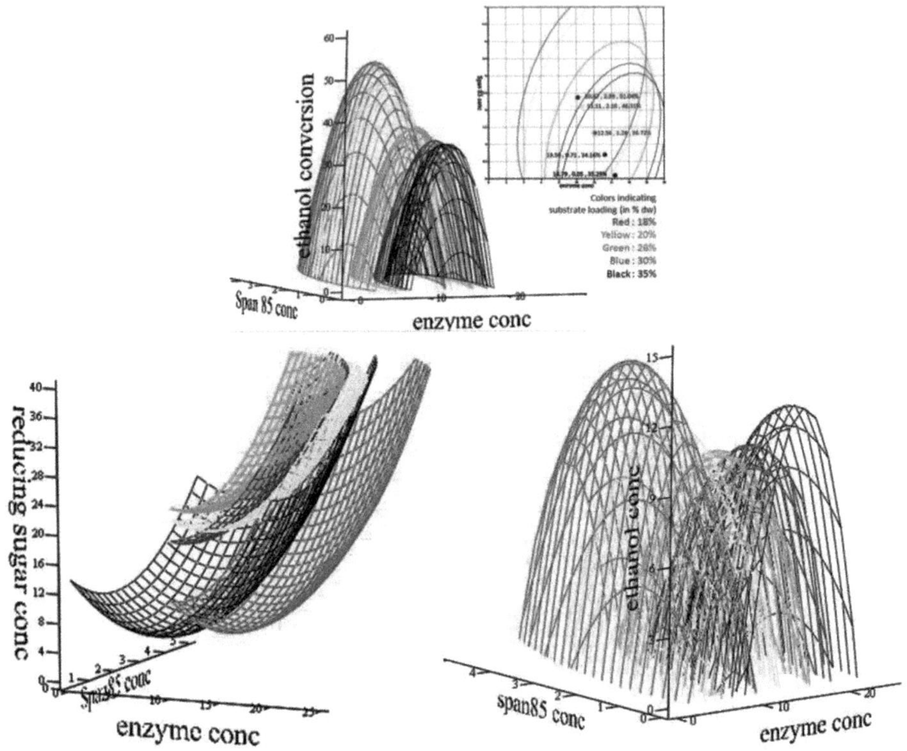

FIGURE 2.2 (See color insert.) Profile of cellulose conversion, reducing sugar concentration and ethanol concentration that predicted by the obtained mathematical model.

Surfactant addition was applied for reducing the negative effect of lignin impurities as aforementioned. Increasing substrate loading will require a higher dose of surfactant addition. Thus, it is very reasonable that increase in resulted reducing sugar will be obtained as surfactant doses are increased. Small discrepancies were observed in the minimum value were not exactly at the point of origin (0,0). In analyzing resulted ethanol concentration and ethanol conversion *Saccharomyces cerevisiae* or yeast performance were included. The resulted mathematical model gave predictions that aligned with a theory about surfactant effect to yeast by showing an optimum value. Span 85 doses lower than optimum value were indicating that the amount of surfactant could not overcome the negative effect of lignin impurities, thus lowering the cellulase performance. On the other hand, Span 85 doses higher than optimum values caused harmful effects to yeast by its increasing toxicity as previously explained. The effect mentioned at last cannot be observed if the observed process is only hydrolysis-based on resulted reducing sugar. Of course, this only happens if the processes were conducted in similar duration and temperature with the one that was used in this study, 48 h hydrolysis, 60 h fermentation in 25–28°C.

Interesting alignment between statistical conclusion with bioprocess theory and facts gave confidence for conducting model verification. In verification step, several random conditions were calculated to give ethanol concentration in fermentation broth over 8%v. were used. This 8%v. limit was adjusted to the lowest accepted ethanol concentration in fermentation broth to be directly sent to distillation unit in one of the existing Indonesian ethanol production factories. The calculated results of depended variables were further called predicted values. List of tested random conditions and their respective predicted values are shown in Table 2.2. These random conditions were then sent to experimental test in the laboratory to obtain the, so-called, actual values. As can be observed in Table 2.2, the obtained actual values were very close to the predicted ones. Thus, the mathematical model had satisfyingly proved as a reliable and useful prediction tool. As also can be seen in the series of tested variables, there are several conditions that can be used for obtaining ethanol concentration over 8%v. that adapt substrate loading and enzyme loading simultaneously. A suitable combination of substrate and enzyme loading was required and these can be fulfilled by RSM analysis. The mathematical model involves the term X_1X_2 in the equation which represents the mutual effect of substrate loading and enzyme loading. By involving this term, the sugar concentrations as feed for fermentation were adjusted to obtain ethanol concentration in fermentation broth over 8%v. As pulp substrate loading increased, the enzyme loading was adjusted at lowest possible concentration. This adaptation condition will be very useful in large scale/commercial application as a precaution if substrate or enzyme supply chain were interrupted. These results cannot be obtained from conventional analysis approach by observing the effect of one variable by keeping the other variables at a constant value.

There will be a certain doubt if the statistical results do not align with background facts and theories. In the worst case, a model also cannot give satisfying predictions. In this problem, second order polynomial equation may not be suitable empirical model for the observed phenomena. The options for still using the same experimental data are very limited since no variables were set constant to be analyzed in a conventional way. For avoiding waste of time and experimental resources, adequate reference studies of the similar or related fields were required. Another problem is

TABLE 2.2 List of Random Conditions and Their Respective Predicted Values that are Tested in Laboratory Experiments. The Results from Laboratory were Called Actual Values.

Independent variables			Dependent variables					
Substrate loading (%dw)	Enzyme (FPU/g substrate)	Span 85 concentration (%vol.)	Resulted reducing sugar (%dw)		Resulted ethanol concentration (%vol.)		Ethanol conversion (%)	
x1	x2	x3	Actual	Predicted	Actual	Predicted	Actual	Predicted
36.5	14.75	1	21.76	20.66	10.48	10.07	28.71	27.19
23	12.5	1	15.63	16.50	8.23	8.41	35.78	36.20
21.25	11	1	12.10	14.05	8.96	8.89	41.67	39.00
20	10.25	1	10.51	11.01	7.77	8.91	38.85	40.12
31	12	1	24.75	21.39	8.81	9.07	33.88	35.32
26	13.25	1	21.98	23.37	8.25	8.24	26.61	27.46
Coefficient of determination (R^2)			0.8880		0.9189		0.9163	

that the optimum condition cannot be found. The most recommended way for finding it was to add more experiments, with broader variable range, which are designed also in CCRD and repeat the RSM analysis sequences. Usually, the problem comes from unsuitable selection of variable range in which the optimum value does not lie in the range. Since RSM analyzing process is only dealing with empirical approach, it will not be easy to create solid correlation between the results with scientific theory and fundamentals and thus, difficult to plan scientific improvement for the observed process.

2.3.2 MODIFICATION OF EFB MATHEMATICAL MODEL FOR IN PALM TRUNK ENZYMATIC HYDROLYSIS PROCESS

RSM had been demonstrated for giving significant contribution to a biochemical engineer in determining optimum conditions. However, obtaining optimum conditions from laboratory experiments is not an easy task, especially in a bioprocess such as bioethanol from cellulose. It had been mentioned that for completing one experiment, 108 h are required and excluding preparations. The possibility of using obtained equation for predicting similar or closely related raw materials seems very worthy to be examined. This study continued by implementing the equation from hydrolysis and fermentation of palm oil EFB pulp to the process that uses palm oil trunk.

As mentioned earlier, palm oil trunks have become a potential raw material due to the regeneration of palm oil plantation that actually happened in these recent years in Indonesia. Cross sectional profile of palm oil trunk can be seen in Figure 2.3. As part of monocotyledon plants, palm oil trunk does not have cambium with less structured arrangement of its fibers. By excluding the barks, palm oil trunk structure can be classified as parenchyma that lies inside the circle and vascular bundle at the outsides.

Vascular bundles have more structured fibers than parenchyma. Botanists believe this is due to its function for trunk outer protection. Parenchyma, on the other hand, is having softer structure and high water adsorption capacity which is very useful in transporting nutrient. It is also having a function as nutrient storage; thus, a small amount of starch and plant sap were found in the parenchyma. Based on this structure and chemical composition, an attempt to direct utilization of parenchyma as feed for enzymatic hydrolysis process, without prior pulping, was conducted but gave unsatisfied results. In this study, palm oil trunk from Malingping, Indonesia was chopped and milled to 50 meshes. Some of the resulted material went to mechanical separation for obtaining vascular bundle and parenchyma part. The other part was not separated and both non-separated and vascular bundle part went through a kraftpulping process that has a similar condition to EFB. The pulp of whole trunks had 6.96%wt. lignin impurities and the vascular bundle had 13.70%wt. Later, both pulps were then separately used as raw material for bioethanol production process. A similar condition in hydrolysis and fermentation process for EFB pulp was applied to both the pulps, 48 h hydrolysis followed by 60-h fermentation in 25–28°C. Fermentation processes were conducted in the same condition, by adding 1%wt. of powdered yeast, 0.3%wt. of urea, and 0.1%wt. of NPK. Observed variables were following the selected condition for EFB as shown in Table 2.3 with the similar goal, obtaining ethanol concentration in fermentation broth over 8%v. The resulted reducing sugar concentrations from hydrolysis of whole trunk pulps and its comparison with hydrolysis of EFB pulp are shown in Figure 2.4.

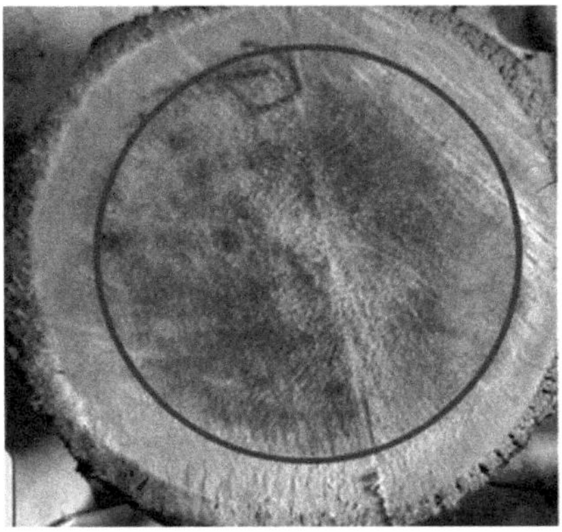

FIGURE 2.3 Cross-sectional area of palm oil trunk. The part that is inside the circle is parenchyma and the one outside the circle is vascular bundle.

Data was represented in a histogram with selected condition numbers as abscise. The numbers were referred to the similar optimum condition as mentioned in Table 2.3. The reducing sugar from palm oil EFB hydrolysis showed better results than resulted reducing sugar from palm oil trunks. The result of the correlative test between trunks and EFB in high-loading substrate enzymatic hydrolysis based on their resulted reducing sugar is also shown in Figure 2.4.

TABLE 2.3 Variables for Hydrolysis Whole Trunk Pulp and Vascular Bundle Pulp.

Optimum condition number	Substrate loading (%dw)	Enzyme (FPU/g substrate)	Span 85 conc. (%vol.)
1	20	10.25	1
2	21.25	11	1
3	23	12	1
4	26	12.5	1
5	31	13.25	1
6	36.5	14.75	1

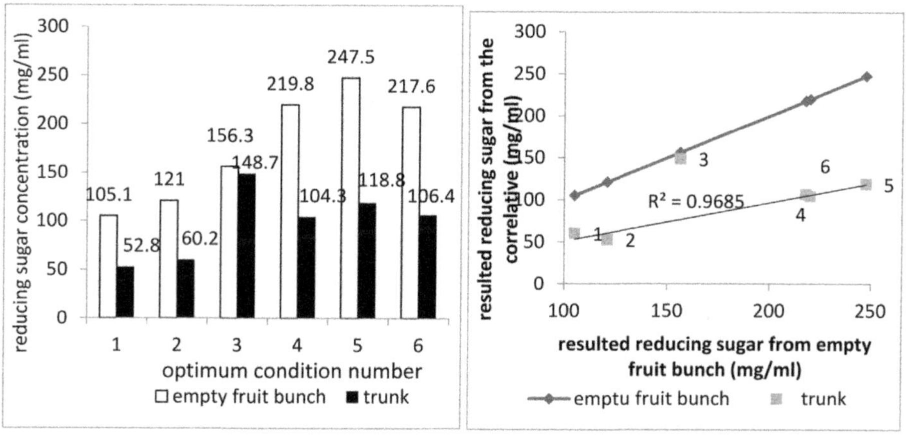

FIGURE 2.4 Resulted reducing sugar obtained from palm oil trunks hydrolysis.

The perfect correlation was simulated by correlating the EFB results to itself as showed in the graph as a dashed line. Square dots represent the resulted reducing sugar from palm oil trunks hydrolysis arranged correlatively to the ones obtained from palm oil EFB in the same optimum operation condition. By excluding data obtained from condition number 3, the calculated correlative coefficient was 0.9993. This result supported the conclusion that optimum conditions for high-loading substrate enzymatic hydrolysis of EFB were similar with optimum conditions for trunk pulp.

Data presented in Figure 2.5 were the comparison of resulted ethanol concentration between palm oil EFB and palm oil trunk. Since all operating conditions of fermentation were set similar for all, the difference was solely caused by the difference of hydrolysis process and outside disturbances.

No sterilization prior fermentation was performed in order to reduce operational cost if this similar process applied in commercial scale. Unsterilized fermentations also allowed the enzymatic hydrolysis continuation during the fermentation, which also called simultaneous saccharification fermentation. However, contamination possibilities need to be considered as outside disturbance.

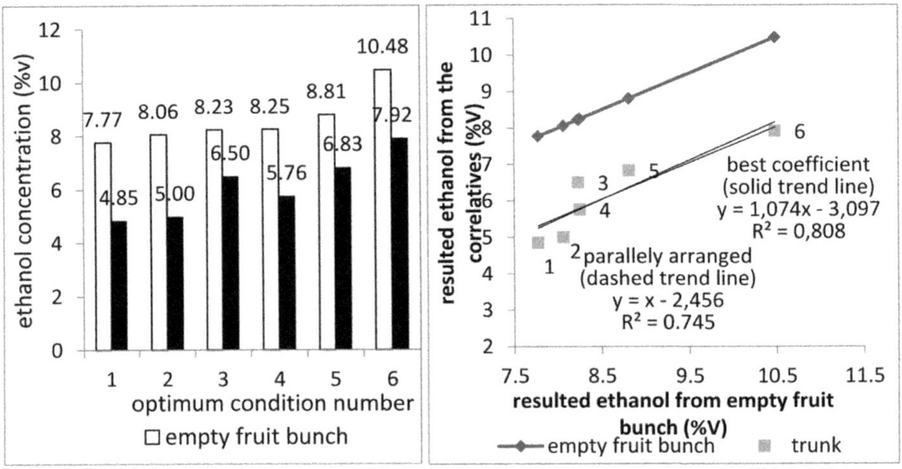

FIGURE 2.5 Ethanol obtained from palm oil trunks hydrolysis and continued by fermentation.

It is shown in Figure 2.3 that the resulted ethanol from trunks was lower than those obtained from EFB. This is aligned with the resulted reducing sugar reported above. In EFB pulp, 1%v. of Span 85 could effectively reduce the negative effect of lignin residue up to 9.34%wt. As previously described, the hydrophobic part of the surfactant binds through hydrophobic interactions to lignin on the lignocellulose fibers and the hydrophilic head group of the surfactant prevents unproductive binding of cellulase to lignin. The content of lignin residue in palm oil trunk pulp was only 6.96%wt. Thus, lower cellulase performance in trunk pulp, which is indicated by lower resulted ethanol concentration compared to EFB pulp, was not caused by lignin impurities. It could be concluded that the reason for lower resulted ethanol concentration was more dominantly caused by the higher crystalline cellulose content in whole trunk pulp. Later, it is discussed that there were two factors that affected cellulase performance, lignin residue and cellulose crystalline in the pulp.

The result of the correlative test between trunks and EFB in high-loading substrate enzymatic hydrolysis based on their resulted ethanol concentration is also shown in Figure 2.5. Unfortunately, none of the tested conditions gave ethanol concentration over 8%v. in the fermentation broth. However, no data was excluded and the best correlative coefficient was calculated as 0.808. Aligning this conclusion with a correlative test using sugar concentration, it is acceptable to assume that optimum conditions for high-loading-substrate enzymatic hydrolysis for EFB and those for trunk pulp were similar.

The trend line's slope for trunks was 1.074 which is very close to 1. It means that the trunks trend line can be considered as parallel to the dashed line for EFB. Arranged trend line to meet the slope equal to 1 reduced the correlative coefficient to 0.7451 and still considered acceptable. From the arranged trendline equation, there was a constant value obtained, −2.456%v. The mathematical equation for predicting resulted ethanol concentration from palm oil trunks can be derived from Equation 2.1.

$$Y_2 = (\mathbf{4.1471 - 2.456}) + 0.0434X_1 - 0.4240X_2 + 10.8375X_3 - 0.04732X_1^2 - 0.2798X_2^2 - 1.5456X_3^2 + 0.2541X_1X_2 - 0.6038X_1X_3 + 0.7078X_2X_3$$

$$Y_2 = \mathbf{1.6904} + 0.0434X_1 - 0.4240X_2 + 10.8375X_3 - 0.04732X_1^2 - 0.2798X_2^2 - 1.5456X_3^2 + 0.2541X_1X_2 - 0.6038X_1X_3 + 0.7078X_2X_3$$

The Equation 2.2 is as follows:

TABLE 2.4 Comparison Obtained Experimental Ethanol Concentration from Trunks Pulp and Empty Fruit Bunch Pulp to Their Correlative Prediction by Mathematical Model.

Optimum condition number	Substrate loading (%dw)	Enzyme (FPU/g substrate)	Resulted ethanol concentration from trunks (%v.)		Resulted ethanol concentration from EFB (%v.)	
			obtained	Predicted (Eq. 2.2)	Obtained	Predicted (Eq. 2.1)
1	20	10.25	4.85	5.55	7.77	8.01
2	21.25	11	5.00	5.63	8.06	8.09
3	23	12	6.50	5.65	8.23	8.11
4	26	12.5	5.76	5.78	8.25	8.24
5	31	13.25	6.83	6.61	8.81	9.07
6	36.5	14.75	7.92	7.61	10.48	10.07
Coefficient of determination			0.7451		0.9189	

The predicting results of Equation 2.2 are shown in Table 2.4. The coefficient of determination for Equation 2.2 was much lower than the prediction made by Equation 2.1 for EFB but is still usable in brief prediction. Thus, using RSM model that is obtained from a certain substrate was very promising to be applied to similar substrates. Increasing more experimental number hopefully will give better correlations. Similar correlation test was conducted for vascular bundle pulp. Unfortunately, the correlation of EFB optimum conditions to vascular bundle was not as satisfying as the correlation between EFB and whole trunks. Resulted reducing sugar of vascular bundle pulp gave acceptable correlation with sugars obtained from EFB pulp from respective optimum conditions as shown in Figure 2.6. However, the correlation of resulted ethanol concentration between EFB and vascular bundle gave unacceptable results. The correlative coefficient was only 0.219, as shown in Figure 2.7, which is much lower than the previous study result (0.7451) which used whole trunk.

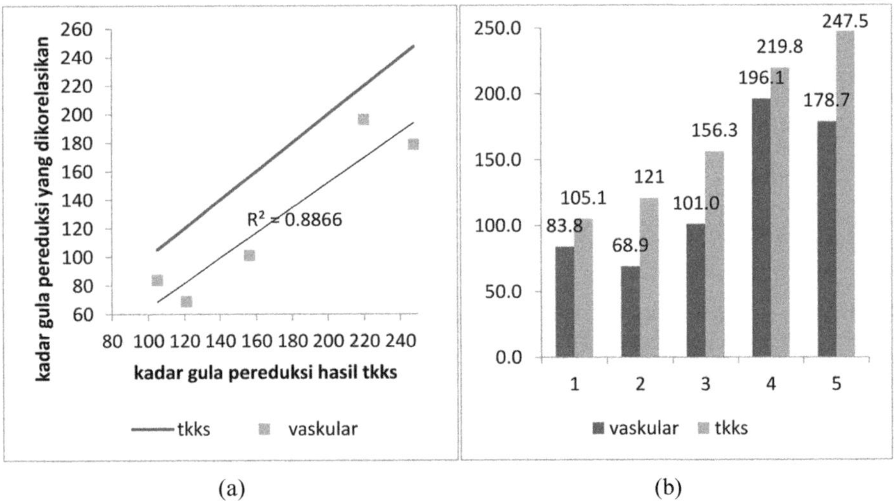

FIGURE 2.6 Ethanol obtained from palm oil vascular bundle hydrolysis and continued by fermentation.

In a previous study, parts of palm oil trunks were separated mechanically to its vascular bundle and parenchyma then each part was measured for its hemicelluloses content and cellulose crystallinity index (Lamaming et al., 2014a). The results show that vascular bundle and parenchyma had hemicelluloses content 25.47 and 35.57%wt., respectively, and cellulose crystallinity index 76.78 and 69.70, respectively. Cellulose crystalline index of palm oil EFB was not found from previous studies. However, in another previous study on Miscanthus, the correlation between hemicelluloses content and crystalline cellulose in its natural existence was shown (Xu et al.,2012). The higher hemicelluloses content in one observed part of plants, the lower cellulose crystallinity index. These previous studies on palm oil trunk and Miscan thus were supporting a conclusion about the correlation between hemicelluloses content and celluloses crystallinity index.

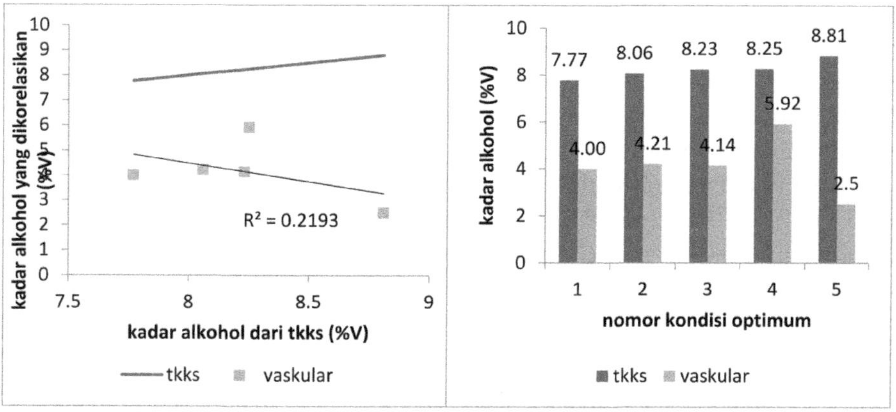

(a) (b)

FIGURE 2.7 (a) Correlation curve (b) alcohol level on vascular bundle.

In another previous study, effects of noncellulosic sugar in natural plants to cellulose crystallinity were specified only to galactoglucuronoxylan (Smith et al.,1998). Three monocotyledons: Italian ryegrass, pineapple, and onion were measured for their galactoglucuronoxylan content and the results were correlated to cellulose crystallinity index. The results showed that higher galactoglucuronoxylan content will lead to higher crystallinity index when the cellulose naturally biosynthesized. Glucuronogalactoxylan has similar properties with starch due to its water solubility and branched molecular structure. Starch was found in palm oil trunks but not in EFB. This fact could strengthen our previous conclusion that palm oil trunks have higher cellulose crystallinity index than its EFB. Not only because it had lower hemicelluloses contented but also because of the existence of starch in palm oil trunks.

Hemicelluloses content in the parenchyma, EFB, whole trunks, and vascular bundle were 35.57, 35.3, 31.8 and 25.47% wt., respectively (Ebtesam et al., 2005; Singh et al., 1999; Yang et al., 2004; Lamaminget et al., 2014). Cellulose crystallinity index for vascular bundle and parenchyma were 76.78 and 69.70, respectively. It will be reasonable to assume that cellulose crystallinity index in EFB was close to parenchyma, that is, 69.70, due to no significant difference between hemicelluloses content in EFB and parenchyma, and the index for the whole trunk would be between 69.70 and 76.70. Ordering from the highest cellulose crystallinity index to the lowest in its natural existence would bevascular bundle (76.78)> whole trunk > EFB = parenchyma (69.70). Since both raw materials were gone through kraft pulping process with similar operation conditions, the cellulose crystallinity index in resulted pulp gave similar tendencies. These facts support the experimental finding that the order of enzymatic hydrolysis performance from the lowest was vascular bundle < whole trunk< EFB.

These results showed the limitation of RSM in searching for a general model that can be applied to similar or closely related conditions. In case of bioethanol production from palm plantation biomass waste, the general model cannot be obtained even if the biomass came from one specific species. However, scientific and experimental facts were aligned with the RSM model which opens windows of opportunities for using RSM in this field. One promising possibility was establishing clear raw material characteristics, including lignin impurities and cellulose crystallinity index as the term for the existing RSM model works satisfyingly.

KEYWORDS

- **information technology—IT**
- **IT emerging technology**
- **bioinformatics**
- **response surface methodology (RSM)**
- **enzymatic hydrolysis process**

REFERENCES

1. Ebtesam, E.; Hussein, H.; Baghdadi, H. H.; El-Saka, M. F. Comparison between Biological and Chemical Treatment of Wastewater Containing Nitrogen and Phosphorus. *J. Ind. Microbiol. Biotechnol.* **2005,** *32,* 195–203.
2. Facioli, N. L.; Barrera-Arellano, D. Optimisation of Enzymatic Esterification of Soybean Oil Deodoriser Distillate. *J. Sci. Food Agric.* **2001,** *81*(12), 1193–1198.
3. Lamaming, J.; Hashim, R.; Sulaiman, O.; Sugimoto, T.; Sato, M.; Hiziroglu, S. Measurement of Some Properties of Binderless Particleboards Made from Young and Old Oil Palm Trunks. *Measurement* **2014a,** *47,* 813–819.
4. Lamaming J.; Hashim, R.; Leh, C. P.; Sulaiman, O.; Sugimoto, T. Proceedings of the Second International Conference on Advances in Bioinformatics, Bio-Technology and Environmental Engineering-ABBE 2014b. ISBN: 978-1-63248-053-8.
5. Pena, M. A.; Fierro, J. L. G. Chemical Structures and Performance of Perovskite Oxides. *Chem. Rev.* **2001,** *101*(7), 1981–2018.
6. Singh, G.; Huan, L. H.; Leng, T.; Kow, D. L. Oil Palm and the Environment: A Malaysian Perspective; Malaysian Oil Palm Growers' Council: Kuala Lumpur, 1999a.
7. Singh, M.; Sétáló, G.; Guan, X.; Warren, M.; Toran-Allerand, C. D. Estrogen-Induced Activation of Mitogen-Activated Protein Kinase in Cerebral Cortical Explants: Convergence of Estrogen and Neurotrophin Signaling Pathways. *J. Neurosci.* **1999b,** *19*(4), 1179–1188.
8. Vander Wiel, S. A.; Tucker, W. T.; Faltin, F. W.; Doganaksoy, N. Algorithmic Statistical Process Control: Concepts and An Application. *Technometrics* **1992,** *34*(3), 286–297.
9. Xu, N.; Zhang, W.; Ren, S.; Liu, F.; et al. Hemicelluloses Negatively Affect Lignocellulose Crystallinity for High Biomass Digestibility under NaOH and H_2SO_4 Pretreatments in Miscanthus. *Biotechnol. Biofuels* **2012,** *5*(1), 58.
10. Yang, H.; Yan, R.; Chin, T.; Liang, D. T.; Chen, H.; Zheng, C. Thermogravimetric Analysis—Fourier Transform Infrared Analysis of Palm Oil Waste Pyrolysis. *Energy Fuels* **2004,** *18*(6), 1814–1821.

CHAPTER 3

MACROMOLECULES VISUALIZATION: AN EMERGING TOOL OF INFORMATICS INNOVATION

HERU SUSANTO[1,2,*], LEU FANG-YIE[1], TEUKU BEUNA BARDANT[1], CHIN KANG CHEN[3], and ANDRIANOPSYAH MAS JAYA PUTRA[1]

[1]*Department of Computer Science and Information Management, Tunghai University, Taichung, Taiwan*

[2]*Computational Science, Research Center for Chemistry, The Indonesian Institute of Sciences, Serpong, Indonesia*

[3]*School of Business and Economics, University of Brunei, Bandar Seri Begawan, Brunei*

[*]*Corresponding author. E-mail: heru.susanto@lipi.go.id, susanto.net@gmail.com*

CONTENTS

ABSTRACT

Scientists have been using a lot of different scientific methods to analyze their research or experiments, hoping to find a cure to every disease that they find, analyzing organisms, further research on the human body, and in a way to develop an advanced medical treatment to people ever since the early 13th century. Gregor Mendel (1822–1884) was one of the scientists that proved a successful experiment in his research of a pea plant, where he recorded every single observation only through a notebook without any use of modern computers or tablets.

The timescale during the 13th century shows a very long time to consume the research activity in the manual way without any technology. However, since technology has been made advanced, scientists can now record thousands of analyzed data in a short period of time. As technology kept growing in the 1940s, it was seen as an important deal for scientists. Charles Babbage (1791–1871) invented a new way of performing mathematical calculations automatically and built a more advanced computing device. After that, computer scientists learned a lot about advanced computers, hardware, software, or databases in order to be able to perform simple operations in the laboratory.

Along with her partner, Ada Lovelace (1815–1852) developed sorts of computer programs to work with numbers as well as programming languages such as Smalltalk, Pascal, and C++ that can sync with Charles invention.[10]

During the mid-1990s, computational and quantitative analysis in science fields exploded where scientists begin to use science programming for their research. Such example will be the deoxyribonucleic acid (DNA) sequence where scientists are able to disentangle the beauty of thousands and millions of genome and nucleotides in a matter of time. As they see the technology is getting more and more important for their lab research, biologists need to cooperate in learning deeper toward the technology world. As they realize that in order to manage, store, and analyze more visualized data and perform statistics results, they are able to produce bioinformatics results that are seen as reliable data that can be used to other scientists' field. This is where bioinformatics was born.

As technology era improved year by year, biologists are now able to develop algorithms or advanced procedures to further understand different types of data collections such as nucleotide, amino acid sequence, protein domains, DNA sequence, pattern recognition, as well as the structure of organisms. It is clear that computer informative system plays an important role in helping scientists or biologists to perform their research in order to help further predictions for the future.[10]

3.1 INTRODUCTION

Computer technology is being widely used not only by common workers such as accountants, lawyers, secretaries, teachers, or any other private or government firms but also by the researchers, especially those who are working in the laboratory fields. There are a few examples of computers science that uses the informatics data, such as bioinformatics, biomedical informatics, chemoinformatics, ecoinformatics, geoinformatics, health informatics, neuroinformatics, social informatics, and veterinary informatics.[18] In the next subsection, we discuss about IT-emerging technology as driver and enabler of sciences area.

3.1.1 INFORMATICS FOR SCIENCES

It refers to the study of the science of processing, gathering information or data, storing, retrieving, and the organizing of the information used. In this new era, technology has been made advanced. It continually becomes more and more powerful in terms of their input of technology which has enabled data to be processed in just a few seconds. If technology has not been made in advance, it may take several years to process a data which is very time-consuming to all the users.[18] Bioinformatics is basically the compilation and management of sciences, data analysis in genomics, proteomics, and in the field of biological sciences. The tools or the software used in the bioinformatics includes the analysis of quality check, trimming, base calling, assembly, as well as gene prediction[2].

These researchers use computer informatics to process lab work or the lab data. They not only work in lab with hands on tube or thermometer doing experiments but also need computers in order to store their findings as well as to compute their end results. It really helps the lab researchers to store their data in order to be used in the future, helps them to understand more of their studies, be able to perform accurately with the correct use of data, and be able to determine the results in a split second.[18]

3.1.1.1 BIOINFORMATICS

Claverie and Notredame[7] have stated that it is commonly heard that highly trained professionals used computers to carry out their tasks. However, biologists also use the computers as much as they do other manual task.

Biologists use computers not only for writing down some memos and sending out e-mails but also to deal with very specific problems. As a whole, these tasks are called bioinformatics; simply referred as molecular biology's computational branch.[7]

According to Al-Ageel et al.,[1] bioinformatics is an interdisciplinary methodology by making use of all information that has been gathered and data that are displayed in order to analyze biological data.

The use of computer appliances helps the biologists to analyze more of their data easily, able to store as much data or findings that they have into the computer, do a wide research regarding their tasks, as well as to retrieve any of the biological information. The study of bioinformatics basically involves a biologist. In a way, they use computers to analyze a deoxyribonucleic acid (DNA) sequence as well as to retrieve advanced study of the molecular genetics.[18]

3.1.1.2 EXAMPLES

Bioinformatics helps biologists to identify more quickly, easily, and most importantly, they are able to browse through the web to search appropriate tools or software for their work fields. In a way, it saves time and the process can be made in such a short time without having to wait for many years for the analysis to complete. A comprehensive database and software saves time and their efforts in searching the appropriate tools.[20]

One of the accomplished users in the bioinformatics field is the Human Genome. Scientists use the computer to process and to map down the understanding of the aspects of human life. For example, a scientist or even a normal human being is unable to explain why children resemble the face of their parents? In what way the resemblance came from? But now, scientists are able to identify and predict how a child will look like by determining the different DNA, height, or even the hair color. Aside from looking at human genes, scientists are able to carry out more advanced experiments in terms of finding a cure to diseases in the human body. Sometimes diseases can occur from past generations to their generations in a way diseases are inherited. Therefore, lab scientists are able to identify the cure for the diseases to prevent them from being inherited to the next generations. In this way, it helps the generations to stay healthier and prescribed medicine as what has been consulted by the doctors.[18]

3.1.2 PAST, PRESENT, AND FUTURE OF INFORMATICS FOR SCIENCES

3.1.2.1 PAST

The Chinese merchants during the 14th century used a form of biometrics technology for identification purposes. They took the children's fingerprints by using ink for identification. This was found by a European explorer named Joao de Barros.

In 1890, the technology of body mechanics and measurements was studied by Alphonse Bertillon. The police used the "Bertillonage method" because it could help to identify criminals. Unfortunately, the method was no longer in use because there had been a false identification of some subjects. The method was removed in favor of fingerprinting and they brought back the method used by Richard Edward Henry of Scotland Yard.

By 20th century, Karl Pearson from the University College of London had made an important finding in the field of biometrics. He studied about the correlation and statistical history which he referred as animal evolution. His works include the method of moments, chi-squared test, correlation, and Pearson system of curves.

In the 1960s and 1970s, signature biometric authentication procedures were advanced. However, the biometric field remains firm until the security agencies and military further developed and investigated the biometric technology more than just the fingerprints recognition.

3.1.2.2 PRESENT

To get the Food and Drug Administration (FDA) into approving a drug requires the protracted process which averagely takes 12 years between the lead identification and the approval of the FDA. The FDA approves the existing drug and uses it to treat different diseases other than what is intended, taking the Viagra as an example where the original purpose is to treat the heart disease. However, it was repurposed to treat the erectile dysfunction. This saves time and money.[21]

Different approaches need to be taken into account in order to repurpose the drug. One of them is to compare the signature of molecule in disease and the signatures observed in cells of the animal or the people that have been tested with different drugs. If there are anticorrelated signatures found, the drug's administration for that disease can at least lessen the symptom or they can even give treatment for the condition. One of the examples of computational approaches is the one that is used by Sirota et al. (2011). They use it to discover drug cimetidine for anti-ulcer as a mean to cure lung adenocarcinoma and approve the off-label usage in vivo by making an animal as the model of the lung cancer.[21]

Other than using the computational methods that have been mentioned above, there is also the experimental approach to drug repositioning. One of the examples is when Nygren et al. screened 1600 known compounds to be compared with two different colon cancer cell lines by using the connectivity map data in order to analyze their findings into more details and by identifying the mebendazole that has the ability to have therapeutic effect in colon cancer.[21]

3.1.2.3 FUTURE

The system in biometric technology consists verification and authentication. Cognitive biometric system was developed to use brain response, stimuli, face perception, and mental performance for search of high-security area.[3] This technology has been advanced that it has created a variety of recognition systems, for example, developed biometric strategies such as retinal scan, face location, hand geometry, iris scan, DNA earprint, and fingerprint. In the coming future, these biometric technologies will be the number one choice to solve any threats in the world and able to increase the management of information security.

3.2 INFORMATICS: IMPACTS FOR SCIENCES

3.2.1 ANALYZING GENE TO PROTEINS

Fiannaca et al.[9] have mentioned that the working molecules of a cell is symbolized by the protein, yet studying the proteins' functions is not sufficient to comprehend the cell machinery completely. This is because it is more essential to have the interactions among the proteins as the cell's biological activity cannot be identified by the functions of the proteins. The interactions between a group of proteins help to control and to give support among the proteins during the biological activities and this is called a protein complex, a functional module of the cell. RNA polymerase and DNA polymerase are the examples of the protein complexes.[9]

According to Buehler and Rashidi,[4] the word "Bioinformatics" is referred to organizing the work, analyzing them, as well as predicting the complex data that have arose from the modern molecular and biochemical techniques as living organisms and the different human studies life are very complex. However, the meanings tend to vary from one's point of view to another. Some define that it depends on the information flow concept in the systems that refer to the biological transmission of genetically encoded information from genes to proteins and from the blueprint to the machinery of life. The bioinformaticians try to comprehend the saying that the genes codes for the physiological characteristics for the disease. Bioinformaticians can assist in explaining the genetic complexity mechanism and the evolutionary relationships among the organisms through the creation, annotation, and biological databases' mining. Bioinformatics is relying on the assumption that quantifiable relationships are present not only between the genes activity and their existence inside the genome but also between the genes sequence and the proteins structure and role. The techniques of database mining are used by the bioinformaticians in order to study the protein complexes, pathways of metabolic, and the networks of the gene.[4]

Buehler and Rashidi[4] also stated that gene sequence and protein structures are the important elements of bioinformatics. Genes are the units of life that are heritable and its molecular sources of information are the genomes of contemporary organisms. It is very important for the practicability of the organism to replicate the information of genetic from one generation to other generations correctly. It is also crucial for the biological evolution, for example, the alteration of organisms in adapting the changes in the environment. Mutation can cause carriers the disease or even death. New traits are also found in some cases which will provide advantages in the reproduction of the organism. Some of the mutations are selected to be retained naturally while the majority of the mutations are removed; therefore, there are differences in our genomes that form a species. This diversity is characterized by the bioinformatics by comparing the genetic sequences and protein structure directly.[4]

Buehler and Rashidi[4] have mentioned that one of the uses of bioinformatics is to study evolution. As evolutionary changes are difficult to be studied in real time, bioinformatics uses systems such as the Linnaean taxonomic system to enable to name, rank, as well as classify the organisms and paleontology to study the fossil records. Some microevolution such as the rapid mutation rate of the human immunodeficiency virus (HIV) and influenza (flu) can be studied in real time and thus help the viruses in escaping the immune detection.[4]

Studying the mutation in living things, according to Buehler and Rashidi,[4] helps a better understanding of the pathogenesis disease and rising infection. Mutations allow the molecular biologists to track the traits plus record the genomes as they play the role of genetic markers. Geneticists use animal models in medicine to analyze and cure the human disease by testing toxicity and effectiveness of the new drugs. However, animal models cannot represent the human subjects in studies such as the human behavior.[4]

Buehler and Rashidi[4] claimed that it is crucial to comprehend the source, quality, and biological significance of the data which it relies on which are the genes and proteins sequences and structures. Bioinformaticians use raw data to discover the connection between the genes, genomes, and proteins. The raw data used are the nucleic acid sequences and protein structures. In accurately interpreting their biological significance, the quality and the precision of the data is essential. For example, when the scientists assess the mutation of genes that is engaged in cancer, in order to obtain the sequence of information in coding the protein and regulating the component, it is important to clone the genes that are involved in cancer. The restriction fragment length polymorphisms are used to discover the chromosomal DNA fragment in the individuals that have an effect on. In cloning a gene, a fragment that is not included in the genome will be put into a small customized vector DNA to get its nucleotide sequence as inserting the secluded fragment into the small customized vector DNA will be resulting in recombinant DNA. Another technique to amplify the DNA is by using the polymerase chain reaction (PCR).[4]

Bioinformatics technology has successfully created opportunities for biologists in shaping modern biological research. This is true according to Charles Darwin's theory of evolution. The cloning of human DNA can be done by the PCR which is the simplest and effective method, and it also prevents the difficulty of obtaining large amounts of human tissues for the purpose of DNA sequencing. Every sequence of dozens of building blocks able to encodes genetic information. The DNA can also be used for studying gene expression and synthesize large amounts of protein for analysis. Another use of bioinformatics technology is that it enables to understand human responses when taking drugs, which is so-called pharmacogenomics. It can be noticeable to those who are susceptible to drugs.[4]

In addition, biometrics technology provides better understanding on the relationships between genes and diseases, such as potassium channel mutations and heartbeat regulation, for example, long QT syndrome. As more genome projects or the full DNA sequence completed, more information was obtained. The information then transformed from the molecular biology laboratory into computer labs, which in return, making the information more accessible and free from public databases. This will open up the role of genetic networks such as embryonic development, the structure and the function of proteins, memory, and aging.[4]

3.2.2 SOFTWARE CATEGORY

BioInfoKnowledgeBase is one of the data used to assist the researchers with numerous tools, software, methods, and databases in biological information. It is easily accessible for the biologists to carry out their research since the databases provide a wide range of biological information such as the nucleotide, protein, bacteria, cyanobacteria, fungi, and other biological terms.[20]

3.2.2.1 GENOME ANNOTATION

Genome annotation refers to the process of locating genes and genome coding to determine what each gene can do. The genome annotation uses tools name such as Apollo genome annotation, ASAP, COFECO, GoGene, IBM Genome, TIGR software tools, KAAS, KOBAS, MADAP, MICheck, Sequin, and a lot of tools are used in the genome annotation software. Under the genome annotation, another subcategory can be found such as gene prediction, genome assembly, and ontology analysis.[19]

3.2.2.2 GENOME ANALYSIS

Genome analysis compare features such as the DNA sequence, structural variation, and gene expression which uses tools such as Bioconductor, bioDAS, Bioverse, CARGO, BRIGEP, IslandPath, Sockeye, and Taverna as a few of the sample tool names which can be found under the use of genome analysis. The subcategories under genome analysis are metagenome analysis, genome alignment, and genome browser.[19]

3.3.2.3 GENE ANALYSIS

Gene analysis is the overall study of genetics and molecular biology which involve tools such as WebGestalt, PANTHER tools, GeneTrail, GeneCodis, COFECO, ToppGene Suite, Endeavor, Babelomics, MirZ, and the SVC. The subcategories are gene regulation, transcription element analysis, splicing, expression analysis, and pathway analysis.[19]

3.2.2.3 NUCLEOTIDE ANALYSIS

Nucleotide analysis is the process of imperiling the DNA and RNA to a wide use of analytical methods in order to understand each of their features. Tools such as Alias Server, AMOD, BioMart, DINA Melt, Feature Extract, Gendoo, FIE2.0, Gene Set Builder, Geno CAD, BCM search launcher sequence utilities, MKT, UGENE, Virtual Ribosome, and the Pegasys are used for workflow management for bioinformatics. The subcategories are motif identification, nucleotide and genome alignment, single-nucleotide polymorphism (SNP) identification, and the structure and sequence detection.[20]

Among the most useful genetic markers in tracking down genes that contribute to diseases and disorders are single base-pair variations in the genomes of the human population. A single base-pair site where variation is found in at least 1% of the population is called an SNP (pronounced "snip"). A few million SNPs occur in the human genome, about once in 100–300 base pairs of both coding and noncoding DNA sequences. At present, they can be detected by very sensitive microarray analysis or by PCR. Once an SNP is identified that is found in all affected people, researchers focus on that region and sequence it. In nearly all cases, the SNP itself does not contribute directly to the disease in question by altering the encoded protein; in fact, most SNPs are in noncoding regions. Instead, if the SNP and a disease-causing allele are close enough, scientists can take advantage of the fact that crossing over between the marker and the gene is very unlikely during gamete formation. Therefore, the marker and gene will almost always be inherited together, even though the marker is not part of the gene. SNPs have been found that correlate with diabetes, heart disease, and several types of cancer, and the search is on for genes that might be involved.

Forensic scientists use an even more sensitive method that takes advantage of variations in length of genetic markers called short tandem repeats (STRs). These are tandem repeated units of two to five nucleotide sequences in specific regions of the genome. The number of repeats present in these regions is highly variable from person to person (polymorphic), and even for a single individual, the two alleles of an STR may differ from each other. For example, one individual may have the sequence ACAT repeated 30 times at one genome locus and 15 times at the same locus on the other homolog, whereas another individual may have 18 repeats at this locus on each homolog. These two genotypes can be expressed by the two repeat numbers: 30,15 and 18,18.

In a murder case, for example, this method can be used to compare DNA samples from the suspect, the victim, and a small amount of blood found at the crime scene. The forensic scientist tests only a few selected portions of the DNA—usually 13 STR markers. However, even this small set of markers can provide a forensically useful genetic profile because the probability that two people (who are not identical twins) would have exactly the same set of STR markers is vanishingly small. The Innocence Project, a nonprofit organization dedicated to overturning wrongful convictions, uses STR analysis of archived samples from crime scenes to revisit old cases. As of 2013, more than 300 innocent people have been released from prison as a result of forensic and legal work by this group.

Genetic profiles can also be useful for other purposes. A comparison of the DNA of a mother, her child, and the purported father can conclusively settle a question of paternity. The largest such effort occurred after the attack on the World Trade Center in 2001; more than 10,000 samples of victims' remains were compared with DNA samples from personal items, such as toothbrushes, provided by families. Ultimately, forensic scientists succeeded in identifying almost 3000 victims using these methods. Just how reliable is a genetic profile? The greater the number of markers examined in a DNA sample, the more likely it is that the profile is unique to one individual. In forensic cases using STR analysis with 13 markers, the probability of two people having identical DNA profiles is somewhere between one chance in 10 billion and one in several trillion. (For comparison, the world's population is between 7 and 8 billion.) The exact probability depends on the frequency of those markers in the general population. Information on how common various markers are in different ethnic groups is critical because these marker frequencies may vary considerably among ethnic groups and between a particular ethnic group and the population as a whole. With the increasing availability of frequency data, forensic scientists can make extremely accurate statistical calculations. Thus, genetic profiles are now accepted as compelling evidence by legal experts and scientists alike.

3.2.2.4 PROTEIN ANALYSIS

Protein analysis is the process of determining the different types of acids in the protein structure especially occurring in the human backbone. The biologist uses a number of tools such as the AACompldent, InterProScan, and the

MassSearch. The subcategories are three-dimensional (3D)-motif analysis, secondary structure, transmembrane segment, modeling and simulations, structure visualization, and the functional pattern recognition.[20]

3.2.2.5 EVOLUTIONARY ANALYSIS

The evolutionary analysis is basically identifying the gene family, gene discovery, as well as the origin ofgenetic diseases. It uses tools such as the Gblocks, MEGA, EEEP, BAMBE, Arlequin, Joes site' phylogeny programs, Weighbor, Tree Editors, Signature, Puzzleboot, POWER, Orthologue Search Service, and the OGtree.[19]

3.2.3 DATABASE CATEGORY

3.2.3.1 NUCLEOTIDE DATABASES

Nucleotide databases are basically the collection of several sources such as the GenBank, RefSeq, TPA, and PDB which are involved in the gene, genome, and the biomedical research data. It uses tools such as the Genome Database for Rosaceae, HGVbase, MGIP, Projector 2, and the TIGR software tools. It is divided into subcategories of sequence submission and retrieval system and the genome sequence databases.[20]

3.2.3.2 PROTEIN DATABASES

Protein databases use a 3D structural data of proteins and the nucleic acids. It uses tools such as swiss-2D page, SCOP, MPDB, HSSP, FSSP, DSSP, and BioMagResBank.[19]

3.2.3.3 BACTERIAL DATABASES

Bacterial databases basically store and analyze the bacterial isolate which is linked to isolate records which uses tools such as database of magnetotactic bacteria, database of natural luminescent bacteria, extra train, ICB database, HOBACGEN; homologous bacterial genes database, and the VFDB.[19]

3.2.3.4 CYANOBACTERIAL DATABASES

Cyanobacterial databases are the database to collect all cyanobacteria for photosynthesis such as the information, gene information, gene annotations, and the mutant information. It uses tools such as cTFbase, CyanoBase, CyanoClust database, CyanoData, CyanoMutants, Cyanosite, and the SynechoNET.[19]

3.2.3.5 FUNGAL DATABASES

Fungal databases are where fungal organisms are kept to be used for further research in the medicine, agriculture, and the industry. It uses tools such as the CORTbase, UNITE, phytopathogenic fungi and oomycete database, MycoBank, MyCoRec, and the dimorphic fungal database.[19]

The databases and the software as per stated above consists of more than 1000 bioinformatics tools which are widely accessible to biologists or the researchers to carry out their extraction and data analysis. The database contains six main sections, "Home," "About Us," "Software," "Databases," "Search," and "Team". These sections show a brief information about the use of the database, the role of the institute, search findings, and present the group who developed this database.[20]

3.3 COMPUTATIONAL TOOLS

Computers and computational methods now have a very important device that can be used for scientific applications thus these methods are becoming vital role in sciences. The applications are covering from the use of computers in molecular biology which is called the classical bioinformatics to complex physiological systems, that is, numerical models. The number of advanced computing tools has been increased; therefore, this enables the models of the physiological system to be more developed in details.[6]

Buehler and Rashidi[4] stated that science experiments are not only involved in test tubes, liquids, and microscope. Experimenting is examining one idea that involves observing set of samples as well as testing it in a certain period of time. Clearly, planning and organization are needed in testing the idea.[4]

This is where the computers come in handy. It is mentioned by Buehler and Rashidi[4] that the computers are essential in an experiment in terms of designing, executing, and analyzing. Computers help in modernizing almost all activities in the laboratory such as the use of word processing, analyzing with spreadsheet, and assessing the internet. Scientists are able to independently carry out their work with the high computational power.[4]

Buehler and Rashidi[4] claimed that the computers help the scientists in calculating and measuring the cells as well as recording the cells' activities. Sophisticated computers that are equipped with cell-sorting machine, video cameras, and digital oscilloscopes help the scientists to accurately record the data from the experiment. Large number of data can be controlled, performed, manipulated, and stored in the computers.[4]

Processors are needed to be inserted in the machine in this modern science, for example, with the spectrophotometers which are used in assessing the absorption of the light. Immediate experiments are allowed by the computer control in order to perform the experiments for example in observing any chemical composition changes in the test solution. The small window controls the microprocessor by putting one or more codes or command text that is allowed to be typed in on view from a short menu. The accurate experiments are not only caused by the computers but also depend on the use of the quality of the instrument.[4]

Before the introduction of computers, the accuracy and the high-quality measurements were already attained in the science world. The accurateness of the experiment is determined by the quality of the instrument that is used by the scientist and not just by the computer. For example, high-quality instrument is needed to cut the sample of frozen cell in the electron microscopy or measuring the neuron's electrical activity.[4]

The scientists work differently since the introduction of internet as it is the vital advanced technical. The communication among the researchers has been improved and the collaboration in the field works has also encouraged. The effectiveness of efforts of research has increased since the existence of data management systems, for examples, the National Center for Biotechnology Information and European Bioinformatics Institute. The specialized and specific management system specifies the biological data. One of the examples is the Protein Data Bank (PDB). The PDB is the storage of the data of the proteins where the relationships among the molecules that are stored are also available. Links of related data should be given by the data management system.[4]

The internet has the interactive mode which gives the benefit to science. Remote computers are used to run the internet to access the interactive tasks. Majority of the computing task use PCs as PCs are less expensive, big capacity, and they can also interconnect the local and nonlocal networks which enable them to gather the parallel computing processors. The internet needs to be linked to switches, routers, and fiber-optic cables. It also has a massive parallel computing which is similar to the supercomputers.

3.4 EXAMPLES OF BIOINFORMATICS TECHNOLOGY

The bioinformatics technology can also be used for information security. It is important because it has the ability to verify and authenticate information effectively. It is a type of software which could detect identification of a person's unique physiological and biological traits. This will open a wide range of opportunities to develop more security and safety from any threats or criminals in the world. This technology has a vital role in preventing any incidents because of its effective tools to detect criminals. The use of biometric technologies provides a beneficial and reliable verification to those who are travelling abroad, the bioinformatics technology will help strengthen the security of passport, visa, and other identifications. The reason behind this action is because of the concerning terrorism in the world lately. Therefore, the main focus of people is based on their security. The next discussion will be elaborated based on its biometrics tools, in addition to its benefits and limitations.

3.4.1 TOOLS OF BIOMETRICS EQUIPMENTS

3.4.1.1 SIGNATURE VERIFICATION TECHNIQUE

The signature recognition is a tracking people's dynamic signature. It can be measured in terms of the direction, the pressure, acceleration, and the length of the strokes, number of strokes, and their duration. The advantage of this technology is that the fraud cannot steal any information on how to write and copy the signature similar to the one that writes it previously. Other devices such as traditional tablets capture 2D coordinated pressure. This technology is mainly used for e-business and other applications where signature is accepted for personal authentication.

The limitations of using signature verification technique are that the signature can look different from the usual user signature. This is because of the digitalized signature, it changes slightly compared to how it was written on paper. Moreover, the user does not see what they had already written. They have to look back at the computer monitor to see the signature. This is a disadvantage to those inexperienced users.

3.4.1.2 VOICE RECOGNITION

Voice recognition has the feature of speech that is different between individuals. Kumar and Walia[13] argue that there are different styles of spoken input: text-dependent, text-prompted, and text-independent. Most applications use text-dependent input, which involves enrollment of more than one voice passwords. It is used whenever there is concern about imposters. This can avoid the imposters from unlocking the passwords, therefore, security of the user is guaranteed.

The voice recognition, however, can also lead to a disadvantage. This is because of performance degradation. Voice may change due to aging which needs to be inscribed by the recognition system. Furthermore, the results may change from behavioral attributes of the voice, verification, and enrollment on another telephone.

3.4.1.3 FACIAL RECOGNITION

Facial recognition has the ability to detect facial image of a person's individual characteristics. The advantages of biometric identification on photos are that it does not require a direct relation with the subject, it is confidential, provide unique and stable biological parameters of a person, and its biometric database are complete and widespread. Furthermore, another benefit is if there is a problem of image processing, it can be transferred to the professionals automatically. It will enable to process and specify image control points, which strengthen the accuracy of the system.

The limit of this technology is that the features are resistant to hairstyle changes, people wearing the glasses, and other transformations.

3.4.1.4 PATTERN RECOGNITION

Pattern recognition was traditionally found in the 1950s; the measurement of data shows results in a lacking form of formula which therefore takes a lot of time to analyze the results without the use of advanced technology computation. This, therefore, slows down the process and the development is decreased.[15]

With the help of advanced technology, biologists are able to compute data mining, analyze datasets, syntax, statistics structure, recurrent patterns, machine learning, developed effective modeling data, and desired outputs.

Pattern recognition basically involves matching and detecting pattern sequences as well as covers the theory and methods for clustering refer to how do objects form into natural groups, dimensionality reduction, and classification.[15]

Example of pattern recognition is the diagnosis of breast cancer; pattern recognition is used to identify specific medicines to cure the molecules hidden in the cancer cells. When they can retrieve the cure, they are able to use it for medical purposes, which are more certified after they have used the pattern recognition test. Patient diagnosed with cancer is now able to consult their doctors on medical instructions. Cancer can be anything, from brain cancer to bone cancer and basically may form anywhere in the human body; therefore, scientists do play an important role here in finding the right cures, doing screening test, and treatment for the patient. As we all know, cancer treatment uses technologies such as chemotherapy, radiotherapy, hormone therapy, biological therapy, and transplant.[15]

There are three ways in which pattern recognition works. First, the clustering methods where Gaussians mixture is used in order the data to be fitted well. Second, the dimensionality reduction, it is where the data are in line with a linear analysis. Last, the classification where the boundary is made to separate two different classes of data. The cluster data basically used in performing a large cluster data, for example, experimenting genes which involves different types of genes and cells and this data is used in order to retrieve uncharacterized genes, similarity, patient samples, and most importantly to discover disease inherited if any. Therefore, the cure can be identified as early as they can to prevent diseases to spread into a difficult situation. There are two types of clustering the procedures: the first is the partitional and the second is hierarchical. The partitional is used to fit a number of data into one of the methods such as the k-means and the Gaussian mixture models. The hierarchical clustering used to group certain objects in the data.[15]

However, the cluster method may not produce a better or an objective result as it may produce confusion between the clusters and the classes as the groups have the same label; classes may overlap sometimes and unsupervised classification may occur. By using the pattern recognition, biologists may face challenges in performing the data, such as storage may be limited and may not fit the whole data that they needed to perform. The works need to be packed down which therefore may cause the data to scatter. Some biolab may not have enough budgets to invest advanced software or databases, which is likely to decrease their research activity and may slow down their experiment research fields. Cells of a human body or even organisms may change at times; biologists depend on advanced computerized system to further their research. Therefore, scientists need to put a lot of investment to make improvements in their computer systems.

As such example, using the pattern recognition, scientists should employ a more skilled biologist that is able to perform the research by using the pattern recognition system or database. In a way, this shows that computer plays an important role in employing skilled labor that is able to use computer system to perform daily tasks. If biologists are unable to use computer system, this may cause difficulty in the laboratory and therefore, this may slow down the process and may cause error if computer systems are not used properly.[15]

The data represented are still performed in a very good way although sometimes data may perform slight error probably due to storage error. Sometimes it is impossible for a biologist to retrieve a better result as what they expected. Biologists may use common feature vector such as zero mean, unit standard deviation, log, square root, and add or multiply elements. Sometimes complicated data results may occur such as binary of 0/1, ordinal 1/2/3, qualitative of red, blue, green, or sequential data of ACTGAATA in biologists term which basically shows a loss of information to some part of the research and it needs to convert into reliable information data. These are few examples of a mislead data perform by biologists in the bioinformatics fields.[15]

3.4.1.5 FINGERPRINTING

Fingerprint recognition gives a wide opportunity for security purposes. "A chance of two users having the same identification in the biometrics security technology system is nearly zero." One of the technologies that can be used is Siemens ID Mouse fingerprints. The results of scanning the fingerprints will produce a very high-quality image while remaining sub-pixel geometric accuracy. Afterward, the image will be kept stored and processed to extract the pixel information. According to Mazumdar and Dhulipala, a fingerprint sensor captures the digital image of a fingerprint that is called live scan. The live scan is digitally processed to create a biometric template which can be stored and collected for matching. Furthermore, it also has matching algorithms which can be used to solve a problem. Stored templates of fingerprints can be used against candidate's fingerprints for authentication and verification purpose. The process starts either using original image with the candidate image to make comparison. Pattern-based algorithm compares the basic fingerprint, such as its arc, whorl, and loop. The template includes the size, type, and orientation of pattern found in the fingerprint image. This can also be differentiated with the template to determine which matches with the match score generator.

Despite its ability and advantages of this bioinformatics technology, there is a probability that it may lead to drawback, for example, if the user uses fingerprints for security reason, and an incident occurred that could cause loss of his or her fingernails, it will be difficult for the verification process.

3.4.1.6 IRIS SCAN

Iris scan is also one of the bioinformatics technologies. It is found to be reliable and accurate for the authentication process. It is another biometric that can scan the iris pattern quickly from the left and right. The scan can be used for identification and verification purposes. The high variance of appearance makes biometric well organized. The iris scan is an accurate measurement which avoids the difficulty of forging and using as an imposter person. Its other advantages are as follows:

1. It is intrinsic isolation and protection from the external environment.
2. It is extremely data-rich physical structure.
3. It is a genetic property—no two eyes are the same. The characteristic that is dependent on genetics is the pigmentation of the iris, which defines its gross anatomy and color.
4. Its stability over time: the impracticality of surgically modifying it without risk to physiological and vision responses to light, which gives a natural test against artifice.

In terms of its iris detection, the irises are detected even if the image is in constructions, visual noise, and different levels of illuminations. The lighting reflections, eyelids, and eyelashes barriers are eliminated. Based on security, it is an important technology for security system. In the verification process, it will first verify and check if the user data that was put in is correct or not, for example, username and password. However, on the second stage, which is the identification stage, the system tries to find who the subject is without any input information. Therefore, the use of iris scan is also a part of bioinformatics tools which detects out biological data processing and analysis.

3.5 BENEFITS AND LIMITATIONS OF BIOINFORMATICS TECHNOLOGY

Computational approaches have been very successful in facilitating, extending, and complementing experimental investigations including bioinformatics. There are benefits and limitations when it comes to using bioinformatics as they have become an indispensable component of many areas of biology. One of the benefits is that bioinformatics has quick sequencing capabilities which aid in saving time as it would be impossible for projects such as Human Genome Project to take place when it requires sequencing and storing of 3 billion nucleotides. If the human genome was to be analyzed using human brain instead of using bioinformatics, it would have taken many generations to complete. The second one is that with the help of advanced technology, it will eventually be possible to sequence DNA of individual patients which would provide physicians with new ways to identify diseases and cure them effectively (personalize medications). Moreover, bioinformatics can store a massive storage in an orderly manner (hard disk drive and indexing) and provide researchers with the storage space they require to store the data they collect. This is very helpful to researchers when they want to compare and analyze the DNA as they are already well organized.

Moreover, bioinformatics techniques such as image and signal processing allow extraction of useful results from large amount of raw data in experimental molecular biology likewise in structural biology, the use of bioinformatics assists in the simulation and modeling of DNA, RNA, and protein structures as well as molecular interactions. Similarly, in the field of genetics and genomics, bioinformatics help in sequencing and annotating genomes and their observed mutations. Bioinformatics plays a role in the textual mining of biological literature and the development of biological as well as gene ontologies to organize and query biological data. Additionally, they also play an important part in the analysis of gene and protein expression and regulation.

The other benefit of bioinformatics is that the tools of bioinformatics aid in the correlation of genetic and genomic data and more broadly in the understanding of evolutionary aspects of molecular biology. At a more integrated level, this bioinformatics tools helps to evaluate and catalog the biological pathways and networks that are an essential part of systems biology.

Computational models of system-wide properties could provide a basis for experimentation and discovery which will therefore result in not only explicit understanding of how organisms are built but also the capability to exhibit specific traits, to discover the casualty of diseases, and to foresee organisms' responses to changes in the environment. As a result, diseases such as HIV, acquired immunodeficiency syndrome, malaria, and so forth can be avoided and treated for the conservation of the environment. In the agriculture field, bioinformatics is also beneficial. Bioinformatics alters the genetic makeup of a certain crop and eventually results in a higher yield and a much more improved quality of food production.

In spite of all the benefits of bioinformatics, there are also limitations to it. The most evident drawback of using bioinformatics is that they are very costly as it requires computers and other technology tools. There is also a possibility of losing all the data collected due to a virus. Similarly, there is a chance for the algorithm to make an error in rare instances and a possibility of sequence matching due to a chance. String matching is also the major problem of the computational biology, text processing, and pattern recognition.

The other limitation of using bioinformatics is that biologists will generally have a much larger boundary of their own department of expertise and contribute most of their time on the computer instead of spending their time at the bench. In addition, analyzing other people's data will be much more commonplace since the concept of ownership will also change because of bioinformatics. Because of this change, scientists will be encouraged, if not forced, to concentrate more on the quality of data annotation and actively participate in their improvement. Last, the key to successful bioinformatics is close face to face teamwork between the biologist, biostatistician, and bioinformaticist. However, this is not always achievable due to the distances between institutions.

3.6 TRANSLATIONAL BIOINFORMATICS

Translational bioinformatics has become a significant aspect in the area of biological advancements. Thus, it has become the third major domain of informatics.[5] According to the American Medical Informatics Association, translational bioinformatics is defined as "...the development of storage, analytic, and interpretive methods to optimize the transformation of increasingly voluminous biomedical data, and genomic data, into proactive, predictive, preventive, and participatory health. Translational bioinformatics includes research on the development of novel techniques for the integration of biological and clinical data and the evolution of clinical informatics methodology to encompass biological observations. The end product of translational bioinformatics is newly found knowledge from these integrative efforts that can be disseminated to a variety of stakeholders, including biomedical scientists, clinicians, and patients" (Tenenbaum, 2016).

There are several reasons as to why translational bioinformatics is on reoccurring demand for the last decade:

First, the information technology (IT) infrastructures developed for translational research allows data managers and statisticians to acquire the data and advanced services required for them to accomplish their tasks.[8]

Second, there is an increase in tools that help in the assessment of molecular states. We can now measure a tremendous number of molecules at the same time. For example, the gene expression microarray has enabled us to determine the number of diseases and found novel subgenotypes of diseases in RNA across thousands of genes.[5] Furthermore, the tools are easily available to the public.

Last, the IT required may assist citizens to have a better healthcare approach as there will be access to prevention strategies, diagnoses, and therapy beforehand as a result from the anticipated results of translational research.[8]

As translational bioinformatics consists of evolved and advanced methods, tools, and services that have improved the connection between researchers and healthcare, there are challenges along the way.

One of the challenges is that semantic interoperability is hard to achieve. This is because, within health information systems, clinical data is often changed due to the continuous use of health informatics standards.

Another challenge is that biomolecular databases are becoming progressively large, complicated, and interoperated. This increases the possibility of risks in the growth, placement, and sharing of the database. Data warehousing can be a solution as it supports the reuse of gathered clinical data that is to be utilized for a different purpose and thus increases the quality of clinical data and restricts repetitions and errors.

3.7 CONCLUSIONS

Advanced technologies have occupied a lot in this modern world and basically have been part of the human life, from education, regular institutes, departments, workplace to business. It really plays an important role in acquiring the technology world so they are able to engage in ambitious projects. It gives an opportunity to people especially the scientists to have such an advancement that will help them to improvise their tasks or studies and basically made the human life much easier. Technology and computer advancement improves the scientist's world, in terms of gaining faster results, easy access to the internet world, able to process reliable data, and gather as much information as they can. However, in acquiring the technology world, it is also quite expensive to invest such a big amount from thousands to billions or even millions in the investment of computers, update software, databases. Without technology advancement, people may experience difficulty and may not be able to stay up to date. Basically, we are living in the world of vice versa. Therefore, it is important as well to have technology as part of our lives in order to increase, improve the standard of living of the people which enables rapid and a cost-effective generation.

Moreover, the use of biometrics technology can have its own advantages and also consequences. The unique features of bioinformatics technology is to provide high accuracy and validity, which not even humans are capable of doing. In the future, as IT will continue to advance, this will provide improvement for a biologist to make work much convenient and less time-consuming. Not just in biology area, but almost all parts of the area, such as businesses, social, and sciences.

KEYWORDS

- **information technology—IT**
- **IT emerging technology**
- **deoxyribonucleic acid (DNA)**
- **bioinformatics**
- **genomics**
- **proteomics**
- **short tandem repeats (STRs)**

REFERENCES

1. Al-Ageel, N.; Al-Wabil, A.; Badr, G.; AlOmar, N. Human Factors in the Design and Evaluation of Bioinformatics Tools. *Procedia Manuf.* **2015,** *3*, 2003–2010.
2. Altman, R. B. Genome Analysis. Bioinformatics Challenges for Personalized Medicine. *Bioinf. (Oxford, England)* **2011,** *27*(13), 1741–1748.
3. Bhattacharyya, D.; Ranjan, R.; Farkhod Alisherov, A.; Choi, M. Biometric Authentication: A Review. *Int. J. u- and e- Service Sci. Tec.* **2009,** *2*(3), 13–27.

4. Buehler, K. L.; Rashidi, H. H. (Eds.) *Bioinformatics Basics—Applications in Biological Science and Medicine;* Taylor & Francis Group: Florida, 2005.
5. Butte, A. J. Translational Bioinformatics: Coming of Age. *J. Am. Med. Inf. Assoc.* **2008,** *15*(6), 709–714.
6. Cannataro, M.; Santosb, R. W.; Sundnesc, J. Biomedical and Bioinformatics Challenges to Computer Science: Bioinformatics, Modeling of Biomedical Systems and Clinical. *Procedia Comput. Sci.* **2011,** *4*, 1058–1061.
7. Claverie, J. M.; Notredame, C. Finding Out What Bioinformatics Can Do for You. In *Bioinformatics—A Beginner's Guide;* Claverie, J. M., Ed.; Wiley India Pvt. Ltd.: New Delhi, 2009; p 9.
8. Daniel, C.; Albuisson, E.; Dart, T.; Avillach, P.; Cuggia, M.; Guo, Y. Translational Bioinformatics and Clinical Research Informatics: Medical Informatics, e-Health, 2013. http://link.springer.com/chapter/10.1007/978-2-8178-0478-1_17.
9. Fiannaca, A.; Rosa, M. L.; Urso, A.; Rizzo, R.; Gaglio, S. A Knowledge-Based Decision Support System in Bioinformatics: An Application to Protein Complex Extraction. *BMC Bioinf.* **2013,** *14*(1), S5. https://doi.org/10.1186/1471-2105-14-S1-S5.
10. Gopal, S. *Bioinformatics: A Computing Perspective;* McGraw Hill: New York, 2009; p 445.
11. Jason, H. M.; Folkert, W. A.; Scott, M. W. Genetics and Population Analysis: A Review. Bioinformatics Challenges for Genome-Wide Association Studies. *Bioinformatics* **2010,** *26*(4), 445–455. http://bioinformatics.oxfordjournals.org/content/26/4/445.full#abstract-1.
12. Kazimov, T.; Mahmudova, S. The Role of Biometric Technology in Information Security. *Int. Res. J. Eng. Technol.* **2015,** *2*(3), 1509–1513.
13. Kumar, S., Walia, E. Analysis of Various Biometric Techniques. *Int. J. Comput. Sci. Inf. Technol.* **2011,** *2*(4), 595–1597.
14. Marguerat, S. RNA-Seq: From Technology to Biology. *Cell. Mol. Life Sci.* **2009,** *67*, 569–579. DOI: 10.1007/s00018-009-0180-6 (accessed Sept 11, 2009).
15. Ridder, D. Pattern Recognition in Bioinformatics. *Brief. Bioinform.* **2013,** *14*, 402–410.
16. San Diego State University. Bioinformatics, Biological & Medical Informatics, n.d. http://informatics.sdsu.edu/bioinformatics/.
17. Shehab, S. A.; Keshk, A.; Mahgoub, H. Fast Dynamic Algorithm for Sequence Alignment Based on Bioinformatics. *Int. J. Comput. Appl.* **2012,** *37*(7), 54–61. http://research.ijcaonline.org/volume37/number7/pxc3876636.pdf.
18. Shelly, G. B. *Discovering Computers, Complete: Your Interactive Guide to the Digital World;* Cengage Learning: United States, 2012.
19. Singh, D. BioInfoKnowledgeBase, 2010. http://webapp.cabgrid.res.in/BIKB/edb_home.html (accessed Sept 18, 2010).
20. Singh, D. BioInfoKnowledgeBase: Comprehensive Information Resource for Bioinformatics Tools. *Am. J. Bioinf.* **2015,** 2(4), 28–33.
21. Tenenbaum, J. D. Translational Bioinformatics: Past, Present and Future. *Genom. Proteom. Bioinform.* **2016,** *14*(1), 31–41. http://www.sciencedirect.com/science/article/pii/S1672022916000401.

PART II
Advanced Dielectric Materials

CHAPTER 4

CHEMICAL MODIFICATION OF DIELECTRIC ELASTOMERS

CHRIS ELLINGFORD, CHAOYING WAN*, LUKASZ FIGIEL, and TONY MCNALLY

International Institute for Nanocomposites Manufacturing (IINM), WMG, University of Warwick, CV4 7AL, United Kingdom

Corresponding author. E-mail: Chaoying.wan@warwick.ac.uk

CONTENTS

ABSTRACT

Dielectric elastomers (DEs) can be used as energy transducers for actuation and energy-harvesting applications. This is due to their excellent mechanical properties, including strain, response and flexibility. However, the dielectric properties such as dielectric permittivity (ε) of non-polar elastomers must be improved before they can be used effectively in real-world applications for energy harvesting, especially for nonpolar DE. The examples of nonpolar DE are polydimethylsiloxane (PDMS) and styrene-butadiene-styrene (SBS). They can be modified extrinsically through blending with other polymers or through compositing with nanoparticles or by intrinsic modification through chemical reactions. Chemical modification creates the opportunity to incorporate polar molecules into the structure of a nonpolar DE, increasing the polarity of the material, and thus, the dielectric properties. This chapter investigates four main methodologies used to carry out these chemical modification reactions: hydrosilylation, thiol-ene click chemistry, azide click chemistry, and atom transfer radical polymerization. The effects on the dielectric and mechanical properties are analyzed for whether these systems could represent viable materials for inclusion in energy-harvesting systems.

4.1 INTRODUCTION

Dielectric elastomers (DEs) are a class of electroactive polymers proposed as promising candidates for actuators, as reported in the literature.[1] They are able to act as electromechanical transducers in which electrical energy passed through the DE induces a deformation of the structure, causing movement.[2] In recent years, DEs have gathered great interest in their potential applications as energy generators[3] by deformation of the elastomer structure to induce an electrical charge—a reversal of the energy input into the DE for actuation purposes. The tensile deformation of the flexible elastomeric structure causes opposite sides of the DE to be stretched and compressed simultaneously, creating polarization across the DE structure and generating a charge.[4] The charge generated is harvested by placing the DE between two compliant electrodes in a circuit.[4,5]

DEs have good physical properties aside from energy transduction. They are stable and electromechanically sensitive, cheap, easy to fabricate, and possess a potential simplicity in their implementation as the elastomers can easily be sandwiched between a pair of electrodes for energy harvesting.[6–9] However, DE requires significant improvement of their electromechanical properties to improve the efficiency of energy transduction within the system before their use in energy-harvesting applications can become widespread.[1]

Typically, there are three main challenges that must be overcome: improvement of the dielectric permittivity (ε) of the DE to increase the potential difference generated upon deformation with a minimal increase in dielectric loss (ε′) to prevent loss of energy in the form of heat, maintaining a high electrical breakdown strength (E_b) to prevent electrically induced failure of the DE, and maintaining a high tensile strength (T) alongside a good Young's modulus (E) and a maximum elongation at break (λ_{max}) for a strong, stretchable DE. Table 4.1 shows a summary of electrical and mechanical properties for the DE of styrene-butadiene-styrene (SBS), styrene-ethylene-butadiene-styrene (SEBS), polydimethylsiloxane (PDMS), and ethylene propylene diene monomer (EPDM) and their comparison to polyvinylidene fluoride (PVDF), a piezoelectric polymer, and ceramic piezoelectric materials barium titanate ($BaTiO_3$) and lead zirconium titanate (PZT) with the repeated units of the polymer backbones for SBS, SEBS, PDMS, EPDM and PVDF as shown in Figure 4.1.

TABLE 4.1 Comparison of Dielectric and Mechanical Properties for Materials of Interest.

Material	Class	ε	ε′	E_b ($V \cdot \mu m^{-1}$)	d_{33} ($pC \cdot N^{-1}$)	T (MPa)	E (MPa)	λ_{max} (%)
SBS[10]	Dielectric elastomer (DE)	3.90	3.0×10^{-4}	65	–	16.4	5.0×10^{-1}	1350
SEBS[11,12]	DE	2.45	5.0×10^{-4}	25	–	27.1	25.4	518
PDMS[13,14]	DE	2.50	2.0×10^{-4}	80	–	1.1	9.0×10^{-1}	200
EPDM[15]	DE	3.00	1.0×10^{-3}	20	–	1.0	9.0×10^{-2}	600
PVDF[16]	Semicrystal-line polymer	13	1.8×10^{-2}	160	20	290	1.1×10^{3}	450
$BaTiO_3$[17–19]	Ceramic	1700	1	38	191	59.0	1.2×10^{5}	–
PZT[18,20–22]	Ceramic	1300	5.0×10^{-2}	120	289	83.0	6.3×10^{4}	–

The DEs are not as stiff as PVDF and ceramic piezoelectric materials and possess a much greater flexibility compared to PVDF, allowing deformation of the structure, making them suitable for flexible and deformable energy-harvesting applications. The E_b value reported for SBS and PDMS is high with a similar value to PZT and PVDF, showing that DE could be used without the high risk of an electrically induced failure. However, it is clear that the ε values of all the DE are significantly lower than that of PVDF and the ceramic piezoelectric materials and do not exhibit a piezoelectric coefficient, for example, d_{33}, which ultimately prevents their use for energy-harvesting applications in their current form.

Thus, for the properties of these materials to become sufficient for large-scale use, there are idealized properties which have been suggested by Madsen et al. First, the ε should be at least 10 while the tan δ should be less than 5%, where tan $\delta = \frac{\varepsilon'}{\varepsilon}$. Second, the T should be greater than 2 MPa with a λ_{max} greater than 200% and an E of around 1 MPa. Finally, the E_b should be greater than 50 V·μm^{-1}.[1] However, this combination of properties is difficult to achieve as targeting the improvement of the ε tends to result in less favorable mechanical properties and a reduction in the E_b with an increase in the ε'.[1]

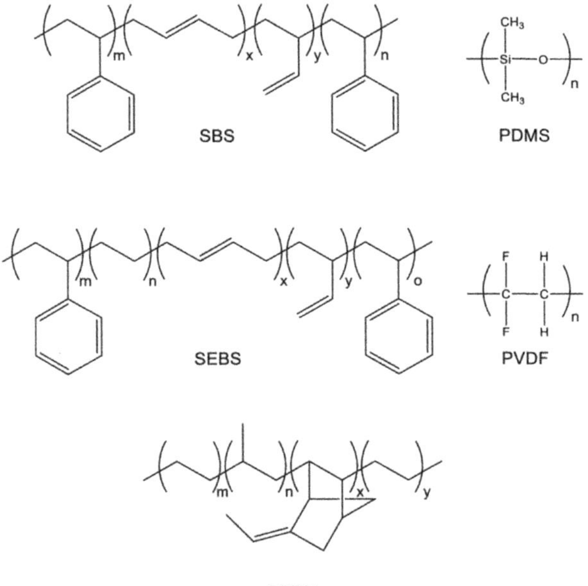

FIGURE 4.1 Examples of dielectric elastomers (DEs) and their chemical structures: SBS, SEBS, PDMS, EPDMS, and the semicrystalline polymer PVDF.

4.2 CHEMICAL MODIFICATION OF DIELECTRIC ELASTOMERS

To date, three approaches have been attempted to improve the properties of DE. Extrinsically, by blending different polymers to combine the advantageous properties they exhibit or the addition of ceramic piezoelectric nanoparticles to increase the dielectric properties and intrinsically by chemical modification of the backbone of the polymer to increase the polarity of the structure.

Chemical grafting of organic dipoles such as cyanopropyl and nitrobenzene onto the backbone of the DE increases the polarity of the structure, leading to an increase in the ε of the elastomer.[23–25] Chemical modification introduces no issues with dispersion and compatibility[26] and the resulting elastomeric systems tend to be more stable upon continued activation.[27] However, the addition of polar groups does not come without drawbacks. It is noted that the flexibility and E_b of the elastomer typically decreases, impacting the electromechanical properties[28] as well as simultaneously increasing water sensitivity and the glass transition temperature (T_g) of the resulting elastomers.[29,30]

PVDF has been modified using chemical modification techniques with examples including the grafting of bulky side groups such as styrene[31] or by synthesizing block copolymers using PVDF.[32] Grafting PVDF with 39 wt.% of styrene resulted in a large increase in the ε from 13 for pure PVDF to 80 and resulted in only a small increase in ε′ from 4.0×10^{-3} to 2.4×10^{-2}.[31] The synthesis of a PVDF-co-2-hexaethylmethacrylate block copolymer resulted in an increase in the ε from 13 for pure PVDF to 45. It also led to a decrease in the tan δ from 3.0×10^{-3} to 2.0×10^{-3} at 105 Hz due to the favorable interactions between the fluoride groups on PVDF and the hydroxyl groups on 2-hexaethylmethacrylate, making the resulting matrix more homogenous.[32]

Research into this area is still in its infancy, with more focus being applied to exploring the potential of this field. So far, research has mainly focused on applying the techniques of hydrosilylation, thiol-ene click chemistry, azide click chemistry, and atom transfer radical polymerization (ATRP) to modify elastomers and investigate their dielectric properties for either actuation or energy-harvesting applications.

4.2.1 HYDROSILYLATION

Hydrosilylation chemical grafting reactions are typically carried out with silicone-based elastomers, primarily PDMS, or poly (methyl hydride) siloxane (PHMS) where Si–H bonds are added to a vinyl group (Fig. 4.2).[33] A nonfunctionalized silicone-based elastomer typically has a dielectric constant of 2.3–2.8[34] and an E_b of approximately 80 V·μm^{-1}.[14] A significant number of organic molecules have been grafted onto siloxane chains including ethers,[35] esters,[35] epoxides,[35–37] carbonates,[38] amines,[35] and aromatic rings[39,40] although no dielectric data has been reported for these modifications.

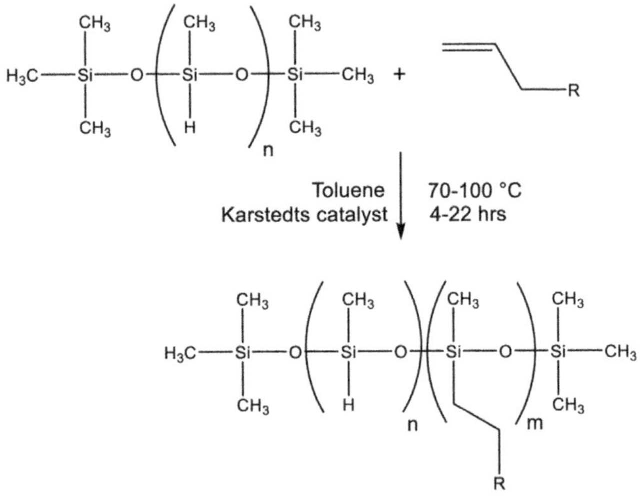

FIGURE 4.2 General reaction scheme and conditions for hydrosilylation on silicone-based polymers.

Typical reaction conditions for hydrosilylation involve reacting PHMS with a vinyl-terminated organic molecule using Karstedt's catalyst while under an inert atmosphere due to the water sensitivity of the reaction. The reactions are heated to 70–100°C and left to react for between 4 and 22 h.[14,41] The thin films are formed by evaporation of the solvent on a nonstick surface and cross-linking is achieved by heating to a high temperature for less than an hour.[14] Due to the requirement for an inert atmosphere and dry conditions during the hydrosilylation step of modified elastomer film formation, an increase in the complexity and the cost in carrying out the reaction for large-scale synthesis compared to other potential reaction methodologies is somewhat observed. When reported, the thicknesses of the thin films formed are between 100 and 150 μm.[42,43]

The ε of pure PDMS increased from 2.5 to 4.2 when the grafting level of allyl cyanide reached 8.0 mol.%[43] followed by an increase up to 15.9 with a grafting level of 89 mol.%.[23] As shown in Figure 4.3, the ε results show a linear relationship with the grafting level of allyl cyanide between different works. The impact of cross-linking the elastomers has no significant effects on the dielectric properties reported between the cross-linked and the uncross-linked results, as previously reported for polymers in the literature,[44] allowing the direct comparison of cross-linked and uncross-linked results.

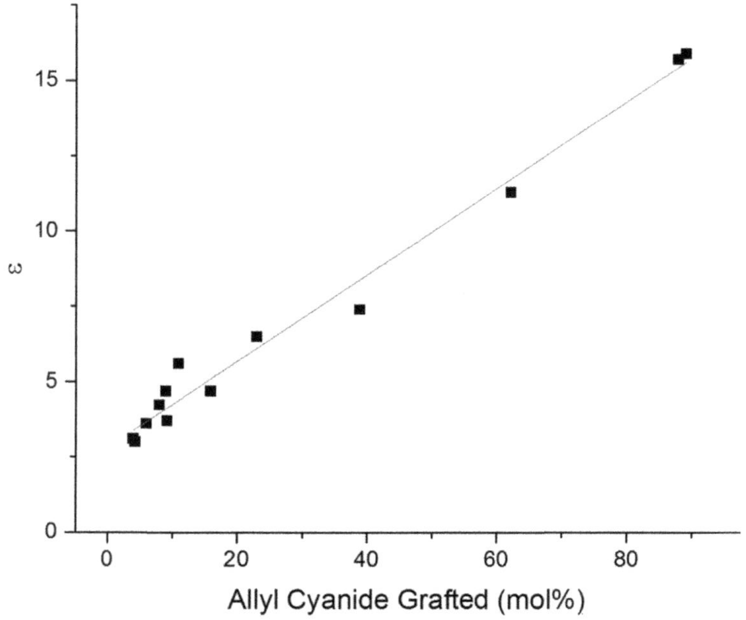

FIGURE 4.3 Linear correlation between allyl cyanide content and the increase in ε.

However, by increasing the dipole content of allyl cyanide the ε′ was shown to increase exponentially regardless of the level of dipole grafting, by increasing from 2.0×10^{-4} for pure PDMS to 4.2×10^{-3} at 8 mol.% grafting[43] and up to a very high value of 2.5 at 89 mol.% grafting (Fig. 4.4).[23] This demonstrates that to produce an elastomer usable in energy-transducing applications, there is an optimum level of grafting to be found for each dipole, which leads to the desired increases in ε, with minimal effect on the ε′.

It is also seen that the incorporation of allyl cyanide affects the mechanical properties and E_b of elastomers. Blending allyl cyanide-modified PHMS (CNATS-993, Fig. 4.5) with pure PDMS[14] and grafting modified PHMS to PDMS[45] were carried out to improve the mechanical and electrical properties exhibited without impacting the ε. CNATS-993 acted as a plasticizer when used as a filler, resulting in an elastomer with a lower E compared to the pure PDMS. The elastomer displayed a reasonable ε of 7.0 and a low ε′, but the E_b was reduced from 80 to 20 $V \cdot \mu m^{-1}$.[14]

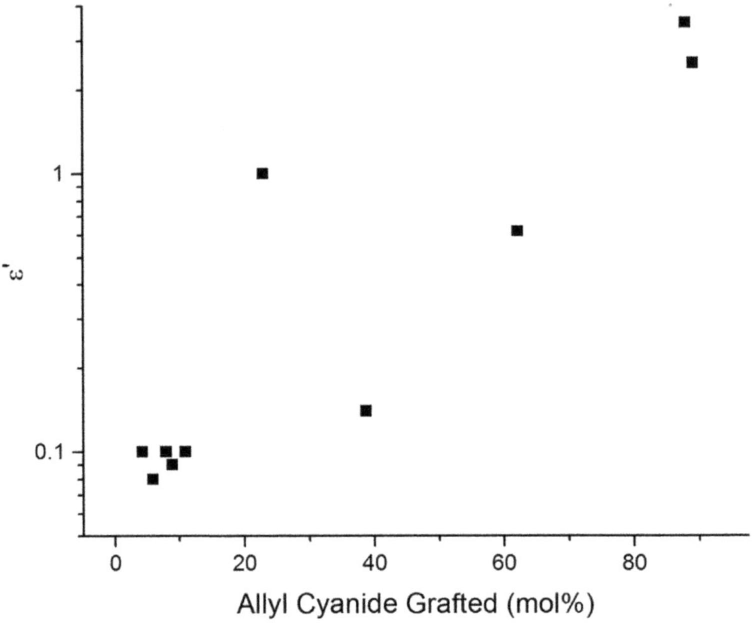

FIGURE 4.4 Graph showing the correlation between ε′ and grafted content of allyl cyanide.

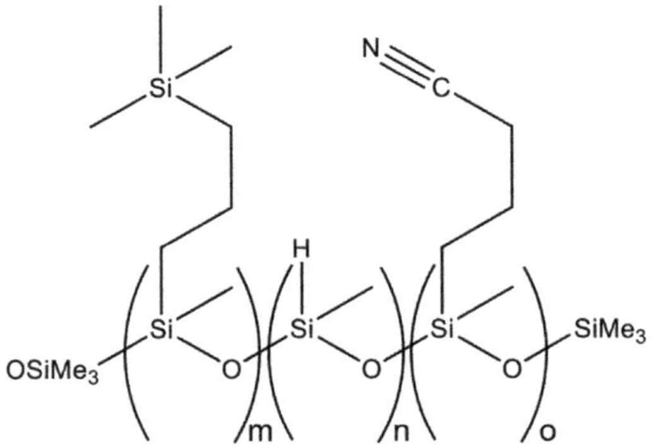

FIGURE 4.5 Chemical structure of CNATS-993.

By grafting allyl cyanide-modified PHMS to PDMS in a 1:2 ratio, a small increase in the ε from 2.5 to 4.9 was observed. The mechanical properties were greatly affected by the incorporation of the polar group, reducing the T and E_b from 3.20 MPa and 49 V μm^{-1} for the pure PDMS sample to 0.49 MPa and 29 V μm^{-1}, respectively.[45] The sections of high-density cyanopropyl groups significantly disrupt the structure of the elastomer, as the cyanopropyl groups are not spread evenly throughout the elastomer structure, resulting in a weaker product compared to the pure PDMS.

Changing the organic dipole grafted onto less polar allyl chloride[46] still impacts the properties exhibited by the elastomers. A maximum of 16.1 mol.% of allyl chloride was grafted onto PHMS-g-PDMS, resulting in an ε of 4.7 and a low ε′ of 4.5×10^{-3}.[42] The ε is similar to the ε for 8.0 mol.% grafting of allyl cyanide as the increase in polarity within the elastomeric structure is the main factor for increasing in the ε.[43] The impact on the mechanical properties and E_b is very low and produces quite promising results. The E_b remained high with a value of 94.4 V·μm^{-1} and the elastomer strong but a good elasticity modulus with a T of 2 MPa, an E of 1 MPa and a λ_{max} of 130%.[43] It is therefore feasible that a higher doping level of allyl chloride could result in a greater ε than what has been reported, and it could have less of an impact on the mechanical properties and E_b in comparison to allyl cyanide.

Increasing the polarity of the grafted dipole from chloro to trifluoro should increase the ε of the elastomer obtained. Using cyclic siloxane rings of 1,3,5-tris (3,3,3-trifluoropropylmethyl)-1,3,5-trimethylcyclotrisiloxane (TFP-TMC), a ring-opening reaction with cyclic PDMS was performed to form a trifluoropropyl-modified siloxane elastomer.[41] While a hydrosilylation reaction was not used to attach the trifluoropropyl group, the results for the siloxane are valid for comparison in this section as it is the same product that would be obtained if the trifluoropropyl groups were grafted onto PHMS through hydrosilylation.

A 57.5 mol.% grafting level of trifluoropropyl groups within the elastomer resulted in an ε of 6.4,[41] comparable with a 23.0 mol.% grafting level of cyanopropyl groups to PDMS.[24] However, the ε′ is two orders of magnitude lower when modifying with trifluoropropyl, meaning the efficiency of energy harvesting is better than that for allyl cyanide.

Interestingly, changing the group between chloro and trifluoropropyl does not appear to have a big effect on the ε, even though the polarity increases. A grafting of 28 mol.% trifluoropropyl resulted in an ε of 5.1,[41] whereas at a 16-mol.% grafting of allyl chloride, the ε was 4.7.[42] This is a small increase in ε for an increase in polarity and 12 mol.% grafting level.

The mechanical properties of 57.5 mol.% trifluoropropyl-grafted siloxane elastomers were severely impacted. A low strength of 5.0×10^{-2} MPa and an E of 1.8×10^{-2} MPa were reported, forming a gel-like elastomer. Reducing the grafting level to 52.9 mol.% did not significantly impact the dielectric properties of the material, but did increase the tensile strength to 1.3×10^{-1} MPa alongside a doubled λ_{max}. No change in the elasticity modulus was observed.[41] However, the elastomer is still too soft to use for energy-harvesting applications.

A different set of work focused on grafting N-allyl-N-methyl-p-nitroaniline onto different commercially available siloxane chains, the dielectric and mechanical properties were influenced Table 4.2). N-allyl-N-methyl-p-nitroaniline is readily described as a "push–pull" dipole, increasing compatibility with the silicone matrix.[47]

PDMS DMS-V31 and DMS-V41, molar masses of 28,000 and 62,700 g·mol^{-1}, respectively, were grafted with up to 13.4 wt.% of N-allyl-N-methyl-p-nitroaniline, increasing the ε for both elastomers to 5.98 for DMS-V31 and 5.4 for DMS-V41. However, by incorporating N-allyl-N-methyl-p-nitroaniline, the ε′ of the elastomers increased by an order of magnitude with respect to the unmodified siloxane chains.[47,48]

DMS-V31 has a shorter chain length compared with DMS-V41, leading to a greater cross-linking density. The impact of this is seen in the *T* of DMS-V31 compared with DMS-V41, 1.7 MPa compared with 0.9 MPa, respectively, despite 13.4 wt.% of N-allyl-N-methyl-p-nitroaniline grafted. However, the dipole does affect the elastic properties of both of the elastomers, with the *E* of DMS-V31 decreasing from 1.7 to 0.3 MPa and for DMS-V41 decreasing from 0.95 to 0.25 MPa. This demonstrates that the increased cross-linking density has a minimal impact on the *E* and the grafted dipole is the key factor for reduction in the elasticity modulus. As expected, the E_b also decreases sharply upon increasing the grafted dipole concentration from 130 and 80 V·µm^{-1} to 40 and 30 V·µm^{-1} for DMS-V31 and DMS-V41, respectively.[47,48]

Elastosil RT625 and Sylgard 184 were also investigated with grafted N-allyl-N-methyl-p-nitroaniline, both at a grafting degree of 10.7 wt.%. Even though the grafting degree was lower, the ε for Elastosil RT625 was between DMS-V31 and DMS-V41, whereas the ε for Sylgard 184 was the highest of all four siloxanes. Both of these modified elastomers were mechanically weaker than DMS-V31 and DMS-V41 with tensile strengths of 0.1 and 0.5 MPa, respectively. Elastosil RT625 elastomer is therefore too soft for energy-harvesting applications. However, while the Sylgard 184 elastomer was relatively soft, it also had the best elasticity modulus with an E of 0.85 MPa.[49] This, coupled with the relatively high E_b and good ε, means it is the best of the four PDMS candidates to continue with for further modification. However, the lack of strength is a disadvantage. Blending of Sylgard 184 and DMS-V31 could produce an elastomer that, when modified, has both good electrical and mechanical properties.

TABLE 4.2 Comparison of Dielectric and Mechanical Properties of Various Forms of PDMS Modified with N-allyl-N-methyl-p-nitroaniline.

Polymer backbone	Organic dipole	Grafting level (wt.%)	Cross-linked	Yield (%)	ε	ε′	E_b (V μm^{-1})	E (MPa)	T (MPa)	λ_{max} (%)
DMS-V31[47,48]	N-allyl-N-methyl-p-nitroaniline 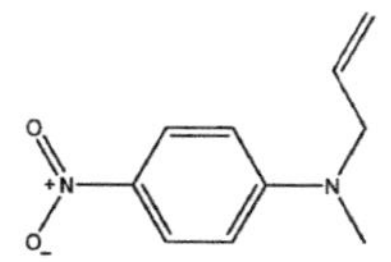	13.4	Yes	–	5.98	0.1	40.0	0.30	1.7	–
DMS-V41[48]	N-allyl-N-methyl-p-nitroaniline 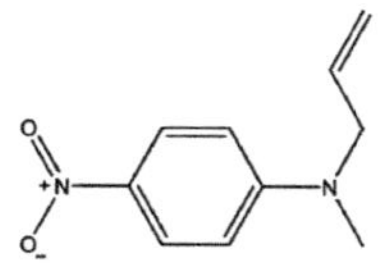	13.4	Yes	–	5.40	0.1	30.0	0.25	0.9	–
Elastosil RT625[49]	N-allyl-N-methyl-p-nitroaniline 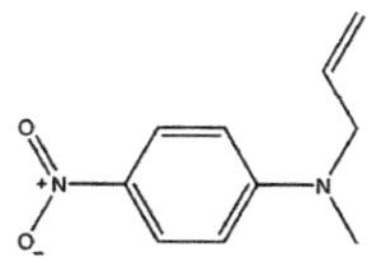	10.7	Yes	–	5.56	0.1	30.7	0.14	0.1	175
Sylgard 184[49]	N-allyl-N-methyl-p-nitroaniline 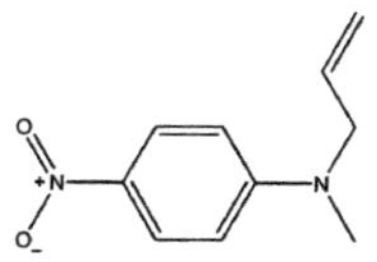	10.7	Yes	–	6.15	0.1	61.1	0.85	0.5	390

The effect of polarity from the grafting of different functional groups to siloxane chains has been investigated by Racles et al. to determine whether or not there is a direct correlation between the dipole moment of the side-chain group and the observed ε.[43] Allyl aldehyde, allyl glycidyl ether, 4-aminopyridine, allyl cyanide, and Disperse Red 1 were grafted onto PHMS-g-PDMS at a doping level of 8 mol.%, while chloropropane-thiol and 3-mercaptopropionic acid were grafted using thiol-ene click chemistry (see Section 4.2.2 for more detail on thiol-ene click reactions) using vinyl-modified siloxane chains (PVMS-g-PDMS). The dipole moments for the grafted organic dipoles were calculated using density functional theory and molecular mechanics.[43]

The general trend observed was an increase in the dipole moment of the organic molecule grafted increasing the ε. The lowest polarity molecule resulted in an ε of 3.8, which went up to a value of 7.4 when higher polarity groups were grafted. However, only a weak correlation was found (Fig. 4.6). Some of the elastomers displayed unexpected values of ε if the ε was dependent only on the dipole moment. The dipole moment of polar group Disperse Red 1 is 10.40 D with the modified elastomer exhibiting an ε of 5.4, whereas 4-aminopryidine had a dipole moment of 5.67 D but showed a higher ε of 7.4. A similar disagreement toward the general trend is between allyl glycidyl ether and chloropropane-thiol. Both have the same ε of 3.8 but a different dipole moment, 2.19 and 3.22 D, respectively.[43] This shows that other factors, not just polarity, affect the resulting dielectric properties for the modified elastomers. This is from a combination of steric factors and water sorption ability of the organic dipoles, as the ions from the water affect the dielectric properties.[50] The effect of this is further seen in the differences in ε′ between 4-aminopyridine and Disperse Red 1. The ε′ for Disperse Red 1 is 2.2×10^{-1}, but it rises greatly to 5.9 for 4-aminopyridine.[43] Therefore, 4-aminopyridine-modified elastomers are unsuitable for energy-transducing applications; however, elastomers grafted with Disperse Red 1 could be suitable candidates for further research if the mechanical properties and E_b of the elastomers satisfy the desired requirements.

An interesting adaptation to the modification of PDMS is to form an elastomer with polar organic groups on the cross-linker instead of on the main chain. Several ready modified cross-linkers can be bought and used directly in reactions to form elastomers, removing an entire synthetic procedure and overcoming the challenges from hydrosilylation as the water sensitivity element of the reaction is negated. However, the maximum achievable degree of grafting is significantly lower than chemical modification of the elastomer backbone.[51]

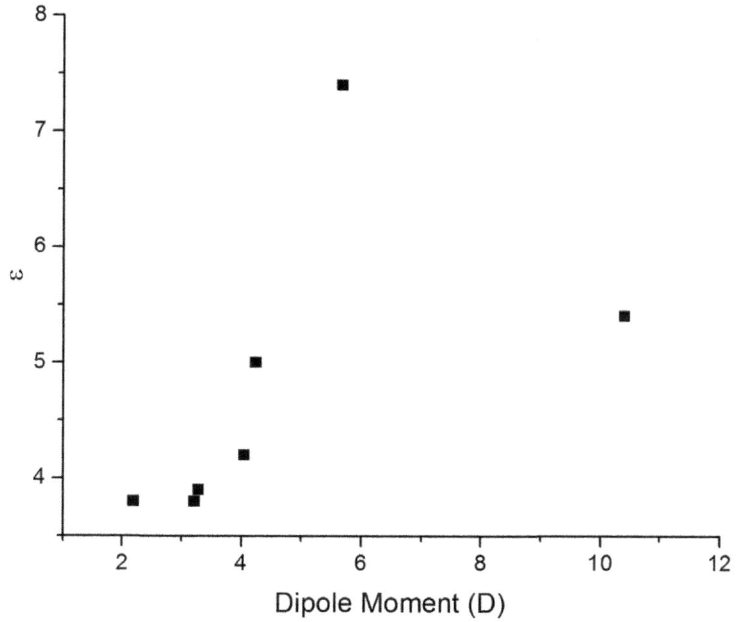

FIGURE 4.6 Graph showing the weak correlation between dipole moment of organic dipole grafted and the ε of the elastomer.

The grafting of cyanopropyl groups onto the cross-linker resulted in the biggest increase in ε for PDMS, from 2.5 to 3.7. This result was unsurprising as cyanopropyl had the highest dipole moment of all of the organic groups. Surprisingly, the addition of the chloropropyl group resulted in a smaller increase in ε than the aminopropyl group on the cross-linker, even though the polarity of chloropropyl is greater.[51] This was attributed to the increase in free volume from the larger size of the chloro group;[51] however, no water sorption experiments for the elastomers were reported.

Cross-linkers modified with phenyl and cyanopropyl resulted in the best mechanical properties with a *T* of 1.2 MPa and an *E* of 0.68 MPa for phenyl, and a *T* of 2.1 MPa and an *E* of 0.95 MPa for cyanopropyl.[51] The good mechanical properties of cyanopropyl and phenyl-modified elastomers compared to the modification with methyl, chloropropyl, and aminopropyl is likely due to inter- and intramolecular interactions arising from the addition of the cyanopropyl group or additionally π–π stacking between the phenyl-modified cross-linkers. Chemical modification of future elastomers could involve both grafting to the chain and to the cross-linker, as long as the cross-linker increases the strength of the material instead of disrupting the structure.

4.2.2 THIOL-ENE CLICK CHEMISTRY

Thiol-ene click grafting reactions can be carried out on any elastomers in which a vinyl group is present such as SBS or polyvinylmethylsiloxane (PVMS) in which an S–H bond is added across the double bond (Fig. 4.7). Typically, unmodified SBS has an ε of 3.90 and a good E_b of 65 $V \cdot \mu m^{-1}$.[10] Typical reaction conditions for the modification of elastomers involve dissolving a photoinitiator such as DMPA, elastomer and organic molecule in solvent, and then irradiation of the solution with ultraviolet (UV) light. The reaction is carried out without the need for an inert atmosphere, unless the organic molecule is at the risk of becoming oxidized.[52] The grafting level for thiol-ene click chemistry is determined by how long the solution is exposed to UV light, making the modification easily controllable.

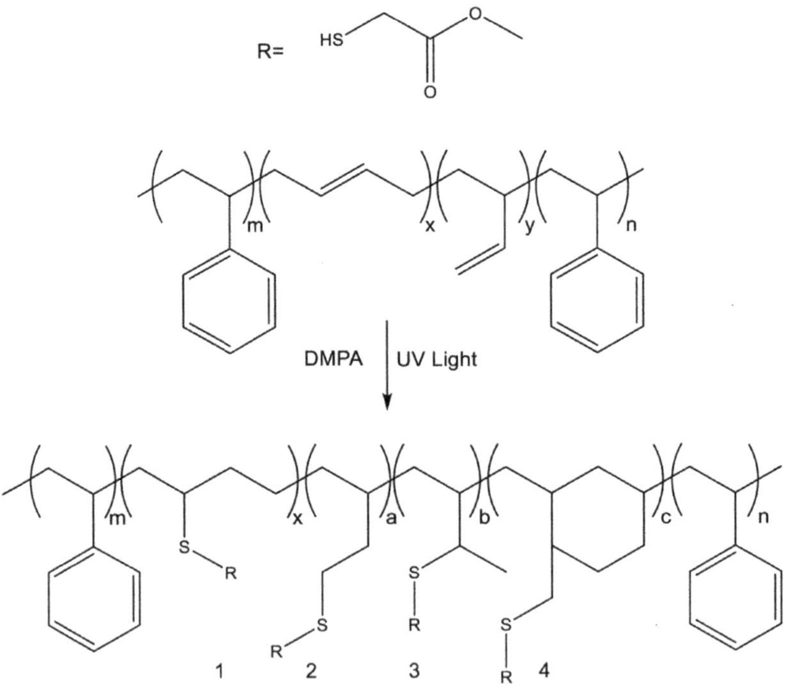

FIGURE 4.7 Thiol-ene click reaction scheme for modification of SBS in which there are four different ways the thiol can bond.

A number of different polar groups have been grafted onto butadiene-based polymers including mercaptan groups with amine,[53,54] carboxylic acid,[53–55] ester,[54,55] and cyclic ether functionalities.[53,55] However, very little work has been done on investigating the resulting dielectric properties of the elastomers. The use of click chemistry for chemical modification of elastomers is a desirable concept, especially when the industrial applications are concerned. This is due to the inherent relative ease that click chemistry reactions provide owing to the simple reaction procedures, simple purification steps, high product yields, and fast reaction times.

Methyl thioglycolate and thioglycolic acid, an ester and carboxylic acid, respectively, were grafted onto SBS; one to the pristine SBS and the other to regular SBS (Table 4.3). Pristine SBS has a regular alignment of polymer chains, increasing the crystallinity of the elastomer compared with the regular SBS,[56] affecting the macroscopic properties of the elastomer.[57] The difference in ε of the two SBS elastomers is immediately seen before modification, as regular SBS has an ε of 3.9[10] and pristine SBS has a lower ε of 2.2.[52]

The high-level grafting of thioglycolic acid to pristine SBS of 83 mol.%, resulted in an increase in ε from 2.2 to 7.2, with an acceptable ε′ of 3.0×10^{-1}.[52] Grafting of methyl thioglycolate onto regular SBS resulted in an elastomer with significantly better dielectric properties, with an ε of 12.2 as well as a reduced ε′ of 7.0×10^{-2} with respect to using pristine SBS.[10] However, it is not clear if the difference in ε is due to structural differences between the elastomers or due to the different functional groups on the organic dipoles.

The mechanical properties of the two elastomers are also different. The modified regular SBS has a high strength, with a T of 3 MPa while simultaneously having a good elasticity with an E of 0.34 MPa and a λ_{max} of 1400%. However, the mechanical properties are poorer for the modified pristine SBS. The strength of the material is less, with a T of 1.1 MPa, and the elastomer is less stretchy and more brittle with an E of 3.3 MPa and a λ_{max} of 300%. From these results, modification of regular SBS appears to be the better option for future modifications. However, the E_b for both elastomers was quite low, 15.7 V·μm^{-1} for regular SBS and 16.0 V·μm^{-1} for pristine SBS. The similarity indicates that it is the incorporation of the grafted dipole reducing the $E_{b,}$ rather than the differences in the structure of the SBS used.

A different approach for using thiol-ene click chemistry has been to synthesize siloxane chains from cyclic monomers containing vinyl groups to give double bond functionality in the elastomer after a ring-opening polymerization.[30,43] For example, chloropropane-thiol and 3-mercaptopropionitrile were grafted onto PVMS-g-PDMS. This work demonstrated that a polar group with a larger dipole moment has the greatest effect on the dielectric permittivity. However, other factors such as the water sorption ability of

TABLE 4.3 Electrical and Mechanical Properties of Elastomers Modified Using Thiol-ene Click Chemistry.

Polymer backbone	Organic dipole	Grafting level (mol.%)	Ultraviolet (UV) light exposure (min)	ε	ε′	E_b (V·µm^{-1})	T (MPa)	E (MPa)	λ_{max} (%)
Regular SBS[10]	Methyl thioglycolate 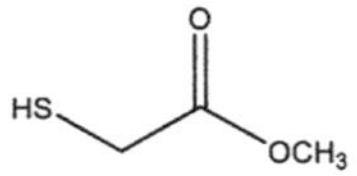	81	40	12.2	7.0×10^{-2}	15.7	3.0	0.34	1400
Pristine SBS[52]	Thioglycolic acid 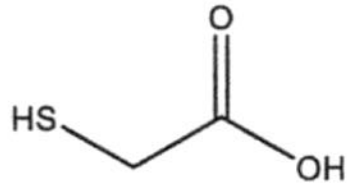	83	30	7.2	3.0×10^{-1}	16.0	1.1	3.30	300
PVMS-*g*-PDMS[43]	Chloropropane-thiol 	8	2 × 15	3.8	2.1×10^{-2}	–	–	–	–
PVMS-*g*-PDMS[43]	3-Mercaptopropionic acid 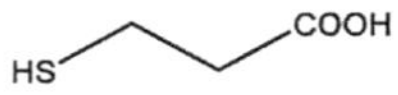	8	2 × 15	3.9	2.0×10^{-2}	–	–	–	–
PVMS-*g*-PDMS[30] uncross-linked	3-Mercaptopropionic acid 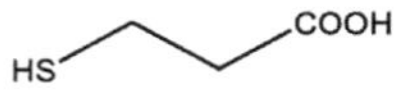	100	5	18.4	~1	–	–	–	–
PVMS-*g*-PDMS[58] cross-linked	3-Mercaptopropionic acid 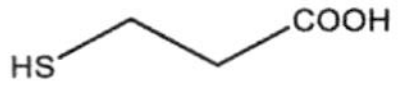	100	4	17.4	~5	15.6	0.4	0.16	320

the organic molecule had secondary influences on the ε and ε′, leading to unexpected results (see Section 4.2.1 for further information).[43]

A different work has investigated grafting 3-mercaptopropionitrile to PVMS at high grafting levels. 3-mercaptopropionitrile is an analogous molecule to allyl cyanide and can be compared to the results in Section 4.2.1. This led to increases in the ε to similar values for both the uncross-linked and cross-linked elastomer, 18.4 and 17.4, respectively, agreeing with the previous studies which suggest cross-linking has no direct impact on the dielectric properties of elastomers.[44] These increases are similar to the ε with allyl cyanide (15.9 at 89 mol.% grafting) showing that the additional sulfur group does not have any significant impacts on the structure of the elastomer. Similarly, the ε′ of the elastomers is very high with an approximate value of 1 for the uncross-linked elastomer[30] and an approximate value of 5 when it is cross-linked.[58] These high values correlate to the reported ε′ for allyl cyanide of 2.5 and follow the general trend in the literature that as the grafting level of the organic dipoles is increased, the ε′ increases exponentially.

The mechanical properties of the cross-linked elastomer are also very poor, perhaps to be expected with such a high grafting level of 3-mercapto-propionitrile. The strength of the elastomer is low, with a *T* of only 0.4 MPa and a very low *E* of 0.16 MPa. The E_b of the elastomer is also reduced, 15.6 V·μm^{-1}. This coupled with the poor mechanical properties means that the silicone elastomer in its current form with 100 mol.% grafting is not suitable for energy-transducing applications.

4.2.3 AZIDE CLICK CHEMISTRY

Azide click chemistry reactions have been carried out to chemically modify silicone-based elastomers by a copper-catalyzed cycloaddition reaction between an azide and alkyne group, forming a 1,4-disubstituted product (Fig. 4.8). A number of molecules have been grafted onto the cross-linker used in PDMS elastomers using this reaction, including molecules with aromatic rings, fluorinated aromatic systems, and ferrocene sandwich systems; however, no dielectric data was recorded for these.[25]

Typical reaction conditions for azide click chemistry involve dissolving the azide-modified PDMS (AMS) and alkyne-modified organic molecules in a dry solvent with Et_3N as a base and CuI as a catalyst. The reaction was stirred for 17 h at 40°C after which the product was obtained as an oil in a quantitative yield.[59] This click chemistry reaction does take longer to form the modified films compared to thiol-ene click chemistry, as the reaction proceeds through a different route; a thermally induced rather than a UV-induced pathway. However, similar to the thiol-ene click reaction, the reaction does produce a high yield of product with no difficult purification techniques required.

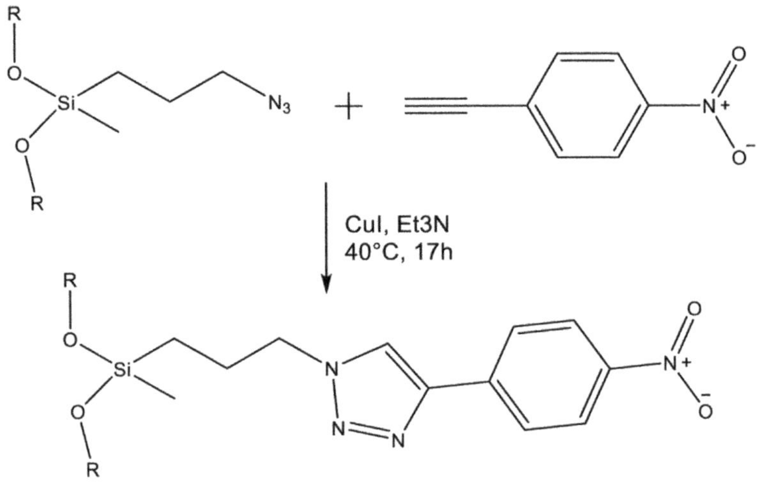

FIGURE 4.8 Reaction scheme for azide click chemistry to form a 1,4-disubstituted product.

Nitrobenzene was grafted onto PDMS-g-AMS at different grafting levels and with different lengths of PDMS spacer molecules as part of the elastomer backbone to vary the distance between the nitrobenzene groups.[59] From Table 4.4, grafting 51 mol.% of nitrobenzene to PDMS increased the ε to 5.1. However, by increasing the nitrobenzene content yet further to 100 mol.% grafting, the value of ε did not increase any further. This is in contrast to results seen in other works and is possibly due to a smaller increase in the density of nitrobenzene groups within the elastomer as a longer PDMS spacer molecule is used. However, this change in grafting level does result in an increase in tan δ from 4.0×10^{-2} to 6.0×10^{-1}. This is expected as increasing the grafting level should increase tan δ and shows that no mistake has been made in determining the ε of both elastomers. Both sets of work show a good E_b, 69.2 and 60.5 $V \cdot \mu m^{-1}$, respectively, suggesting that these silicone-based elastomers would be resistant to electrically induced failures. However, no mechanical properties are reported for the determination of the strength and elasticity of the elastomer.[59]

TABLE 4.4 Properties for Elastomer Systems Modified by Azide Click Chemistry.

Polymer backbone	PDMS spacer length ($g \cdot mol^{-1}$)	Organic dipole	Grafting Level (mol.%)	ε	Tan δ	E_b ($V \cdot \mu m^{-1}$)
PDMS-g-AMS[59]	1200	Nitrobenzene 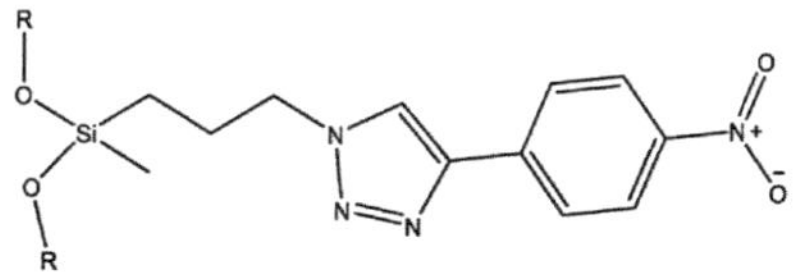	51	5.1	4.0×10^{-2}	69.2
PDMS-g-AMS[59]	1200	Nitrobenzene 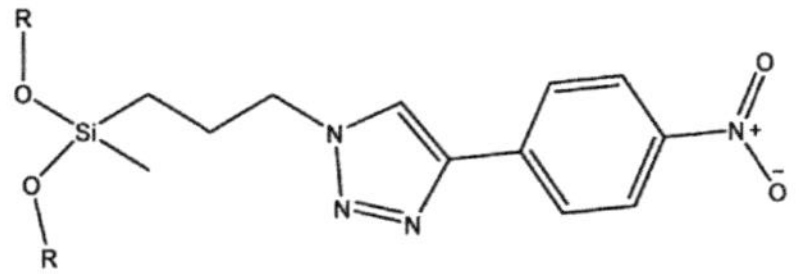	100	5.0	6.0×10^{-1}	60.5
PDMS-g-AMS[59]	580	Nitrobenzene 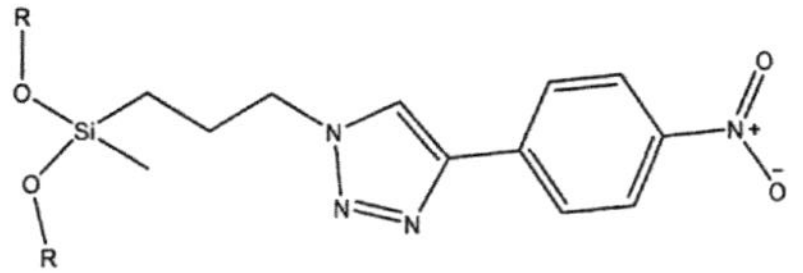	42	7.3	1.5×10^{-1}	64.1
PDMS-g-AMS[59]	580	Nitrobenzene 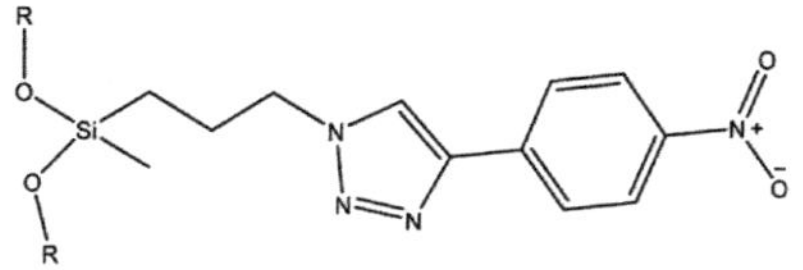	100	8.5	9.0×10^{-1}	65.0

TABLE 4.4 *(Continued(*

Polymer backbone	PDMS spacer length ($g \cdot mol^{-1}$)	Organic dipole	Grafting Level (mol.%)	ε	Tan δ	E_b ($V \cdot \mu m^{-1}$)
PDMS[25]	28,000	Nitrobenzene (on cross-linker) 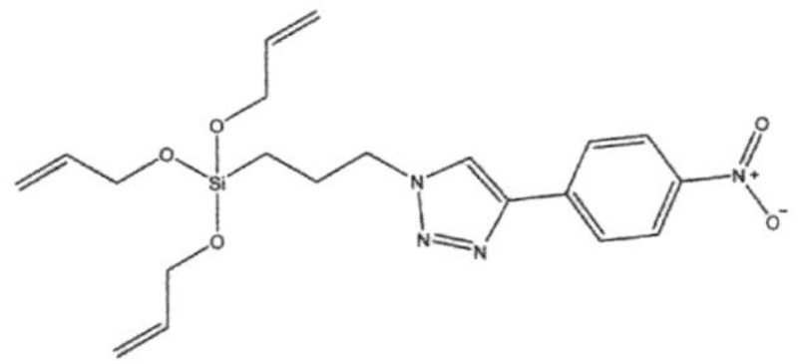	0.25 wt.%	3.1	1.0×10^{-3} (ε′)	–
PDMS[61]	6000	Nitroazobenzene (on cross-linker) 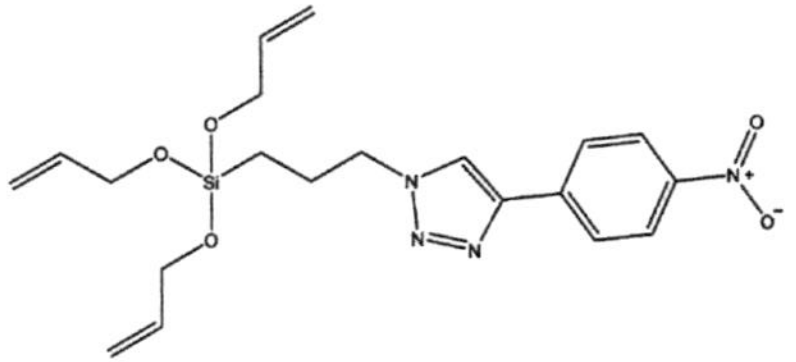	1.35 wt.%	3.2	5.2×10^{-4}	124.2

By reducing the PDMS spacer length from 1200 to 580 $g \cdot mol^{-1}$, the ε increases due to an increase in the density of the organic dipole groups within the chain. At a 42 mol.% grafting level, the ε is already higher than when a longer PDMS spacer chain is used, reporting a value of 7.3. A further increase in the grafting level to 100 mol.% leads to an increase in ε to 8.5. This contrasts with the changes in ε seen with longer PDMS spacer chains used; however, it does put the observed results in line with the literature expectations. The tan δ values are slightly higher for the shorter PDMS spacer lengths, increasing from 1.5×10^{-1} at 42 mol.% to 9.0×10^{-1} at 100 mol.% grafting. The E_b values remain high, 64.1 and 65.0 $V \cdot \mu m^{-1}$, respectively. In fact, the E_b values for all of the modified silicone elastomers are higher than the reference PDMS elastomer, 55.4 $V \cdot \mu m^{-1}$.[59] This is due to the purification steps undertaken in the work once the films are formed to remove the metallic copper impurities,[59] as these can affect the dielectric properties of the films formed.[60]

The systems incorporating nitrobenzene- and nitroazobenzene-modified cross-linkers using azide click chemistry have been reported. As expected, the increases in ε are much lower when organic dipoles are grafted onto the cross-linker used to form the elastomer compared to the modification of the main silicone chain, also seen in Section 4.2.1. Both nitrobenzene and nitroazobenzene, a mesogenic-type organic dipole (see Section 4.2.4 for more information on mesogens), increased the ε from 2.8 to 3.1[25] and 3.2,[61] respectively. This is a large increase considering that only a 0.25 wt.% loading of nitrobenzene-modified cross-linker was used.[25] Different lengths of PDMS chains are used in these works; however, an increased chain length appeared to have no detrimental impact on the ε in this case. The ε′ for using nitrobenzene is low, 1.0×10^{-3}, approximately an order of magnitude increase compared to a purely azide-modified cross-linker in the elastomer. For nitroazobenzene, the increase in tan δ was from 6.0×10^{-4} to 5.2×10^{-4}, a small increase for a larger loading of the organic dipole. At this concentration, the E_b increased from 110 to 124.2 $V \cdot \mu m^{-1}$. The nitroazobenzene increased the breakdown strength of the structure either through intermolecular interactions from the nitro groups or by stabilization of charge through the aromaticity of the dipole.[61] However, at higher grafting levels of nitroazobenzene, the ε remains the same but a decrease in E_b is observed to 57.0 $V \cdot \mu m^{-1}$ at 3.60 wt.% loading. This shows that at 1.35 wt.%, the threshold at which the organic dipole has a negative effect on the E_b has not been reached yet;[62] however, it is lower than 2.25 wt.%.[61]

4.2.4 ATOM TRANSFER RADICAL POLYMERIZATION (ATRP)

ATRP is a versatile copper-mediated living radical polymerization technique[63] used for the polymerization of highly controlled block copolymers resulting in a low polydispersity index (PDI) for the product, with the PDI tending to be 1.[64,65] The reaction proceeds through the activation of a radical initiator, in which a carbon–halide bond is typically broken to generate a free radical. This free radical reacts with vinyl groups on the monomer units to propagate the reaction, forming a polymer (Fig. 4.9).[63]

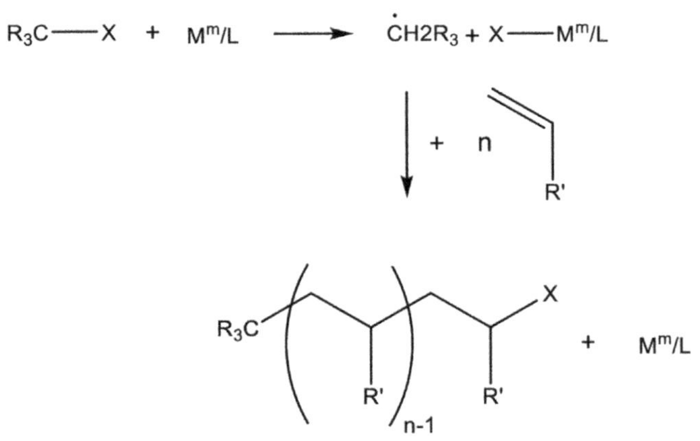

FIGURE 4.9 Reaction scheme for atom transfer radical polymerization (ATRP).

Typical reaction conditions for ATRP involve a reaction under nitrogen between the monomer, CuBr, 1,1,4,7,10,10-hexamethyltriethylenetetramine, anisole, and methyl-2-bromo-2-methylpropionate at 90°C for 10 h. Purification steps for this reaction involve using an alumina column to remove the copper catalyst followed by precipitation to obtain the purified polymer.[66]

This reaction has been used to incorporate rod-like mesogen groups as side chains or as part of the elastomer backbone (used interchangeably with the term liquid crystals) within the elastomer structure to form materials known as liquid crystalline elastomers (LCEs). LCEs were first hypothesized as potentially exhibiting actuation properties in 1975[67] and first experimentally proven in 1997.[68]

Mesogens are small, "rod-like" organic molecules,[69] which are typically 2 nm long and 0.5 nm thick[70] to create a crystal phase in an LCE, known as a liquid–crystalline monodomain.[71] The examples of mesogens are shown in Figure 4.10 with azo groups,[72] ferrocenyl groups,[73] and aniline groups.[74]

More specifically, the order of alignment of the mesogens within the LCE influences the properties exhibited by these crystalline monodomains. The monodomains can exist in four orientations.

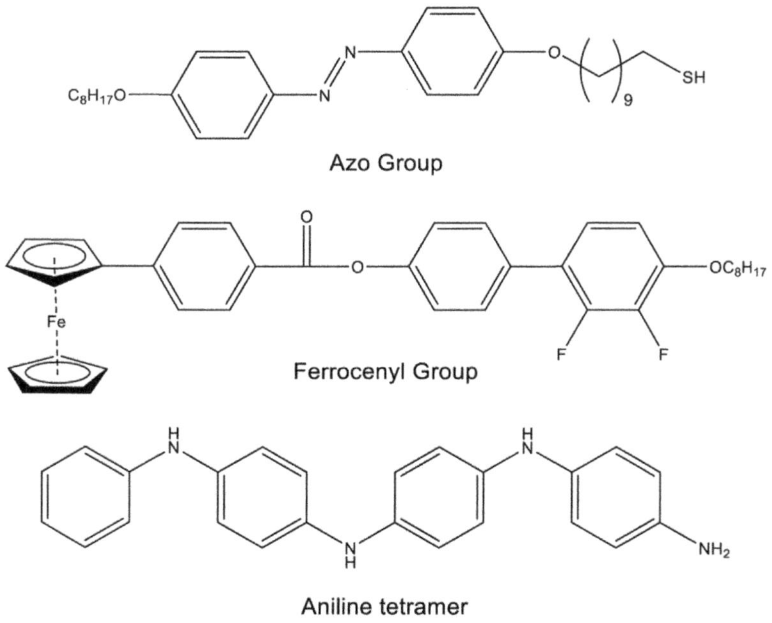

FIGURE 4.10 The examples of Azo (top), ferrocenyl (middle), and aniline (bottom) containing mesogens.

First, there is the isotropic phase, in which there are no long-ranged positional or orientational orders exhibited by the mesogens.[69] Typically speaking, this type of ordering is found in LCE at higher temperatures where the kinetic energy of the mesogens is such that it can overcome any intermolecular interactions.[69] On the other extreme, there is the crystalline phase in which long-range positional and orientation order are observed, typically for LCE at lower temperatures. In the crystalline phase, the mesogens will have regular lattice sites and will be aligned in a common direction,[69] making them anisotropic.[75]

The two intermediate phases, referred to as mesophases, are the nematic and smectic phase. The nematic phase does not display any long-range positional order; however, the majority of mesogens have a common directionality in their orientation.[69] Typically speaking, most LCEs exist between the isotropic and nematic phase.[76] The smectic phase is where the mesogens have formed regular layers, which can slide over each other easily and also have a common directionality.[69]

The mesogens can be incorporated within the polymer backbone as block copolymers or as side-chain groups (Fig. 4.11) in one of the four ways. The first method involves a competitive reaction between the cross-linker of the elastomer and the mesogen units to become side-chain groups[75] through a catalyzed reaction between the vinyl groups and typically Si–H bonds, making this approach very desirable for the modification of polysiloxanes.[77] However, the drawback of using this "one-pot" synthetic method is the resulting elastomers are difficult to purify.[69,78]

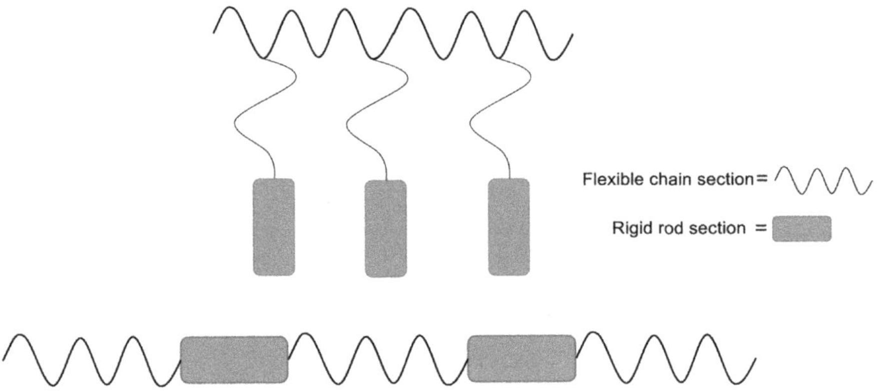

FIGURE 4.11 Top: Incorporation of mesogens as side-chain units. Bottom: Incorporation of mesogens in the polymer backbone.

The second synthetic route involves an already modified chain which is then cross-linked by using multifunctional cross-linkers to generate the elastomeric network.[75] This allows the modified individual polymer chains to be purified to remove impurities and low molecular weight by-products prior to cross-linking.[78]

The third method of incorporating mesogens is to chemically modify the polymer so that the mesogens and cross-linkable groups are already attached to the backbone,[75] followed by the application of UV light, resulting in a cross-linking reaction.[79]

The final method uses modified monomeric units with mesogens in which the polymer is then synthesized and cross-linked through radical polymerization through vinyl-end groups on the monomers.[75] One example of this is given by Lub et al. where they used UV light to carry out the polymerization and cross-linking simultaneously.[80]

In LCE, the orientation of mesogen units can be manipulated during synthesis to induce a stronger common directionality[69] or to completely change the direction of their alignment.[66] One of the three techniques can be used. First, the application of an electric field to the LCE causes the mesogens to align parallel to the electric field followed by cross-linking of the elastomeric chains to lock the mesogens into place.[81] The alternative methods of manipulation are to apply a magnetic field to cause the mesogens to align orthogonally to it[82] or to chemically treat the surface. The orientation of the mesogens can then be locked into place by cross-linking of the elastomeric chains.[76] The advantage of this is that the LCE can exhibit a much stronger dielectric response as the polarities of the LCE are enhanced.

Interestingly, other polymers such as nylon-11 can exhibit a similar crystalline region giving rise to dielectric properties.[83] The polar amine groups in odd-numbered nylons allow hydrogen bonding between themselves, creating various polarized mesophase crystals.[84] Investigations of their dielectric properties have been conducted and were found to display an ε of 2.5,[85] similar to that of PDMS.

Most studies have focused on the actuation of the LCE upon a thermal[86,87] or photo stimulus,[87,88] when incorporating mesogens as side-chain groups. However, some dielectric studies have been carried out with respect to LCE by forming a block copolymer between the mesogens and PDMS or poly (n-butyl acrylate) (PBA) (Table 4.5).

An example synthesis from Madsen et al. for the formation of monomeric 11-(4-cyano-4′-biphenyloxy)undecyl methacrylate (11CBMA) is achieved through a multistep reaction where 4′-cyano-4-hydroxybiphenyl, K_2CO_3, and KI are heated under reflux in dry acetone for 30 min. 11-bromo-1-undecanol is added and refluxed for a further 12 h before the removal of the solvent and purification by recrystallization to obtain pure 11-[(4-cyano-4′-biphenylyl)oxy]undecanol. This product is then dissolved with N,N′-dicyclohexylcarbodiimide and 4-dimethylaminopyridine in dry dichloromethane (DCM), and stirred for 30 min. A solution of methacrylic acid dissolved in DCM is added dropwise before further stirring of the reaction mixture for 17 h. The precipitate was then removed and solvent evaporated and the remains were purified using flash chromatography with silica gel to obtain pure 11CBMA.[66]

The synthesis of 11CBMA results in a high overall product yield of 89.1%. The formation of the PDMS-g-P11CBMA$_n$ diblock copolymer LCE also results in a high yield of 81.1%.[66] PDMS diblock copolymers containing 40 units of 11CBMA to form PDMS-g-P11CBMA$_{40}$ showed a good increase in the ε to 6.42, compared with 2.38 of the pure PDMS used in the study. The

TABLE 4.5 Properties of Elastomer Systems Modified Using ATRP and the Alignment Methodologies for the Mesogens.

Polymer backbone	Organic dipole (X)	Yield of X (%)	Yield of polymer (%)	Mesogen alignment method	Mesogen alignment for 11CBMA	Mesogen alignment for PDMS-g-P11CBMA$_n$	ε	Tan δ
PDMS-g-X[66]	11CBMA$_{40}$	89.1	81.1	None	Smectic A	Undefined	6.43	2.4×10^{-2}
PDMS-g-X[66]	11CBMA$_{40}$	89.1	81.1	Homogeneous alignment	Smectic A	Undefined	5.37	2.4×10^{-2}
PDMS-g-X[66]	11CBMA$_{40}$	89.1	81.1	Homeotropic alignment	Smectic A	Undefined	7.29	2.4×10^{-2}
X-g-PBA-g-X[89]	11CBMA$_{53}$	–	69.6	Thermal annealing	Smectic	Undefined	7.82	3.2×10^{-2}
X-g-PBA-g-X[89]	11CBMA$_{43}$	–	69.6	Thermal annealing	Smectic	Undefined	6.56	2.5×10^{-2}
X-g-PBA-g-X[89]	11CBMA$_{43}$	–	69.6	Solvent annealing (xylene)	Smectic	Undefined	5.50	3.0×10^{-2}
X-g-PBA-g-X[89]	11CBMA$_{43}$	–	69.6	None	Smectic	Undefined	4.19	3.0×10^{-2}

tan δ also remained low at 2.4×10^{-2}. The alignment of the mesogens was in a smectic-A fashion before the formation of the diblock copolymer. However, when the PDMS-g-P11CBMA$_{40}$ diblock copolymers were formed, the phase structure of mesogens became undefined.[66]

The literature reports have shown that the alignment of mesogens can be modified upon application of an electrical or magnetic field or by chemical treatment of the LCE.[81,82] Homogenous alignment of mesogens was achieved by rubbing the surface of the LCE in one direction, ensuring their orientation is parallel to the film surface; however, the smectic-A phase structure was still lost. This alignment led to a decrease in the ε value from 6.43 to 5.37 as the mesogens have lower dipole polarizability in this orientation while maintaining a very similar tan δ value.[66]

Homeotropically aligned mesogens, perpendicular to the surface of the film, can be attained by chemical treatment of the LCE. This increased the ε from 6.43 to 7.29 with a very similar tan δ reported once again. As seen with homogenously aligned LCE, the homeotropically aligned LCE also had an undefined phase structure suggesting that when the mesogens are the part of the diblock copolymer, microdomain structures are more difficult to form.[66] As the ε has varied values depending on the orientation of the mesogens within the LCE, with tan δ remaining constant, it shows that the orientation of the mesogens within an energy-harvesting device would be important to achieve the maximum potential for energy generation from it. No mechanical or E_b properties were reported for these LCEs, so while the dielectric properties can be deemed promising, the practicality of use for these elastomers is still in question.

11CMBA has also been used to form a triblock copolymer LCE with PBA in the form of P11CBMA$_n$-g-PBA-g-P11CBMA$_n$. Once again, the mesogens had a smectic microphase structure that disappeared upon the formation of the triblock copolymer, indicating that it is from the constrictions of forming a copolymer, rather than a direct effect from PDMS or PBA. For the LCE P11CBMA$_{53}$-g-PBA-g-P11CBMA$_{53}$, a maximum ε was recorded of 7.82 when the film was formed by thermal annealing, with a low tan δ of 3.2×10^{-2}. Thermal annealing of the elastomers provides an uniformity in the direction of alignment for the mesogens, resulting in a high ε.[89]

This impact of annealing on the dielectric properties was further explored by investigating the effect of thermal annealing and solvent annealing compared with no annealing of P11CBMA$_{40}$-g-PBA-g-P11CBMA$_{40}$ films. Each of these techniques affected the uniformity of alignment of the mesogens within the elastomer in different ways, resulting in different values of ε. When no annealing of the LCE film was carried out, it resulted in the lowest ε reported with a value of 4.19, as the alignment of mesogens was the least ordered resulting in the lowest dipole moment.[89] When the elastomer was annealed using solvent annealing, xylene produced the best dielectric result with an increase in the ε by 5.50[89] showing that the slow evaporation rate of xylene[90] and the effect the solvent has on the nanophase structure, resulted in an increased polarizability of the elastomer[91] from an increased order of mesogen alignment.

The best processing method for improving the ε was through thermal annealing of the elastomer film. This resulted in a further increase in the ε to 6.56 and a slight lowering of tan δ to 2.5×10^{-2}, showing that thermal annealing is the best technique to create a uniform alignment of mesogens.[89] However, the rate of heating and cooling was only described as "controlled," with no physical rates given in the work.[89] Potentially, had the thermal annealing been conducted at a slower rate, an even higher order of alignment for the mesogens could have been obtained, increasing the polarizability and hence, the ε even further. Therefore, the method of annealing and the conditions under which annealing takes place when using LCE should be considered for optimization of the dielectric properties. Once again, there were no investigations into the mechanical properties or E_b of the elastomers formed, leaving unanswered questions about the practical applicability of these systems.

4.2.5 OTHER REACTIONS

Other organic dipole molecules have been grafted onto elastomers over the years using different techniques to those in previous sections with their dielectric properties investigated, as stand-alone works. 2,2′-[[4-[(4-nitrophenyl)azo]phenyl]imino] bisethanol (DR19) was covalently grafted onto PDMS in a condensation cure reaction.[92] The maximum grafting level of DR19 reported was 13.2 wt.%, which increased the ε from 2.72 to 4.88. The ε is less than when the highly polar DR1 was grafted onto PDMS at a lower grafting level (see Section 4.2.1), even though the core structure and polarity of the molecules are similar (Fig. 4.12). This indicates that there are other factors influencing the ε such as water sorption[43] and steric hindrance versus chain interactions affecting the free volume available within the elastomer.[92]

There are big differences in the mechanical and electrical properties when DR19 is grafted at a 10.3 and a 13.2 wt.% level. At the lower doping level, the strength of the elastomer is greater with a T of 1.15 MPa, while also exhibiting a very low E of 0.37 MPa, and a λ_{max} of 525%. Once the doping increased to 13.2 wt.%, the strength of the elastomer decreases with a T of 0.90 MPa and a positive increase in E to 0.73 MPa as well as a λ_{max} of 225% was seen. Of course, there is also a drop in the E_b by increasing the grafting level, from 89.4 V·μm^{-1} at 10.3 wt.% to 56.7 V·μm^{-1} at 13.2 wt.%[92] as the higher grafting level of polar moieties increases the chance of a short-term breakdown of the structure.[62] The mechanical properties and E_b are overall best when the grafting level is 13.2 wt.% due to the increase in the elasticity modulus. Given the difference in ε between the two elastomers and the silicone elastomers grafted with DR1 in Section 4.2.1, it would be interesting to compare the difference in mechanical and electrical properties of these elastomers to determine which organic dipole would be most suitable for energy-harvesting applications.

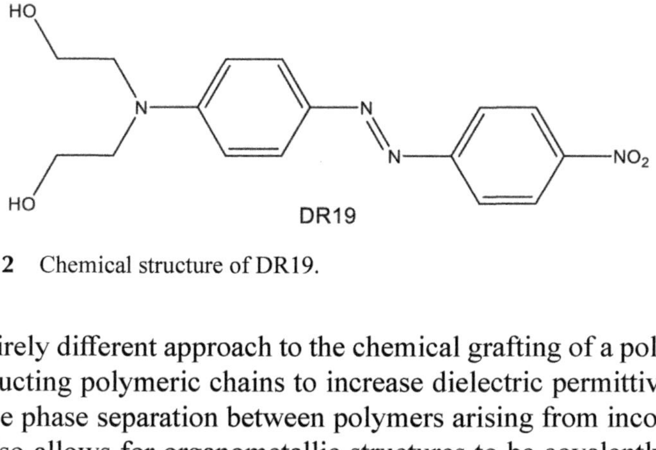

FIGURE 4.12 Chemical structure of DR19.

An entirely different approach to the chemical grafting of a polymer is to graft conducting polymeric chains to increase dielectric permittivity and to help reduce phase separation between polymers arising from incompatibilities.[93] It also allows for organometallic structures to be covalently attached between the two polymer chains to increase the dielectric properties, while, at the same time, preventing agglomeration of metal nanoparticles within the elastomer which reduces the E_b.[94]

An example of this involves grafting polyaniline (PA) to maleic anhydride (MA), which is already attached to SEBS, through ring-opening reactions.[94] Another work has grafted PA onto polyurethane (PU) through a copper phthalocyanine (CuPc) macrocyclic ring.[93] Both sets of work incorporated PA due to its ability to increase the dielectric constant through the percolative phenomenon, where a sharp rise in conductivity is observed for nanocomposites by the formation of PA nanodomains in the elastomer structure.[95] The addition of CuPc additionally increases the ε of the native elastomer.

The grafting of 2 vol.% PA to SEBS resulted in an increase of ε to approximately 5.5 from 2 at 104 Hz. However, if the doping level was increased further to 2.1 vol.%, the ε increased to 10 due to the incomplete grafting of PA resulting in the free polymer within the SEBS elastomer. This causes the PA chains to form a network with each other as the percolation threshold has been reached, increasing the ε and the overall conductivity of the elastomer.[94] This effect is also seen in the difference in tan δ between the two grafting levels, in which the increase of 0.1 vol.% of PA results in an increase from 0.10 to 0.40.[94]

The mechanical properties and E_b of the elastomers before reaching the percolation threshold are acceptable. The modified elastomer has a high strength with a T of 5 MPa and a good E of 1.6 MPa and λ_{max} of 650%. The E_b of the elastomer remains high compared to the unmodified SEBS-g-MA elastomer at 110 $V \cdot \mu m^{-1}$. However, when the grafting level is increased to 2.1 vol.%, the E_b drops drastically to 65 $V \cdot \mu m^{-1}$ as the percolation threshold is reached.[94] No mechanical properties were reported at this grafting level.

Another set of work grafted CuPc and PA onto polyurethane (PU) chains. CuPc is a macrocyclic molecule, commonly used in dyes, which can display a good conductivity when polymerized.[96] Grafting 23 vol.% of CuPc molecules to PU increases the ε from approximately 9–30 alongside a small rise in the ε′ from 0.05 to 0.15. However, by grafting these macrocyclic molecules to the elastomer, the flexibility of the elastomer drops significantly, with an increase in E from 20 (at 1 Hz) to 60 MPa (at 1 Hz) as a direct effect from the addition of the macrocyclic rings.[93]

PA was also grafted onto the PU–CuPc to form an elastomer of PU-g-CuPc-g-PA. In total, 14.4 vol.% of PA was grafted, increasing the ε value from 30 to 105 as the percolation threshold had been crossed. At the same time, the increase in the ε′ was only up to 0.28, a small increase for such a large gain in ε. However, the flexibility of the elastomer decreased further with E increasing to 90 MPa at 1 Hz,[93] making the elastomer too rigid for energy harvesting. No E_b results were recorded for either of the systems.

4.3 CONCLUSION

So far, only chemical modification of siloxane-based elastomers and the impact this has on the dielectric and mechanical properties have been investigated in any great depth. This leaves many DE materials still to be investigated to determine whether chemical modification can ever achieve dielectric properties that surpass pure PVDF and are more in line with modified PVDF systems. It would seem that grafting small, highly polar groups to elastomers would yield the best results for ε. However, there are usually large increases in ε′, especially when increased water sorption is a factor and so are poorer mechanical properties. This raises the question of whether designing organic dipoles to have specific groups for polarity and interactions with each other could improve the dielectric, mechanical, and E_b properties. Far more work is required to understand which properties of the organic dipoles influence the overall ε for designing and predicting how these would affect the elastomer.

The DE with the most practical applicability had a similar ε to that of pure PVDF. Methyl thioglycolate grafted onto SBS at a 81 mol.% level resulted in an ε of 12.2 and ε′ of 7.0×10^{-2} [10] (for PVDF: ε of 13 and ε′ of 1.8×10^{-2}).[16] Differences in the mechanical properties of the two materials are noted. PVDF has a high strength (*T* of 290 MPa) and low elasticity (*E* of 1.1×10^3 MPa, λ_{max} of 450%),[16] whereas the modified SBS elastomer has a much lower, although still strong, tensile strength of 3 MPa and a very flexible and stretchy structure with an *E* of 0.34 MPa and a λ_{max} of 1400%.[10] In works where the ε was reported to be higher than pure PVDF, the ε′ rose to values much greater than 5% of the ε.

KEYWORDS

- **dielectric elastomer**
- **hydrosilylation**
- **thiol-ene click chemistry**
- **azide click chemistry**
- **atom transfer radical polymerization**

REFERENCES

1. Madsen, F. B.; Daugaard, A. E.; Hvilsted, Skov, A. L. *Macromol. Rapid Commun.* **2016,** *37*, 378–413.
2. Pelrine, R.; Kornbluh, R. *Dielectric Elastomers as Electromechanical Transducers*, Elsevier: Amsterdam, 2008; pp 3–12. DOI: http://dx.doi.org/10.1016/B978–0–08–047488–5.00001–0.
3. Zhou, J.; Jiang, L.; Khayat, R. E. *Soft Matter* **2015,** *11*, 2983–2992.

4. Biggs, J.; Danielmeier, K.; Hitzbleck, J.; Krause, J.; Kridl, T.; Nowak, S.; Orselli, E.; Quan, X.; Schapeler, D.; Sutherland, W.; Wagner, J. *Angew. Chem. Int. Ed.* **2013,** *52,* 9409–9421.
5. Koo, C. M. *Electroactive Thermoplastic Dielectric Elastomers as a New Generation Polymer Actuators;* INTECH Open Access Publisher: South Korea, 2012.
6. Lee, J.; Kwon, H.; Seo, J.; Shin, S.; Koo, J. H.; Pang, C.; Son, S.; Kim, J. H.; Jang, Y. H.; Kim, D. E.; Lee, T. *Adv. Mater.* **2015,** *27,* 2433–2439.
7. Amjadi, M.; Pichitpajongkit, A.; Lee, S.; Ryu, S.; Park, I. *ACS Nano* **2014,** *8,* 5154–5163.
8. Pelrine, R.; Kornbluh, R.; Pei, Q.; Joseph, J. *Science* **2000,** *287,* 836–839.
9. Shankar, R.; Ghosh, T. K.; Spontak, R. J. *Adv. Mater.* **2007,** *19,* 2218–2223.
10. Sun, H.; Jiang, C.; Ning, N.; Zhang, L.; Tian, M.; Yuan, S. *Polym. Chem.* **2016,** *7,* 4072–4080.
11. Grigorescu, R. M.; Ciuprina, F.; Ghioca, P.; Ghiurea, M.; Iancu, L.; Spurcaciu, B.; Panaitescu, D. M. *J. Phys. Chem. Solids* **2016,** *89,* 97–106.
12. AzoMaterials, Styrene-Ethylene-Butylene-Styrene Based Thermoplastic Elastomer—SEBS TPE 45A. http://www.azom.com/article.aspx? ArticleID=873 (accessed December 2016).
13. Banerjee, S. S.; Ramakrishnan, I.; Satapathy, B. K. *Polym. Eng. Sci.* **2016,** *56,* 491–499.
14. Risse, S.; Kussmaul, B.; Krüger, H.; Kofod, G. *Adv. Funct. Mater.* **2012,** *22,* 3958–3962.
15. A. R. Coatings, Specifications for Cured EPDM Rubber—Advanced Rubber Coatings, http://www.advancedrubbercoatings.com/site/1584080/page/765223 (accessed Dec 2016).
16. Goodfellow, Polyvinylidenefluoride (PVDF) Material Information, http://www.goodfellow.com/E/Polyvinylidenefluoride.html (accessed December 2016).
17. Dent, A. C.; Bowen, C. R.; Stevens, R.; Cain, M. G.; Stewart, M. *J. Eur. Ceram. Soc.* **2007,** *27,* 3739–3743.
18. Baur, C.; Apo, D. J.; Maurya, D.; Priya, S.; Voit, W. In *Polymer Composites for Energy Harvesting, Conversion, and Storage*; Lai, L., Wong-Ng, W. and Sharp, J. Ed.; American Chemical Society: Washington, DC, 2014; Vol. 1161, Ch. 1, pp 1–27.
19. MatWeb-LLC, Channel Industries 600 Barium Titanate Piezoelectric, http://www.matweb.com/search/datasheet.aspx? matguid=b030bd8a82774bff8e485986efe77af5&ckck=1 (accessed Jan 2017).
20. Udayakumar, K. R.; Schuele, P. J.; Chen, J.; Krupanidhi, S. B.; Cross, L. E. *J. Appl. Phys.* **1995,** *77,* 3981–3986.
21. MEMSnet, Material: Lead Zirconate Titanate (PZT). https://www.memsnet.org/material/leadzirconatetitanatepzt/ (accessed Jan 2017).
22. MatWeb-LLC, Channel Industries 5800 Lead Zirconate Titanate Piezoelectric. http://www.matweb.com/search/datasheet.aspx? matguid=5e080b454b194836862ddb0d882bb194 (accessed Jan 2017).
23. Racles, C.; Alexandru, M.; Bele, A.; Musteata, V. E.; Cazacu, M.; Opris, D. M. *RSC Adv.* **2014,** *4*(37), 620–37,628.
24. Racles, C.; Maria, C.; Beatrice, F.; Dorina, M. O. *Smart Mater. Struct.* **2013,** *22,* 104004.
25. Madsen, F. B.; Dimitrov, I.; Daugaard, A. E.; Hvilsted, S.; Skov, A. L. *Polym. Chem.* **2013,** *4,* 1700–1707.
26. Yang, D.; Ruan, M.; Huang, S.; Wu, Y.; Li, S.; Wang, H.; Ao, X.; Liang, Y.; Guo, W.; Zhang, L. *RSC Adv.* **2016,** *6,* 90,172–90,183.
27. Skov, A. L.; Vudayagiri, S.; Benslimane, M. *Proc. SPIE* **2013,** *8687,* 86,871I–86,871.

28. Dünki, S. J.; Ko, Y. S; Nüesch, F. A.; Opris, D. M. *Adv. Funct. Mater.* **2015,** *25,* 2467–2475.
29. Chen, T.; Qiu, J.; Zhu, K.; Li, J.; Wang, J.; Li, S.; Wang, X. *J. Phys. Chem. B* **2015,** *119,* 4521–4530.
30. Dunki, S. J.; Tress, M.; Kremer, F.; Ko, S. Y.; Nuesch, F. A.; Varganici, C.-D.; Racles, C.; Opris, D. M. *RSC Adv.* **2015,** *5,* 50,054–50,062.
31. Thakur, V. K.; Tan, E. J.; Lin, M.-F.; Lee, P. S. *Polym. Chem.* **2011,** *2,* 2000–2009.
32. Thakur, V. K.; Tan, E. J.; Lin, M.-F.; Lee, P. S. *J. Mater. Chem.* **2011,** *21,* 3751–3759.
33. Marciniec, B. *Hydrosilylation: A Comprehensive Review on Recent Advances;* Springer Science & Business Media: Netherlands, 2008.
34. Mark, J. E. *Polymer Data Handbook,* 2nd ed.; Oxford University Press: New York, 2009.
35. Marciniec, B.; Guliński, J.; Kopylova, L.; Maciejewski, H.; Grundwald-Wyspiańska, M.; Lewandowski, M. *Appl. Organomet. Chem.* **1997,** *11,* 843–849.
36. Matisons, J. G.; Provatas, A. *Macromolecules* **1994,** *27,* 3397–3405.
37. Coqueret, X.; Lablache-Combier, A.; Loucheux, C. *Eur. Polym. J.* **1988,** *24,* 1137–1143.
38. Zhu, Z.; Einset, A. G.; Yang, C.-Y.; Chen, W.-X.; Wnek, G. E. *Macromolecules* **1994,** *27,* 4076–4079.
39. Coqueret, X.; Wegner, G. *Organometallics* **1991,** *10,* 3139–3145.
40. Chien, L. C.; Cada, L. G. *Macromolecules* **1994,** *27,* 3721–3726.
41. Dascalu, M.; Dunki, S. J.; Quinsaat, J.-E. Q.; Ko, Y. S.; Opris, D. M. *RSC Adv.* **2015,** *5,*104,516–104,523.
42. Madsen, F. B.; Yu, L.; Daugaard, A. E.; Hvilsted, S.; Skov, A. L. *RSC Adv.* **2015,** *5,* 10,254–10,259.
43. Racles, C.; Cozan, V.; Bele, A.; Dascalu, M. *Des. Monomers Polym.* **2016,** *19,* 496–507.
44. Böhm, M.; Soden, W. v.; Heinrich, W.; Yehia, A. A. *Colloid Polym. Sci.* **1987,** *265,* 295–303.
45. Racles, C.; Bele, A.; Dascalu, M.; Musteata, V. E.; Varganici, C. D.; Ionita, D.; Vlad, S.; Cazacu, M.; Dunki, S. J.; Opris, D. M. *RSC Adv.* **2015,** *5,* 58, 428–58, 438.
46. Collings, P. J.; Hird, M. *Introduction to Liquid Crystals: Chemistry and Physics;* Taylor and Francis: London, 1997.
47. Kussmaul, B.; Risse, S.; Kofod, G.; Waché, R.; Wegener, M.; McCarthy, D. N.; Krüger, H.; Gerhard, R. *Adv. Funct. Mater.* **2011,** *21,* 4589–4594.
48. Björn, K.; Sebastian, R.; Michael, W.; Guggi, K.; Hartmut, K. *Smart Mater. Struct.* **2012,** *21,* 64005.
49. Risse, S.; Kussmaul, B.; Kruger, H.; Kofod, G. *RSC Adv.* **2012,** *2,* 9029–9035.
50. Zou, C.; Fothergill, J. C.; Rowe, S. W. *IEEE Trans. Dielectr. Electr. Insul.* **2008,** *15,* 106–117.
51. Bele, A.; Cazacu, M.; Racles, C.; Stiubianu, G.; Ovezea, D.; Ignat, M. *Adv. Eng. Mater.* **2015,** *17,* 1302–1312.
52. Tian, M.; Yan, H.; Sun, H.; Zhang, L.; Ning, N. *RSC Adv.* **2016,** *6,* 96,190–96,195.
53. Hordyjewicz-Baran, Z.; You, L.; Smarsly, B.; Sigel, R.; Schlaad, H. *Macromolecules* **2007,** *40,* 3901–3903.
54. Justynska, J.; Hordyjewicz, Z.; Schlaad, H. *Polymer* **2005,** *46,* 12,057–12,064.
55. ten Brummelhuis, N.; Diehl, C.; Schlaad, H. *Macromolecules* **2008,** *41,* 9946–9947.
56. Stoddart, A. *Nat. Mater*. **2013,** *12,* 282.
57. Distefano, G.; Suzuki, H.; Tsujimoto, M.; Isoda, S.; Bracco, S.; Comotti, A.; Sozzani, P.; Uemura, T.; Kitagawa, S. *Nat. Chem.* **2013,** *5,* 335–341.

58. Dunki, S. J.; Nuesch, F. A.; Opris, D. M. *J. Mater. Chem. C* **2016,** *4,* 10,545–10,553.
59. Madsen, F. B.; Yu, L.; Daugaard, A. E.; Hvilsted, S.; Skov, A. L. *Polymer* **2014,** *55,* 6212–6219.
60. Bostrom, J. O.; Marsden, E.; Hampton, R. N.; Nilsson, U. *IEEE Electr. Insul. Mag.* **2003,** *19,* 6–12.
61. Madsen, F. B.; Daugaard, A. E.; Hvilsted, S.; Benslimane, M. Y.; Skov, A. L. *Smart Mater. Struct.* **2013,** *22,* 104,002.
62. Li, S.; Yin, G.; Chen, G.; Li, J.; Bai, S.; Zhong, L.; Zhang, Y.; Lei, Q. *IEEE Trans. Dielectr. Electr. Insul.* **2010,** *17,* 1523–1535.
63. Matyjaszewski, K. *Macromolecules* **2012,** *45,* 4015–4039.
64. Min, K.; Gao, H.; Matyjaszewski, K. *J. Am. Chem. Soc.* **2005,** *127,* 3825–3830.
65. Ding, H.; Park, S.; Zhong, M.; Pan, X.; Pietrasik, J.; Bettinger, C. J.; Matyjaszewski, K. *Macromolecules* **2016,** *49,* 6752–6760.
66. Zhang, C.; Wang, D.; He, J.; Liu, M.; Hu, G.-H.; Dang, Z.-M. *Polym. Chem.* **2014,** *5,* 2513–2520.
67. Gennes, P. G. d. *C. R. Seances Acad. Sci. Ser. B* **1975,** *281,* 101.
68. Bergmann, G. H. F.; Finkelmann, H.; Percec, V.; Zhao, M. *Macromol. Rapid Commun.* **1997,** *18,* 353–360.
69. Menzel, A. M. *Phys. Rep.* **2015,** *554,* 1–45.
70. Gennes, P. G. d.; Prost, J. *The Physics of Liquid Crystals;* Clarendon Press: Oxford, 1995.
71. Brömmel, F.; Kramer, D.; Finkelmann, H. Preparation of Liquid Crystal Elastomers. In *Liquid Crystal Elastomers: Materials and Applications;* de Jeu, W. H. Ed.; Springer Berlin Heidelberg: Berlin, Heidelberg, 2012; Vol. 250, pp 1–48.
72. Duan, J.; Wang, M.; Bian, H.; Zhou, Y.; Ma, J.; Liu, C.; Chen, D. *Mater. Chem. Phys.* **2014,** *148,* 1013–1021.
73. Loubser, C.; Imrie, C.; van Rooyen, P. H. *Adv. Mater.* **1993,** *5,* 45–47.
74. Yusuf, Y.; Ono, Y.; Sumisaki, Y.; Cladis, P. E.; Brand, H. R.; Finkelmann, H.; Kai, S. *Phys. Rev. E* **2004,** *69,* 21710.
75. Ohm, C.; Brehmer, M.; Zentel, R. *Adv. Mater.* **2010,** *22,* 3366–3387.
76. Küpfer, J.; Finkelmann, H. *Makromol. Chem. Rapid Commun.* **1991,** *12,* 717–726.
77. Finkelmann, H.; Rehage, G. *Makromol. Chem. Rapid Commun.* **1980,** *1,* 733–740.
78. Jiang, H.; Li, C.; Huang, X. *Nanoscale* **2013,** *5,* 5225–5240.
79. Beyer, P.; Braun, L.; Zentel, R. *Macromol. Chem. Phys.* **2007,** *208,* 2439–2448.
80. Lub, J.; Broer, D. J.; van den Broek, N. *Liebigs Ann.* **1997,** *1997,* 2281–2288.
81. Küupfer, J.; Finkelmann, H. *Macromol. Chem. Phys.* **1994,** *195,* 1353–1367.
82. Lubensky, T. C.; Mukhopadhyay, R.; Radzihovsky, L.; Xing, X. *Phys. Rev. E* **2002,** *66,* 11702.
83. Mathur, S. C.; Scheinbeim, J. I.; Newman, B. A. *J. Appl. Phys.* **1984,** *56,* 2419–2425.
84. Zhang, Z.; Litt, M. H.; Zhu, L. *Macromolecules* **2016,** *49,* 3070–3082.
85. Wu, G.; Yano, O.; Soen, T. *Polym. J.* **1986,** *18,* 51–61.
86. Greco, F.; Domenici, V.; Desii, A.; Sinibaldi, E.; Zupancic, B.; Zalar, B.; Mazzolai, B.; Mattoli, V. *Soft Matter* **2013,** *9,* 11,405–11,416.
87. Fu, S.; Zhang, H.; Zhao, Y. *J. Mater. Chem. C* **2016,** *4,* 4946–4953.
88. Yang, Y.; Zhan, W.; Peng, R.; He, C.; Pang, X.; Shi, D.; Jiang, T.; Lin, Z. *Adv. Mater.* **2015,** *27,* 6376–6381.
89. Zhang, C.; Wang, D.; He, J.; Liang, T.; Hu, G.-H. Dang, Z.-M. *Polym. Adv. Technol.* **2014,** *25,* 920–926.

90. Hofmann, H. E. *Ind. Eng. Chem.* **1932,** *24,* 135–140.
91. Huang, W.-H.; Chen, P.-Y.; Tung, S.-H. *Macromolecules* **2012,** *45,* 1562–1569.
92. Zhang, L.; Wang, D.; Hu, P.; Zha, J.-W.; You, F.; Li, S. T.; Dang, Z. M. *J. Mater. Chem. C,* **2015,** *3,* 4883–4889.
93. Huang, C.; Zhang, Q. M. *Adv. Mater.* **2005,** *17,* 1153–1158.
94. Stoyanov, H.; Kollosche, M.; McCarthy, D. N.; Kofod, G. *J. Mater. Chem.* **2010,** *20,* 7558–7564.
95. Paul, B. K.; Mitra, R. K. *J. Colloid Interface Sci.* **2006,** *295,* 230–242.
96. Epstein, A.; Wildi, B. S. *J. Chem. Phys.* **1960,** *32,* 324–329.

CHAPTER 5

TRANSPARENT DIELECTRIC MATERIALS

LUMINITA IOANA BURUIANA*, ANDREEA IRINA BARZIC, and CAMELIA HULUBEI

"Petru Poni" Institute of Macromolecular Chemistry, 41A Grigore Ghica Voda Alley, 700487 Iasi, Romania

Corresponding author. E-mail: luminitab25@yahoo.com

CONTENTS

ABSTRACT

Materials with dielectric behavior represent a widespread topic in scientific community. Latest trends in this research are focused on enhancing transparency since in certain applications this is an essential requirement. This chapter deals with current approaches in adapting the dielectric features of macromolecular compounds for achieving the targeted performance. Synthesis and processing routes used in obtaining modern dielectrics are described. Further, some insights on rendering intelligent behavior to this category of materials are presented. The practical importance of transparent dielectrics in various sectors of industry is overviewed.

5.1 INTRODUCTION

The dielectric materials have found utilization in numerous aspects in our everyday life, starting from electronic devices, food packaging, biomaterials, and products for agriculture.[1–4] The behavior of these compounds in the presence of electromagnetic radiations is very important since, in the majority of cases, the transparency and/or permittivity delimit their applicability.

The dielectric properties of materials can be tuned through chemical and physical procedures. Dielectric constant (*k*) is very important in practice since it also describes the system behavior when it is exposed to electromagnetic radiations. This parameter is related to the compound ability to hold electric charge for long periods of time. Depending on the pursued application, scientists were focused on two large research directions:

- Development of low-*k* materials
- Achievement of high-*k* materials

The literature[5–8] reports several classifications of dielectrics based on their permittivity values. In some cases, a low-*k* material is identified when its permittivity is lower than 4.2. Other studies suggest that the classification must be made in regard to some references. Therefore, for low-*k* materials, the comparison is made with a traditional dielectric, namely silicon dioxide, for which the dielectric constant is 3.9. Below this value ($k<3.9$), one achieves materials with low permittivity. For high-*k* compounds, silicon nitride is the threshold; thus, such materials must have $k>7$. A more complex sorting of dielectrics considers that they can be rated based on their properties,[6] namely insulating materials having $k<13$, electrical storage materials with $k>12$

(useful for capacitor purposes and involving high breakdown voltage), and ferroelectrics, piezoelectrics, and pyroelectrics (for narrowband electrical filters). The majority of cases simply consider that one may group dielectrics based on silicon oxide value (k=3.9) and, for this reason, here the discussion is made following this criterion. A schematic representation of the most important dielectric materials is displayed in Figure 5.1.

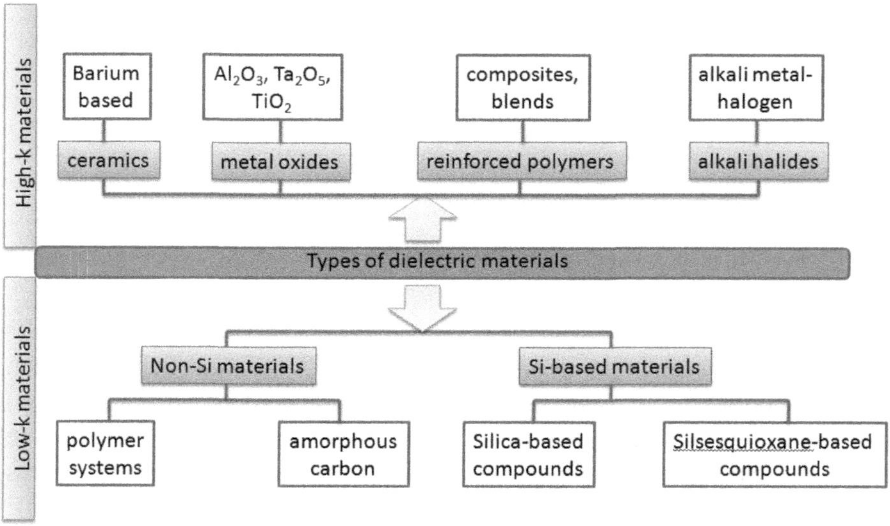

FIGURE 5.1 The main categories of dielectric materials divided based on their permittivity values.

Among the dielectric compounds reported in the literature, a great deal of attention was ascribed to polymers owing to their good processability under various shapes and low cost.[9] Their properties can be easily tuned, this being one of the most important advantages of the common dielectrics. This raised the interest of scientists to develop macromolecular compounds with advanced physical performance in regard to classical insulators.[10–12] Having all these in view, this chapter is particularly focused on transparent dielectric polymers, without excluding other types of materials.

Based on the typical chemical features of each polymer class, one may distinguish specific mechanical, thermal, optical, and dielectric properties. There are multiple facile strategies to control the dielectric constant of dielectrics, which are also applicable to polymers. Figure 5.2 schematically depicts all methodologies to adapt the value of permittivity of polymers to the demands of pursued application.

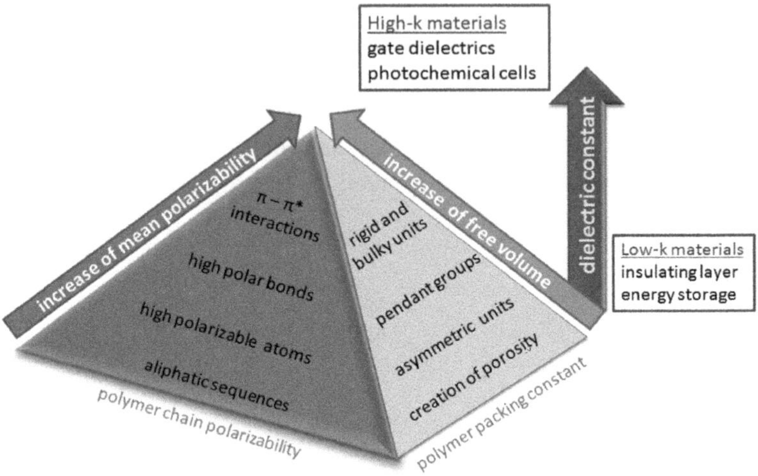

FIGURE 5.2 Schematic representation of the factors which affect the dielectric constant of a polymer.

This property depends on two major factors:[5–8] the ratio between mean molecular polarizability and van der Waals volume (denoted here with φ) and packing ability of chains (K_p). The dielectric constant is a monotone increasing function with respect to the product of these two parameters, as described in Equations 5.1 and 5.2:

$$k = \frac{\sqrt{1+2K_p\varphi}}{\sqrt{1-K_p\varphi}} \tag{5.1}$$

$$\varphi = \frac{\alpha}{V_w} \tag{5.2}$$

where α is the mean molecular polarizability and V_w is the van der Waals volume.

The dielectric constant expresses the intensity of the polarization mechanisms. The most important ones mentioned in literature[13] are:

- Electronic polarization—displacement of the center of positive and negative charges
- Atomic polarization—deformation of atomic position in molecules
- Orientational polarization—alignment of dipoles during exposure to electric fields, resulting a net polarization in the direction of electric field oscillation

A successful approach to adapt the φ factor consists in adjusting both molecular polarizability and van der Waals volume. Introduction in polymer backbone of different sequences or atoms with low polarizability, such as aliphatic groups or fluorine, is beneficial for lowering *k* values. Conversely, polymers that contain aromatic rings, high polarizable atoms (sulfur, iodine, and bromine), and bonds (Si–F, S–C, double/triple) in their structure have a strong polar character and, implicitly, high dielectric constant. Moreover, the chain geometry should be discussed to determine if a polymer is polar or nonpolar. In other words, the symmetrical disposal of dipoles along the chains might cause the cancellation of dipole moments and diminishing of the corresponding polarization mechanism, which is seen in a lower dielectric constant. The influence of van der Waals volume is seen in polymer packing constant. The latter can be quantified through the fractional free volume, as presented in Equation 5.3:

$$V_{free} = \frac{V_w - V_0}{V_w} \tag{5.3}$$

where V_{free} is the fractional free volume and V_0 is the occupied molar volume of a repeat unit.

In order to enhance the free volume, one may introduce in polymer structure flexible bridging units, bulky and pendant groups, which reduce chain packing density. Another strategy is to create a porous morphology through physical methods, which diminishes the dielectric constant by reduction of material's density (occurrence of micro/nano-voids by constitutive/subtractive porosity) and insertion of air whose relative permittivity is about one. A higher fractional free volume will determine a lower number of polarizable groups per unit volume. A proper solution to achieve high free volume consists in replacement of hydrogen with fluorine atoms that lead to lower dielectric constant because fluorine occupies higher volume.

As a general remark, one may state that polar polymers at low frequencies (of about 60 Hz) present dielectric constants values between 3 and 9, whereas at high frequencies (of about 100 Hz) exhibit dielectric constants in the range of 3–5. When dealing with nonpolar polymers, the *k* values are independent of the alternating current frequency since the electron polarization is effectively instantaneous; therefore, the dielectric constants values are lower than 3.

On the other hand, for certain applications, dielectric material must not absorb radiation in a certain wavelength range. *Optical transparency* is related to transmission of radiation from visible spectral range. In practice, the capacity of polymers and other low-molecular-weight materials to transmit radiations must be investigated in a wide range of spectrum, including infrared and ultraviolet regions.[14] A mandatory condition for a compound to allow passing of infrared radiations is to contain mobile electrons and phonons. Most polymers absorb ultraviolet radiations and, consequently, in this spectral domain are not transparent. For this reason, the structure might be affected because the absorbed energy has a similar order of magnitude with the strength of some constituting chemical bonds. In order to solve this drawback, one could add some stabilizers that have the capacity to absorb ultraviolet radiation and enable neutralization of the energy. The transparency is affected by both bulk and surface characteristics of the material.

Regarding the first issue, one can mention that amorphous polymers are known to be transparent since the lack of long-range order (polymer chains are not able to arrange themselves) reduces the possibility of diffusion phenomena. The presence of crystalline domains determines scattering of incident radiation, which is noticed when light passes from amorphous to crystalline zones. The addition of fillers or colorants may generate opacity. In the cases when additive size is higher than the wavelength of visible radiation (0.4–0.7 μm), scattering is produced in detriment of transparency and thus haze may appear.[14] Transmittance is somehow related to the refractive index so that particular fillers are able to maintain a certain degree of transparency (translucent). There are situations in which the dye is dissolved in the macromolecular matrix, resulting, in the end, a colored transparent polymer. The modification in material's microstructure caused by degradation or oxidation, or diffusion of specific components, could lower light transmission. When discussing polymer solutions in organic solvents, one should consider the spectral features of the solvent (overlapping of absorption bands must be avoided) and the amount of solved polymer (absorption must be kept in the measuring scale of the apparatus). For polymers in the solid phase, particularly films, it is observed that transmittance is also affected by thickness. Therefore, several polymers are able to transmit incident radiation as thin films.

Surface scattering represents one of the major factors responsible for the loss of optical transparency. Scattering phenomena taking place at the polymer surface might be induced by scratches or flaws arising in the stage of film preparation. Another optical property, widely seen at the polymer surface, is gloss appeared as a result of light reflection. The majority of polymers exhibit low roughness surfaces and high gloss. Haze is caused by light diffraction, namely as the ratio of impinging radiation that is diffracted above 2.5°. When the haze is higher than 30%, the polymer becomes translucent. Moreover, for anisotropic polymers, the phenomenon of double refraction (birefringence) is remarked.

Analyzing the factors that affect both dielectric constant and transparency of materials, one can establish an indirect connection between the two properties. However, there are several aspects, such as temperature (implicitly glass transition temperature—T_g) or radiation wavelengths that impede a general rule formulation. However, few correlations can be established, particularly for polymer systems. A degree of crystallinity determines the reduction of transparency and permittivity. In case of polymer, the more mobile the chains are the higher the dielectric constant will be.[15,16] The presence of chemical groups which are able to absorb visible radiations reduce transparency, while increase the permittivity owing to their polarizable character.[17] The dielectric properties of the composites are mainly affected by the amount of the introduced particles, but the size is also important. The dielectric constant is reported to be higher for nanocomposites than for micro-composites at low frequencies.[18] Thus, if nanofiller size is lower than 0.4–0.7 μm, the material is transparent and will present high-k values, considering the higher density of polarizable nanoparticles. Moreover, the uniform distribution of reinforcement agent inside the matrix is essential for the enhancement of the permittivity[19] and transparency—for the latter, especially, at low amounts of nanofiller to avoid creation of diffusion centers. Figure 5.3 presents a schematic correlation of the main factors influencing the above-discussed physical properties.

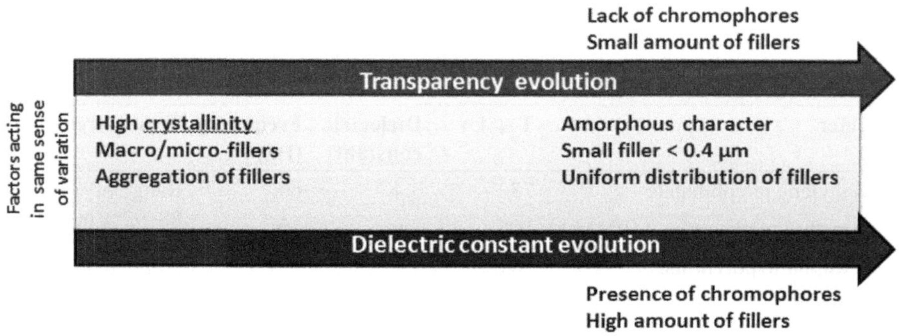

FIGURE 5.3 The main factors influencing the dielectric constant and transparency of polymer materials.

This chapter deals with fundamental aspects of low-k and high-k polymers. The first category can be achieved by reducing the backbone conjugation by using aliphatic, bulky, and noncoplanar sequences. Moreover, the main design procedures to accomplish a specific macromolecular architecture that creates a big free volume for diminishing the dielectric property are presented. On the other hand, the manuscript also describes high-*k* polymers that can be prepared by inserting specific fillers that enhance main chain polarizability and determine higher material density. Several examples of dielectrics with intelligent behavior and their applications are included.

5.2 TRANSPARENT POLYMER MATERIALS WITH LOW DIELECTRIC CONSTANT

Generally, the organic dielectric materials present lower permittivity values compared with inorganic materials due to their hydrophobic nature and low polarizability. Among the transparent polymer films, one can distinguish amorphous ones. In Table 5.1, some examples from this category of materials together with their dielectric and thermal features are given. Thin insulating films—used in a wide variety of components—are usually amorphous or near amorphous in nature, with higher electrical resistance compared to polycrystalline or crystalline thin films.[20] Based on the service temperature (T_g), the polymer films are divided into three classes,[21] namely conventional ($T_g < 100°C$), common high temperature ($100 \leq T_g < 200°C$), and high temperature ($T_g \geq 200°C$). Despite their excellent optical transparency and low dielectric constant, the service temperature is certainly the major limitation of the polymers for microelectronic applications.

TABLE 5.1 Examples of Some Low-k and Transparent Amorphous Polymers.

Polymer	T_g (°C)	Dielectric constant	Frequency (Hz)	Transparency
Polyethylene terephthalate	78	3.4	60	Range of 88–92% in apparent light transmittance
Poly(methyl methacrylate) (PMMA)	105	3.0	10^6	
Fully aliphatic polyimide	>200	2.47	$>10^6$	
Polycarbonate	145	2.93	10^6	
Atactic polypropylene	−20	2.15	10^6	
Polystyrene (PS)	95	2.40	10^6	
Poly(vinyl chloride)	80	2.70	10^6	

On the other hand, there are high-performance polymer films with excellent thermal stability (above 300°C) and low dielectric constant—for example, those derived from fully aromatic polyimides (PI)—which are limited in their optoelectronic applications by the deep colors and the poor optical transmittance.

For a linear dielectric polymer, there is a direct proportionality between the electric moment and the intensity of the applied electric field. In this context, low-*k* materials can be accomplished based on polymer structures with low polarizability. Therefore, fluorinated and non-fluorinated aliphatic macromolecular structures, containing single C–F and C–C bonds with the lowest corresponding electronic polarizability (of about 0.555 $Å^3$ and 0.531, respectively), become suitable for low-*k* applications.[22,23] At the same time, polymers having double and triple bonds in their structures gain an increased π-electrons mobility (C=C, $\alpha = 1.643$ $Å^3$, and C≡C, $\alpha = 2.036$ $Å^3$) and enhanced electronic polarizability. Given the mentioned structural peculiarities, low-*k* polymers exhibit poor adhesion, in contrast with those with aromatic π-bonding configurations, which have more pronounced bonding strength. In addition, it was experimentally proved that increasing the bond length, bonding orientation, as well as discontinuing the chain by inserting single bond atoms or groups of atoms into the polymer skeleton is a suitable approach to lower the *k* value.[20]

The modification of the chemical polymer structure can significantly change the share of the polarization mechanisms to the magnitude of dielectric constant. The main approaches to achieve low-*k* dielectric materials are:

- The reduction of the electronic polarization mechanism involves the diminishing of the electrons mobility along the chains under an applied field. This can be done by introducing of specific atoms, such as fluorine[24] and/or carbon (C),[25] which will produce a material with a lower electronic polarizability.
- The decrease of the orientation and/or the ionic contribution requires the control of the atom's response to an electric field as a function on the atoms nature (Si, C, H, N, F) from the material. The optimal solution relies on the increase of the material's free volume and implicitly by the decrease of polarizable group's density per volume unit with impact on the final of the atomic or dipolar polarization mechanisms.

Porous materials can be also considered as adequate candidates for low-*k* dielectric materials. The general ways to introduce *free volume* in the form of porosity are:

- The obtaining of a low-density network based on selected suitable monomers

- The chemical modification of the polymer structures in order to lower the network density and/or polarizability by introducing space-occupying groups, such as methyl, ethyl, and phenyl groups in the backbone
- The discard of a labile second phase (a porogen or sacrificial place-holder) embedded in the network structure of the deposed films.[26]

It is to be noted that of all nonporous bulk materials, perfluorinated aliphatic polymers, such as poly(tetrafluoroethylene), present the lowest dielectric constant ($k<2$) due to the low polarizability of the C–F and C–C bonds present in its structure.[27] In addition, aliphatic polymers, such as poly(ethylene), have low permittivity ($k<2.25$).[28] Taking into account the earlier described features, one must achieve a compromise between transparency, high thermal stability, and low dielectric constant for development of optical films for optoelectronic industry.

A variety of polymers has been created for their use as the materials with low dielectric constant for the applications in several areas, including microelectronics. Among them, one can enumerate fluoropolymers, heteroaromatic polymers, PI, hydrocarbon polymers without any polar groups, films deposited from the gas phase by chemical vapor deposition, plasma-enhanced chemical vapor deposition, and so forth.[29] The materials that best combine the above-discussed properties are polyether ether ketone (PEEK), poly(arylene ether) (PAEs), polynorbornene, poly(silsesquioxanes), some of heterocyclic polymers, and so forth. In the following paragraphs, a short review of the most significant low-*k* transparent polymers is made.

PEEK represents a thermoplastic matrix resin, which is very interesting for fiber-reinforced composite applications. However, this polymer system has been difficult to produce in high-molecular-weight form. PEEKs are by their nature insoluble and cannot be cast from solution into polymer films. The only known film form of the commercial product PEEK relies on extrusion. The extruded film is difficult to accomplish and is pale to dark amber in color, depending on thickness.[30] Cassidy et al.[31] described a method for producing processable soluble fluorinated PEEK that has the ability to form a transparent film and coating. The fluorinated PEEK film is a low dielectric material with a dielectric constant near 2.4, and possible special utility in electronic applications where high electrical insulation and thermal stability are mandatory. Moreover, high optical transparency fluorinated PEEK film and coating are characterized by outstanding high transparency at wavelengths in the visible domain of the electromagnetic spectrum. Thus, such materials are optimal as materials for solar energy applications.[31] Flexible and tough films based on new poly(aryl ether ketones/sulfones) (PAEKs/PAESs) containing ortho-methyl and pendant trifluoromethyl-substituted phenyl groups were also reported.[32] The resulted polymer films showed good thermal stability with glass transitions between 197 and 235°C, low dielectric constants of 2.67–2.73, and low water uptakes of 0.21–0.40%. Moreover, the polymers revealed good transparency with light transmittance at 450 nm of about 96% and low cutoff wavelength, namely 285 nm. The low light absorption of PAEKs recommends them for the opto-communication wavelengths of 1310 and 1550 nm.[32]

PAEs (or aromatic polyethers) are one of the basic classes of polymers for the plastics, known for their superior performance characteristics in many applications. The performance of optical films based on new PAEs with excellent thermal stabilities (T_d=500°C and T_g=300°C) was reported.[33] The physical features of PAEs were significantly determined by the monomers chemical structure and their catenation (linking position and the type of bisphenol pendants). The synergetic effect generated by the insertion of the flexible ether linkage (bond angle ~122°) and the bulky CF_3 groups together with other structural peculiarities led to molecular conformation and implicitly a specific packing density.[33] All these aspects conducted to a wide range of properties such as increased solubility, good thermal and chemical resistance, decreased color and intermolecular charge transfer complex, high optical transparency of ~90%, and dielectric constant of about 1.87–2.01. These excellent characteristics have imposed the PAEs as polymer substrates with potential applicability for high-speed electronic devices.[34]

Polynorbornene is a dicyclic hydrocarbon polymer, with a stiff saturated backbone and pure hydrocarbon content.[35] As a result of its structural nature, polynorbornene presents T_g (above 350°C),[36] low dielectric constant of approximately 2.2,[37] and good optical transparency. This combination of properties makes it very interesting as compound for interlayer dielectric in electronic packages and on-chip stress buffer applications. Functional polynorbornene with adequate grafted groups became adequate for adhesion purposes.[38]

PIs, among the heterocyclic polymers and the traditional engineering plastics are recognized as high-performance systems. Typically, PIs are synthesized from dianhydride and diamine monomers through various polymerization procedures. These polymers possess good combined properties, including high thermal and chemical stability, high radiation resistance, high mechanical and insulating properties, and in particular, inherent high refractive index.[39] They are utilized in advanced integrated circuits with multiple functionalities.[40,41] The thermal stability and low-*k* value, combined with high mechanical modulus and toughness, low thermal expansion coefficient, and low interfacial stress are the key points in the production of dielectric PIs for microelectronics. The strategies created to control such types of PI morphologies involve the chemical modifications of the polymer skeleton to enhance the solubility, optical and dielectric properties and/or the incorporation of nanoscale pores into PIs, rendering nanoporous PI dielectrics with dimensionally stable microarchitecture.[42] It is well known that the aliphatic structures determine a lower permittivity and enhanced transparency, which result from their low molecular density and polarity and diminished inter- or intramolecular charge transfer.[43] In this context, the polymer chain structure can be adapted by incorporation of alicyclic structures, short side groups as highly electronegative $-CF_3$ or bulky $-SO_2-$ moieties, proper linkages, or by using the copolymerization technique, in order to disrupt the symmetry and recurrence of regularity of the polymer chain and to reduce or inhibit the intra- and/or intermolecular charge transfer complex (CTC) formation with impact on an improved optical transparency.[44] Copolymerization is considered as the most permissive and successful method for systematic effective changes in the polymer properties. Combination of rigid and flexible, aromatic and aliphatic monomers by copolymerization may be used to control the polymer mechanical and thermal properties, according to the specific requirements of high-performance applications.[44]

This leads to the formation of high optical transparent copolyimide (CPI) thin films, with a transmittance greater than 85% at 400 nm and dielectric constant of 2.77–3.38. The results are due to the presence of groups with low polarizability (alicyclic units and 6F bridge) in various CPI structures, which are obtained from combination of stoichiometric 1/1 dianhydride/diamine monomers selected as aliphatic/aromatic pairs (at a molecular ratio of 1/3 per structural CPI units). The corresponding materials may be used as a flexible transparent substrate for micro- and optoelectronics and biomedicine.[45,46]

Silsesquioxanes represent a class of organosilicon compounds containing carbon–silicon bonds, with the chemical formula $[RSiO_{3/2}]_n$, where R=H, alkyl, aryl, alkoxyl, or other organic groups with functional derivatives.[47] Silsesquioxanes are colorless solids, with a cage-like structure, and have outstanding thermal, chemical, mechanical, and electronic properties and are extensively studied as types of dielectric materials, additives, and preceramics. This type of monomers can be partially opened under cross-linking on curing, leading to poly(hydrogen silsesquioxane) or poly(methylsilsesquioxane) (PMSSQ), which are recognized as polymer materials with low dielectric constant. Their refractive index and dielectric constant are the result of the extent to which the Si–O–Si cage structure can be found in the final polymer network structure. The latter can be adapted as a function of the curing process in nitrogen and air atmosphere.[26] Curing at very high temperatures obviously results in the eventual formation of SiO_2 by oxidation under loss of carbon and hydrogen functions. A low dielectric constant nanoporous PMSSQ can be accomplished by using poly(styrene-block-2-vinylpyridine) (PS-b-P2VP) as a template. The miscible hybrid and the narrow thermal decomposition of the PS-b-P2VP lead to nanoporous films with refractive index and dielectric constant which can be tuned by the loading ratio in the range of 1.361–1.139 and 2.359–1.509, respectively.[48] Other porous PMSSQ films were prepared from PMSSQ/amphiphilic block copolymer (ABC) hybrids, and this was followed by spin coating and multistep baking. The ABCs were poly(styrene-block-acrylic acid) (PS-b-PAA) and poly(styrene-block-3-trimethoxysilylpropyl methacrylate) (PS-b-PMSMA). Both intramolecular and intermolecular hydrogen bonding existed in the PMSSQ/PS-b-PAA hybrid led to macrophase separation. The resulted PMSSQ/PS-b-PMSMA nanoporous hybrid thin films presented pores with size less than 15 nm and dielectric constant decreasing from 2.603 to 1.843 with the PS-b-PMSMA loading increase from 0 to 50 wt.%, respectively. Thus, the chemical bonding in hybrid materials is essential for the preparation of low-*k* nanoporous films and provides evidence of pore formation in the PMSSQ thin films.[49]

Hydrophobic PMSSQ aerogels by a surfactant-free method using alkoxysilane with the ionic group were prepared.[50] The obtained aerogels expected to be transparent with good insulation properties. The obtained aerogels have low-density and high visible light transmittance equivalent to those of the materials prepared under the presence of surfactant. In addition, the thermal conductivity is as low as that of conventional silica aerogels and PMSSQ aerogels prepared with surfactant.[50]

Another category of low-*k* transparent dielectrics is xerogels. They are made of interconnected clusters of SiO_2. The dimension of a cluster is smaller (20 nm) than the wavelength of visible light, making these films optically transparent. The permittivity of a silica xerogel is affected by its volume fraction porosity, being very low (of about 1.78). Such compounds are excellent cladding materials for optical waveguides.[51]

Amorphous carbon films were revealed to have the dielectric constant below 3, good transparency, and also a high thermal stability at a temperature of at least about 400°C. Such dielectrics are useful in realizing semiconductor devices.[52]

5.3 TRANSPARENT POLYMER MATERIALS WITH HIGH DIELECTRIC CONSTANT

There are several low-molecular-weight compounds that are known to exhibit high dielectric constant. The most relevant examples are listed in Table 5.2.[5,36]

TABLE 5.2 Examples of Most Important High-k Materials.

Material	Dielectric constant
$PbMgNbO_3 + PbTiO_3$	22,600
$PbLaZrTiO_3$	1000
$BaSrTiO_3$	300
$BaTiO_3$	100–1250
TiO_2	50
Al_2O_3	9
AlN	8
(Bz, Ca, Sr) F_2	7

High-permittivity systems have become important mainly in three areas, memory cell dielectrics, gate dielectrics, and passive components. These materials are known for their high dielectric performance, but when analyzing the transparency they have some limitations. The majority of high-*k* materials are not transparent, this property being more typical for metal oxides. In order to supply the need for transparent dielectrics, such compounds are incorporated in polymers in low amounts. Thus, polymer nanocomposites are the most investigated transparent high-*k* dielectrics.

Kim et al.[53] developed a dielectric film through chemical vapor deposited graphene interlayer and examined the influence of the graphene interlayer on the dielectric performance. The resulted system is highly transparent, being in the form of a flexible film. The latter is a polymer/graphene/polymer "sandwich structure" accomplished by a one-step transfer method. The corresponding dielectric constant of the material was 51, with a dielectric loss of 0.05 at 1 kHz. The carbon-based interlayer in the sample constitutes a space charge layer, that is, an accumulation of polarized charge carriers near the graphene, generating an induced space charge polarization and enhanced permittivity.

The feature of the space charge layer for the graphene film, the sheet resistance of the graphene interlayer, was adapted through thermal annealing, which produced partial oxidation. The dielectric material with bigger sheet resistance owing to the oxidized graphene interlayer exhibited reduced dielectric constant in regard to that of the graphene with smaller interlayer sheet resistance. Oxidizing the graphene interlayer affects the space charge density in the dielectric film, in such a manner that diminished the capacitance. Considering the simplicity of the manufacturing process and high permittivity, as well as the high transparency and flexibility, this material is ideal for applications in plastic electronics.

Tseng et al.[54] reported the preparation of transparent polyimide/graphene oxide (GO) nanocomposites. The colorless and organo-soluble matrix was prepared from an alicyclic dianhydride and aromatic diamine, containing an ether bridge, using N,N-dime thylacetamide, and γ-butyrolactone as solvents. The GO particles were blended with the transparent PI in N,N-dimethylacetamide solution to obtain the nanocomposite films. When inserting only 0.001 wt.% of filler in PI matrix, the material not only exhibits the great resistance to moisture but also displays superior visible light transmission, enhanced mechanical strength, and excellent dimensional stability, simultaneously. The authors did not make reference at the dielectric behavior of the system but given the high permittivity ($\sim 10^6$ at 1 kHz) of GO,[55] it could be assumed that the PI/GO are high-k materials with good transparency. Other GO-based polymer composites[56] were based on a matrix derived from polypyrrole/polyvinyl alcohol (WPPy/PVA) blend. At 50 Hz and 150°C, the permittivity increases from 27.93 for WPPy/PVA (50/50 blend) to 155.18 for nanocomposites with 3 wt.% GO loading. At 50 Hz, the dielectric loss is augmented from 2.01 for WPPy/PVA (50/50 blend) to 4.71 for nanocomposites with 3 wt.% GO. These high-k WPPy/PVA/GO materials have high dielectric performance being suitable for high-frequency capacitors or embedded capacitors.

Barium titanate-based nanocomposites were prepared by Nagao et al.[57] showing that utilization of PMMA as matrix leads to films that have transparency and high permittivities. The nanoparticles were obtained by hydrolysis of a barium/titanium complex alkoxide in 2-methoxyethanol, and then the surface is modified with a silane coupling agent (3-methacryloxypropyltrimethoxysilane) to enhance their interface interactions with PMMA. The insertion of the surface-modified filler into PMMA was made up to a nanoparticle content almost similar to particle close-packing state. The refractive index of the resulted material increased with nanoparticle incorporation, keeping the relative transmittance normalized with PMMA film above 90%. A high refractive index of 1.82 was achieved at a nanoparticle content of 53 vol.%. For the same composition, the dielectric constant was 36 and the dissipation factor was low as 0.05.

Metal oxides inserted in transparent polymers, such as PMMA or PVA, is a good approach to obtain transparent nanocomposites with high-*k* values. Sugumaran et al[58] used TiO_2 to prepare multiphase materials through dipcoating method. The transmittance of the samples was higher than 80% in the visible domain. The optical bandgap was evaluated to be about 3.77 and 3.78 eV, respectively, for PVA–TiO_2 and PMMA–TiO_2 films. High values of permittivity were achieved for both systems, varying between 24.6 and 26.8. Zhu et al.[59] reported the procedure to obtain transparent dielectrics based on hafnium–aluminum oxide (HAO). The average transmittance of the HAO samples was over 88% within the visible radiations region and Al incorporation in HfO_2 can extend the bandgap of HAO. These films with the 2:1 ratio of Hf and Al rendered great dielectric performance, namely the relative permittivity of 12.1. Such materials are also promising for the development of new polymer composites.

5.4 SINGLE/MULTICOMPONENT TRANSPARENT DIELECTRICS WITH INTELLIGENT BEHAVIOR

The continuous upgrade of current technologies involves utilization of materials with multiple functions and sometimes with intelligent behavior. Polymers containing certain structural peculiarities are able to adapt some of their properties in response to external stimuli, such as electric or magnetic fields. In this section, the most significant achievements from the field of smart dielectric materials based on polymers are reviewed.

In the past decades, electroactive polymers (EAPs) have been proved to be a novel and efficient category of intelligent materials.[60] They are able to transform in a reversible manner electrical energy into mechanical output by modification of shape or size as a result of exposure to external stimuli. Among the polymers that exhibit good mechanical properties, which can be controlled through the abovementioned procedure, a great deal of attention was given to elastomers. Several reports reveal the importance of these transparent materials in the development of actuators,[61] tunable lenses,[62] buildings,[63] adhesive for smart windows,[64] and microelectromechanical systems.[65] In addition, the low density and easy operation of EAPs, combined with instant response and big deformation, recommend them as good compounds for artificial muscles.[66,67]

Dielectric elastomer actuator (DEA) is generally made of a soft elastomer (usually electrical insulator) placed between two compliant electrodes. Under high voltages, opposite charges are traveling toward corresponding electrodes, determining the generation of electrostatic forces. The electrostatic force is able to deform the elastomer on the longitudinal direction and consequently produces planar deformation. The actuation performance can be considerably improved if the field strength is enhanced. Therefore, this can be achieved by introducing a big voltage across the DEA or by prestraining the dielectric on the order from 200 to 500%. The literature reports several types of polymers that are able to perform actuation as dielectric layers, some of them having also good transparency. Among them one can mention silicones,[68] acrylates,[69] PVA,[61] PMMA,[64] and polyurethanes.[70] Many efforts have been made to improve the dielectric behavior of elastomeric materials by means of functionalization of polar groups from the polymer structure.[71] At the same time, the generation of huge deformation is based on conductive compliant electrodes, which present Young's modulus much smaller than that of the DEA.[66]

For applications in the biomedical domain, it was reported that DEA based on complaint PVA hydrogel electrodes exhibit remarkable performance. Beside its biocompatible character, PVA electrodes can be obtained with flexible networks through inter- and intramolecular hydrogen bonds. These aspects are favoring the stretchable features of the hydrogel electrodes, while it is still electroactive even after 2960 cycles. Moreover, the problem arising from the slippage of interfaces is avoided by the good adherence of soft PVA hydrogel to the dielectric material. The DEA derived from compliant PVA hydrogel electrodes proved to present significant actuating ability over 5 kV with the mechanism of ion migration. The prepared electrodes consist of about 90 wt.% water, which maintains structural compliance during actuation. The latter is affected by several factors, including applied voltage, ramp speed, electrode elastic modulus, and water content.[61] The addition of salts (LiCl, NaCl, and $CuSO_4$) can lead to high deformations as a result of ion migration; that is strain over 78% was recorded at 3.5 kV. The tunable, long lifetime, and easy manufacturing recommend the DEA based on PVA an ideal material for sensors, artificial muscles, and optical applications.

Moreover, some elastomer actuators also have the implication in the realization of lenses owing to their good transparency.[62] Comparative to classic lens assemblies, current trends are focused on improving the tunable and adaptive character in terms of compactness, efficiency, and flexibility. Shian et al.[62] prepared an elastomer–liquid lens system, which is based on an inline, transparent electroactive polymer actuator. The system relies on a stiff frame, a transparent liquid of fixed volume, and two elastomeric components—one passive and the other electroactive. The latter is a transparent dielectric elastomer covered on both sides with transparent compliant electrodes. The used dielectric is a commercially available acrylic elastomer and the electrodes are made of single-walled carbon nanotube mats. In the rest state, the refractive properties of the liquid and the extent of the membrane bulging are factors that influence the focal length and the numerical aperture of the lens. The second factor is determined by the volume of the liquid found inside the cavity. A higher amount of liquid lowers the focal length, while affects the opposite mode of the numerical aperture of the lens. The focal length of the lens can be constructed to range upon actuation depending on the location of the electroactive membrane. In other words, depending on which membrane is actuated (considering both of them electroactive), the variation of the focal length is noticed—relative to the rest state. Changes in focal length were recorded to be greater than 100% with this system, responding in less than 1 s. However, the transparency of the lens with two electroactive components will be influenced by the additional electrode pair. Considering the fact that the diameter of each membrane is the same, the actuation will also modify the numerical aperture of the lens, which is the reverse of the focal length. The lens' adaptive capability is controlled by the relative difference in curvature of each electroactive part as a result of variation of the internal liquid pressure. The deep analysis of membrane deformation within imposed conditions revealed that by choosing the adequate lens dimensions, even larger focusing dynamic ranges can be obtained.

A polymer system derived from poly(propylene glycol) (PPG), PMMA, and $LiCF_3SO_3$ is the key to obtain a material for smart windows given the resulted transparent, adhesive, and viscoelastic character.[64] The premise, which lied on the basis of the report of Mani and Stevens,[64] relies on the fact that polyelectrolytesrepresent dielectrics with suitable features for energy-efficient smart windows. The preparation of the electrolyte system was carried out by the matrix polymerization method. The initiator, involved in the reaction of methyl methacrylate with PPG doped with $LiCF_3SO_3$, was azobisisobutyronitrile. It must be mentioned that the introduction of $LiCF_3SO_3$ makes the two otherwise incompatible polymers compatible. PPG is noncrystalline and exhibits relatively high ionic conductivity when doped with alkali metal salts.[72] However, this material presents some drawbacks, such as hydrophilic surface, low mechanical stability, and reduced adhesiveness. To overcome these undesired aspects, a good solution was to incorporate PMMA. Various proportions of the salt and PMMA were examined to verify the performance of the system for pursued application. Following the proposed procedure, it was revealed that introduction of PMMA up to 8% leads to a protective coating for the PPG–salt complex, enhancing the hydrophobic character, concomitantly with the improvement of adhesive property of the electrolyte compound. The ionic conductivities of these polyelectrolytes were determined at various temperatures in the range of −20 to 110°C. The resulted dependence describes a conduction mechanism that is following Vogel–Tammann–Fulcher relation. At a high ratio of ether oxygen to alkali metal cation, PMMA causes the enhancement of the glass transition temperature of the system; the aspect is not noticed at a low ratio. The reported material is an ideal candidate for obtaining suitable components used in electrochrome-based smart windows.

Rizzello et al.[73] attempted to construct a model by combining the effects of the electrical dynamics which evaluates the current and displacement and considers the deterioration of the electromechanical actuation owing to the electrical dynamics. The latter can be neglected as the actuation frequency is higher. They extended the electromechanical model from the work of Hodgins et al.[74] with the electrical dynamics of dielectric EAPs. They combined two models: one that presents the electromechanical coupling and the nonlinear viscoelasticity of a dielectric EAP circular actuator and the other which predicts the electrical response of an electrical and mechanical input. The suggested approach led to a complex model, which accounts for the coupling between electrical, mechanical, and viscoelastic dynamics. Based on these ideas, they succeeded in modeling the effects of the electrical dynamics on the electromechanical response of a dielectric EAP circular actuator with a mass–spring load. The motion is determined by the distortion of the EAP provoked by the electrostatic compressive force between two compliant electrodes connected to the polymer surface. To preload the membrane, they used a mass and a linear spring. This enabled the stroke in the out-of-plane direction. The elaboration of mathematical models, which accurately depict the nonlinear coupling between electrical and mechanical dynamics, is essential for the development of high-precision position control algorithms functioning in high-frequency regimes (up to 150 Hz). The understanding of the nonlinear electrical dynamics of the actuator from the circuit is fundamental to obtain desirable features, such as increased modeling precision for high-frequency actuation, self-sensing, or control energy minimization. Rizzello et al.[73] performed numerous experiments, showing that the model can be used to predict both actuator current and displacement. Therefore, their approach allows increasing the overall displacement prediction exactness in regard to actuator models which do not consider electrical behavior.

Qiu and Hu[75] reported the importance of natural polymers derived from cellulose in preparation of smart materials. Most of the cellulose derivatives are transparent dielectrics and gain intelligent behavior in reaction to many environmental stimuli if their structure is properly adapted. Thus, by chemical modifications and physical incorporating/blending, the resulted materials become sensitive to temperature, pH, light, electricity, magnetic fields, and mechanical forces. Such compounds can be processed under various forms, such as nanoparticles, gels, copolymers, and membranes. All *chemical strategies* to achieve such smart materials are focused on modifications at the three alcoholic hydroxyl groups from the anhydroglucose unit. The classical reactions involve esterification, etherification, and oxidation processes. Chemical modifications can be performed both in heterogeneous and homogeneous conditions. Given the crystalline character of this material, its solubility is restricted to a narrow range of solvents, facilitating the reactions in heterogeneous conditions. Most of the chemical reactions take place particularly at the surface layer in heterogeneous conditions. Thus, the gross structure of the cellulosic material is less affected. Inhomogeneous conditions, occurrence of significant changes in the initial supermolecular structure of the sample and the limitation of the completeness of the chemical reaction can be controlled. In this way, well-defined polymeric materials can be prepared. Among the *physical procedures* to obtain smart cellulose systems, one can mention blending with other materials of high-molecular weight or addition of micro/nanoscale fillers. The advantage of these materials is given by renewable nature and outstanding mechanical properties of cellulosic component.[76,77] In the fabrication stage of multiphase/multicomponent materials containing cellulose, the latter plays an important role as matrix, filler, or coating/shell. The practical importance of these transparent dielectrics is exploited in many fields depending on the developed intelligent feature. Some cellulosic materials are able to target tissues or to reach specific intracellular locations if they are adapted to respond to redox potential, pH, light, temperature, and magnetic fields. Such systems are designed in different forms based on cellulose and other biomaterials for drug delivery. In the case of *aggregates* and *hydrogels*, external factors determine self-assembly and post-assembly triggering strategies creating an alternative method for the manipulation of self-assembled architectures of synthetic macromolecular aggregates in releasing of active principles. The assembly process can modify the transmittance and hydrodynamic radius, and medicinal substance loaded in the aggregates can be released while they disassemble in the presence of stimuli action in the surrounding medium. Stimuli-responsive hydrogels, which have the possibility to swell or shrink as a function of external stimuli, have been tested for biomedical purposes owing to their special properties, such as biocompatibility, biodegradability, and biological functionality.[78,79] The mechanism of hydrogels' swelling and shrinking occurs in an analogous manner to that of aggregate assembly, namely intermolecular and intramolecular hydrogen bonding or electrostatic forces are modulated by stimuli presence. Drugs incorporated in such materials can be delivered while they swell to looser structures as the medium conditions are changed. Another category of release cellulosic materials relies on microcapsules, nanoparticles, and membranes. During diffusion of the drug, the macromolecular shell affects the release kinetics.[80] The shell microcapsule swelling or membrane porosity is controlled by external factors such as temperature or pH.[81] Response to magnetic fields can be rendered to these systems if gold particles are modified with smart polymers.[82]

Another area where the intelligent features of cellulosic dielectrics is mechanical-adaptive materials. The basic idea here is to adapt elastic or Young's modulus (which ranges between 20 and 30 GPa) through modulation of the microstructure. This can be easily achieved in case of cellulose nanofibers embedded in polymer matrices. Depending on the reinforcement degree, one can notice the formation of rigid whisker networks in which stress transfer is favored by hydrogen bonding between the whiskers. The tendency of self-association cellulose nanofibers is facilitated by the intense interactions at surface hydroxyl groups.[83] The mechanical characteristics of polymer matrices between "on" and "off" states of hydrogen bonding of cellulose whiskers indicate significant differences. All these aspects recommend utilization of these smart cellulose materials in manufacturing mechanical-adaptive systems.

Natural polymers in combination with metal chloride or conducting polymers are also a good alternative for electroactive components and even for sensors. For example, optical pH sensors can be accomplished by immobilizing compound that is pH indicator on cellulose materials, revealing different colors when pH changes.[84] Chemosensors are obtained when they are grafted on cellulose fibers for the detection of cyanide ions in aqueous media.[84,85]

5.5 APPLICATIONS AND FUTURE PERSPECTIVES

Considerable efforts have been made recently to develop new materials with proper optical and dielectric properties for applications in electronic industries, which represent an essential point of current research and developments in the field of science and engineering. Thus, a real challenge is representing the implementation of transparent high-performance electronic materials, such as semiconductors, electric contacts, or dielectric/passivation layers in transistor and circuit structures. At the same time, an essential goal of the researchers is to achieve the suitable balance between application and specific material properties because the transistor performance and materials properties requirements vary in function of the final product device specifications. From the first research attempts until now, the classes of materials available for transparent electronics applications have presented a continuous development. This field of application was initially based on oxide materials that are both electrically conductive and optically transparent (known as transparent conducting oxides), being used in antistatic coatings, touch display panels, solar cells, flat panel displays, smart windows, and optical coatings as *passive* electrical or optical coatings. Nowadays, recent breakthroughs in this field of research highlight new *active* materials for functional transparent electronics. These systems demand new material sets, in addition to the transparent conductive oxides, such as conducting, dielectric, and semiconducting materials, for various device fabrication.

Insulators are generally used from dielectric materials with low dielectric constant, known as passivation compounds. Their applications are based especially on interlayer dielectric that reduces the resistance–capacitance time delays, fast signal propagation, or power dissipation in the high density. In electronic packaging, dielectric materials ensure the pathways for electronic device connections in multilayer circuit boards. Future direction in this research area will focus on decreasing relative permittivity of these devices in order to reduce the damaging effect of odd and coupling capacitances.[86] Some dielectric materials are applicable in encapsulation of the balls that link the substrate and the matrix from the electronic devices. This dielectric characteristic is also known as underfill and has the role to protect against any network failure and decrease thermal unsuitability between the bridging layers. In order to obtain peculiar properties of the apparatus, the research in this area must be oriented on obtaining high values of the electric permittivity that assure good polarizability for capacitors. These insulator features are useful in propagation and reflection of the electromagnetic waves and in the design of some semiconductor devices, dielectric amplifiers, memory elements, or piezoelectric transducers. All the abovementioned properties provide the nonpolar character of the material, the requirement of becoming a polar one being accomplished by introducing small amount of impurities for storing an important quantity of charges at low applied electric field. In this context, research efforts were made to obtain polymers such as transparent polyimide reinforced with Al_2O_3, $BaTiO_3$, or ZnO_3 that exhibit a higher dielectric constant.[87,88]

Piezoelectric transformers that have power yield below 10 W have been used for backlight inverters of laptop computers. In this context, it is known that the base materials are piezoelectric ceramics with maximum vibration speed of 1 m/s, obtained by rare-earth ion doping. These transformers have several advantages in regard to the conventional Rosen type, meaning high power, isolation resistance, and the connection technology.[89] These devices found new trend of applications in integrated ultrasonic motor and piezoelectric actuator drive system for active vibration control on helicopters. Latter studies present a demonstration regarding the preparation of a compliant, highly transparent, and electrically conductive electrode that, in combination with some dielectric elastomers, permits the manufacture of the adaptive lens. Thus, the utilization of an electrical signal to change the shape of a liquid lens without external actuation was been proposed.[90] Moreover, specific optic properties—high transparency of the electrodes, liquid, and elastomer—allow fabrication of simple and adaptive lenses useful in space-constrained tasks. These lenses show some benefits in comparison with conventional ones, in regard to efficiency, cost, flexibility, and compactness, and, at the same time, few components are necessary to manufacture them. The studies proved that length variation recorded was bigger than 100%, but, it is shown that by selecting proper lens size, even larger focusing dynamic ranges can be obtained.[91]

Dielectric electroactive polymers (DEAPs) represent new types of smart materials, manufactured by conjugating metal electrodes on both surfaces of a polymer core. This feature is useful to associate mechanical and electrical properties, meaning that electric field stimulation causes deformation and instead, a deformation leads to an electrical impact. These categories of dielectric polymers have drawn attention, especially in building industry. Thus, small DEAPs devices can be used to control air/fluid flow in ducts for heating, ventilation, and air-conditioning systems. At the same time, they found applications in large flexible structures such as building envelopes. As this device can be laminated easily and they have great power density, they can be used in the management systems of facades in order to control transparency or ventilation of the building envelopes. Moreover, vibration isolation possibilities and building monitoring purposes are studied as well.[63]

The current tendency in scientific research regarding materials with low dielectric constant useful in the microelectronic industry will continue to expand as the demand for faster processing techniques are in continuous development.[92] Chemical bonds with low polarizability and high porosity are the main factors through which a decreasing of dielectric constant can be obtained. The potential application in microelectronic circuit implies the necessity of satisfying some demands in regard to properties and reliability. In this context, the pore characterization, thin film deposition on porous substrate, specific properties of these films (especially mechanical features) and, also, conduction mechanism are the future steps to follow in order to improve low-*k* materials performance in innovative fundamental and applied science. Moreover, the transparency of such dielectrics is exploited in the creation of advanced alignment layers for nematic molecules. This is used in liquid crystal display technology, where PIs are often used. Besides their high transparency and low permittivity, their surface needs to be optimized to achieve the proper anisotropy for nematic orientation.[93]

Some recent researches evaluated optoelectronic properties of some modified polysulfones and their blends with some conductive polymers. The achieved results highlighted the extremely important role played by the control balance between the refractive indexes of the chemical conformation and the optical parameters in order to obtain the desired properties in specific electronic applications.[94,95] These phosphorus-modified polysulfones appear as interesting transparent materials due to their high transmittance (about 85% transmission over the whole visible spectral range) and optical energy values higher than 3.26 eV, having at the same time, a relatively low dielectric constant (between 2 and 6). The studies performed emphasize their utility in microelectronics, indicating that transparency, conductivity, and implicitly, electron interactions represent main features to improve electrical performance.

Polymer-based composites with excellent dielectric performance are very known area in the material science field. Polymers present some advantages from this point of view, such as low dielectric loss, easy processing, and low cost, but they also have some drawback including very low dielectric constant.[96] All of the drawbacks challenged the researchers to find some ingenious ways to improve these characteristics. Thus, a key issue is to increase the dielectric constant of polymers while retaining other excellent performances. A strategy is to manufacture a capacitor by adding some conductive filler into polymers. Among these fillers, functionalized carbon nanotubes exhibit a unique combination of mechanical, electrical, and thermal properties. The system obtained formed by flexible dielectric PS and multi-walled carbon nanotubes show a stable high dielectric constant and low dielectric loss in the wide frequency range. Such type of composites could be used as high-energy-density capacitors and flexible high-*k* components such as organic printed circuit boards-embedded dielectric devices.

On the other hand, polymer-based electronics display wide potential for printed polymer electrical circuits, flexible electronics, intelligent labels, or large area displays.[97] Introducing novel technology for coating organic polymers (such as PS) would bring great benefits due to their ability to attend to different electronic applications, being considered promising electro-optical materials. Owing to its transparent property, PS has opened new application possibilities in this field, especially photonic applications. From this point of view, the literature noticed some investigations regarding transparent conductive polymers with different shapes and morphologies prepared by coating with a nanostructured Au layer.[98] This new studies presented several advantages in comparison with other current metal deposition technologies, that is, the obtained process did not require specific templates and could be directly implemented on different polymer morphologies. At the same time, the obtained gold-coated PS substrate revealed attractive properties for optoelectronic applications, such as good optical transparency and excellent electrical conductivity. Therefore, this research brings out a versatile and easy gold coating technology that could be subsequently extended to other polymer types in order to obtain materials with suitable properties for optoelectronic devices.

5.6 CONCLUSIONS

This chapter makes a short review of the most significant features of low-*k* and high-*k* polymers. Thus, obtaining low-*k* polymers by reducing the macromolecule skeleton conjugation and using aliphatic, bulky, and non-coplanar sequences was summarized. The main design procedures to accomplish a specific macromolecular architecture that creates a big free volume for diminishing the dielectric property were also presented. At the same time, the chapter also describes high-*k* polymers prepared by introducing specific fillers that increase the main chain polarizability and, consequently, determine higher material density. Some examples of dielectrics with intelligent behavior and their applications are overviewed. At the end, the chapter highlights the main application and future trend of both categories of polymers.

ACKNOWLEDGMENT

This work was supported by a grant of the Romanian National Authority for Scientific Research and Innovation, CNCS—UEFISCDI, Project PN-II-RU-TE-2014–4–2976, no. 256/1.10.2015.

KEYWORDS

- **polymer**
- **transparency**
- **dielectric constant**
- **applications**

REFERENCES

1. McCoul, D.; Hu, W.; Gao, M.; Mehta, V.; Pei, Q. *Adv. Electron. Mater.* **2016,** *2,* 1500407.
2. Thomas, S.; Joseph, K.; Malhotra, S. K.; Goda, K.; Sreekala, M. S. *Polymer Composites, Macro- and Microcomposites;* Wiley: Germany, 2012; Vol. 1.
3. Fathpour, S.; Jalali, B. *Silicon Photonics for Telecommunications and Biomedicine;* CRC Press: New York, 2011.
4. Robertson, G. L. *Food Packaging: Principles and Practice*, 3rd ed.; CRC Press: Boca Raton, 2012.
5. Singh, R.; Ulrich, R. K. *Electrochem. Soc. Interface* **1999,** *8,* 26.
6. Joshi, D. R. *Engineering Physics;* Tata McGraw Hill: New Delhi, 2010.
7. Huang, X.; Zhi, C. *Polymer Nanocomposites: Electrical and Thermal Properties;* Springer: New York, 2016.
8. Shamiryan, D.; Abell, T.; Iacopi, F.; Maex, K. *Mater. Today* **2004,** *7,* 34.
9. Ahmad, Z. Polymeric Dielectric Materials. In *Dielectric Material;* Silaghi, M. A., Ed.; InTech: Croatia, 2012; pp 1–26.
10. Cosutchi, A. I.; Hulubei, C.; Stoica, I.; Dobromir, M.; Ioan, S. *e-Polym.* **2008,** *8,* 778.
11. Simpson, J. O.; St. Clair, A. K. *Thin Solid Films* **1997,** *308–309*, 480.
12. Damaceanu, M. D.; Musteata, V. E.; Cristea, M.; Bruma, M. *Eur. Polym. J.* **2010,** *46,* 1049.
13. Martinez-Vega, J. *Dielectric Materials for Electrical Engineering;* Wiley: U.K., 2010.
14. Meeten, G. H. *Optical Properties of Polymers;* Elsevier: New York, 2007.
15. Effect of Structure on the Dielectric Constant, 2008. https://www.doitpoms.ac.uk/tlplib/dielectrics/structure.php (accessed April 2017).
16. Callinan, T. D.; Parks, A. M. Conference on *Electrical Insulation,* 1959. DOI: 10.1109/EIC.1959.7533355.
17. Steybe, F.; Effenberger, F.; Gubler, U.; Bosshard, C.; Günter, P. *Tetrahedron* **1989,** *54,* 8469.
18. Barber, P.; Balasubramanian, S.; Anguchamy, Y.; Gong, S.; Wibowo, A.; Gao, H.; Ploehn, H. J.; Loye, H. C. *Materials* **2009,** *2,* 1697.
19. Zhang, L.; Shan, X.; Bass, P.; Tong, Y.; Rolin, T. D.; Hill, C. W.; Brewer, J. C.; Tucker, D. S.; Cheng, Z. Y. *Sci. Rep.* **2016,** *6.* DOI: 10.1038/srep35763.
20. Gupta, T. *Copper Interconnect Technology;* Springer: London, 2009.
21. Pyshkin, S. L.; Ballato, J. *Optoelectronics—Materials and Devices;* InTech: Croatia, 2015.

22. Morgan, M.; Ryan, E. T.; Zaho, J.; Hu, C.; Ho, P. S. *Annu. Rev. Mater. Sci.* **2000,** *30,* 645.
23. Hess, D. W. *J. Electrochem. Soc.* **2003,** *150*(1), S-1.
24. Kitoh, H.; Mroyama, M.; Sasaki, M.; Iwasawa, M.; Kimura, H. *Jpn. J. Appl. Phys.* **1996,** *35,* 1464.
25. Gill, A.; Patel, V. *J. Electrochem. Soc.* **2004,** *151,* 133.
26. Kohl, P. A. *Annu. Rev. Chem. Biomol. Eng.* **2011,** *2,* 379.
27. Rosenmayer, C. T.; Bartz, J. W.; Hammes, J. *Mater. Res. Soc. Symp. Proc.* **1997,** *476,* 231.
28. Tapaswi, P. K.; Choi, M. C.; Jung, Y. S.; Cho, H. J.; Seo, D. J.; Ha, C. S. *J. Polym. Sci. Part A*. **2014**, *52*(16), 2316..
29. Maier, G. *Prog. Polym. Sci.* **2001,** *26,* 3.
30. Critchley, J. P.; Knight, G. J.; Wright, W. W. *Heat-Resistant Polymers: Technologically Useful Materials;* Plenum Press: New York, 1983; p 173.
31. Patrick, E.; Cassidy, G. L.; Tullos, A. K. St. Clair. Low Dielectric Poly(phenylene ether ketone) Film. US Patent 4,902,769A, 1990. http://www.boedeker.com/peek_p.html.
32. Shang, C.; Zhao, X.; Li, J.; Liu, J.; Huang, W. *High Perform. Polym.* **2012,** *24,* 692.
33. Colquhoun, H. M.; Williams, D. J. *Acc. Chem. Res.* **2000,** *33,* 189.
34. Lee, C. C.; Huang, W. Y. *Polym. J.* **2010,** *43,* 180.
35. Heitz, W. *Pure Appl. Chem.* **1995,** *67,* 1951.
36. Treichel, H.; Withers, B.; Ruhl, G.; Ansmann, P.; Wurl, R.; Muller, C.; Dietlmeier, M.; Maier, G. Low Dielectric Constant Materials for Interlayer Dielectrics. In *Low-k and High-k Materials;* Nalwa, H. S., Ed.; Academic Press: Oxford, UK, 1999.
37. Grove, N. R.; Kohl, P. A.; Bidstrup-Allen, S. A.; Shick, R. A.; Goodall, B. L.; Jayaraman, S. *Mater. Res. Soc. Symp. Proc.* **1997,** *476,* 3.
38. Grove, N. R.; Kohl, P. A.; Allen, S. A.; Jayaraman, J.; Shick, R. *J. Polym. Sci. Part B* **1999,** *37,* 3003.
39. Sroog, C. E. *Prog. Polym. Sci.* **1991,** *16,* 561.
40. Ioan, S.; Hulubei, C.; Popovici, D.; Musteata, V. E. *Polym. Eng. Sci.* **2013,** *53,* 1430.
41. Cosutchi, A. I.; Hulubei, C.; Buda, M.; Botila, T.; Ioan, S. *Rev. Roum. Chim.* **2007,** *52,* 665.
42. Ree, M.; Yoon, J.; Heo, K. *J. Mater. Chem.* **2006,** *16,* 685.
43. Mathews, A. S.; Kim, I.; Ha, C. S. *Macromol. Res.* **2007,** *15,* 114.
44. Ni, H. J.; Liu, J. G.; Wang, Z. H.; Yang, S. Y. *J. Ind. Eng. Chem.* **2015,** *28,* 16.
45. Hulubei, C.; Popovici, D.; Bruma, M. Process for Synthesys of Transparet Flexible Polyimide Materials. Patent No. RO131123-A2, 2016.
46. Buruiana, L. I.; Barzic, A. I.; Stoica, I.; Hulubei, C. *J. Polym. Res.* **2016,** *23,* 217.
47. Cordes, D. B.; Lickiss, P. D.; Rataboul, F. *Chem. Rev.* **2010,** *110,* 2081.
48. Yang, C. C.; Wu, P. T.; Chen, W. C.; Chen, H. L. *Polymer* **2004,** *45,* 5691.
49. Chang, Y.; Chen, C. Y.; Chen, W. C. *J. Polym. Sci. Part B: Polym. Phys.* **2004,** *42,* 4466.
50. Hayase, G.; Nagayama, S.; Nonomura, K.; Kanamori, K.; Maeno, A.; Kaji, H.; Nakanishi, K. *J. Asian Ceram. Soc.* **2017,** *262,* 1.
51. Jain, A.; Rogojevi, S.; Ponoth, S.; Agarwal, N.; Matthew, I.; Gill, W. N.; Persans, P.; Tomozawa, M.; Plawsky, J. L.; Simonyi, E. *Thin Solid Films* **2001,** *398,* 513.
52. Bencher, C. D. Removable Amorphous Carbon CMP Stop. US Patent 6,541,397 B1, 2002.
53. Young, J.; Lee, K. J.; Lee, W. H.; Kholmanov, I. N.; Suk, J. W.; Kim, T. Y.; Hao, Y.; Chou, H.; Akinwande, D.; Ruoff, R. S. *ACS Nano* **2014,** *8,* 269.

54. Tseng, I. H.; Liao, Y. F.; Chiang, J. C.; Tsai, M. H. *Mater. Chem. Phys.* **2012,** *136,* 247.
55. Kumar, K. S.; Pittala, S.; Sanyadanam, S.; Pai, P. *RSC Adv.* **2015,** *5,* 14,768.
56. Deshmukh, K.; Ahamed, M. B.; Pasha, S. K.; Deshmukh, R. R.; Bhagat, P. R. *RSC Adv.* **2015,** *5,* 61,933
57. Nagao, D.; Kinoshita, T.; Watanabe, A.; Konno, M. *Polym. Int.* **2011,** *60,* 1180.
58. Sugumaran, S.; Bellan, C. S. *Optik—Int. J. LightElectron Opt.* **2014,** *125,* 5128.
59. Zhu, L.; Gao, Y.; Li, X.; Sun, X. W. *J. Mater. Res.* **2014,** *29,* 1620.
60. Bar-Cohen, Y. *J. Spacecr. Rockets* **2002,** *39,* 822.
61. Xu, C.; Li, B.; Xu, C.; Zheng, J. *J. Mater. Sci.: Mater. Electron.* **2015,** *26,* 9213.
62. Shian, S.; Diebold, R. M.; Clarke, D. R. *Opt. Express* **2013,** 21, 8669.
63. Berardi, U. *Intell. Build. Int.* **2010,** *2,* 167.
64. Mani, T.; Stevens, J. R. *Polymer* **1992,** *33,* 834.
65. Brochu, P.; Pei, Q. *Macromol. Rapid Commun.* **2010,** *31,* 10.
66. Biggs, J.; Danielmeier, K.; Hitzbleck, J.; Krause, J.; Kridl, T.; Nowak, S.; Orselli, E.; Quan, X. N.; Schapeler, D.; Sutherland, W.; Wagner, J. *Angew. Chem. Int. Edit.* **2013,** *52,* 9409.
67. Bar-Cohen, Y.; Wallmersperger, T. *Smart Struct. Mater.* **2009.**
68. Zhang, X. Q.; Wissler, M.; Jaehne, B.; Broennimann, R.; Kovacs, G. *Proc. SPIE Int. Soc. Opt. Eng.* **2004,** *5385,* 78.
69. Pelrine, R.; Kornbluh, R.; Pei, Q. B.; Joseph, J. *Science* **2000,** *287,* 836.
70. Venkataswamy, K.; Ard, K.; Beatty, C. L. *Polym. Eng. Sci.* **1982,** *22,* 961.
71. Kussmaul, B.; Risse, S.; Wegener, M.; Kofod, G.; Kruger, H. *Electroact. Polym. Actuators Dev. (Eapad)* **2012,** *83,* 400, 1.
72. Watanabe, M.; Ikeda, J.; Shinohara, I. *Polym. J.* **1983,** *15,* 175.
73. Rizzello, G.; Hodgins, M.; Naso, D.; York, A.; Seelecke, S. *Smart Mater. Struct.* **2015,** *24,* 94003.
74. Hodgins, M.; Rizzello, G.; Naso, D.; York, A.; Seelecke, S. *Smart Mater. Struct.* **2014,** *23,* 104,006.
75. Qiu, X.; Hu, S. *Materials* **2013,** *6,* 738.
76. Siqueira, G.; Bras, J.; Dufresne, A. *Polymers* **2010,** *2,* 728.
77. Khalil, H. P. S. A.; Bhat, A. H.; Yusra, A. F. I. *Carbohydr. Polym.* **2012,** *87,* 963.
78. Prabaharan, M.; Mano, J. F. *Macromol. Biosci.* **2006,** *6,* 991.
79. Sannino, A.; Demitri, C.; Madaghiele, M. *Materials* **2009,** *2,* 353.
80. Barzic, A. I.; Nechifor, C. D.; Stoica, I., Dorohoi, D. O. *J. Macromol. Sci. Part B* **2016,** *55,* 575.
81. Fang, A.; Cathala, B. *Colloids Surf., B* **2011,** *82,* 81.
82. Gaharwar, A. K.; Wong, J. E.; Müller-Schulte, D.; Bahadur, D.; Richtering, W. *J. Nanosci. Nanotechnol.* **2009,** *9,* 5355.
83. de Susa Lima, M. M.; Borsali, R. *Macromol. Rapid Commun.* **2004,** *25,* 771.
84. Schueren, L. V. D.; Clerck, K. D.; Brancatelli, G.; Rosace, G.; Damme, E. V.; Vos, W. D. *Sens. Actuators B Chem.* **2012,** *162,* 27.
85. Isaad, J.; Achari, A. E. *Tetrahedron* **2011,** *67,* 4939.
86. Ahmad, Z. Polymeric Dielectric Materials. In *Dielectric Material;* Silaghi, M. Al., Ed.; InTech: Rijeka, Croatia, 2012; pp 3–26.
87. Xie, S. H.; Zhu, B. K.; Li, J. B.; Wei, X. Z.; Xu, Z. K. *Polym. Test.* **2004,** *23,* 797.
88. Liu, L.; Liang, B.; Wang, W.; Lei, Q. *J. Compos. Mater.* **2006,** *40,* 2175.
89. Uchino, K.; Priya, S.; Ural, S.; Vazquez Carazo, A.; Ezaki, T. High power piezoelectric transformers—Their applications to smart actuator systems. In *Developments in*

Dielectric Materials and Electronic Devices; Nair, K. M.; Guo, R.; Bhalla, A. S.; Hirano, S.; Suvorov, D., Eds.; The American Ceramic Society: Indianapolis, USA, 2005; p 383.
90. Shian, S.; Diebold, R. M.; McNamara, A.; Clarke, D. R. *Appl. Phys. Lett.* **2012,** *101,* 61101.
91. Shian, S.; Diebold, R. M.; Clarke, D. R. *Opt. Express* **2013,** *21,* 8669.
92. Lee, H. S.; Lee, A. S.; Baek, K. Y.; Hwang, S. S. Low Dielectric Materials for Microelectronics. In *Dielectric Material;* Al. Silaghi, M., Ed.; InTech: Rijeka, Croatia, 2012; p 59.
93. Barzic, A. I.; Stoica, I.; Popovici, D.; Vlad, S.; Cozan, V.; Hulubei, C. *Polym. Bull.* **2013,** *70,* 1553.
94. Buruiana, L. I.; Avram, E.; Popa, A.; Musteata, V. E.; Ioan, S. *Polym. Bull.* **2012,** *68,* 1641.
95. Buruiana, L.; Avram, E.; Musteata, V. E.; Filimon, A. *Mat. Chem. Phys.* **2016,** *177,* 442.
96. Yang, C.; Lin, Y.; Nan, C. W. *Carbon* **2009,** *47,* 1096.
97. Mildner, W.; Hecker, K. Roadmap for Organic and Printed Electronics. In *Polymer Electronics—A Flexible Technology;* Gardiner, F.; Carter, E., Ed.; iSmithers Rapra: UK, 2009; pp 1–12.
98. Trachtenberg, A.; Vinod, T. P.; Jelinek, R. *Polymer* **2014,** *55,* 5095.

CHAPTER 6

HIGH-T_C SUPERCONDUCTING Bi CUPRATES: CHASING THE ELUSIVE MONOPHASE

T. KANNAN and P. PREDEEP*

LAMP, Department of Physics, National Institute of Technology, Calicut, Kerala, India

Corresponding author. E-mail: ppredeep@gmail.com

CONTENTS

ABSTRACT

Bismuth-based cuprates offer superconductors above liquid nitrogen temperature. However, even after decades of their existence, the synthesis of phase-stable Bi cuprates superconductors is still a challenging problem. Glass precursor technique which has been shown as a different but highly efficient synthesis route for these types of superconducting ceramic also requires optimized critical synthesis conditions. In this chapter, various aspects of this problem are discussed and tricky issues of keeping stoichiometry during reaction, identifying optimum annealing temperature through Avrami index, and other factors are reviewed critically.

6.1 INTRODUCTION TO CERAMIC SUPERCONDUCTORS

The research in superconductivity often looked like something similar to *alchemy*, chasing an elusive dream of realizing materials that allow the current to flow through them forever without any resistance at room temperature. This ambitious, at the same time often desperate, scientific chase for a dream material, along with its still fully unknown mechanism, makes superconductivity a mysterious[61] and weird one. In this context, those in the field of superconductivity research may well remember the tremendous hype and hope of realizing room temperature superconductors during the 1980s. However, this dream still continues to be a mirage to the scientific community as a whole. It was like the hype created during the advent of nanotechnology in recent times, and the fascinating possibilities about superconductivity filled the air during the 1980s when it was hyped that the scientific community was at the brink of developing the long-cherished dream of room temperature superconductivity. However, the difference is that while nanotechnology fast matured theoretically and experimentally and opened up numerous application areas from fairness creams to biotech and medicine, room temperature superconductivity still continues to be an evasive dream for the scientists. The superconductivity research had so far made large strides in achieving reasonably realizable transition temperatures (T_c) that made many important applications possible. One of the life-saving applications of superconductivity comes in the form of *whole-body scanning* machine. On the other end, bullet trains at super speeds form a monumental achievement of superconductivity research outside the laboratory.[62] The huge promise of solutions to our planet's vows and desires, starting from energy, healthcare, ultrafast communication, quantum computing, and much more, is sure to sustain on the superconductivity research at a high pitch. There do promise in the air and there are silver linings. Research in the field, it seems, is currently preoccupied in consolidating the gains of the current high-temperature superconductors—the ceramics, popularly known as perovskites.

It looked odd to find that the most successful high-temperature superconductors, the yttrium barium copper oxides (YBCOs), popularly known as 1–2–3 compounds,[67] and the bismuth strontium calcium copper oxide (BSCCOs) or Pb–BSCCO (PBSCCO), are just ceramics that are best insulators at room temperature. At reasonably low temperatures—at liquid nitrogen temperature—they become perfectly superconducting.

The interest in ceramic superconductors was triggered by the discovery of superconducting transition at 35 K in lanthanum barium copper oxide ceramic system,[11] $La_{2-x}Ba_xCuO_4$. This T_c soon took a big stride to 92 K for 123 systems. The unique aspect of these two systems was that they all contained rare-earth elements.

The timeline of superconductivity research saw a quantum jump in 1987 when Michel et al.[42] reported the first high-temperature T_c oxide ceramic system that was based on Bi–Sr–Cu–O and did not contain any rare-earth components. The rare-earth-free Bi–Sr–Cu–O perovskites got a fillip after 1 year when Maeda et al.[36] added calcium to Bi–Sr–Cu–O to increase the T_c to 85 K. This triggered new rare-earth-free systems in a short while with reports of achieving T_c 110 K by Chen et al.[16] for Bi–Sr–Ca–Cu–O system. Since then, various efforts in improving the T_c of such systems continued and reports of T_c reaching 125 K for the Ti–Ba–Ca–Cu–O system by Parkin et al. presented interesting and encouraging data. However, the synthesis of Bi systems is not as straightforward as that of rare-earth systems such as YBCO. Rare-earth-free systems are plagued by stray phases of lower T_cs and getting a monophase superconducting phase in such systems remains a challenge even now. Many efforts are being reported by varying the elemental ratios and dopants such as Pb and so forth with partial success. Even for those compositions that were reported to have monophase in Bi systems, reproducibility requires too many precautions and accuracy in controlling the elemental ratios. Failure rates were also very high.

In the course of this elusive perusal for finding easy and reproducible solutions to this problem of achieving superconducting monophase in rare-earth-free BSCCO perovskites, new method such as *glassy precursor* route was widely pursued technique. In comparison with the conventional solid-state sintering technique, the glassy route proved to be more efficient and realizable. The objective of this chapter is to provide an introduction to glassy precursor route for developing stable Bi perovskites and to present certain interesting parameters and optimizing factors that we observed during the work carried out in this area. It would not be out of context to have a look back at the evolution of the Bi compounds for superconducting property development before embarking into the presentation of such interesting data.

6.2 EVOLUTION OF BSCCO SYSTEMS

The field of high-temperature superconductivity (HTSC) era got a fillip, with the processing of YBCO[67] in the year 1987. This means YBCO compound gave the hitherto highest T_c and it was rather easy to synthesize with good phase stability. This instilled great expectations and hope toward developing superconducting materials, showing superconductivity at practically higher temperatures. The discovery of bismuth-based cuprates with a T_c of 110 K by Maeda et al.[36] in the year 1988 and thalium-based cuprates with a T_c of 120–125 K by Sheng and Hermann, and finally the mercury-based cuprates with a T_c of 135 K created a new hope for HTSC based on cuprates. That mercury, the first known superconductor that had become the key element for the best high-T_c cuprates after 82 years, can be credited to the art and handy work of nature and science. Figure 6.1 shows the rise in the highest critical temperature since the discovery of superconductivity. Most of the known cuprate superconductors structurally belong to a single family and are closely related to each other.[31]

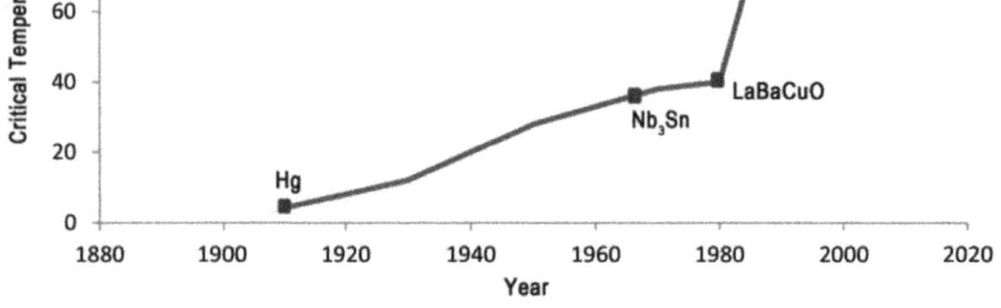

FIGURE 6.1 History of rise of critical temperature with year.

Bismuth-based cuprates are attractive HTSC materials because the grain alignment of these materials is along c-axis, which, in turn, increases the amount of critical current.[35] Bi–Sr–Ca–Cu–O system has three different phases: $Bi_2Sr_2Ca_0Cu_1O_{4+x}$ (2201), T_c=7–20 K; $Bi_2Sr_2Ca_1Cu_2O_{8+x}$ (2212), T_c=80–90 K; and $Bi_2Sr_2Ca_2Cu_3O_{12+x}$ (2223), T_c=110 K. Unit cells for the different phases are mainly pseudo-tetragonal with lattice parameters, a≈b≈5.4 Å and c≈24.4, 30.9, and 37 Å for Bi-2201, Bi-2212, and Bi-2223, respectively.

Bismuth-based cuprates are attractive because of high-T_c phase 2223. The preparation of 2223 as a single-phase material is difficult because different phases are stable only within a small temperature range. It is difficult to grow large and pure crystals based on BSCCO system and it becomes exceedingly difficult in the case of 2223, a three-layered bismuth cuprate. The complexity of BSCCO phase diagram[65] is that all the members in the system melt congruently and the primary phase field always contains the equilibrium of different phases. This means that all members of the system do not melt simultaneously at the same temperature making them incompatible and therefore the primary phase of the system tries to adjust by developing an equilibrium that complicates the synthesis process. Before embarking into the various aspects of BSCCO ceramic superconductors, it is better to have a brief description of crystal structure of BSCCO system and the roles of different planes in the cause for superconductivity.

6.3 CRYSTAL STRUCTURE

Crystal structure of these materials is perovskite in nature as shown in Figure 6.2. The structures of HTSC cuprates are related to that of perovskite.

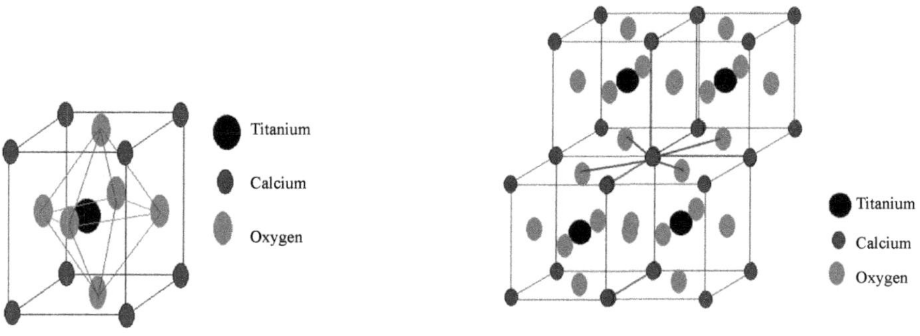

FIGURE 6.2 Perovskite structure based on $CaTiO_3$.

A typical example is $CaTiO_3$. Figure 6.2 shows a unit cell consisting of five atoms, calcium atoms being at the corners of the cube, a titanium atom at the center of the cube, and oxygen atom at centers of the cube faces. Titanium site is located between six oxygen atoms forming a perfect octahedron. The calcium site is surrounded by 12 equidistant oxygen atoms that form a cuboctahedron.[49]

As a first approximation, crystal structures of high-temperature superconductors of the cuprate type, including the Bi-2223 (Fig. 6.3a) and 2212 (Fig. 6.3b) compounds, as a rule, consist of two blocks, that is, the superconducting block $CuO_2[ACuO_2]_{k-1}$ and the charge reservoir block BO[CO] BO (where O—oxygen, A=Ca, Y; B=Sr, Ba; and C=Bi, Ta, Hg).[15]. In the structures, the CuO_2 planes formed by the O_4 squares centered by copper atoms and shared by vertices serve as charge carriers. Upon a change in the oxidation state of cations, the charge reservoir block plays the role of a

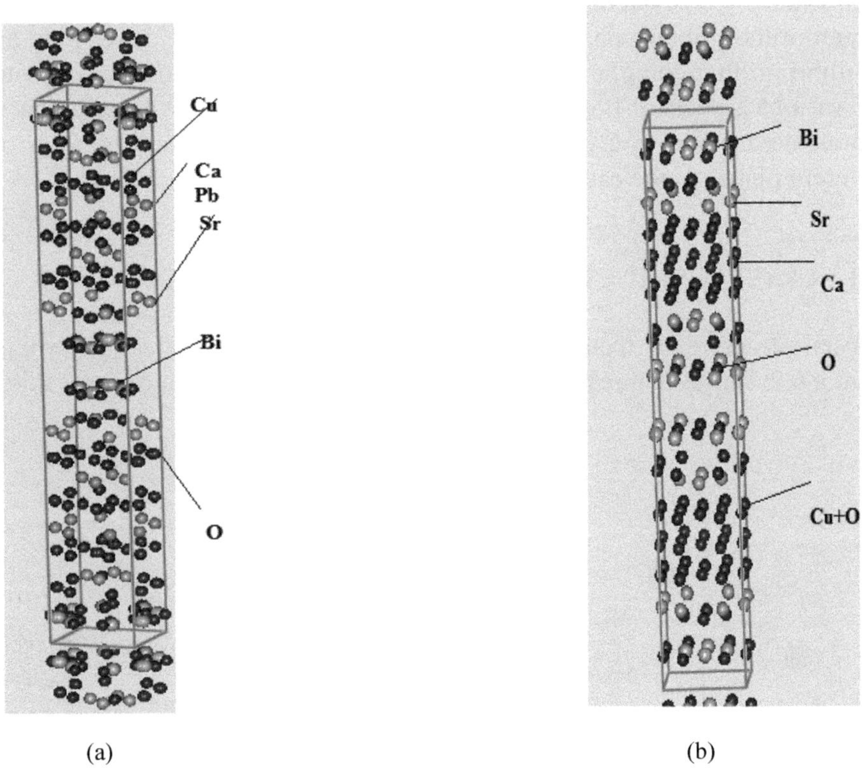

FIGURE 6.3 Crystal structures of (a) 2223 and (b) 2212.

source of charge carriers transferred to the superconducting block.[57]. CuO_2 layers or planes are the metallic, superconducting layers containing delocalized charge carriers[10] which are essentially holes. The arrangement of Cu and O atoms in [001] plane of the perovskite structure contains Ti^{2+} and O^{2-} ions. The Bi–O layers are believed to donate holes to the superconducting CuO_2 planes. The Ca^{2+} cation layers are insulating and donate electrons to the CuO_2 planes. The SrO layers apparently serve primarily as barriers that isolate groups of CuO_2 planes from each other. The layers of atoms of Ca^{2+} and Bi^{2+} are often referred to as the charge reservoir layers since they control charge concentrations on the CuO_2 layers. An important feature which distinguishes the bismuth high-temperature superconductors from the mercury- and thallium-based high-temperature superconductors whose T_c is greater than 110 K is the inverse dependence of the lattice parameters a and b on the number of planes in the superconducting block.[57]

6.4 GLASS CERAMIC METHOD: ADVANTAGES AND DISADVANTAGES

The glass precursor technique has several advantages. Most important is that one can get various components mixed at the molecular level in the melt.[54] One of the salient features of this method was that uniformity and high density coupled with almost zero porosity can be achieved by this method. The second stage of the synthesis from glass involves crystallization that gives control over the formation of highly oriented grains of the ceramic materials.[20] This means that high critical current density, one of the major deciding parameters of quality of superconductivity, could be achieved. In addition, there are many exciting advantages to this glass precursor route of BSCCO formation. They could be summarized as:

- Controlled thermal annealing of the supercooled melt will lead to unique metastable crystalline phases having unusual characteristics.
- The crystallization starts from a perfectly homogeneous glassy state and the final product will have real chemical homogeneity.
- Microstructural control is possible by designing appropriate annealing temperatures and timings.

Besides, in the application level, the glassy route has the advantage of manufacturing ceramic superconductors in various shapes and forms such as wires, coils, fibers, films, and so forth, as the glass phase allows such maneuvering. It is not that glassy route is all rosy and not without flaws. There are certain aspects that demand care and attention in this method as well. One such aspect is that the chemical components of the system should not be volatile even at a few hundred degrees above the melting temperature. Another critical aspect relates to glass preparation technique itself—the reaction vessel should be nonreactive with the melt. Now, various techniques, such as container-free melting techniques, solid-state reaction,[52] organic precursor,[12] solgel technique,[70] and so forth, are available to bypass this latter problem. Considering the versatile advantages over other methods, the glassy route holds much promise in the efforts toward developing easy and reproducible monophase rare-earth-free superconducting Bi perovskites.

Glass ceramic method for fabrication of the 2223 phase in the PBSCCO system has been carried out by many researchers in the field of HTSC. Glass ceramic method can offer better HTSC materials with many advantages[1,8] such as pore-free, highly dense, homogeneous structure with strong grain connection and shape than the material prepared through conventional solid-state reaction technique. Factors which characterize the microstructure of the ceramic core are texture, quality of boundaries, strength of coupling, grain size, ceramic density, existence of cracks, and so forth.[53] Moreover, this technique provides[8] strongly connected grains with good grain growth having large current-carrying capability. High-T_c superconductors prepared using the melt-quenching method is christened by Komatsu et al. as "high-T_c superconducting glass ceramic." A great stride has been made in the field of practical application of high-temperature superconducting layered cuprates since the discovery of these materials.

A successful synthesis of high-T_c perovskites depends on the purity of the starting materials, choice of the oxides, carbonates or hydrates, the homogeneity of the mixture of powders, and the heating schedule. Bi-2223, the highest T_c phase in BSCCO and PBSCCO series, is exceptionally tricky to get in its wholesome form. Attempts to synthesize Bi-2223 by sintering the stoichiometric mixtures of oxides and carbonates at 1123 K resulted in the formation of Bi-2212 as majority phase and Bi-2223 as minority phase. A good number of investigators[2,3,9,16,21,23–27,29,30,39,43,58,63,69] had reported the integration of lead in the BSCCO system (PBSCCO) coupled with extended sintering time to get increased amount of high-T_c phase (2223). Lead is known[59] to partially substitute into bismuth oxide planes of the crystal structure of the 110 K phase. The effect of this substitution is to stabilize the structure and facilitate its formation as a nearly pure phase.[43] High-resolution analytical electron microscopy reveals[3] that Pb atoms are located in the Bi–O layers of the structure with an atomic ratio of Pb to Bi of 1:9. Bi in the +3 oxidation state and lead in the +2 state have similar electronic configuration of $6s^2$,$6p^0$ which might have helped lead atoms to be easily incorporated into the structure. However, the other side of the picture is that the larger size of Pb ions (1.2 Å) in comparison with Bi ions (0.96 Å) can cause a distortion in the crystal lattice.[9] Positive aspect of the Pb addition is that the liquid formed due to Pb can increase the volume fraction of 2223. The absence of the necessary amount of liquid in the solid phase may result in the formation of a mixed layered structure, including syntactic intergrowths of 2223, 2212, and 2201 phases. The only single-phase Bi-2223 samples reported[49,65] so far was prepared by precise control of the processing conditions and partial substitution of Bi by Pb. Doping of Pb in the BSCCO system can strongly influence the crystallization kinetics, superconducting grain formation, and superconducting properties which, in turn, help easy grain growth and crystallization of bulk samples.[2] Further Pb doping introduces[2] a modified higher T_c with a double transition followed by a rearranged electronic structure of 2223 sample. In order to optimize the processing of 2223 ceramics, Pb solubility as a function of the temperature and cation ratio has been examined[30] in detail by considering the Gibbs phase rule for heterogeneous phase equilibrium. It has been found that the maximum amount of equilibrium phase increases by 1 due to the additional component PbO. Investigations[30] of the Ca:Cu ratio variations x=2, 3, 4, 5, 6, 8 and the Pb content y=0, 0.2, 0.4, 0.6, 0.8, 1.0 in $Bi_2Pb_ySr_2Ca_{x-1}Cu_xO_z$ revealed that under the standard conditions, the best composition for 2223 phase is Bi:Pb:Sr:Ca:Cu=1.4:0.34 :1.91:2.03:3.06.[19,32] These findings suggested Endo et al.[19] to have a stringent control of stoichiometry in the processing techniques of 2223 phase. If the Bi percentage is increased beyond the Pb solubility limit, two plumbates, such as Ca_2PbO_4 and the phase $Pb_4Sr_5CuO_{10}$ better known as 451, could arise within the concentration range of 2212 and 2223. For the equilibrium of 451 with 2212 and 2223, the Pb-to-Bi ratio and Sr-to-Ca ratio should be 4:1 and 1:1, respectively.[40] The solubility of Pb in 2223 depends on the temperature and the maximum solubility is at 1123 K.[26,27,37,39] Oxygen partial pressure can also control the lead solubility, which increases with a decrease in oxygen partial pressure.[32] Lead also plays a major role in the structural transformation in the Bi-based cuprates, where the structure transforms from tetragonal to orthorhombic with the increase of Pb content.[23–25] All the above results show that the substitution of Bi by Pb can enhance the processing window and stabilization of crystal structure of 2223. Thus, a number of problems that are of interest at present for researchers and designers of High-T_c superconducting materials are extremely wide.

Experimental results in the processing of PBSCCO system are influenced by many factors, including the processing conditions to the sensitivity of phase assemblages, the sluggish kinetics of the phase transformation, and the very narrowly spaced phase stability fields of high-T_c phases. Mutual solid-state compatibilities between 2212 and 2223 phases are extensive. This is due to their similar structure and also they belong to the same homologous family, and more importantly, 2212 phases is a precursor for the formation of the 2223 phase.[49] Moreover, it is to be remembered that the high free energy of the 2223 phase, when compared with 2212 due to its longer c-axis, makes it a slow runner during phase formation.[58]

The variation in the Ca-to-Cu ratio also has a telling effect on the formation 2223 in PBSCCO system. Ca and Cu diffusion can control the kinetics of 2223 phase formation. The presence of Pb enhances[2] the diffusion of Ca and Cu during insertion of an additional layer of Ca–O and Cu–O into the unit cell of 2212 and thereby accelerates the growth of 2223 phase. According to Wong-Ng et al.,[66] the chemical content of Ca plays a major role and recommended an excess amount of Ca and Cu for the formation of considerable quantities of high-T_c phase. Chen et al.[16] suggested that the best composition should be stoichiometric in Ca and Cu but with a small excess of Pb. Wada et al.[63] argued that when the sample was deficient in Sr and Ca and rich in Bi and Cu, 2223 could be formed without difficulty. The main issue is that no clear boundaries of the chemical regions of the 2223 structure have been determined thus far and has remained a lasting problem in the synthesis of 2223. The strong influence of the chemical composition on the formation of the 2223 phase is well established and commonly accepted through various reports.[22,37,38] The 2223 phase exhibits a very slender variation in the Sr and Ca content in comparison with the 2212 phase which is in the range of Sr:Ca$\approx$1.9:2.1–2.2. Contradicting results exist in the case of bismuth. Bi content reported by Grivel et al.[22] is 2, whereas other researchers[11,55,56] observed a much higher Bi content of almost Bi or Bi+Pb=2.2–2.5. However, a variation of T_c has not been observed. Bansal et al.[9] have reported that an increase in the Pb content up to 0.3 can cause an increase in transition temperature which is due to the transfer of charge between the Bi–O and Cu–O layers through holes created by the substitution of Bi ion by Pb ion in the bismuth layer. On the contrary, it is predicted that higher enrichment with Pb, 0.35 and 0.5, can decrease T_c due to the decrease in contact area among high-T_c grains and increase the transition width.

The synthesis of the metal-substituted Bi-2223 phase requires[29] extensive heat treatment because the conditions for the formation of high-T_c superconductors (Bi-2223) are challenging and temperature at which the reaction takes place for the formation 2223 has a complex pattern which forces the other phases to come into sight.

Komatsu et al.[33] were the first to succeed in the preparation of superconducting Bi-based ceramics with a T_c (0)=92 K using melt-quenching method. They reported that melt-quenched samples such as $Bi_{1.5}SrCaCu_2O_x$ and $Bi_{0.5}Pb_{0.5}SrCaCu_2O_x$ were amorphous and showed glass transition and crystallization. It is necessary to have several complicated heating processes to create a physically and chemically stable high-T_c superconductor because the microstructure of a high-T_c superconductor based on BSCCO system, which has been prepared through glass ceramic route is anisotropic and nonhomogenous. It is very important to follow a regular sequence in the heat treatment of these materials. Difficulty in removing the impurity phases such as $(SrCa)_2CuO_3$ or Ca_2PbO_4 despite the sophisticated heat treatment process remains a challenge.[29] Although the excessive Pb may also lead to the formation of Ca_2PbO_4 as an impurity phase, it is more effective than Cu in synthesizing pure devitrified 2223.[69] However, the addition of Pb increases[21] the formation of Ca_2PbO_4 between 1103 and 1113 K.

The evaporation of Bi and Pb is to be avoided by selecting a proper melt temperature and time, while complete melting is to be ensured. Temperature above 1573 K could result[46] in evaporation of volatile elements such as Bi and Pb, and the final composition will not be same as the initial one. A better knowledge of Bi-2223/melts equilibrium is required[21] for the optimization of melt processing.

However, the majority of all these reports, leaving a few, suffer from a shortcoming that the chemical composition of the glassy material after the quenching step was not analyzed. Therefore, it is difficult to draw any further conclusion that which of the hypotheses is most promising for all in the light of results discussed in various reports. PBSCCO compounds are more prone to defect formation than it was believed earlier. The presence of syntactic intergrowths of several compositionally controllable polytypes will make it difficult to get an ideal polycrystal with a structure free of any defects. Single crystals of a pure polytype can in principle be obtained but only with great difficulty.[35]

The formation of 2223 phase of PBSCCO system is a function of annealing temperature applied during the thermal cycle which demands the accuracy of annealing and melting temperature, which is not being given any importance by many authors. It is reported that the optimum annealing temperature for the formation of the 2223 phase was achieved in between 1098 and 1138 K.[32] Varying the annealing temperature is found to decrease the ratio of high-temperature phase.

It has been reported after studying the effect of numerous thermal treatments and ambient conditions on the doping and transport properties of 2212 that temperature higher than 773 K is required to bring on major changes in the structure and in the electronic features of these bismuth-based cuprates. On the other hand, it is also reported that some modifications can also be induced in these materials at much lower temperature. The reason for this inconsistency is not being explained clearly. The need for longer heat treatment duration seems to indicate that not only thermodynamic equilibrium conditions but also the kinetics of the process plays a vital role. It is interesting to note that[14] crystal size can also influence the experimental results due to the kinetics of crystallization. The abovementioned facts on the processing techniques reveal that some significant factors might have been overlooked in the past.

A number of studies have been done between the years 1987 and 2000 on bismuth-based cuprates with lead doping for obtaining a crystal with a high percentage of 2223. For this, we could see a variety of starting compositions, doping, sintering temperatures, times of sintering, different atmospheric conditions, pelletizing pressure, and so on.

Bismuth-based 2223 has a slow rate of phase formation and an exceedingly thin range of temperature stability.[46] It is reported[47] that four or five consecutive sintering of moderate duration of 40 h and each of them being followed by transitional grinding were required for the formation of a stable 2223. Morgan et al.[44] suggested a mechanism of the formation of 2223 compound out of the composition $Bi_{1.7}Pb_{0.4}Sr_{1.6}Ca_{2.4}Cu_{3.6}O_x$ based on a three-layered copper oxide packets formed from precursors of Ca_2PbO_4, Ca_2CuO_3, and 2212 through dissolution in small liquid droplets enriched with Bi and Pb and the temperature for the appearance of the liquid phase was found to be 1098 K. In another experiment, Morgan et al.[45] investigated the ceramics of composition $Bi_{1.84}Pb_{0.34}Sr_{1.9}Ca_2Cu_{3.1}Ag_{0.02}O_x$ alloyed with Ag and showed that the liquid phase appears at a lower temperature 1095 K and it was enriched in Bi, Pb, and Ag. The X-ray diffraction (XRD) studies revealed an accelerated growth of perfect, more distinct, and coarser plates of the 2223 phase at 1099–1101 K.

The main limitation of Bi-2223 is its anisotropy and difficulty in producing high critical current densities obtained on some very short pressed samples. This is due to the granular nature of Bi-2223 and short coherence length of high-temperature superconductors. Arshad et al.[4] reported that volume fraction of 2223 phase at 1118 K increases with time; the maximum value of the 2223 phase was obtained at 120 h. It has been observed that the formation of the high-T_c phase is remarkably enhanced at the temperature of the endothermic peak of differential thermal analysis curve. The best result has been obtained in the sample sintered for 24 h at 1128 K. This also indicates that at 1128 K, the large volume fraction of Bi-2223 phase with $T_c = 113$ K is obtained in short time and as the sintering time increased, it decomposed into $Bi_2Sr_2CaCu_2O_x$ and other phases. When the samples were heat-treated for longer durations, decomposition of Bi-2223 phase accompanied by morphology changes and decrease in magnetic and transport properties was noticed,[5] which may be due to the cation losses or changes in oxygen content. This may also be due to processing conditions and techniques and material composition, and a consensus on the reason for this has not been reached among the researchers. Longtime sintering cannot totally recover induced defects which have been caused by the destruction of already formed microstructure due to intermediate grinding. However, phase amount and morphology of grain boundaries could be increased by enhanced sintering time and conditions as per the findings of Bunescu et al.[13] Young et al.[29] also recommended several sophisticated heating processes to synthesize a physically and chemically stable high-T_c Bi superconductor. The high-T_c phase can be synthesized only after prolonged sintering and seldom results in the phase purity even though the addition of Pb increases Bi-2223 volume fraction. However, Polasek et al.[48]claimed that the phase assemblage is expected to take place at a temperature below the Bi-2223 equilibrium range (1123–1163 K) and sintering for short periods will give better critical current density values.

It is to be noted that in the melt-quenching technique followed by glass ceramic route, selection of temperature and time continues to be a challenge for PBSCCO compounds. Selected melt temperature must be[46] as low as possible to avoid the evaporation of Bi and Pb but ensuring a complete melt. Melt temperatures above 1573 K will lead to deviation from initial chemical composition.[46,51] The selection of melt temperature for glass ceramic method is very crucial as the material used for fabrication is volatile. Ramanathan et al.[51] synthesized $Bi_2Sr_2Ca_{1.9}Cu_4O_x$ that represents a Cu-rich composition. The resulting composition was $Bi_{1.6}Sr_{1.31}Ca_{0.87}Cu_{3.08}O_x$ after melting at 1423 K for 30 min in an alumina crucible. Long process time of 30 min at 1423 K might have caused[46] the evaporation of Bi which, in turn, caused the deviation from stoichiometry of other elements as well. Tampieri et al.[60] pointed out that the melt process is highly dependent on Bi_2O_3 content that lowers the total melt temperature. Bi_2O_3 and CuO are recognized as the necessary constituents to form a glassy state in bismuth cuprates even though they are not glass-forming agents. For the formation of 2223, Matsubara et al.[41] reported that Al_2O_3 crucibles have shown better results than Ni crucibles. The formation of 2223 whiskers has been enhanced due to contamination from Al_2O_3 crucibles.

It is important to note that high amount of Bi-2223 phase in the sample does not guarantee a high-quality product. There is an intricate relationship among phase assembly, chemical composition, crystal structure, coupling of grains, and properties of grain boundary. A sample excellent in its structural and electrical properties may be weak in their magnetic properties and vice versa.[6]

The main drawback of glass ceramic technique is that in many cases, the crystallization process may not progress to completion and a fraction of residual glass may remain. The presence of residual glass in superconductors processed through glass ceramic route possesses a barrier for critical current density j_c. The coupling between superconducting crystals at the grain boundaries in the PBSCCO ceramics is very weak.[68]

Various reports discussed above give an idea of the multiple, complex, and different observations made in the PBSCCO systems which make it nearly impossible to suggest a technique and homogeneity range for the processing of high-T_c 2223 phase. It is amazing to note that how large variations of initial compositions in the six elements in the PBSCCO system have been used with the aim to engineer an optimal 2223 PBSCCO superconductor. This complicates the logical conclusions of the chemical region and processing conditions of the 2223 phase. Wang et al.[64] suggested in the beginning of the oxide superconductive material that considerable optimization of preparative conditions is required in the processing of cuprate-based superconductors. In spite of a great progress achieved in the field of investigation of high-T_c superconductors, further studies of the fine structure and exact stoichiometry of the ceramics, including doped ones, are required in combination with measurements of superconducting and other physical properties, which is necessary for the optimization of the processing conditions and superconducting properties of these materials.

In the background of the limitations in achieving single-phase Bi systems and the advantages of the glass ceramic route over the usual solid-state sintering technique, a novel and simple approach has been developed to optimize the processing conditions. This uses Avrami index[28] as a crucial parameter that helps in this direction.

6.5 AVRAMI INDEX: A NEW TOOL FOR PROCESSING HIGH-TEMPERATURE CERAMIC SUPERCONDUCTORS

As already explained, phase purity is found to be evasive in Pb-doped bismuth cuprates even after enhanced heat treatment conditions, different stoichiometric conditions, and substitution of different elements. The activation energy of crystallization and dimensionality of crystal growth are the two major factors other than sintering conditions and chemical composition which can play a key role in obtaining pure Bi-2223. Crystallization of these glassy materials is a function of time and temperature change because of their metastable state. Knowledge of crystallization mechanism is noticeably essential for the optimization of processing parameters required for the processing of high-T_c superconductors prepared through glass ceramic route. It is explained[28] that correlation of Avrami index and the activation energy of crystallization are factors important for optimizing the processing conditions of PBSCCO system to the superconducting grain formation. Considerable information about the type of crystallization taking place within the sample can be inferred from the computed value of the Avrami index.

Figure 6.4 shows the XRD of quenched samples of three different compositions of Pb-doped BSCCO system, B_1, B_2, and B_3. The XRD of the melt-quenched samples shows amorphous nature for the $Bi_{1.6}Pb_{0.4}Sr_2CaCu_2O_x$ (B_1), $Bi_{1.6}Pb_{0.4}Sr_2Ca_{1.5}Cu_{2.5}O_x$ (B_2), and $Bi_{1.6}Pb_{0.4}Sr_2Ca_3Cu_4O_x$ (B_3). Two types of microstructures, glassy and crystalline, are being noticed in B_3 Figure 6.4c. The XRD of the other two samples B_1 (Fig. 6.4a) and B_2 (Fig. 6.4b) showed a broad halo around $2\theta=30°$ and for B_3, a few crystalline peaks are also obtained with a broad halo around $2\theta=30°$. This indicates a long-range atomic arrangement and also the periodicity of the three-dimensional network in the quenched sample.

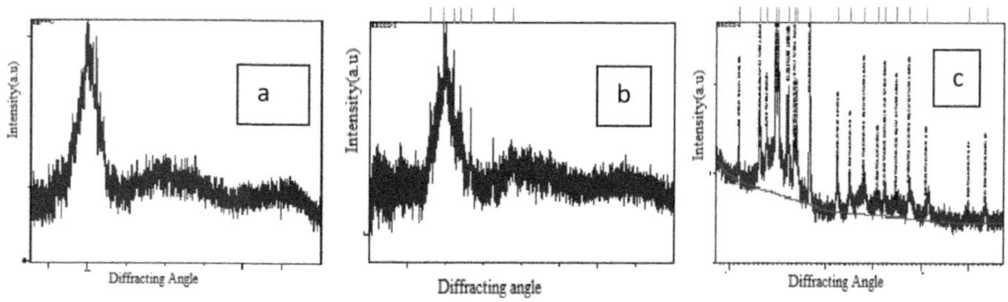

FIGURE 6.4 X-ray diffraction pattern of quenched $Bi_{1.6}Pb_{0.4}Sr_2Cu_nCa_{n-1}O_w$, at 1403 K where (a) n=2, (b) n=2.5, and (c) n=4.

Energy-dispersive X-ray analysis revealed the chemical content deviation of elements which can be attributed to the higher melting temperature and longer processing time as pointed out by Nilsson et al.[46].

Differential scanning calorimetry (DSC) has been used to study the thermal behavior of the glassy precursor by heating the samples through different heating rates 5, 10, 15, and 20 K/min. DSC thermogram for n=2, 2.5, and 4 heated in argon atmosphere at different heating rates are shown in Figure 6.5. It is seen that the glass transition temperature T_g and crystallization temperature T_x shifted to a lower temperature region with the increase of Ca-to-Cu ratio. The relative increase of bismuth which is a good glass-forming agent might have reduced the glass transition temperature. The shift of crystallization temperature to the high-temperature region by increasing rate of heating can be attributed to the change in the nucleation rate of different phases.

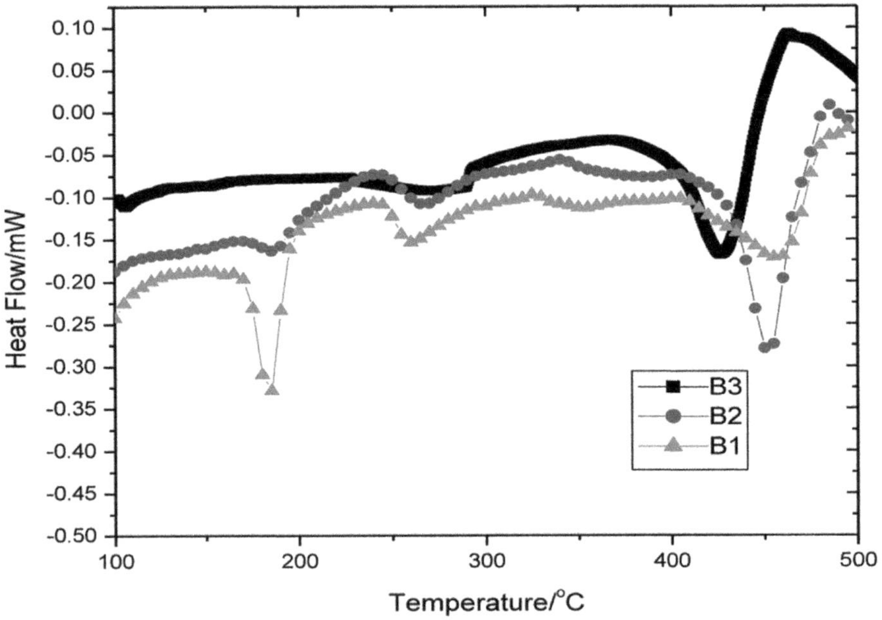

FIGURE 6.5 (See color insert.) Differential scanning calorimetry thermogram of different compositions B_1, B_2, and B_3 at a heating rate of 10 K/min.

The values obtained by different methods are agreeing with each other and are shown in Table 6.1. It can be seen that the activation energy decreased with the increase of Ca-to-Cu ratio. The activation energy of Pb-doped B_2 is almost equal to that of $Bi_2Sr_2Ca_2O_{10+\delta}$ (336.46 kJ/mol).[7] The rearrangement of distinct atoms are required for the crystallization of glass system[17] and each atom must overcome the binding energy with its immediate neighbors to take its lattice position of primary crystals. The increase of activation energy can be credited to the high interaction of atoms. The above facts show that B_1 and B_2 are much more stable than B_3. The atomic diffusion in the system will become delicate with the increase of E_c. The difficulty in atomic diffusion decelerates the nucleation and growth of crystals.[1] The presence of CaO crystallites[46] in the glass, which increases with the increase of Ca, may be the reason for the decrease in activation energy.

TABLE 6.1 Activation Energy for Crystallization (kJ/mol) Calculated by Different Models in Non-isothermal Calorimetry.

Composition	Kissinger method (kJ/mol)	Augis Bennet (kJ/mol)	Mahadevan approximation (kJ/mol)	Matusita–Saka (kJ/mol)
B_1	740.95	731.16	753.59	751.67
B_2	343.87	333.94	356.49	341.22
B_3	199.71	188.6	212.04	263.00

The evaluation of Avrami index is performed from Matusita–Saka method by using Osawa's equation

$$n\left[-\ln(1-x)\right]= p\ln\alpha-1.052\frac{qE_c}{RT}+\text{constant} \tag{6.1}$$

"x" is the fractional crystallization and "α" is the heating rate. T is the temperature in degree celsius. Avrami index has been evaluated for a range of temperature under the exothermic peak and the statistical average is shown in the report.

Graphs drawn between ln[−ln(1−x)] and ln(α) for different composition at constant temperatures are shown in Figure 6.6 and slopes of these give the value of q, Avrami index. Fractional crystallization "x" is evaluated using the first half of the exothermic peak of DSC thermogram from the ratio of area between a defined temperatures ranges to the total area under the graph. Considerable information about the type of crystallization taking place within the sample can be inferred from the computed value of the Avrami index. The values of Avrami index for different compositions are tabulated in Table 6.2. It can be concluded that a large number of nuclei already exists in the specimen[50] if the Avrami index does not change with temperature and p is taken as equal to q. A change in the value of p with temperature confirms

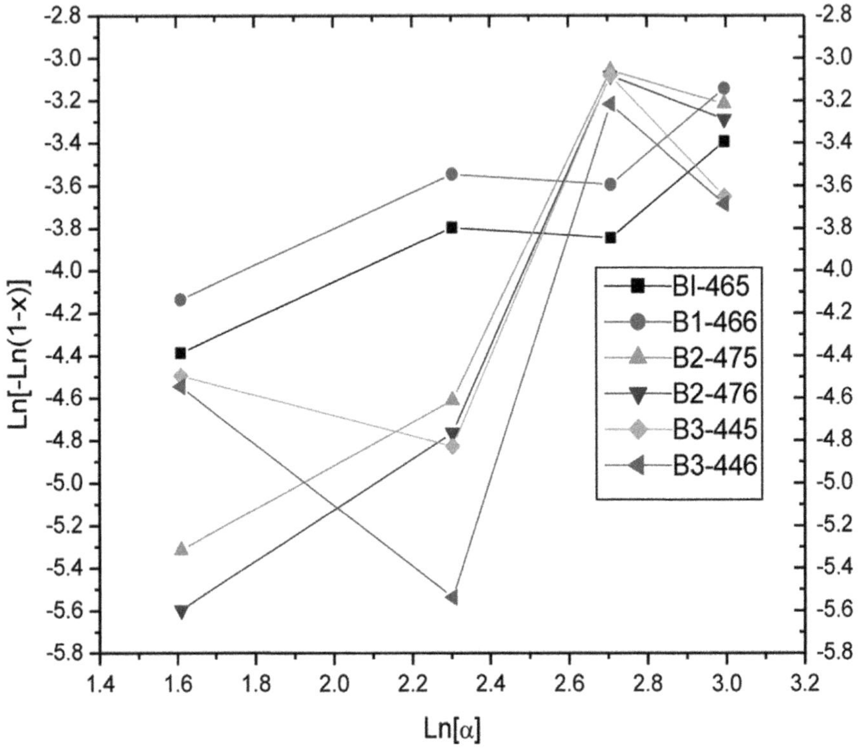

FIGURE 6.6 **(See color insert.)** Evaluation of Avrami index for all the samples.

TABLE 6.2 Variation of Avrami Index with Temperature.

B_1	Temperature (°C)	469	470	471	472	473
	Avrami index	0.98	1.12	1.32	1.61	2.18
B_2	Temperature (°C)	475	476	477	478	479
	Avrami index	1.71	1.89	2.14	2.58	4.57
B_3	Temperature(°C)	442	443	444	445	446
	Avrami index	0.85	0.86	0.87	0.89	0.93

the absence of preexisting nuclei in the sample[50] Hence, p is taken as equal to $q-1$. The values of p calculated for different samples from Table 6.2 are as follows: $p=1$ for B_3, $p=2$ for B_1, and $p=3$ for B_2. The above results clearly indicate that the crystal growth in B_1 is due to two-dimensional and surface nucleation,[50] whereas that of B_2 is dominated by three-dimensional along with two-dimensional growth and surface nucleation and to end with, B_1

crystal growth is dominated by surface nucleation. The presence of preexisting nuclei and the low activation energy of B_3, when compared with the other two samples, are the possible reason for the growth of superconducting grain in it even though it has been sintered for 24 h. Doped Bi perovskite with less activation energy has been easily converted into a superconductor, whereas there is no superconducting grain formation for that with higher activation energy. On the other hand, composites with multidimensional growths have been easily converted into 110 K phase. The superconducting property could not be improved for composition with surface nucleation, even after enhancing the duration of sintering. From the above, it has been seen that Avrami index can be effectively used as a guide in the process for single-phase Bi cuprates through glass ceramic route.

6.6 SINTERING TEMPERATURE: A CRITICAL PARAMETER FOR SYNTHESIS OF CERAMIC SUPERCONDUCTORS

It is observed that the accuracy of the annealing and melting temperature was not finding much importance in many reports even though the formation of 2223 phase in PBSCCO system is a function of annealing temperature. It is found[34] that the best annealing temperature for the formation of the 2223 phase was achieved in between 1098 and 1138 K. Varying the annealing temperature is found to change the ratio of high-temperature phase to the low-temperature phase. The complexity in thermodynamic equilibrium conditions and kinetics of crystallization necessitates the prolonged heat treatment for these materials for obtaining phase purity. It is seen that the selection of melt temperature is crucial in glassy precursor route as the different reactants have different melting points. The relative increase of Bi in the quenched sample can reduce the sintering temperature considerably. Further, it is seen that even a temperature change within 5 K plays a major role in the processing of Bi superconductors as it has been found that a major phase change is taking place even in this range.

Figure 6.7 shows the XRD pattern of a superconducting sample sintered at different temperatures. At 1113 K, all the peaks got vanished, which confirms the destruction of already formed grains. With the increase of heat treatment temperature, the full width at half maximum of the peak decreases indicating an increase in particle size and crystallinity of the sample. Growth kinetics is enhanced with the increase of heat treatment conditions, which, in turn, helped the formation of large grains.

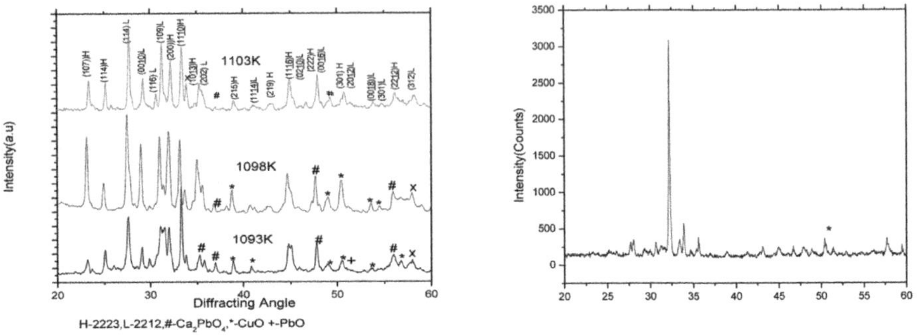

FIGURE 6.7 **(See color insert.)** Diffraction pattern of the composition (n=4) sintered at different temperatures (a) 1093, 1098, and 1103 and (b) 1113 K.

The resistance–temperature characteristics of n=4 sintered at different temperatures is shown in Figure 6.8. The sample sintered at 1103 K has shown a double transition at 112 and 90 K with T_c (0) 78 K. The increase in transition width may be due to the poor intergrain connectivity as evidenced from scanning electron microscopy micrograph and this can be improved by sintering the sample for a longer duration. A two-step transition in a sample shows the presence of 2212 and 2223 phases. The sample sintered at 1098 K has shown T_c at 82 K and T_c (0) at 72 K, where the sample at 1093 K has shown the T_c at 78 K and T_c (0) at 70.5 K. The sample sintered at 1113 K has shown an irregular transport property, which has not been included, as the grains got disturbed due to melting. It is seen that the selection of melt temperature is crucial in glassy precursor route as the different reactants have different melting points.

Ensuring exact stoichiometry as initial composition after quenching is very difficult in PBSCCO system because of the proximity of ionic radius of different atoms, which is helping them to share their spaces. It is obvious[57] that point defects formed upon the replacement of calcium cations by heavy cations in the calcium layers should have a noticeable effect on the superconducting properties of bismuth-based high-temperature superconductors. The replacement of calcium coordinated by eight oxygen atoms of the CuO_2 planes results in the appearance of local fluctuations of the charge carrier concentration and correspondingly a decrease in the critical temperature T_c. These defects with a size of the order of the coherence length in the case of their statistical distribution should be treated as pinning centers with local variation in the critical temperature. When defects are orderly structured, we should expect manifestations of strong pinning effects of the columnar defect type. It is to be noted that magnetic properties for materials, which look like very good in XRD, can have a poor performance due to the low-level amorphous impurity phases possibly concentrated on grain boundaries.

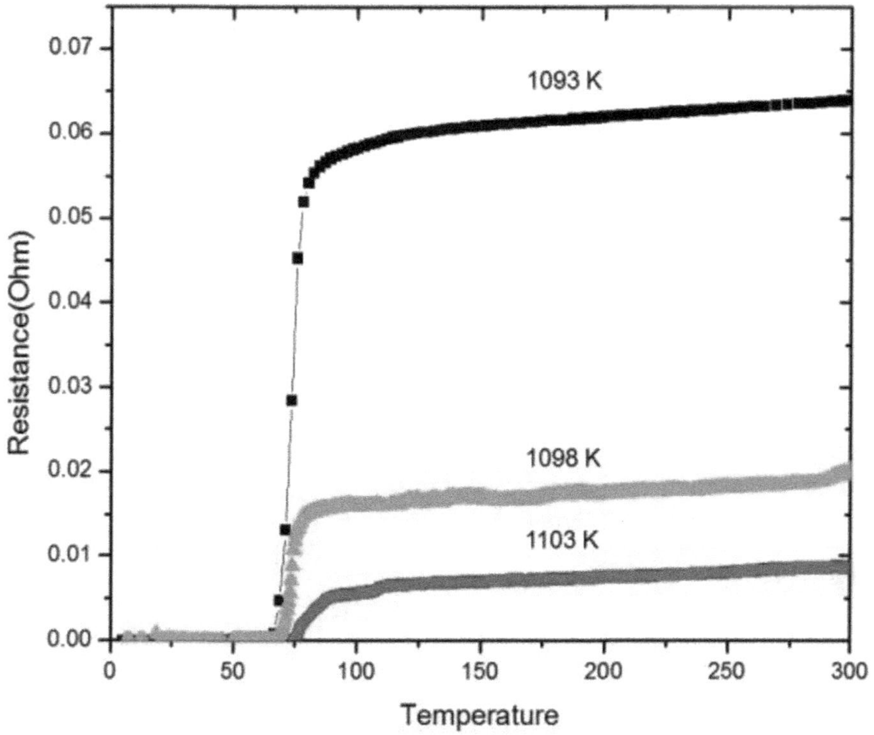

FIGURE 6.8 (See color insert.) Variation of resistance with temperature composition (n=4) sintered for 24 h at different temperatures.

The presence of residual secondary phases[18] and residual glasses is the major current limiting factor in the processing of PBSCCO ceramic superconductors. Processing of a physically and chemically stable high-T_c superconductor[29] necessitates several sophisticated heating techniques because the microstructure of a high-T_c superconductor is anisotropic and nonhomogenous. Thermodynamically, the stability of 2212 over a wide range of temperature in the phase equilibria makes its detection effortless, whereas that of 2223 is poor because of its stability in a narrow temperature range in the phase equilibria. Specifically, extensive heat treatment is required to synthesize the metal-substituted Bi-2223-phase high-T_c superconductor because the condition for the formation of Bi-2223 phase is demanding and needs prolonged heat treatment, and the reaction for the formation of 2223 phase has an intricate pattern, which forces the other phases to appear. The formation of secondary phases can be controlled only after ensuring exact stoichiometry as initial composition after quenching and determination of dimensionality of crystal growth, using Avrami index. The above strategy can considerably reduce the time, energy, and the effort for the processing bismuth-based ceramic superconductors in future.

KEYWORDS

- **glass ceramic**
- **melt quenching**
- **Avrami index**
- **monophase**
- **crystalization kinetics**

REFERENCES

1. Aksan, M. A.; Yakinici, M. E.; Balci, Y. Thermal Analysis Study of $Bi_2Sr_2Ca_2Cu_{3-x}Er_xO_{10+\delta}$ Glass Ceramic System. *J. Therm. Anal. Calorim.* **2005**, *81*, 417–423.
2. Alamgir, A. K. M.; Yamada, H.; Harada, N.; Osaki, K.; Tada, N. Effects of Pb Doping on the Microstructure and Superconductivity of Bulk BSCCO 2212. *Appl. Supercond. IEEE Transact.* **1999**, *9*(2), 1864–1867.
3. Arima, H. T.; Shimizu, K.; Otsuka, Y.; Y. Murata; Kawai, T. Observation of the High-T_c Phase and Determination of the Pb Position in a Bi-Pb-Sr-Ca-Cu Oxide Superconductor. *Jpn. J. Appl. Phys.* **1989,** *28*(1), L57–L59.
4. Arshad, M.; Qureshi, A. H. Time and Temperature-Based Study for the Production of High T_c Phase by Sol-Gel Technique in Pb-BSCCO System. *J. Therm. Anal. Calorim.* **2006,** *83*(2), 415–419.
5. Badica, P.; Adica, G.; Alexeev, A. F.; Gridosova, T. Y.; Morozov, V. V. Structural Modifications of Superconducting Phases in Bi (Pb)-Sr-Ca-Cu-O System. *Phys. C Supercond.* **1996,** *259*(1), 92–96.
6. Badicca, P.; Addica, G. Bi-2223 Freeze-Dried Ceramic: Specific Features, Related Problems and Search for New Solutions. *J. Optoelectro. Adv. Mater.* **2003,** *5*(4), 1029–1040.
7. Balci, Y. Crystallisation Kinetics of Glass Ceramic Superconductors and Investigation of the Electrical Conductivity Properties. Ph.D. Thesis, Firat University, Elazig, Turkey, 1997.
8. Balci, Y.; Ceylan, M.; Yakinci, M. E. An Investigation on the Activation Energy and the Enthalpy of the Primary Crystallization of Glass–Ceramic Bi-Rich BSCCO High T_c Superconductors. *Mater. Sci. Eng. B* **2001,** *86*(1), 83–91.

9. Bansal, N. P. NASA Contractor Report. 1989, 185,184.
10. Bardeen, J.; Cooper, L. N.; Schrieffer, J. R. Theory of Superconductivity. *Phys. Rev.* **1957,** *108*(5), 1175.
11. Bednorz, J. G; Muller, K. A. Electronic Structure of High T_c Superconductor $Sr_{0.2}La_{1.8}CuO_4$. *Phys. Rev. Lett.* **1987,** *58*, 908.
12. Bhat, V.; Ganguli, A. K.; Nanjundaswamy, K. S.; Ram, R. M.; Gopalakrishnan, J.; Rao, C. N. R. Approaches to the Synthesis of High-T_c Superconducting Oxides in La–Ba–Cu–O and Y–Ba–Cu–O Systems. *Phase Transit. Multinatl. J.* **1987,** *10*(1–2), 87–95.
13. Bunescu, M. C.; Adica, G.; Badica, P.; Vasiliu, P.; Nita, P.; Mandache, S. SEM Studies on BSCCO Superconducting Ceramic Produced by Spray Frozen, Freeze Drying Technique. *Phys. C Supercond.* **1997,** *281*(2), 191–197.
14. Cardona, M.; Thomsen, C.; Liu, R.; von Schnering, H. G.; Hartweg, M.; Yan, Y. F.; Zhao, Z. X. Raman Scattering on Superconducting Crystals of $Bi_2\ (Sr_{1-x}Ca_x)_{n+2}Cu_{n+1}O_{(6+2n)+\delta}$ (n=0, 1). *Solid State Commun.* **1988,** *66*(12), 1225–1230.
15. Cava, R. J.; Hewat, A. W.; Hewat, E. A.; Batlogg, B.; Marezio, M.; Rabe, K. M.; Krajewski, J. J.; Peck, W. F., Jr.; Rupp, L. W., Jr. Structural Anomalies, Oxygen Ordering and Superconductivity in Oxygen Deficient $Ba_2YCu_3O_x$. *Phys. C Supercon.* **1990,** *165*(5), 419–433.
16. Chen, Y. L.; Stevens, R. 2223 Phase Formation in Bi (Pb)–Sr–Ca–Cu–O: III, the Role of Atmosphere. *J. Am. Ceram. Soc.* **1992,** *75*(5), 1160–1166.
17. Cima, M. J.; Jiang, X. P.; Chow, H. M.; Huggerly, J. S.; Flemings, M. C.; Drody, H. D.; Laudise, R. A.; Johnson, D. W. Influence of Growth Parameters on the Microstructure of Directionally Solidified $Bi_2Sr_2CaCu_2O_y$. *J. Mater. Res.* **1990,** *5*(9), 1834–1849.
18. Deng, H.; Hua, P.; Wang, W.; Dong, C.; Chen, H.; Wu, F.; Wang, X.; Zhou, Y.; Yuan, G. Phase Transformation and Critical Current Density of (Bi, Pb)-2223/Ag Superconducting Tapes by a Low Temperature Low Oxygen Pressure Post-Annealing Method, Physica-C339, 2000; pp 181–194.
19. Endo, U.; Koyama, S.; Kawai, T. Composition Dependence on the Superconducting Properties of Bi-Pb-Sr-Ca-Cu-O. *Jpn. J. Appl. Phys.* **1989,** *28*, L190.
20. Esteve, D.; Martinis, J. M.; Urbina, C.; Devoret, M. H.; Collin, G.; Monod, P. Revcolevschi, A. Observation of the AC Josephson Effect Inside Copper-Oxide-Based Superconductors. *Europhys. Lett.* **1987,** *3*(11), 1237.
21. Gomes, G. G., Jr.; Bispo, E. R.; Polasek, A.; Neves, M. A.; Rizzo, F.; Amorim, H. S.; Ogasawara, T. Quantitative Analysis of Phases Present in Partially Melt Processed (Bi, Pb)-2223 Superconductor International Conference on Advanced Materials, Rio de Janeiro, Brazil, 2009.
22. Grivel, J. C.; Flukiger, R. Formation Mechanism of the Pb Free $Bi_2Sr_2Ca_2Cu_3O_{10}$ Phase. *Supercond. Sci. Technol.* **1998,** *11*, 288.
23. Iwai, Y.; Hoshi, Y.; Saito, H.; Takata, M. Influence of the Oxygen Partial Pressure on the Solubility of PbO in the $(BiPb)_2Sr_2CaCu_2O_{8+\delta}$ Superconducting Oxides. *Phys. C Supercond.* **1990,** *170*(3), 319–324.
24. Jeremie, A.; Alami-Yadri, K.; Grivel, J. C.; Flükiger, R. Bi, Pb (2212) and Bi (2223) Formation in the Bi-Pb-Sr-Ca-Cu-O System. *Supercond. Sci. Technol.* **1993,** *6*(10), 730.
25. Jeremie, A.; Grivel, J. C.; Flükiger, R. Superconducting and Crystallographic Properties of Bi, Pb (2212). *Phy. C Supercond.* **1994,** *235*, 943–944.
26. Kaesche, S.; Majewski, P.; Aldinger, F. Phase Relations and Homogeneity Region of the High Temperature Superconducting Phase $(Bi, Pb)_2Sr_2Ca_2Cu_3O_{10+d}$. *J. Electron. Mater.* **1995,** *24*(12), 1829–1834.

27. Kaesche, S.; Majewski, P.; Aldinger, F. Phase Equilibria in the System Bi_2O_3–PbO–SrO–CaO–CuO with Special Regard to the $(Bi, Pb)_{2+x}Sr_2Ca_2Cu_3O_{10+d}$ Phase. *Int. J. Mater. Res.* **1996,** *87*(7), 587–593.
28. Kannan, T.; Predeep, P. Importance of Avrami Index in the Processing of Bismuth Based High Temperature Superconductors. *J. Alloys Compd.* **2013,** *579*, 65–70.
29. Ki, Y. S.; Lee, M. S. The Effect of a Large Amount of Ag Introduced into the $Bi_{1.84}Pb_{0.34}Sr_{1.91}Ca_{2.03}Cu_{3.06}O_{10+\delta}$ (110 K Phase) High-T_c Superconductor. *Supercond. Sci. Technol.* **2006,** *19*(12), 1253.
30. Kijima, N.; Endo, H.; Tsuchiya, J.; Sumiyama, A.; Mizuno, M.; Oguri, Y. Reaction Mechanism of Forming the High-T_c Superconductor in the Pb–Bi–Sr–Ca–Cu–O System. *Jpn. J. Appl. Phys.* **1988,** *27*(10A), L1852.
31. Kirtley, J. R.; Tsuei, C. C.; Verwijs, C. J. M.; Harkema, S.; Hilgenkamp, H. Angle-Resolved Phase-Sensitive Determination of the In-Plane Gap Symmetry in $YBa_2Cu_3O_{7-\delta}$. *Nat. Phys.* **2006,** *2*(3), 190–194.
32. Kocabas, K.; Bilgili, O.; Yasar, N. The Effect of Sintering Temperature in $Bi_{1.7}Pb_{0.2}Sb_{0.1}Sr_2Ca_2Cu_3O_y$ Superconductors. *J. Supercond. Novel Magn.* **2009,** *27*, 643–650.
33. Komatsu, T.; Ohki, T.; Hirose, C.; Matusita, K. Superconducting Properties of Glass-Ceramics in the Bi–Sr–Ca–Cu–O System. *J. Non-Cryst. Solids* **1989,** *113*(2), 274–281.
34. Koyama, S.; Endo, U.; Kawai, T. Preparation of Single 110 K Phase of the Bi–Pb–Sr–Ca–Cu–O Superconductor. *Jpn. J. Appl. Phys.* **1988,** *27*(10A), L1861.
35. Kumakura, H.; Togano, K.; Kase, J.; Morimoto, T.; Maeda, H. Superconducting Properties of Textured Bi-Sr-Ca-Cu-O Tapes Prepared by Applying Doctor Blade Casting. *Cryogenics* **1990,** *30*(11), 919–923.
36. Maeda, H.; Tanaka, Y.; Fukutomi, M.; Asano, T. A New High T_c Oxide Superconductor Without a Rare Earth Element. *Jap. J. App. Phys.* **1988,** *27*, L209–L210.
37. Majewski, P. Phase Diagram Studies in the System Bi–Pb–Sr–Ca–Cu–O–Ag. Supercond. *Sci. Technol.* **1997,** *10*(7), 453.
38. Majewski, P.; Hettich, B.; Schulze, K.; Petzow, G. Preparation of Unleaded $Bi_2Sr_2Ca_2Cu_3O_{10}$. *Adv. Mater.* **1991,** *3*(10), 488–491.
39. Majewski, P.; Kaesche, S.; Aldinger, F. Fundamentals of the Preparation of High-T_c, Superconducting $(Bi, Pb)_{2+x}Sr_2Ca_2Cu_3O_{10+}\delta$ Ceramics. *Adv. Mater.* **1996,** *8*(9), 762–765.
40. Majewski, P.; Su, H. L.; Aldinger, F. Engineered Flux Pinning Centres in Pb-Doped High Temperature Superconducting "$Bi_2Sr_2CaCu_2O_8$" Ceramics. *J. Mat. Sci.* **1996,** *31*(8), 2035–2042.
41. Matsubara, I.; Funahashi, R.; Ogura, T.; Yamshita, H.; Tsuru, K.; Kawai, T. Growth Mechanism of $Bi_2Sr_2CaCu_2O_x$ Superconducting Whiskers. *J. Cryst. Growth* **1994,** *141*(1), 131–140.
42. Michel, C.; Hervieu, M.; Borel, M. M.; Grandin, A.; Deslandes, F.; Provost, J.; Raveau, B. Superconductivity in the Bi-Sr-Cu-O System. *Z. Phys. B Condens. Matter*, **1987,** *68*(4), 421–423.
43. Mizuno, M.; Endo, H.; Tsuchiya, J.; Kijima, N.; Sumiyama, A.; Oguri, Y. Superconductivity of $Bi_2Sr_2Ca_2Cu_3Pb_xO_y$, (x=0.2, 0.4, 0.6). *Jpn. J. Appl. Phys.* **1988,** *27*(7), L1225–L1227.
44. Morgan, P. E. D.; Housley, R. M.; Porter, J. R.; Ratto, J. J. Low Level Mobile Liquid Droplet Mechanism Allowing Development of Large Platelets of High-T_c "Bi-2223" Phase Within a Ceramic. *Phys. C* (Amsterdam) **1991,** *176*(1–3), 279–284.

45. Morgan, P. E. D.; Piché, J. D.; Housley, R. M. Use of a Thermal Gradient to Study the Role of Liquid Phase During the Formation of Bismuth High Temperature Superconductors. *Phys. C* (Amsterdam) **1992,** *191*(1–2), 179–184.
46. Nilsson, A.; Gruner, W.; Acker, J.; Wetzig, K. Critical Aspects on Preparation of Bi-2223 Glassy Precursor by Melt-Process. *J. Noncryst. Solids*, **2008,** *354*, 845.
47. Pierre, L.; Schneck, J.; Mosin, D.; Toledano, J. C.; Primot, J.; Daguet, C.; Savary, H. Role of Lead Substitution in the Production of 110-K Superconducting Single-Phase Bi-Sr-Ca-Cu-O Ceramics. *J. Appl. Phys.* **1990,** *68*(5), 2296–2303.
48. Polasek, A.; Saléh, L. A.; Borges, H. A.; Hering, E. N.; Marinkovic, B.; Assunção, F. C. R.; Serra, E. T.; de Oliverira, G. S. Processing of Bulk Bi-2223 High-Temperature Superconductor. *Mater. Res.* **2005,** *8*(4), 391–394.
49. Poole, C. P., Jr. *Handbook of Superconductivity;* Academic Press: San Diego, 2000.
50. Predeep, P.; Saxena, N. S.; Kumar, A. Crystallization and Specific Heat Studies of $Se_{100-x}Sb_x$ (x=0, 2 and 4) Glass. *J. Phys. Chem. Solids* **1997,** *58*(3), 385–389
51. Ramanathan, S.; Li, Z.; Ravi-Chandar, K. On the Growth of BSCCO Whiskers. *Phys. C Supercond.* **1997,** *289*(3), 192–198.
52. Rao, C. N. R.; Gopalakrishnan, J. *New Directions in Solid State Chemistry;* Cambridge University Press: UK, 1989.
53. Romanov, E. P.; Blinova, Y. V.; Sudareva, S. V.; Krinitsina, T. P.; Akimov, I. I. Mechanism of Formation, Fine Structure, and Superconducting Properties of High-Temperature Superconductors and Superconducting Composites. *Phys. Met. Metallography* **2006,** *101*(1), 27–44.
54. Ruvalds, J. Theoretical Prospects for High-Temperature Superconductors. *Supercond. Sci. Technol.* **1996,** *9*(11), 905.
55. Sastry, P. V. P. S. S.; West, A. R. Novel Synthetic Pathway to Bi (Pb)-2223 Phase with Variable Ca:Sr Ratio, $Bi_{1.7}Pb_{0.3}Sr_{4-x}Ca_xCu3O_y$: $1.85 \leq x \leq 2.4$. *J. Mater. Chem.* **1994,** *4*(4), 647–649.
56. Sato, K. I.; Shibuta, N.; Mukai, H.; Hikata, T.; Ueyama, M.; Kato, T. Development of Silver-Sheathed Bismuth Superconducting Wires and Their Application. *J. App. Phys.* **1991,** *70*(10), 6484–6488.
57. Shamray, V. F.; Mikkhailova, A. B.; Mitin, A. V. Crystal Structure and Superconductivity of Bi-2223. *Crystallogr. Rep.* **2009,** *54*(4), 584–590.
58. Shi, D.; Boley, M. S.; Chen, J. G.; Xu, M.; Vandervoort, K.; Liao, Y. X.; Zangvil, A.; Akujieze, J.; Segre, C. Origin of Enhanced Growth of the 110 K Superconducting Phase by Pb Doping in the Bi-Sr-Ca-Cu-O System. *Appl. Phys. Lett.* 1989, *55*(7), 699–701.
59. Takano, M.; Takada, J.; Oda, K.; Kitaguchi, H.; Miura, Y.; Ikeda, Y.; Mazaki, H. High-T_c Phase Promoted and Stabilized in the Bi, Pb-Sr-Ca-Cu-O System. *Jpn. J. Appl. Phys.* **1988,** *27*(6), L1041.
60. Tampieri, A.; Landi, E.; Bellosi, A. Kinetic Study of the Formation of $Bi_{1.8}Pb_{0.2}Sr_2Ca_2Cu_3O_x$ Ceramic Superconductor. *Mater. Chem. Phys.* **1993,** *34*(2), 157–161.
61. Thomas, F. Structural and Superconducting Properties of High T_c Superconductors, ISBN 87-550-2476-9; 87-550-2477-7.
62. Tinkham, M. *Introduction to Superconductivity*; Courier Dover Publications: New York, 2012.
63. Wada, T.; Suzuki, N.; Maeda, A.; Uchida, S. I.; Uchinokura, K.; Tanaka, S. Preparation of High-T_c (110 K) Superconducting Phase by the Annealing of Low-T_c (80 K) Superconductor $Bi_2(Sr_{0.5}Ca_{0.5})_{3-x}Cu_2O_y$. *Jpn. J. Appl. Phys.* **1988,** *27*(6A), L1031.

64. Wang, H. H.; Carlson, K. D.; Geiser, U.; Thorn, R. J.; Kao, H. C. I.; Beno, M. A.; Monaghan, M. R.; Allen, T. J.; Proksch, R. B. Comparison of Carbonate, Citrate, and Oxalate Chemical Routes to the High-T_c Metal Oxide Superconductors, Lanthanum Strontium Copper Oxide, $La_{2-x}Sr_xCuO_4$. *Inorg. Chem.* **1987,** *26(*10), 1474–1476.
65. Wong-Ng, W.; Cook, L. P.; Kearsley, A.; Greenwood, W. Primary Phase Field of the Pb-Doped 2223 High-T_c Superconductor in the (Bi, Pb)-Sr-Ca-Cu-O System. *J. Res. Nat. Inst. Stand. Technol.* **1999,** *104*, 277–290.
66. Wong-Ng, W.; Freinman, S. W. High T_c Superconducting Bi-Sr-Ca-Cu-O Glass Ceramics: A Review. *Appl. Supercond.* **1994,** *2*(3), 163–180.
67. Wu, W. K.; Ashburn, J. R.; Torng, C. J.; Hor, P. H.; Meng, R. L.; Gao, L.; Huang, Q.; Chu, C. W. Superconductivity at 93 K in a New Mixed-Phase Y-Ba-Cu-O Compound System at Ambient Pressure. *Phys. Rev. Lett.* **1987,** *58*(9), 908–910.
68. Yanagisawa, E.; Diederich, D. R.; Kumakura, H.; Togano, K.; Maeda, H.; Takahashi, K. Properties of Pb-Doped Bi-Sr-Ca-Cu-O Superconductors. *Jpn. J. Appl. Phys.* **1988,** *27*(8), L1460–L1462.
69. Yurchenko, I. A. Representations About Nature of Processes Occurring in the Material at Different Stages of Bi (Pb)–Sr–Ca–Cu–O Ceramic Manufacture. *J. Mater. Sci.* **2006,** *41*(11), 3507–3520.
70. Zhuang, H. R.; Kozuka, H.; Sakka, S. Preparation of Superconducting Bi–Sr–Ca–Cu–O Ceramics by the Sol-Gel Method. *J. Mater. Sci.* **1990,** *25*(11), 4762–4766.

CHAPTER 7

PIEZOELECTRIC MATERIALS FOR NANOGENERATORS

YUNLONG ZI

Department of Mechanical and Automation Engineering, Chinese University of Hong Kong, Shatin, N.T., Hong Kong, China

**Corresponding author. E-mail: yunlongzi@mae.cuhk.edu.hk*

CONTENTS

ABSTRACT

The development of the Internet of Things and big data requires data collection and actuation through trillions of devices, each of which requires energy supply. To address the issue of the battery on the limited lifetime, nanogenerators have been developed for over 10 years to power these widely-distributed devices. Piezoelectric materials provide a possibility to conduct energy harvesting in the nanoscale. All types of piezoelectric materials, including binary III-V or II-VI compounds, 2D materials, perovskite materials, and polymer ferroelectric materials, are all developed as piezoelectric nanogenerators. The applications of these nanogenerators are demonstrated in smart wearable technologies, biomedical detection, self-powered sensing, and so on, which set the foundation for future functional smart systems.

7.1 INTRODUCTION

As excellent materials to convert ambient mechanical energy/signals into electrical energy/signals, piezoelectric materials have demonstrated promising performances for nanogenerators and self-powered sensors. This chapter mainly introduces piezoelectric materials for nanogenerators, starting from fundaments of piezoelectric effect to different types of piezoelectric materials and nanogenerators. Different kinds of piezoelectric materials, including binary II–VI, III–V, two-dimensional (2D) materials, perovskite materials, and polymer materials, are introduced, including the mechanism for piezoelectricity, material characterizations, electrical performances, and applications as nanogenerators.

7.1.1 PIEZOELECTRIC EFFECT

As fast development of the instrument, micro devices, the Internet of Things (IoT), and so forth, materials based on the piezoelectric effect have been demanded for mechanical energy harvesting, sensing, detection, and computation. The piezoelectric effect was originally discovered for over a century[1] and it has been rapidly developed for several decades. This is the fundamental type of dielectric material for electromechanical coupling-related studies.

The mechanisms of piezoelectricity in different materials are not exactly same, but they are usually related to the variation of the polarization density in dielectric materials under the external strain, or the introduced strain due to the varied polarization density as affected by the external electric field (the converse piezoelectric effect). Here, we mainly discuss the former one, which is commonly called (direct) piezoelectric effect. Different from other common dielectric materials, there are usually dipole moments already existing in piezoelectric materials. The positive and negative sites in the dipole moments are usually atoms (such as H(+) and F(−) in polyvinylidene fluoride [PVDF]) or ions (such as Zn^{2+} and O^{2-} in zinc oxide [ZnO]) in the materials. When external mechanical strain is applied in the piezoelectric material, the magnitude/direction of the dipole moments changes, and then the additional potential difference is built due to the polarization density P,[2] which can be expressed as:

$$P_i = d_{ijk}\, T_{jk} \text{ or } \{P\} = [d]\, \{T\} \tag{7.1}$$

where $[d]$ is the piezoelectric coefficients matrix and $\{T\}$ is the stress matrix. They can be further expressed as:

$$\begin{bmatrix} P_1 \\ P_2 \\ P_3 \end{bmatrix} = \begin{bmatrix} 0 & 0 & 0 & 0 & d_{15} & 0 \\ 0 & 0 & 0 & d_{24} & 0 & 0 \\ d_{31} & d_{32} & d_{33} & 0 & 0 & 0 \end{bmatrix} \begin{bmatrix} T_1 \\ T_2 \\ T_3 \\ T_4 \\ T_5 \\ T_6 \end{bmatrix} \tag{7.2}$$

Usually, when the piezoelectric material is only subject to normal pressure, for example, normal stress in third axis, then:

$$P_3 = d_{33}\, T_3 \tag{7.3}$$

Therefore, also known as piezoelectric modulus, d_{33} is usually considered as the most important piezoelectric coefficient in piezoelectric materials. Piezoelectric coefficients of various piezoelectric materials are summarized in Table 7.1.

TABLE 7.1 Piezoelectric Coefficients of Some Piezoelectric Materials as Reported.

Materials	d_{33} (pC/N)	Materials	d_{33} (pC/N)
$BaTiO_3$	190[28]	GaN	3.1[29]
Lead zirconate titanate 5H	593[28]	2D-MoS_2	3.73*[23]
Lead magnesium niobate–lead titanate	1200[30]	2D-$MoTe_2$	9.13*[23]
ZnO	9.2[31]	Polyvinylidene fluoride (PVDF)	33[32]
CdS	32.8[33]	Poly (vinylidenefluoride-co-tetrafluoroethylene)	38[27]

*For 2D materials, the reported piezoelectric coefficient is referring to d_{11}.

7.1.2 PIEZOELECTRIC NANOGENERATORS

In recent years, portable and wearable electronics, which can be used as sensors, actuators, detectors, and so forth, have grown rapidly. Each of these electronics requires only a very tiny amount of power consumption (usually in μW–mW level); however, their number is huge, as the development of the IoT. Traditionally, energy storage units (batteries and supercapacitors) are used to power these electronics. However, these energy storage units have limited lifetime, and their disposal may result in huge environmental issues. To provide a sustainable and environmentally friendly solution, nanogenerator, as a device that converts ambient small-scale energy to electricity, was first proposed in 2006 by Wang et al. at Georgia Tech.[3] According to them, piezoelectric nanogenerator (PENG) is the first and an effective one that can convert mechanical energy to electric power, utilizing piezoelectric effect in materials. In the past decade, multiple materials, especially nanomaterials, have been developed for PENG, such as ZnO nanowires,[3] zinc sulfide (ZnS) nanowires,[4] cadmium sulfide (CdS) nanowires,[5] gallium nitride (GaN) nanowires,[6] molybdenum disulfide (MoS_2) 2D materials,[7] barium titanate ($BaTiO_3$) nanowires,[8] and PVDF nanofibers.[9] Different integrations and fabrication methods are invented to effectively harvest different mechanical energy, including vertical integrated nanogenerator (VING),[10–12] lateral integrated nanogenerator (LING),[12,13] and other types of integrations such as microfiber–nanowire hybrid structure.[14]

7.1.3 PIEZOELECTRIC MATERIALS

Many binary compound semiconductor materials, especially groups II–VI and III–V whose structure is wurtzite (WZ) or zinc blende (ZB), show piezoelectric effect. Typical materials include ZnO, CdS, and GaN. To synthesize these piezoelectric nanomaterials, hydrothermal synthesis can be utilized as a facile and cost-effective method;[15] alternatively, various vapor deposition methods, such as chemical vapor deposition (CVD), can be used.[16–19] Even though these II–VI and III–V materials have been focused and studied for decades for their applications as nanogenerators, their output is limited due to their relatively low d_{33} and semiconducting property. To improve their outputs, Schottky barriers[3] or p–n homostructures[20] are usually formed. Furthermore, by coupling piezoelectricity and semiconducting property, these materials can also be used in piezotronics[21] and piezo-phototronics[22] for purposes of electronic tuning, sensing, and human–machine interface.

Before experimental demonstration in 2014,[7] 2D materials, including MoS_2, molybdenum diselenide ($MoSe_2$), molybdenum ditelluride ($MoTe_2$), tungsten disulfide (WS_2), tungsten diselenide (WSe_2), and tungsten ditelluride (WTe_2), have been theoretically predicted to have a piezoelectric effect.[23] Successful demonstration of nanogenerator made by 2D materials enables the possibility of energy harvesting in nanometer-to-micrometer size, as well as sensing through mechanisms of piezotronics[7] and piezo-phototronics.[24]

Traditional piezoelectric materials are usually with perovskite crystal structures, such as $BaTiO_3$, lead zirconate titanate ($Pb[Zr_xTi_{1-x}]O_3$, $0 \leq x \leq 1$, PZT), and lead magnesium niobate–lead titanate ($[1-x]Pb(Mg_{1/3}Nb_{2/3})O_{3-x}PbTiO_3$, $0 \leq x \leq 1$, PMN–PT). These materials have been demonstrated to have high piezoelectric output performance. The high-output signals mainly come from the high d_{33}. These materials can also be synthesized as nanomaterials for nanogenerators using solid-state chemical reaction[8] or hydrothermal method.[25]

Ferroelectric polymers can also be used as ideal piezoelectric materials for nanogenerators with naturally high flexibility. PVDF has been demonstrated as an efficient material for electricity generation and self-powered sensing.[9,26] PVDF nanofibers can be prepared by electrospinning process. Copolymers including PVDF, such as poly(vinylidenefluoride-co-tetrafluoroethylene) (P[VDF-TFE]), can be used to improve the piezoelectric response by increasing the crystallinity.[27]

7.2 SEMICONDUCTING BINARY COMPOUNDS

In semiconductors, a lot of binary compounds with WZ or ZB crystal structures show the property of piezoelectricity. The uniqueness of these group materials is that, due to their semiconducting behavior, the piezoelectricity can be coupled with varieties of different properties, toward a broad range of applications in electronics,[21] optoelectronics,[34] touch sensors,[35] chemical sensors,[36] and so forth. However, since the resistivity of these materials is not high compared with insulating materials, the Schottky barriers are usually demanded for better electrical output performance in nanogenerators.[3,14] In these materials, ZnO nanowire was the first material used in nanogenerators, as discovered by Wang et al., Georgia Tech.[3] In this section, we will discuss several common semiconducting binary compound piezoelectric materials.

7.2.1 II–VI AND III–V BINARY COMPOUNDS

7.2.1.1 ZnO

ZnO, as well as most of the piezoelectric binary semiconductor materials, has two main crystal structures, hexagonal WZ and cubic ZB.[37] The WZ structure is most common and most stable one in the ambient environment. In ZnO as well as most of the other piezoelectric materials, the positive or negative centers are usually formed by ions, atoms, or radicals. For example, in ZnO material with WZ structure, the zinc ions (Zn^{2+}) take the positive charge and the oxygen ions (O^{2-}) take the negative charge. Thus, without any strain, in a single crystal unit cell, which has a pyramid shape, the centers of positive and negative charges overlap with each other. With a stress applied on the crystal in the *c*-axis, the unit cell is deformed, and the center of the negative charge, as defined by the positions of the four oxygen ions in the corners, becomes dislocated. Thus, it is not overlapped with the positive charge center, which is defined by the position of the zinc ion in the center. Hence, the dipole moment in each single unit changes and the piezoelectric polarization density varies accordingly.[21] Due to the varied polarization density, the piezopotential is created. The simulation of distributions of a single ZnO nanowire under compress and tensile strains is described by Gao et al., showing piezopotential difference of 0.4 V in a ~1-μm length strained nanowire.[38,39]

ZnO can be synthesized into various nanostructures including nanowires, nanorods, nanobelts, nanoparticles, and so forth. The most typical nanostructure is the nanowires. The ZnO nanowires used in PENGs can be synthesized by hydrothermal[15] or vapor deposition methods.[16] To growth-aligned ZnO nanowires, substrates (silicon, glass, or others) can be pre-deposited with a thin layer of ZnO film or nanoparticles as the seed layer.[40,41] The aqueous zinc salts such as zinc nitrate and zinc acetate are used as the precursor.[42] The pre-seeding layer can also be conducted in conjunction with lithography method to control the nucleation sites for growth.[43]

The piezoelectricity of the ZnO nanowires enables several interesting research topics. Especially, PENG was first developed based on ZnO nanowire arrays, with the nanowires deformed by atomic force microscopy (AFM) Si/Pt probes.[3] The mechanism of this AFM measurement is illustrated in Figure 7.1, with a piezoelectric CdS nanowire as an example. A Schottky barrier is formed between Pt and CdS.[44,45] As deflected by the AFM probe, the CdS nanowire exhibits a positive piezopotential side and a negative piezopotential side. When the Pt probe is first contacted with the positive potential side, the Schottky barrier is reversely biased, and thus, there is nearly no current output (Fig. 7.1a). When the probe slides to the other side of the nanowire, CdS with the negative potential is contacted, resulting in the positive-biased barrier, with current created to rebalance the existing piezopotentials (Fig. 7.1b).

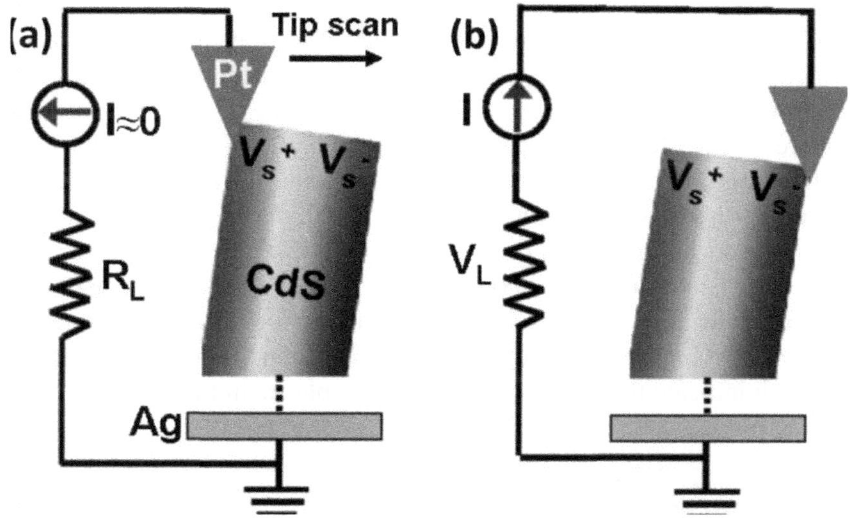

FIGURE 7.1 The piezoelectric output voltage as measured by atomic force microscopy (AFM) probe, while the probe contacting the (a) stretched and (b) released surfaces of a nanowire.

Source: Reproduced with permission from ref 5. © 2008 AIP Publishing LLC.

The AFM measurement on the ZnO nanowire can result in ~8-mV voltage output. A serial of nanogenerator research was further performed based on ZnO nanowires, such as the nanogenerator based on nanowire–fiber hybrid structure,[14] vertical integrated nanogenerator (VING) for ultrasonic wave energy harvesting,[46] and lateral integrated nanogenerator (LING).[12,13] PENG based on ZnO nanowires has achieved an output voltage of 58 V, corresponding to power density 0.78 W/cm^3,[47] with various applications demonstrated for radio frequency transmission[48] and other electronic devices.[47]

7.2.1.2 p-TYPE DOPED ZnO

ZnO has been widely applied in nanogenerators as well as other devices. However, most of these devices are based on intrinsically n-type ZnO materials. In order to optimize the performance and broaden the application of PENGs, p-type ZnO materials are highly demanded, especially p-type ZnO nanowires. P-type ZnO materials have been previously investigated through various methods.[49–51] Here, we introduce a representative work to develop p-type doped ZnO nanowires with dopant antimony (Sb).[52]

Previously, to synthesize the ZnO nanowires, 25 mM zinc nitrate [$Zn(NO_3)_2$] was used as the source of zinc. The dopant source, Sb, was provided by antimony acetate [$Sb(CH_3COO)_3$], with the concentration tuned from 0.2 to 2% molar ratio relative to zinc. The grown Sb-doped ZnO nanowires are as shown in Figure 7.2a, in which at 0.2% doping concentration, the length could be up to 60 μm. Due to the possible energy barrier provided by the Sb dopants, the nanowire length decreases significantly as the doping concentration is increased.

For doped materials, it is common to observe voids in the as-synthesized materials under high-resolution transmission electron microscopy (HRTEM).[53] In Figure 7.2b, they are also observed in Sb-doped ZnO nanowires, which indicate the presence of Sb atomic planes in the lattice.

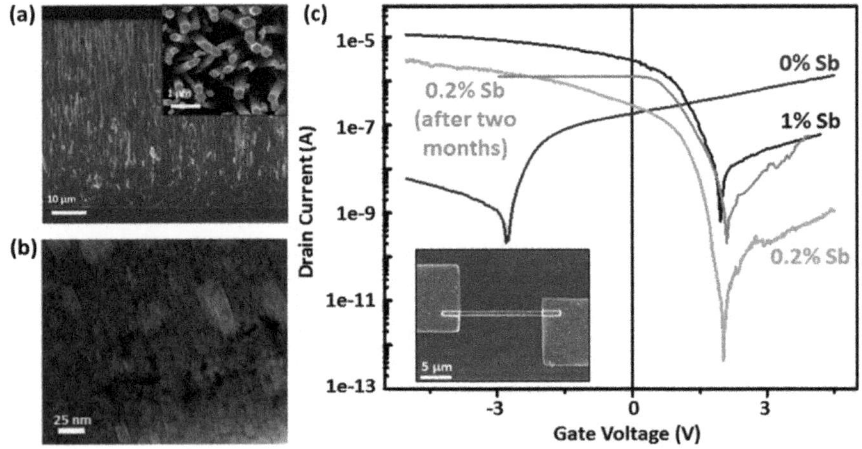

FIGURE 7.2 © 2013, American Chemical Society. p-type doped ZnO nanowires. (a) Scanning electron microscopy (SEM) picture of the p-type nanowires prepared with 0.2% Sb doping concentration, with inset showing the top-view SEM; (b) transmission electron microscopy (TEM) picture of the nanowire, which shows the voids due to the Sb doping as the red arrows; (c) I_{DS}–V_G curves of single p-type ZnO nanowire field-effect transistors (FETs) as a function of doping level. Inset shows SEM image of the nanowire FET, with faked colors.

Source: Reproduced with permission from ref 52. © 2013, American Chemical Society.

The single-nanowire field-effect transistors (NWFETs) with back-gate configuration were fabricated to characterize the carrier properties of the undoped and doped ZnO nanowires, as shown in inset of Figure 7.2c. As we can see, the undoped ZnO NWFET exhibits very typical n-type characteristics, with a threshold gate voltage of −2.8 V. For both 0.2 and 1% doped ZnO NWFETs, p-type characterizations are observed, with threshold gate voltages of 2.1 and 2 V, respectively. After 2 months, the threshold voltage of the 0.2% doped NWFET is still about 2.1 V, demonstrating the successful doping and stable electrical performance.

With the successful p-type doping method demonstrated for ZnO, several further research can be enabled. For example, n–p homojunction ZnO materials can be fabricated, and nanogenerators based on these homojunction materials show much higher output performance.[20] This doping method can also be applied for materials toward self-powered sensing,[20] piezotronics,[52] and optoelectronics.[54]

7.2.1.3 ZnS

ZnS with WZ structure can also be utilized as a material to build nanogenerators. The electric power generation mechanism of ZnS is similar to that of ZnO. Lu et al. prepared ZnS nanowires and measured the piezoelectric output performance of ZnS nanowire arrays for the first time.[4] To synthesize ZnS nanowire arrays, the ZnO–ZnS nanowire heterostructure arrays were first grown using a vacuum tube furnace. And then, the ZnS nanowires were obtained by removing ZnO parts with dilute KOH solution. Figure 7.3a shows the topography profile (red curve, solid square) and the voltage output performance (blue curve, hollow square) of ZnS nanowire arrays using AFM, which is similar to the previous report for ZnO nanowires.[3] The voltage output measured demonstrated the processes of charge accumulation and release while deforming ZnS nanowires using the AFM probe. The scanning electron microscopy (SEM) image of ZnS nanowires is shown in Figure 7.3b, exhibiting the high-density nanowires.[4,55] By scanning over nanowire arrays with an area of 20×20 μm^2, the voltage output performance as recorded by the AFM system is shown in Figure 7.3c. As we can see, the uniformly distributed peak-like voltage output is demonstrated with the magnitude of about 2 mV. Since the as-synthesized ZnS nanowires are shorter and the piezoelectric coefficient is lower,[56] the output voltage of this sample is smaller than that of ZnO.

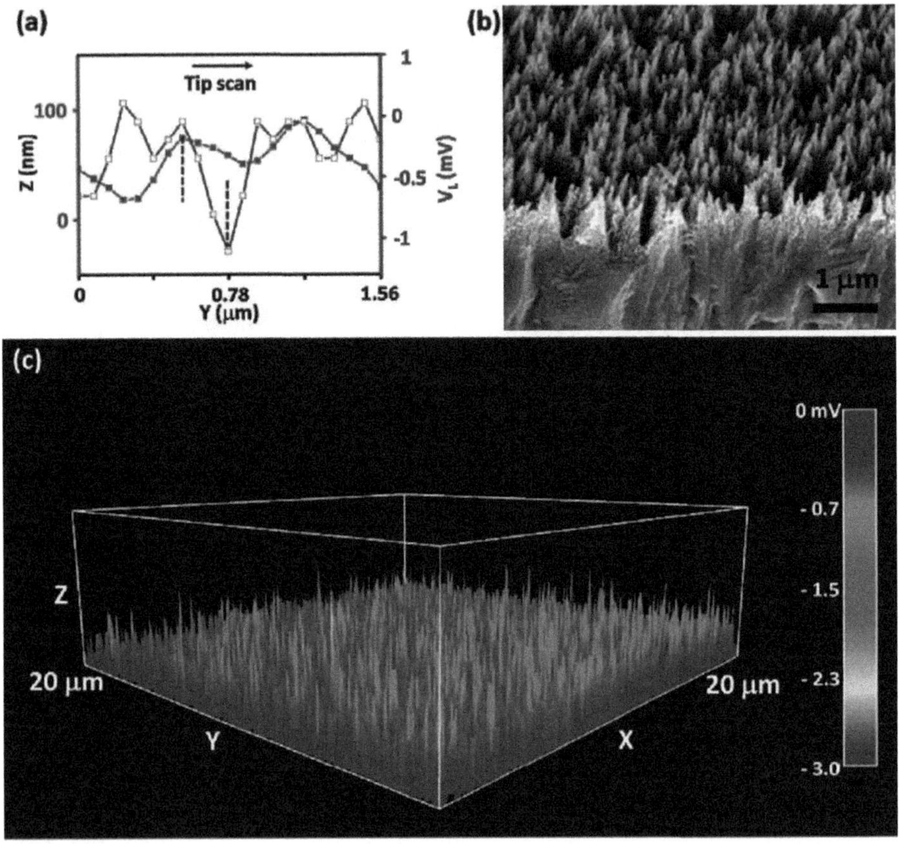

FIGURE 7.3 ZnS nanowire arrays as nanogenerators. (a) As measured in AFM across the nanowire arrays, the topography profile (red) and voltage (blue); (b) the tilt view nanowire arrays under SEM; and (c) the 3D voltage profile as measured by AFM tip scanning.

Source: Reproduced with permission from ref 4. © 2009, American Chemical Society.

7.2.1.4 CdS

Lin et al. reported CdS nanowire arrays as nanogenerators.[5] Initially, the CdS nanowires are grown by the hydrothermal method. Figure 7.4a shows the SEM image of the as-grown CdS nanowires with diameter of 150 nm and length of several micrometers. The CdS nanowires have two phases coexisted in a single nanowire, as characterized by HRTEM as shown in Figure 7.4b. The two phases are ZB and WZ crystal structures, with directions identified as ZB[111] and WZ[0001], respectively. Both of phases contribute to the piezoelectricity performance of the nanowire.

As tested by AFM method as described above, the output voltage is measured as only about 0.5 mV. To improve the voltage output performance, the authors also synthesized pure-WZ-phase CdS nanowires using physical vapor deposition method. The topography profile and 3D voltage output as measured by AFM probe are shown in Figure 7.4c and d, respectively. From the newly grown nanowire arrays, the output voltage can approach ~3 mV, which is much higher than that produced by ZB–WZ double-phase CdS nanowires. High-quality CdS nanowires can also be potentially used for piezo-phototronic applications.[57]

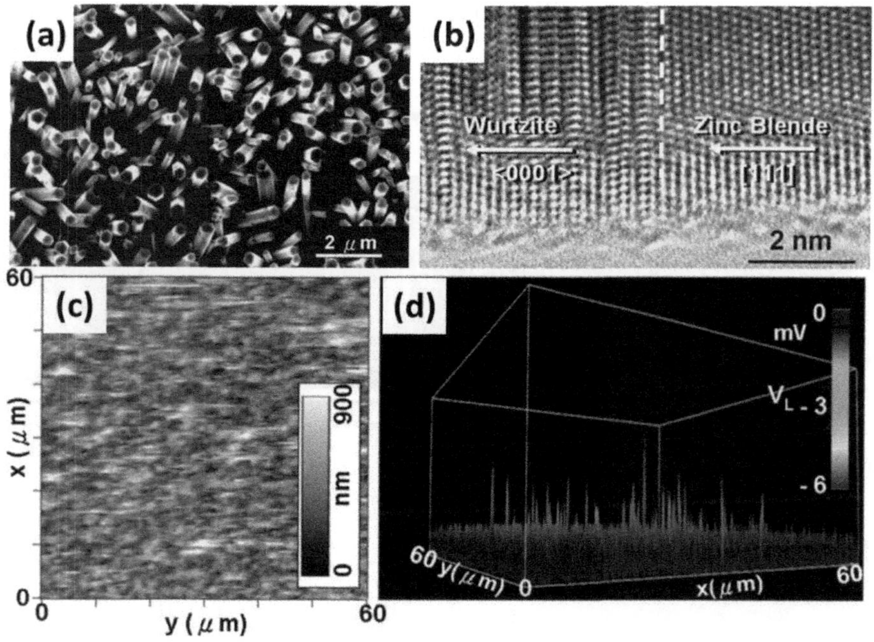

FIGURE 7.4 CdS nanowire arrays as nanogenerators. (a) SEM image of as-grown CdS nanowire arrays; (b) high-resolution transmission electron microscopy (HRTEM) image showing zinc blende and wurtzite (WZ) structure homojunction; (c) the topography profile as measured by AFM; and (d) the 3D voltage profile as measured by AFM.

Source: Reproduced with permission from ref 5. © 2008 AIP Publishing LLC.

7.2.1.5 GaN

Huang et al. first reported GaN nanowires for high-output nanogenerators.[6] Initially, the GaN nanowire was grown by vapor–liquid–solid method. The high-density GaN nanowire arrays grown on the (0001) sapphire substrate was shown in Figure 7.5a, with diameter of 25–70 nm and length of 10–20 μm. The nanowire growth directions have threefold symmetry due to the orientation of the grown substrate. The crystal structure of GaN nanowires is exhibited as WZ structure, with consistent lattice spacing, as shown in Figure 7.5b. Similar to other early PENG studies, the output performance of the GaN nanowires are also characterized by AFM probes. The 3D voltage profile and voltage output by a line scan were shown in Figure 7.5c and d, respectively. The negative output voltage of average absolute value of ~20 mV was demonstrated, and some of the output voltage peaks were up to hundreds of millivolt. This value is much higher than that of ZnO, ZnS, and CdS nanowires as described above. This is mainly due to long length of nanowires synthesized.

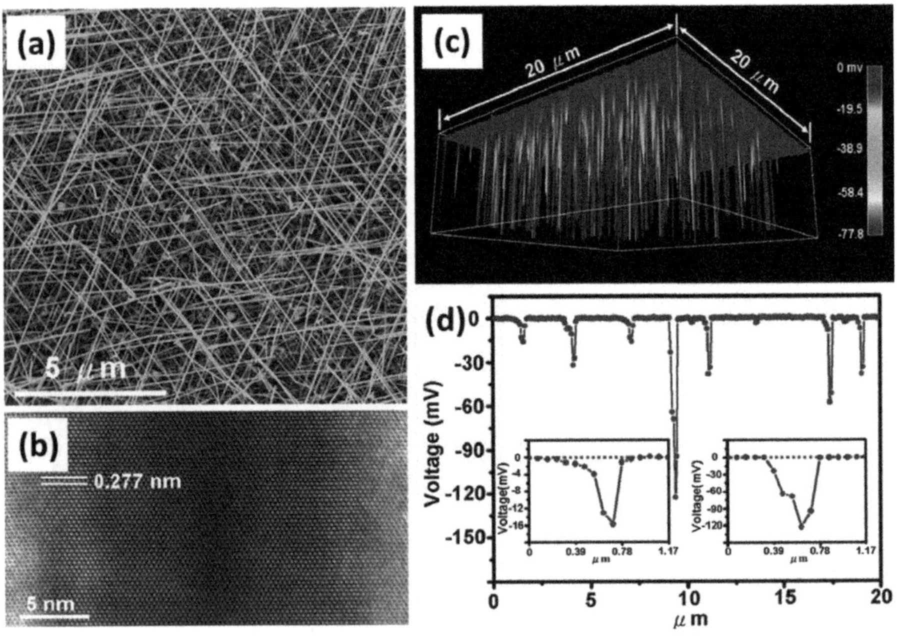

FIGURE 7.5 GaN nanowire arrays as nanogenerators. (a) SEM image showing the GaN nanowires grown on the sapphire substrate; (b) HRTEM image of a single GaN nanowire showing WZ structure; (c) 3D voltage output as scanned by AFM probe; and (d) voltage output by a single AFM probe line scan.

Source: Reproduced with permission from ref 6. © 2010, American Chemical Society

GaN nanowire arrays are also fabricated to be nanogenerator devices similar to that made by ZnO nanowires,[58,59] with voltage output up to ~1 V and output power density of 1 nW/cm^2. The high-quality synthesized GaN nanowires with long length are likely to be a promising material for nanogenerators for power generation.

7.2.2 TWO-DIMENSIONAL MATERIALS

The 2D materials, especially the single-atomic-layer boron nitride, MoS_2, $MoSe_2$, WTe_2, and so forth, have been theoretically predicted to have a piezoelectric effect,[23,60,61] which opens the possible applications of 2D materials as nanoscale energy generation and piezotronic devices, with outstanding flexibility. The 2D metal dichalcogenide materials are a big group of piezoelectric semiconducting materials with broad potential applications. Here, Wu et al. have demonstrated the first experiment of piezoelectricity in 2D materials and fabricated the first single-atomic-layer MoS_2 nanogenerator and piezotronic device.[7] The 2D-materials-based nanogenerators have potential applications of self-powered nanodevices, adaptive biosensors, and tunable/flexible nanoelectronic devices.

The 2D PENG was fabricated using MoS_2 monolayer flake (obtained by either mechanical exfoliation or CVD growth) as the piezoelectric materials, on the flexible substrate, such as the polyethylene terephthalate (PET). The electrode area was defined by lithography method to contact two sides of the flake. The SEM image of the fabricated MoS_2 2D nanogenerator was shown in Figure 7.6a, with the photo of the device on the transparent flexible substrate shown in inset. The open-circuit voltage and short-circuit current were produced by applying an external force on the substrate through a linear motor. With the increase of the applied strain on the substrate, the open-circuit voltage and short-circuit current were both increased, as shown in Figure 7.6b, for single-atomic-layer MoS_2 with a dimension of about 5 μm × 10 μm. The peak open-circuit voltage and short-circuit current reached 18 mV and 27 pA, respectively. By connecting the nanogenerator with an external load, with the increase of the load resistance, the output voltage increased, while the output current decreased (Fig. 7.6c). The maximum peak power output reached 55.3 fW at the optimized load resistance of ~220 MΩ, corresponding to the power density of 2 mW/m^2. A very interesting phenomenon is that only the nanogenerators with odd-number atomic layer have effective piezoelectric current output, and this current decreases with the increase of the layer number (Fig. 7.6d). This result is consistent with a previous report.[62]

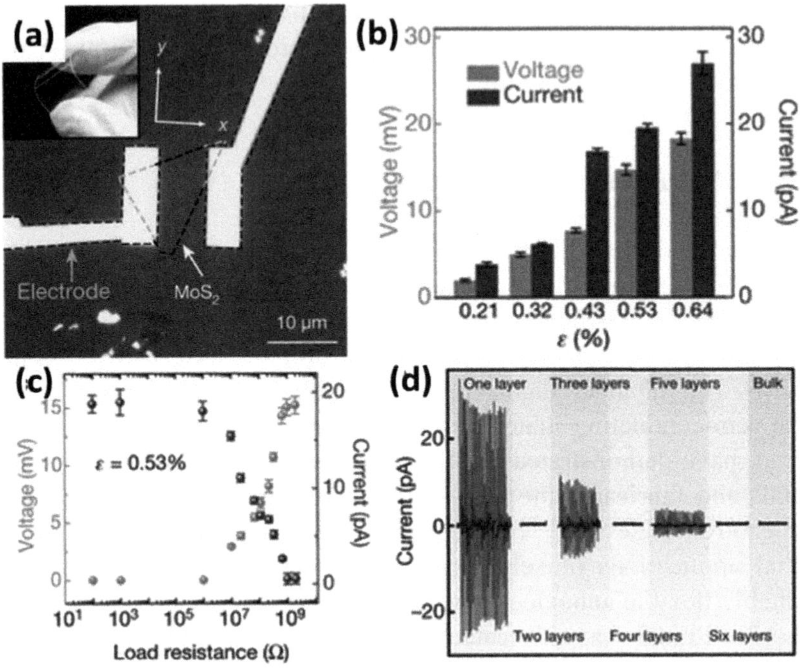

FIGURE 7.6 Single-layer MoS_2 nanogenerator and output performance. (a) The SEM image of the nanogenerator showing MoS_2 area and the contact electrode. The inset shows the image of the device on the flexible substrate. (b) The increase of the output voltage and current with an increase of the strain. (c) The output voltage and current variation with the external load resistance. (d) The current output of MoS_2 with one to six layers and the bulk material.

Source: Reproduced with permission from ref 7. © 2014 Nature Publishing Group.

The discovery and experimental demonstration of piezoelectricity in the 2D materials open a new subfield by utilizing 2D piezoelectric materials for energy harvesting and electronic device tuning. Recent progress by the authors also demonstrates the application of the 2D piezoelectric material for optoelectronic devices (piezo-phototronics).[24] Further development is expected in this topic with broad applications in nanosystems, sensing devices, and nanoelectronics.

7.3 PEROVSKITE MATERIALS

Several materials with perovskite structures have been demonstrated to have higher piezoelectric coefficient than that with WZ structures.[28,31] PZT is a traditional piezoelectric material with outstanding performance. PMN–PT is another piezoelectric material with a high coefficient and broad applications. However, these materials contain lead (Pd), a toxic element, which limits the application areas of these materials. Alternatively, perovskite $BaTiO_3$, which is environment-friendly and biocompatible, can be used to produce piezoelectric outputs. These materials are all demonstrated as outstanding PENGs.

7.3.1 Pd-BASED PEROVSKITE MATERIALS

7.3.1.1 LEAD ZIRCONATE TITANATE

PZT is a very common piezoelectric material with various applications demonstrated. For mechanical energy harvesting, Park et al. have demonstrated a high-efficiency flexible PZT thin-film-based nanogenerator. The fabrication process is through the laser liftoff technology, which allows simple, large-scale transfers of thin films to the flexible substrate. First, through the solgel method, the piezoelectric PZT thin film with a thickness of 2 μm was deposited on the polished sapphire substrate. The sapphire/PZT substrate was then attached to a PET substrate. A laser beam was utilized to irradiate through the sapphire side to vaporize the PZT/sapphire interface. Thus, the PZT/PET film can be separated from the sapphire substrate. And then, 100-nm Au interdigitated electrodes were deposited on the PZT thin films. Finally, the nanogenerator was encapsulated and the piezoelectric property of the film was enhanced under the applied electric field.

The thickness of the PZT film is kept 2 μm during the transfer process. The fabricated PZT-film-based nanogenerator illustrates a high flexibility and outstanding output performance. Under 0.386% of the bending strain, the voltage in the open-circuit condition and the current in the short-circuit condition reach ~200 V and ~1.5 μA, respectively, with a current density of 150 μA/cm^2. With external load, the output current and the voltage are measured to be decreased and increased with the increase of the load resistance, respectively. The optimized peak power density achieves 17.5 mW/cm^2 at the matched load resistance of 200 MΩ.

Such PZT-film-based nanogenerator has demonstrated plenty of applications due to its high-output performance. It was used to simultaneously power over 100 light-emitting diode (LED) bulbs.[63] PZT-based nanogenerator was also implanted in vivo to harvest energy from the motions of internal organs to power cardiac pacemakers.[64]

7.3.1.2 LEAD MAGNESIUM NIOBATE–LEAD TITANATE

PMN–PT thin-film-based nanogenerator is another high-output nanogenerator to enable self-powered pacemaking, as demonstrated by Hwang et al.[65] To prepare this nanogenerator, the metal–insulator–metal (MIM) structure was deposited by Au, PMN–PT, and Au thin films on epoxy/Si substrate. Due to lattice mismatch, Ni film was selected to exfoliate MIM structure from the original substrate and transfer it onto flexible PET substrate. The Au films on both sides were then directly connected to stimulate a cardiac pacemaker, using piezoelectric output generated by applying bending strain on the PET substrate. Under the periodical strain of 0.36%, a 1.7 cm × 1.7 cm area nanogenerator can produce an open-circuit voltage of ~8 V and short-circuit current of nearly 150 μA. The function of the artificial pacemaker was then demonstrated under in vivo environment, successfully stimulating the heart of a rat. Such high current output can also be used to charge coin cell batteries and power 50 green LEDs.

7.3.2 Pd-FREE PEROVSKITE MATERIALS

Lead-free materials are usually preferred for devices considering environmental issues and biocompatibility. $BaTiO_3$ is a typical lead-free biocompatible perovskite piezoelectric material with excellent output characteristics. As reported by Park et al., $BaTiO_3$ thin-film-based nanogenerator arrays were first utilized to generate electricity.[66] The fabrication process of this nanogenerator arrays was illustrated in Figure 7.7a. First, the MIM structure made up of Pt–$BaTiO_3$–Au multilayer thin films was deposited on SiO_2/Si substrate. The cross-sectional SEM image of the structure was shown in Figure 7.7b, with the composition of the $BaTiO_3$ film confirmed by X-ray photoelectron spectroscopy spectrum in the inset. By using inductively coupled plasma–reactive ion etching, the MIM structure was etched to be bridge structures as protected by masks (Fig. 7.7c), followed by wet etching to remove underlying Si layer (cross-sectional SEM as shown in Figure 7.7c inset). And then, the MIM structure can be transferred to a flexible Kapton substrate using polydimethylsiloxane stamps. Epoxy with patterned openings on the contact areas was applied to protect the structure. Finally, interdigital metal (Au) electrodes were deposited as contacts of the thin-film nanogenerator arrays.

With about 1350 MIM structures and effective total area of 13.4 mm^2, the output performance of the nanogenerator arrays was evaluated under strain of 0.4–0.55%. The open-circuit voltage of 0.35 V and peak short-circuit current of 12 nA were demonstrated, as shown in Figure 7.7d. The output power density of ~7 mW/cm^3 was achieved.

Besides the thin-film structures, $BaTiO_3$ nanowires were also developed as nanogenerators.[8] Such biocompatible $BaTiO_3$-based nanogenerators show potential applications as implantable devices and artificial electronic skins.

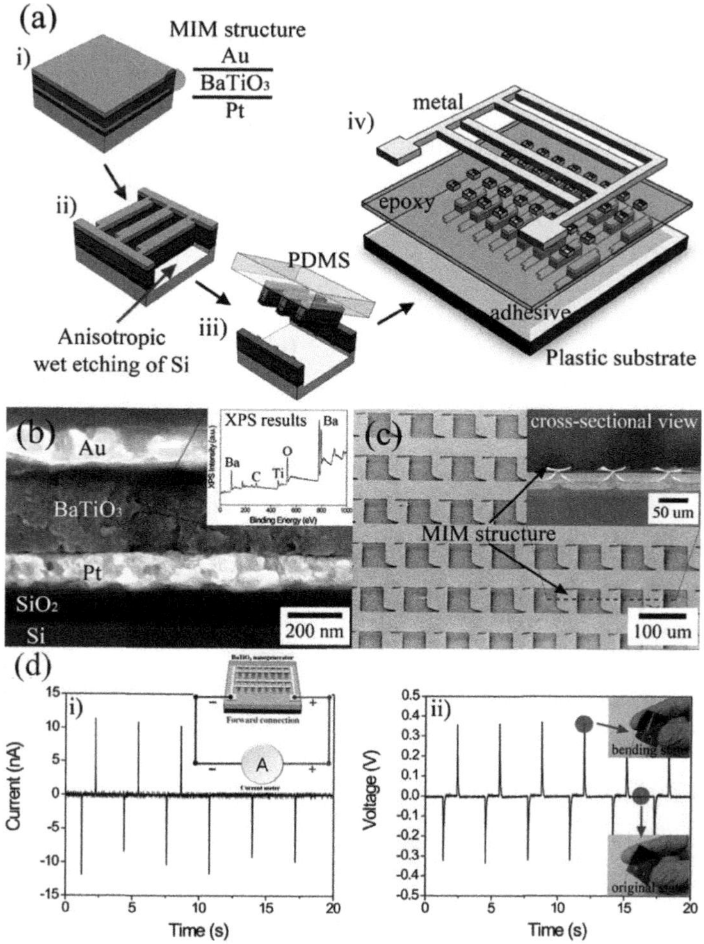

FIGURE 7.7 $BaTiO_3$-based nanogenerator arrays. (a) The fabrication process; (b) the cross-sectional SEM image of a single metal–insulator–metal (MIM) unit nanogenerator, with inset showing X-ray photoelectron spectroscopy spectrum of $BaTiO_3$; (c) SEM image showing the MIM structure "bridge" arrays, with the inset showing underlying Si layer removed; and (d) the short-circuit current and open-circuit voltage outputs of the nanogenerators fabricated on a flexible substrate.

Source: Reproduced with permission from ref 66. © 2010, American Chemical Society

7.4 POLYMER FERROELECTRIC MATERIALS

Since polymers are naturally flexible materials, piezoelectric polymer materials show even broader applications as nanogenerators. PVDF is a typical piezoelectric material for mechanical energy harvesting. Through direct-write technology based on near-field electrospinning, Chang et al. demonstrated PVDF nanofibers as PENGs[9] (Fig. 7.8a). Due to the high electric field (> 107 V/m) applied, the orientations of the dipole moments are naturally aligned during fabrication, forming mainly β-phase. The fabricated single PVDF nanofiber nanogenerator with two electrodes is shown in Figure 7.8b. The typical open-circuit voltage and short-circuit current output reach 30 mV and 4 nA, respectively (Fig. 7.8c and d). The average energy conversion efficiency is estimated at ~12.5%, which is higher than PVDF thin-film-based structures.

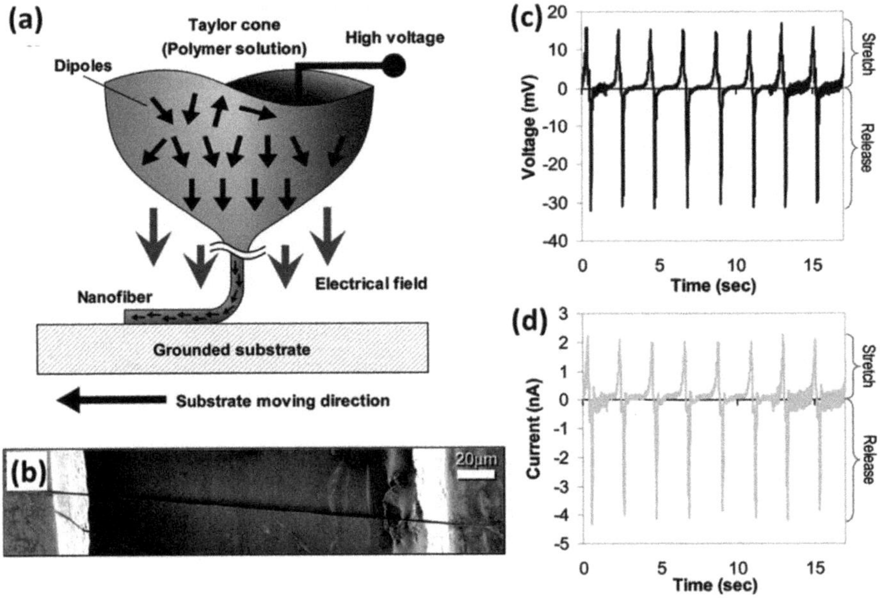

FIGURE 7.8 Polyvinylidene fluoride (PVDF) nanofiber-based nanogenerator. (a) The fabrication process by electrospinning combining direct-write, mechanical stretching, and in situ electrical poling; (b) the SEM image showing a single PVDF nanofiber; (c) the open-circuit voltage; and (d) short-circuit current.

Source: Reproduced with permission from ref 9. © 2010, American Chemical Society.

PVDF can also form copolymers such as poly(vinylidenefluoride-co-trifluoroethylene) (P[VDF-TrFE]) for nanogenerators with higher output performance due to the improved piezoelectric coefficient.[67] Compared to other piezoelectric materials, another advantage of the polymer piezoelectric materials is the ultra-high resistivity, which makes the voltage produced remain for a relatively long time in the open-circuit condition. Such features open the possibility of utilizing the voltage signal for self-powered pressure sensing through the piezoelectric effect and accurate temperature sensing through the pyroelectric effect.[26] Due to the natural flexibility and facile fabrication/processing, PVDF and its copolymers will open plenty of applications, especially the wearable electronics and smart textiles.

7.5 SUMMARY

In this chapter, the piezoelectric materials for the nanogenerators are reviewed. Each material has its own characteristics with unique applications. For the semiconducting piezoelectric materials such as ZnO, CdS, GaN, or 2D MoS_2, it is easy to couple piezopotentials with semiconducting behaviors for sensing and computing through piezotronics and piezo-phototronics. The high electric generation of this type of piezoelectric materials may be achieved through novel device integration methods. For perovskite materials with high piezoelectric coefficients, including PZT, PMN–PT, and $BaTiO_3$, it is easy to achieve high-output performance for large-scale applications. Polymer piezoelectric materials such as PVDF, as naturally highly flexible materials, have potential to be used in smart wearable technologies and self-powered sensing. Piezoelectric materials with potential applications in energy generation, sensing, and computing will pave the foundation for future functional smart systems and the IoT.

KEYWORDS

- **piezoelectric materials**
- **piezoelectric nanogenerators**
- **mechanical energy harvesting**

REFERENCES

1. Katzir, S. The Discovery of the Piezoelectric Effect. *Archive Hist. Exact Sci.* **2003,** *57*(1), 61–91.
2. Birkholz, M. Crystal-Field Induced Dipoles in Heteropolar Crystals II: Physical Significance. *Zeitschrift Für Physik B Condens. Matter* **1995,** *96*(3), 333–340.
3. Wang, Z. L.; Song, J. Piezoelectric Nanogenerators Based on Zinc Oxide Nanowire Arrays. *Science* **2006,** *312*(5771), 242–246.
4. Lu, M.-Y.; Song, J.; Lu, M.-P.; Lee, C.-Y.; Chen, L.-J.; Wang, Z. L. ZnO–ZnS Heterojunction and ZnS Nanowire Arrays for Electricity Generation. *ACS Nano* **2009,** *3*(2), 357–362.
5. Lin, Y.-F.; Song, J.; Ding, Y.; Lu, S.-Y.; Wang, Z. L. Piezoelectric Nanogenerator Using CdS Nanowires. *Appl. Phys. Lett.* **2008,** *92*(2), 022105.
6. Huang, C.-T.; Song, J.; Lee, W.-F.; Ding, Y.; Gao, Z.; Hao, Y.; Chen, L.-J.; Wang, Z. L. GaN Nanowire Arrays for High-Output Nanogenerators. *J. Am. Chem. Soc.* **2010,** *132*(13), 4766–4771.
7. Wu, W.; Wang, L.; Li, Y.; Zhang, F.; Lin, L.; Niu, S.; Chenet, D.; Zhang, X.; Hao, Y.; Heinz, T. F.; Hone, J.; Wang, Z. L. Piezoelectricity of Single-Atomic-Layer MoS_2 for Energy Conversion and Piezotronics. *Nature* **2014,** *514*(7523), 470–474.
8. Wang, Z.; Hu, J.; Suryavanshi, A. P.; Yum, K.; Yu, M.-F. Voltage Generation from Individual $BaTiO_3$ Nanowires Under Periodic Tensile Mechanical Load. *Nano Lett.* **2007,** *7*(10), 2966–2969.
9. Chang, C.; Tran, V. H.; Wang, J.; Fuh, Y.-K.; Lin, L. Direct-Write Piezoelectric Polymeric Nanogenerator with High Energy Conversion Efficiency. *Nano Lett.* **2010,** *10*(2), 726–731.
10. Choi, D.; Choi, M.-Y.; Shin, H.-J.; Yoon, S.-M.; Seo, J.-S.; Choi, J.-Y.; Lee, S. Y.; Kim, J. M.; Kim, S.-W. Nanoscale Networked Single-Walled Carbon-Nanotube Electrodes for Transparent Flexible Nanogenerators. *J. Phys. Chem. C* **2010,** *114*(2), 1379–1384.
11. Choi, M.-Y.; Choi, D.; Jin, M.-J.; Kim, I.; Kim, S.-H.; Choi, J.-Y.; Lee, S. Y.; Kim, J. M.; Kim, S.-W. Mechanically Powered Transparent Flexible Charge-Generating Nanodevices with Piezoelectric ZnO Nanorods. *Adv. Mater.* **2009,** *21*(21), 2185–2189.
12. Xu, S.; Qin, Y.; Xu, C.; Wei, Y.; Yang, R.; Wang, Z. L. Self-Powered Nanowire Devices. *Nat. Nano* **2010,** *5*(5), 366–373.
13. Zhu, G.; Yang, R.; Wang, S.; Wang, Z. L. Flexible High-Output Nanogenerator Based on Lateral ZnO Nanowire Array. *Nano Lett.* **2010,** *10*(8), 3151–3155.
14. Qin, Y.; Wang, X.; Wang, Z. L. Microfibre-Nanowire Hybrid Structure for Energy Scavenging. *Nature* **2008,** *451*(7180), 809–813.
15. Xu, C.; Shin, P.; Cao, L.; Gao, D. Preferential Growth of Long ZnO Nanowire Array and Its Application in Dye-Sensitized Solar Cells. *J. Phys. Chem. C* **2010,** *114*(1), 125–129.
16. Pan, Z. W; Dai, Z. R.; Wang, Z. L. Nanobelts of Semiconducting Oxides. *Science* **2001,** *291*(5510), 1947–1949.
17. Duan, X.; Lieber, C. M. General Synthesis of Compound Semiconductor Nanowires. *Adv. Mater.* **2000,** *12*(4), 298–302.
18. Zi, Y.; Jung, K.; Zakharov, D.; Yang, C. Understanding Self-Aligned Planar Growth of InAs Nanowires. *Nano Lett.* **2013,** *13*(6), 2786–2791.
19. Zi, Y.; Zhao, Y.; Candebat, D.; Appenzeller, J.; Yang, C. Synthesis of Antimony-Based Nanowires Using the Simple Vapor Deposition Method. *Chem. Phys. Chem.* **2012,** *13*(10), 2585–2588.

20. Pradel, K. C.; Wu, W.; Ding, Y.; Wang, Z. L. Solution-Derived ZnO Homojunction Nanowire Films on Wearable Substrates for Energy Conversion and Self-Powered Gesture Recognition. *Nano Lett.* **2014,** *14*(12), 6897–6905.
21. Wang, Z. L. Nanopiezotronics. *Adv. Mater.* **2007,** *19*(6), 889–892.
22. Wang, Z. L. Piezopotential Gated Nanowire Devices: Piezotronics and Piezo-Phototronics. *Nano Today* **2010,** *5*(6), 540–552.
23. Duerloo, K.-A. N.; Ong, M. T.; Reed, E. J. Intrinsic Piezoelectricity in Two-Dimensional Materials. *J. Phys. Chem. Lett.* **2012,** *3*(19), 2871–2876.
24. Wu, W.; Wang, L.; Yu, R.; Liu, Y.; Wei, S.-H.; Hone, J.; Wang, Z. L. Piezophototronic Effect in Single-Atomic-Layer MoS2 for Strain-Gated Flexible Optoelectronics. *Adv. Mater.* **2016,** *28*(38), 8463–8468.
25. Xu, S.; Poirier, G.; Yao, N. PMN-PT Nanowires with a Very High Piezoelectric Constant. *Nano Lett.* **2012,** *12*(5), 2238–2242.
26. Zi, Y.; Lin, L.; Wang, J.; Wang, S.; Chen, J.; Fan, X.; Yang, P.-K.; Yi, F.; Wang, Z. L. Triboelectric–Pyroelectric–Piezoelectric Hybrid Cell for High-Efficiency Energy-Harvesting and Self-Powered Sensing. *Adv. Mater.* **2015,** *27*(14), 2340–2347.
27. Omote, K.; Ohigashi, H.; Koga, K. Temperature Dependence of Elastic, Dielectric, and Piezoelectric Properties of "Single Crystalline" Films of Vinylidene Fluoride Trifluoroethylene Copolymer. *J. Appl. Phys.* **1997,** *81*(6), 2760–2769.
28. Kholkin, A.; Pertsev, N.; Goltsev, A. Piezoelectricity and Crystal Symmetry. In *Piezoelectric and Acoustic Materials for Transducer Applications;* Safari, A., Akdoğan, E. K., Eds.; Springer: Boston, MA, USA, 2008; pp 17–38.
29. Lueng, C.; Chan, H. L.; Surya, C.; Choy, C. Piezoelectric Coefficient of Aluminum Nitride and Gallium Nitride. *J. Appl. Phys.* **2000,** *88*(9), 5360.
30. Sabolsky, E.; James, A.; Kwon, S.; Trolier-McKinstry, S.; Messing, G. Piezoelectric Properties of <001> Textured Pb ($Mg_{1/3}\cdot Nb_{2/3}$) O_3–$PbTiO_3$ Ceramics. *Appl. Phys. Lett.* **2001,** *78*(17).
31. Broitman, E.; Soomro, M. Y.; Lu, J.; Willander, M.; Hultman, L. Nanoscale Piezoelectric Response of ZnO Nanowires Measured Using a Nanoindentation Technique. *Phys. Chem. Chem. Phys.* **2013,** *15*(26), 11113–11118.
32. Nix, E. L; Ward, I. M. The Measurement of the Shear Piezoelectric Coefficients of Polyvinylidene Fluoride. *Ferroelectrics* **1986,** *67*(1), 137–141.
33. Wang, X.; He, X.; Zhu, H.; Sun, L.; Fu, W.; Wang, X.; Hoong, L. C; Wang, H.; Zeng, Q.; Zhao, W. Subatomic Deformation Driven by Vertical Piezoelectricity from CdS Ultrathin Films. *Sci. Adv.* **2016,** *2*(7), e1600209.
34. Pan, C.; Dong, L.; Zhu, G.; Niu, S.; Yu, R.; Yang, Q.; Liu, Y.; Wang, Z. L. High-Resolution Electroluminescent Imaging of Pressure Distribution Using a Piezoelectric Nanowire Led Array. *Nat. Photon.* **2013,** *7*(9), 752–758.
35. Wu, W.; Wen, X.; Wang, Z. L. Taxel-Addressable Matrix of Vertical-Nanowire Piezotronic Transistors for Active and Adaptive Tactile Imaging. *Science* **2013,** *340*(6135), 952–957.
36. Yu, R.; Pan, C.; Chen, J.; Zhu, G.; Wang, Z. L. Enhanced Performance of a ZnO Nanowire-Based Self-Powered Glucose Sensor by Piezotronic Effect. *Adv. Funct. Mater.* **2013,** *23*(47), 5868–5874.
37. Fierro, J. L. G. *Metal Oxides: Chemistry and Applications;* CRC Press: Boca Raton, FL, USA, 2005.
38. Gao, Y.; Wang, Z. L. Electrostatic Potential in a Bent Piezoelectric Nanowire. The Fundamental Theory of Nanogenerator and Nanopiezotronics. *Nano Lett.* **2007,** *7*(8), 2499–2505.

39. Gao, Z.; Zhou, J.; Gu, Y.; Fei, P.; Hao, Y.; Bao, G.; Wang, Z. L. Effects of Piezoelectric Potential on the Transport Characteristics of Metal-ZnO Nanowire-Metal Field Effect Transistor. *J. Appl. Phys.* **2009,** *105*(11), 113707.
40. Wu, W.-Y.; Yeh, C.-C.; Ting, J.-M., Effects of Seed Layer Characteristics on the Synthesis of ZnO Nanowires. *J. Am. Ceram. Soc.* **2009,** *92*(11), 2718–2723.
41. Greene, L.E.; Law, M.; Tan, D. H.; Montano, M.; Goldberger, J.; Somorjai, G.; Yang, P. General Route to Vertical ZnO Nanowire Arrays Using Textured ZnO Seeds. *Nano Lett.* **2005,** *5*(7), 1231–1236.
42. Greene, L. E.; Law, M.; Goldberger, J.; Kim, F.; Johnson, J. C.; Zhang, Y.; Saykally, R. J.; Yang, P. Low-Temperature Wafer-Scale Production of ZnO Nanowire Arrays. *Angew. Chem. Int. Ed.* **2003,** *42*(26), 3031–3034.
43. Wen, X.; Wu, W.; Ding, Y.; Wang, Z. L. Seedless Synthesis of Patterned ZnO Nanowire Arrays on Metal Thin Films (Au, Ag, Cu, Sn) and Their Application for Flexible Electromechanical Sensing. *J. Mater. Chem.* **2012,** *22*(19), 9469–9476.
44. Pierret, R. F. *Semiconductor Device Fundamentals;* Pearson Education, 1996.
45. Park, W. I.; Yi, G.-C.; Kim, J.-W.; Park, S.-M. Schottky Nanocontacts on ZnO Nanorod Arrays. *Appl. Phys. Lett.* **2003,** *82*(24), 4358–4360.
46. Wang, X.; Song, J.; Liu, J.; Wang, Z. L. Direct-Current Nanogenerator Driven by Ultrasonic Waves. *Science* **2007,** *316*(5821), 102–105.
47. Zhu, G.; Wang, A. C.; Liu, Y.; Zhou, Y.; Wang, Z. L. Functional Electrical Stimulation by Nanogenerator with 58 V Output Voltage. *Nano Lett.* **2012,** *12*(6), 3086–3090.
48. Hu, Y.; Zhang, Y.; Xu, C.; Lin, L.; Snyder, R. L.; Wang, Z. L. Self-Powered System with Wireless Data Transmission. *Nano Lett.* **2011,** *11*(6), 2572–2577.
49. Lee, W.; Jeong, M.-C.; Myoung, J.-M. Optical Characteristics of Arsenic-Doped ZnO Nanowires. *Appl. Phys. Lett.* **2004,** *85*(25), 6167–6169.
50. Xiang, B.; Wang, P.; Zhang, X.; Dayeh, S. A.; Aplin, D. P. R.; Soci, C.; Yu, D.; Wang, D. Rational Synthesis of p-Type Zinc Oxide Nanowire Arrays Using Simple Chemical Vapor Deposition. *Nano Lett.* **2007,** *7*(2), 323–328.
51. Yuan, G. D.; Zhang, W. J.; Jie, J. S.; Fan, X.; Zapien, J. A.; Leung, Y. H.; Luo, L. B.; Wang, P. F.; Lee, C. S.; Lee, S. T. p-Type ZnO Nanowire Arrays. *Nano Lett.* **2008,** *8*(8), 2591–2597.
52. Pradel, K. C.; Wu, W.; Zhou, Y.; Wen, X.; Ding, Y; Wang, Z. L. Piezotronic Effect in Solution-Grown p-Type ZnO Nanowires and Films. *Nano Lett.* **2013,** *13*(6), 2647–2653.
53. Yankovich, A. B.; Puchala, B.; Wang, F.; Seo, J.-H.; Morgan, D.; Wang, X.; Ma, Z.; Kvit, A. V.; Voyles, P. M. Stable p-Type Conduction from Sb-Decorated Head-to-Head Basal Plane Inversion Domain Boundaries in ZnO Nanowires. *Nano Lett.* **2012,** *12*(3), 1311–1316.
54. Pradel, K. C.; Ding, Y.; Wu, W.; Bando, Y.; Fukata, N.; Wang, Z. L. Optoelectronic Properties of Solution Grown ZnO N-P Or P-N Core–Shell Nanowire Arrays. *ACS Appl. Mater. Interfaces* **2016,** *8*(7), 4287–4291.
55. Lu, H.-Y.; Chu, S.-Y.; Chang, C.-C. Synthesis and Optical Properties of Well-Aligned ZnS Nanowires on Si Substrate. *J. Cryst. Growth* **2005,** *280*(1–2), 173–178.
56. Xin, J.; Zheng, Y.; Shi, E. Piezoelectricity of Zinc-Blende and Wurtzite Structure Binary Compounds. *Appl. Phys. Lett.* **2007,** *91*(11), 112902.
57. Lin, Y.-F.; Song, J.; Ding, Y.; Lu, S.-Y.; Wang, Z. L. Alternating the Output of a CdS Nanowire Nanogenerator by a White-Light-Stimulated Optoelectronic Effect. *Adv. Mater.* **2008,** *20*(16), 3127–3130.

58. Long, L.; Chen-Ho, L.; Youfan, H.; Yan, Z.; Xue, W.; Chen, X.; Robert, L. S.; Lih, J. C.; Zhong L. W. High Output Nanogenerator Based on Assembly of GaN Nanowires. *Nanotechnology* **2011,** *22*(47), 475401.
59. Wang, C.-H.; Liao, W.-S.; Lin, Z.-H.; Ku, N.-J.; Li, Y.-C.; Chen, Y.-C.; Wang, Z.-L.; Liu, C.-P. Optimization of the Output Efficiency of GaN Nanowire Piezoelectric Nanogenerators by Tuning the Free Carrier Concentration. *Adv. Energy Mater.* **2014,***4*(16), n/a–n/a.
60. Michel, K. H.; Verberck, B. Phonon Dispersions and Piezoelectricity in Bulk and Multilayers of Hexagonal Boron Nitride. *Phys. Rev. B* **2011,** *83*(11), 115328.
61. Michel, K. H.; Verberck, B. Theory of Elastic and Piezoelectric Effects in Two-Dimensional Hexagonal Boron Nitride. *Phys. Rev. B* **2009,** *80*(22), 224301.
62. Li, Y.; Rao, Y.; Mak, K. F.; You, Y.; Wang, S.; Dean, C. R.; Heinz, T. F. Probing Symmetry Properties of Few-Layer MoS_2 and H-BN by Optical Second-Harmonic Generation. *Nano Lett.* **2013,** *13*(7), 3329–3333.
63. Park, K.-I.; Son, J. H.; Hwang, G.-T.; Jeong, C. K.; Ryu, J.; Koo, M.; Choi, I.; Lee, S. H.; Byun, M.; Wang, Z. L.; Lee, K. J. Highly-Efficient, Flexible Piezoelectric PZT Thin Film Nanogenerator on Plastic Substrates. *Adv. Mater.* **2014,** *26*(16), 2514–2520.
64. Dagdeviren, C.; Yang, B. D.; Su, Y.; Tran, P. L.; Joe, P.; Anderson, E.; Xia, J.; Doraiswamy, V.; Dehdashti, B.; Feng, X.; Lu, B.; Poston, R.; Khalpey, Z.; Ghaffari, R.; Huang, Y.; Slepian, M. J.; Rogers, J. A. Conformal Piezoelectric Energy Harvesting and Storage from Motions of the Heart, Lung, and Diaphragm. *Proc. Nat. Acad. Sci.* **2014,** *111*(5), 1927–1932.
65. Hwang, G.-T.; Park, H.; Lee, J.-H.; Oh, S.; Park, K.-I.; Byun, M.; Park, H.; Ahn, G.; Jeong, C. K.; No, K.; Kwon, H.; Lee, S.-G.; Joung, B.; Lee, K. J. Self-Powered Cardiac Pacemaker Enabled by Flexible Single Crystalline PMN-PT Piezoelectric Energy Harvester. *Adv. Mater.* **2014,** *26*(28), 4880–4887.
66. Park, K.-I.; Xu, S.; Liu, Y.; Hwang, G.-T.; Kang, S.-J. L.; Wang, Z. L.; Lee, K. J. Piezoelectric $BaTiO_3$ Thin Film Nanogenerator on Plastic Substrates. *Nano Lett.* **2010,** *10*(12), 4939–4943.
67. Persano, L.; Dagdeviren, C.; Su, Y.; Zhang, Y.; Girardo, S.; Pisignano, D.; Huang, Y.; Rogers, J. A. High Performance Piezoelectric Devices Based on Aligned Arrays of Nanofibers of Poly(vinylidenefluoride-co-trifluoroethylene). *Nat. Commun.* **2013,** *4*, 1633.

PART III
New Insights on Nanotechniques

CHAPTER 8

APPLICATION OF NANOTECHNOLOGY IN CHEMICAL ENGINEERING AND CARBON NANOTUBES: A CRITICAL OVERVIEW AND A VISION FOR THE FUTURE

SUKANCHAN PALIT*

Department of Chemical Engineering, University of Petroleum and Energy Studies, Post Office Bidholi via Premnagar, Dehradun 248007, India, Tel.: +918958728093

**Corresponding author. E-mail: sukanchan68@gmail.com, sukanchan92@gmail.com*

CONTENTS

ABSTRACT

Science and engineering of nanotechnology are moving from one visionary paradigm to another. Nanovision in today's human civilization is the utmost need of the hour. Chemical process engineering and nanotechnology are the two opposite sides of the visionary scientific coin. Carbon nanotubes (CNTs) and its interface with nanotechnology are the challenging areas of scientific research pursuit today. In this treatise, the author with deep and cogent insight presents a well-researched nanotechnology applications paper. Human civilization and scientific endeavor stand in the midst of scientific comprehension and vision. The challenge and vision of this chapter are to elucidate on the vast scientific applications of CNTs and the wide domain of application of nanotechnology in chemical engineering. The success of science and technology, the wide futuristic vision, and the scientific imagination are veritably opening new frontiers of scientific forbearance and scientific candor in the field of nanotechnology today. The vision of this chapter is vast and versatile. The author treads a visionary as well as a weary path of scientific endeavor in nanotechnology and CNT applications.

8.1 INTRODUCTION

Science and technology of nanoscience and nanotechnology are moving forward at a rapid pace surpassing visionary frontiers. Energy engineering and technology are the cornerstones and the pivots of human scientific endeavor and the progress of human civilization. Nanotechnology is revolutionizing the scientific landscape and moving toward the newer world of scientific regeneration and deep scientific rejuvenation. Sustainable development, whether it is energy or environment, are the utmost need of this century. Successful sustainability will veritably change the wide scientific fabric of human scientific endeavor. Today, chemical process engineering, environmental engineering science, and energy engineering are witnessing immense challenges as finding alternate energy sources stand a as major issue in the progress of human civilization; thus, there is a need of the application of nanotechnology and carbon nanotubes (CNTs). CNTs have widespread applications in regard to energy or environment. Science is a huge colossus with definite vision and a definite target of its own. CNTs, long, thin cylinders of carbon, were discovered in 1991 by Sumio Iijima. These are large macromolecules that are unique for their size, shape, and remarkable physical properties. Scientific vision, technological motivation, and the futuristic vision are all leading a long and visionary way toward the true emancipation and true realization of science of nanotechnology. The CNTs can be thought of as a sheet of graphite (a hexagonal lattice of carbon) rolled into a cylinder. These intriguing structures have sparked immense excitement and scientific profundity in recent years and a large amount of research has been dedicated to their deep understanding. Currently, the physical properties are still being discovered and disputed. Nanotubes have a broad range of electronic, thermal, and structural properties that change depending on the different kinds of nanotube (defined by its diameter, length, and chirality, or twist). To make things more robust and interesting, besides having a single cylindrical wall, nanotubes can have multiple walls—cylinders inside the other cylinders. The application of nanotechnology in chemical engineering is vast and versatile. The science of membrane separation processes and nanofiltration are the technological innovations of our times. In this chapter, the author deeply comprehends the scientific potential and scientific success of nanofiltration in separation phenomenon of complex mixtures.

8.2 THE AIM AND OBJECTIVE OF THIS STUDY

Science and engineering of nanotechnology are moving toward a newer visionary realm at a drastic pace. Nanotechnology is a revolutionary domain of scientific research pursuit today. Technological candor, scientific forbearance in nanotechnology applications, and the wide scientific vision will all lead a long and visionary way in the true emancipation and true realization of nanoscience today. CNTs are a vast domain of science. The pivotal aim and objective of this treatise target the scientific potential and the scientific success in nanotechnology applications in chemical engineering and the broader domain of engineering science. In this treatise, the author deeply comprehends the important nanotechnology applications in chemical engineering with the sole and primary objective of furtherance of science and engineering. CNTs and nanotechnology are the vast and versatile domains of scientific research pursuit today. This treatise is an effective eye-opener toward a newer visionary era in the field of nanotechnology research.[3,13,14]

8.3 THE NEED AND RATIONALE BEHIND THIS STUDY

The world of nanotechnology is moving from one visionary paradigm toward another. Technology and engineering science of nanotechnology are challenging the scientific landscape. Mankind stands in the midst of deep scientific imagination and introspection. The science of nanotechnology is ushering in a new era in scientific endeavor today. Environmental engineering disasters, the loss of ecological biodiversity, and the domain of chemical process safety are rigorously opening up new avenues of scientific justification and truth in years to come. Nanotechnology is opening up new vistas of scientific endeavor in chemical engineering. Membrane science and nanofiltration are the areas of scientific endeavor today. Mankind's immense scientific prowess, the vast scientific vision behind environmental engineering, and the futuristic vision of water science and technology will all lead a long way in the true emancipation of nanotechnology nanovision of tomorrow. The world of science and technology is moving toward definite forward directions in regard to application of CNTs in chemical engineering. CNTs are the next-generation advanced materials and there has been a tremendous development in this field. Technology advancements, the wide vision of science, and the success of sustainability are the forerunners toward greater application of nanotechnology in chemical engineering today.[3,13,14]

8.4 WHAT DO YOU MEAN BY NANOTECHNOLOGY?

Nanotechnology is the science and technology of small things, in particular, things that are less than 100 nm in size. One nanometer is 10^{-9} m or about three atoms long. For comparison, a human hair is about 60–80,000-nm wide. Nanostructures, objects with nanoscale features, are not new and they were not first created by man. There are many examples of nanostructures in nature in the way that plants and animals have evolved. Similarly, there are many natural nanoscale materials, such as catalysts, porous materials, certain minerals, soot particles, etc. that have unique properties particularly because of the nanoscale features. Now is the era of engineered nanomaterials and devices. Scientific vision and technological cognizance are in the path of new scientific regeneration and deep introspection in regard to nanotechnology applications. Nanotechnology combines solid-state physics, chemistry, electrical engineering, chemical engineering, biochemistry, biophysics, and materials science.[13,14]

Nanotechnology promises to deliver novel products and processes or enhance the performance of the existing ones across sectors. It encompasses interventions in a range of domains such as water, energy, health, agriculture, and environment that could veritably enable solutions to several development-related problems in developing countries. Several industry-related sectors such as pharmaceuticals, electronics, automobiles, textile, chemicals and manufacturing sector, information technology and communications, as well as biotechnology appear poised to be greatly harnessed from nanotechnology science. Human scientific research pursuit today stands in the midst of deep scientific distress and scientific introspection.[13,14]

8.5 SCIENTIFIC DOCTRINE OF CHEMICAL PROCESS ENGINEERING

Chemical engineering is today witnessing new technology dimensions. Scientific truth, scientific fortitude, and scientific girth are the forerunners toward a newer era in chemical engineering science. Alternate energy resources, renewable technology, and novel separation processes are the present research trends in chemical process engineering. Human civilization is moving toward the domain of nanotechnology and advanced materials. Technological profundity is witnessing an immense test of today's visionary timeframe. Membrane science and nanofiltration are the vast domains of effective and futuristic research pursuit. Chemical process engineering stands in the midst of immense scientific vision and scientific contemplation. The technology of novel separation processes is highly advanced today. Membrane science, industrial wastewater treatment, and drinking water treatment also stand in the midst of immense vision and introspection. Global water shortage and its solutions are urging the scientific domain to move toward newer scientific rejuvenation. Heavy metal and arsenic groundwater contamination are major environmental issues today. Here, the importance of membrane science and nanofiltration also comes into play. Drinking water treatment and desalination are the other vast areas of scientific importance. In developed as well as developing world, arsenic drinking water contamination is in a state of immense disaster. Here, the interface of chemical engineering and nanotechnology comes into the picture. CNTs are challenging the entire scientific fabric of nanoscience and nanotechnology. Carbon nanotubes and advanced materials are ushering in a new era of scientific validation and deep forbearance. Engineered materials and its applications are the other side of the visionary coin of nanotechnology.[3,13,14]

Today, the scientific doctrine of chemical engineering needs to be reenvisioned and reenvisaged with the passage of scientific history, scientific vision, and time. Chemical engineering plays a vast role in nanotechnology emancipation. CNTs are an emerging and promising area of chemical engineering endeavor.

8.6 SCIENTIFIC VISION BEHIND NANOTECHNOLOGY APPLICATIONS

Nanotechnology is surpassing every scientific barrier. Scientific vision, scientific prowess, and deep scientific discernment are of utmost necessity in the avenues of scientific success and rigor today. Nanotechnology and chemical process engineering are two opposite sides of the visionary scientific coin. The future of the synergistic effect of nanotechnology and chemical process engineering is wide and bright. Technology has vast and versatile answers to the intricacies of nanotechnology applications in any branch of science and engineering today. The success, challenge, and vision need to be reenvisioned and reenvisaged with the passage of scientific history and scientific forbearance.[3]

8.7 WHAT DO YOU MEAN BY CARBON NANOTUBES (CNTS)?

CNTs are drastically changing the scientific landscape of both nanotechnology and chemical process engineering. This is a promising and scientifically inspiring area of scientific endeavor. CNTs have veritably attracted the fancy of the scientific domain throughout the world. Technological vision, scientific challenges, and the futuristic vision of nanotechnology and chemical process engineering will all lead a long and visionary way in the true emancipation of CNT application.[13,14] CNT is a tube-shaped material, made of carbon, having a diameter measuring on the nanometer scale. A nanometer is one-billionth of a meter or about 10,000 times smaller than a human hair. CNTs are unique because the bonding between the atoms is very strong and the tubes can have extreme aspect ratios.[13,14] Technological and scientific validation are of utmost importance in the avenues of scientific endeavor today. The challenge and the vision of science of nanotechnology are crossing vast and versatile visionary frontiers. In this treatise, the author pointedly focuses on the scientific vision behind the nanotechnology applications in chemical engineering, especially the CNT application in human society.[13,14]

8.8 RECENT SCIENTIFIC RESEARCH PURSUIT IN THE FIELD OF NANOTECHNOLOGY

Nanotechnology research pursuit is the cornerstone of science and engineering today. CNTs are of utmost importance in the avenues of scientific endeavor in our present-day human civilization. Environmental engineering paradigm, the need for environmental protection, and the futuristic vision of chemical engineering will all lead a long and visionary way in the true emancipation of nanotechnology applications. In this treatise, the author repeatedly urges and points toward the vast scientific success and the deep scientific vision in nanotechnology applications.[13,14]

Ong et al.[7] deeply discussed CNTs, in a review, as next-generation nanomaterials for clean water technologies. Advances in nanoscale science and engineering can be used to find a solution to water quality issues using nanomaterials. Technological vision and scientific profundity are witnessing drastic changes. CNTs are a type of nanomaterial that has been extensively studied in the domain of industrial wastewater treatment. The discovery of CNTs has revolutionized and revitalized the future of nanotechnology.[7] Scientific advancements and scientific vision of CNTs are still in the path of deep introspection. CNTs are recognized as highly efficient absorbents for removing heavy metals from water because of their large surface active sites and controlled pore distribution.[7] Adsorption of heavy metals onto CNTs involves complementary steps that may be attributed to physical adsorption, electrostatic attraction, precipitation, and chemical attraction.[7] This review widens scientific fortitude and scientific vision. This review is a vital eye-opener toward the greater understanding of the application of nanotubes in environment protection and chemical engineering.[7]

Hussain[6] has discussed on the subject of carbon nanomaterials (CNMs) as adsorbents for environmental analysis. Environmental engineering and environmental protection are ushering in a newer eon in the field of science and engineering. Nanomaterials, with bodily structures less than 100 nm in one or more dimension, have widely attracted significant attention from scientists in recent years, mainly due to their unique, attractive, thermal, mechanical, electronic, and biological properties. Their high surface to volume ratio, the possibility of surface functionalization, and favorable thermal features provide the flexibility needed for a broad range of analytical applications.[6] This treatise gives a wide glimpse on CNMs for environmental analysis.[6] These are widely considered as one of the promising material for future engineering applications; they can be used in nanocomposite applications. Scientific subtleties and truth are in the midst of deep introspection and vision. CNMs can be used as nanoabsorbents for liquid- as well as gas-phase adsorption of environmental pollutants because of their special nanoscale adsorbent properties.[6] Adsorption is an important separation process in chemical engineering. Mass transfer and reaction phenomenon are widely encompassed in the success of operation of adsorption. The author in this treatise deeply comprehends CNTs, fullerenes, and adsorption on CNMs.[6] The treatise also deeply elucidates on CNMs for preconcentration of environmental pollutants, solid-phase extraction, CNT as solvent-phase extraction material, fullerenes as solid-phase extraction materials, and the wide world of solid-phase microextraction.[6] The author also has touched upon chromatographic applications of CNMs.[6]

Palit[8] with deep and cogent insight has discussed advanced oxidation processes, nanofiltration, and application of bubble column reactor in a review. The author delineates with deep foresight the vision of nanofiltration and the wide domain of membrane science with the sole objective of furtherance of science and technology. The author has touched upon the fouling phenomenon in membrane technology. This treatise gives a vast understanding of the domain of advanced oxidation processes, particularly ozonation. Application of bubble column reactor in environmental engineering is the other cornerstone of this treatise.[8]

8.9 APPLICATION OF NANOTECHNOLOGY IN CHEMICAL ENGINEERING

Nanotechnology has vast and versatile applications in chemical engineering. The scientific and engineering challenges are immense as human civilization trudges a weary path toward a new century. Industrial wastewater treatment, the world of drinking water treatment, and the futuristic vision of membrane science are the forerunners toward a newer visionary eon of science today. Environmental engineering science is in the path of newer scientific regeneration and wide scientific revamping. Membrane science and nanofiltration are the vast avenues of scientific research pursuit. Ultrafiltration, nanofiltration, and microfiltration are the most challenging areas of scientific endeavor. The grave concerns behind environmental sustainability and the futuristic vision of nanotechnology are the torchbearers toward a newer era of chemical process engineering. Chemical engineering is surpassing wide and vast visionary frontiers. Technology and engineering science have no answers to various research questions of chemical engineering and environmental engineering science. Novel separation phenomenon is one of the few answers toward many water-related issues. Global water challenges today stand in the midst of deep catastrophe and scientific forbearance. Human civilization and human research pursuit are faced with the greatest bane—arsenic and heavy metal groundwater contamination. Desalination and nanofiltration are the only answers toward global water issues. This is a major area of scientific vision in nanotechnology applications. Chemical engineering and environmental engineering paradigms are the two opposite sides of the visionary coin of endeavor.[13,14]

8.10 WATER SCIENCE, MEMBRANE SCIENCE, AND NANOFILTRATION

Water science, membrane science, and nanofiltration are the challenging and ever-growing areas of scientific rejuvenation in the present day-to-day human civilization. Mankind is in the state of immense scientific revamping as science and technology move forward. Filtration is defined as the separation of two or more components from a fluid stream based primarily on the size differences. In conventional usage, it usually refers to the separation of solid immiscible particles from liquid or gaseous streams. The primary role of a membrane is to act as a selective barrier. It should allow passage of certain components and retain certain other components of a mixture. By implication, either the permeating stream or retained phase should be enriched in one or more components.[3] In its widest sense, a membrane could be defined as "a region of discontinuity interposed between two phases" or as a "phase that acts as a barrier to prevent the mass movement but allows restricted and regulated passage of one or more species through it."[3] Technology and engineering science of membranes are highly advanced today. Scientific fortitude, scientific girth, and determination are all leading a long and visionary way in true emancipation of membrane science and novel separation processes.[3] A visionary research work describes membrane as a gaseous, liquid, solid, or combinations of these.[3] Membranes can be further classified by (a) nature of the membrane—natural versus synthetic, (b) structure of the membrane—porous versus nonporous, (c) application of the membrane—gaseous-phase separations, gas–liquid, liquid–liquid, etc., and (d) mechanism of membrane separation—adsorptive versus diffusive, ion-exchange, osmotic, or nonselective membranes.[3,13,14]

Water technology today stands in the midst of deep contemplation and scientific truth. The targets and vision of water technology needs to be reenvisioned and restructured in this century. Global water challenges are the major components of mankind's scientific rigor and progress. Desalination science is another area which needs to be pondered upon with the progress of human civilization. Its scientific rigor encompasses basic fundamentals of chemical engineering and chemical technology. Technological profundity is at stake. Scientific cognizance of the entire domain of chemical engineering needs to be reenvisioned and restructured at this stage of 21st century. Chemical process engineering is a vast and ever-growing area of scientific endeavor. In this treatise, the author pointedly focuses on the scientific success, the deep scientific vision, and the ever-growing scientific discernment in the field of nanotechnology applications in chemical engineering.[3,13,14]

8.11 RECENT RESEARCH ENDEAVOR IN CNTS

Research endeavor in the field of CNTs is surpassing vast and versatile scientific boundaries. Nanotechnology and CNTs are challenging the vast scientific landscape of scientific discernment and profundity today. The landmark achievement of science and technology is in the areas of nanotechnology, nuclear science, and space technology. Science is a huge colossus with a vast vision of its own. CNTs is equally a vast area of scientific research pursuit today.

Ajayan et al.[1] discussed with deep and cogent insight about the applications of nanotubes. CNTs have attracted immense attention to the scientific domain in the 21st century. The small dimensions, strength, and the remarkable physical properties of these structures make them a unique material with a whole range of exquisite applications. In this review, the authors discuss some of the important materials science applications of CNTs.[1] Materials science and metallurgical engineering applications are revolutionizing the scientific landscape of human scientific endeavor. The authors deeply discuss the electronic and electrochemical applications of nanotubes, nanotubes as mechanical reinforcements in high-performance composites, nanotube-based field emitters, and their use as nanoprobes in metrology and biological and chemical investigations, and as templates for the creation of other nanostructures.[1] Validation of materials science is of utmost importance in the progress of human civilization today. Scientific and academic rigor, the vast scientific vision, and the futuristic vision will all lead a long and visionary way to the true realization and emancipation of materials science and nanotechnology. The vast challenges that ensue in realizing some of these applications are deeply focused and discussed from the point of view of manufacturing, processing, and cost considerations. Technological vision and validation of science are deeply challenging the scientific fabric of materials science. The uniqueness of the nanotube arises from its structure and the inherent subtlety in the structure, which is the widely accepted helicity in the arrangement of carbon atoms in hexagonal arrays on their surface honeycomb lattices.[1]

Popov[9] discussed with deep foresight and lucidity about the properties and applications of nanotubes. CNTs are unique tubular structures of nanometer diameter and large length/diameter ratio. The nanotubes may consist of one up to tens and hundreds of concentric shells of carbons with adjacent cell separation of approximately 0.34 nm. The carbon network of the shells is closely related to the honeycomb arrangement of the carbon atoms in the graphite sheets. The vast research questions in the CNTs application still remain unanswered. The challenge and vision of science are growing with the progress of scientific and academic rigor in the field of CNTs.[9] CNTs have immense application as central elements in electronic devices including field-effect transistors, single-electron transistors, and rectifying diodes. This chapter is widely researched and opens vast and versatile avenues in the field of materials science and nanotechnology in the decades to come.[9]

CMP Cientifica Report[4] elucidates, in details, a white paper targeting CNTs. The report elucidates the scientific success, deep scientific potential, and scientific vision of the application of CNTs in society. Technology and science are slowly in the process of immense evolution. Nanotechnology has revolutionized the wide scientific fabric of endeavor and determination. The report delineates the function of CNTs and related structures. The authors described single-walled nanotubes, chirality, multi-walled CNTs (MWCNTs), nanohoms nanofibers, CNT production processes, and nanotube applications.[4] The pivotal point of this report was a wider glimpse of CNTs.[4] CNTs were "discovered" in 1991 by Sumio Iijima of NEC and are effectively long, thin cylinders of graphite, which one will be familiar as the material in a pencil or as the basis of lubricants. Graphite is made up of layers of carbon atoms arranged in a hexagonal lattice. Scientist Dr. Richard Smalley who shared the *Nobel Prize* for the discovery of a related form of carbon called buckminsterfullerene. This technology opened up the wide scientific avenues in CNTs. Validation of science, the vast technological vision, and the futuristic vision of materials science are veritably the forerunners toward a newer eon in the field of nanotechnology and CNTs.[4]

Seetharamappa et al.[10] dealt lucidly in details with CNTs as next-generation electronic materials. The wide current interest in CNTs is a direct consequence of the synthesis of buckminsterfullerene in 1985 and its derivatives thereafter. The discovery that carbon could form stable, ordered structures other than graphite and diamond has stimulated researchers in the world for the search of other allotropes of carbon. This further led to another wonderful and promising finding in 1990 that C_{60} could be produced in a simple arc evaporation apparatus, which is readily available in most laboratories. Technological findings and scientific profundity in CNTs are the most important parameters in the success of nanotechnology today. Sumio Iijima discovered fullerene-related CNTs in 1991 using a single evaporator.[10] After that discovery, the scientific world catapulted into a wider visionary era. Man's immense scientific prowess, the technological validation of nanotechnology, and the scientific success of CNTs totally changed the scientific endeavor of nanovision. The word nanotube is derived from their size, because the diameter of a nanotube is on the order of a few nanometers (approximately 50,000 times smaller than the width of a hair) and can be up to several micrometers in length.[10] A nanotube (also known as bucky-tube) is a member of the fullerene structural family.[10] CNTs are cylindrical carbon molecules with novel properties (outstanding mechanical, electrical, thermal, and chemical properties: 100 times stronger than steel; best field emission emitters; can maintain current density of more than 10^{-9} A/cm^2 and thermal conductivity comparable to that of diamond) which make them potentially useful in a wide variety of applications (e.g., optics, nanoelectronics, composite materials, conductive polymers, sensors, etc.). CNTs are of two types, namely, single-walled CNTs (SWCNTs) and MWCNTs. SWCNTs were discovered in 1993 and most of these have a diameter close to 1 nm, with a tube length that may be many thousands of times larger and up to orders of centimeters.[10] Validation of science and scientific forbearance are the cornerstones of research pursuit in nanotube and nanotechnology research today. Technological paradigm is changing at a rapid pace with the passage of scientific history, scientific vision, and visionary timeframe. This chapter redefines and revisits the success of scientific research pursuit in the field of CNTs with the sole objective of advancement of science.

Baughman et al.[2] described lucidly the route toward application of CNTs. The science of nanotubes is entering into a newer era and a newer domain in the vast and versatile domain of nanotechnology.[2] Many potential applications have been proposed for CNTs including conductive and high-strength composites, energy storage and energy conversion devices; sensors; field emission displays and radiation sources; hydrogen storage media; and nanometer-sized semiconductor devices, probes, and interconnects. The authors with deep vision and scientific truth discuss nanotube synthesis and processing, CNT composites, electrochemical devices, hydrogen storage, nanometer-sized electronic devices, and sensors and probes.[2] The vision, challenge, and targets of nanotechnology science are slowly evolving. Human scientific vision is in a state of immense scientific rejuvenation.[2] The vast exponential increase in patent filings and publications on CNTs indicates growing interest that parallels academic and scientific rigor. Application of CNTs in chemical engineering is slowly and drastically surpassing visionary frontiers. This domain of scientific research is latent yet extremely promising. The vision and targets of engineering science need to be reenvisioned and revamped with the course of scientific history and scientific regeneration.[2]

8.12 VALIDATION OF SCIENCE IN NANOTECHNOLOGY

Nanotechnology is in the path of newer scientific regeneration and scientific rejuvenation. Mankind is in the state of immense scientific distress and deep scientific forbearance. Nanotechnology and chemical engineering are the two opposite sides of the scientific coin today. CNTs' applications are vastly changing the scientific landscape. This discovery is a wonder of science. Technology and engineering science are moving toward wider knowledge dimensions with the passage of scientific history, sagacity, and visionary timeframe. The vast challenges of engineering, technology, and science need to be reenvisioned as forays into nanotechnology revolutionize the scientific frontiers. Validation of the science of nanotechnology is of utmost importance in the path toward scientific emancipation. Energy, environment, food, and water are the vital components of civilization's progress. At such a crucial juncture of scientific history and time, mankind's scientific prowess is in a state of immense quagmire. This treatise widely researches on the scientific fortitude and the scientific cognizance in the field of nanotechnology applications in chemical engineering and technology. Scientific validation in environmental engineering is the other visionary avenue of research pursuit. CNTs and its application are the forerunners toward a newer visionary eon of science and technology. The validation of nanotechnology; thus, needs to be readdressed with the progress of human civilization and scientific rigor.

8.13 THE CHALLENGE OF WATER CRISIS AND WATER SHORTAGE

Global water crisis and related research and development initiatives are changing the face of scientific research pursuit in environmental engineering science. Human civilization stands in the midst of vision and introspection. Drinking water treatment, provision of potable water, and the vast research domain of industrial wastewater treatment are opening up new scientific challenges in research pursuit. Science and technology are huge colossus with a wide vision of its own. Shannon et al.[11] elucidated with deep conscience science and technology of water purification and drinking water issues in the coming decades. One of the challenging issues of future of human civilization throughout the world is the inadequate access to clean drinking water and sanitation. Problems with water are expected to grow alarmingly in the coming decades with water scarcity occurring globally even in water-rich countries. Technology and engineering have few answers toward the success of water technology applications. Lower cost and less energy are the pivotal points of scientific research in water science today. In this chapter, the author elucidates some of the science and technology developed to improve disinfection and decontamination of drinking water as well as efforts to increase water supplies through water recycling and zero-discharge norms. The authors rigorously discussed the topics of disinfection, decontamination, reuse and reclamation, and the wide world of desalination. This is a phenomenal area of scientific research pursuit and this chapter succeeds in the presentation of a wide range of water technology in details. Scientific vision in water technology is surpassing wide and vast visionary boundaries.[11] Heavy metal and arsenic groundwater contamination are banes of human civilization and scientific research pursuit.[11] The challenge of water technology is promising and ever-growing. The authors in this treatise discuss the areas of scientific revalidation in water technology with the sole vision of human scientific progress.[11] The vast problems worldwide associated with the lack of clean, fresh water are well known: 1.2 billion people lack access to pure drinking water; 2.6 billion have little or no sanitation; millions of people die annually—3900 children a day—through diseases transmitted by unsafe water or human excreta.[11,12]

8.14 GLOBAL GROUNDWATER REMEDIATION

Heavy metal and arsenic groundwater remediation are the upshot of scientific research in environmental engineering today. Technological sagacity and scientific validation are the cornerstones of human scientific and academic rigor. Groundwater contamination and its removal stand in the midst of deep scientific contemplation and the technological wonders. Science of heavy metal remediation is in the path of scientific revival with newer innovations and rapid discoveries. Developing countries such as India and Bangladesh are in the threshold of an immense scientific disaster. Here comes the need of sound remediation technologies. Hashim et al.[5] delineated lucidly remediation technologies for heavy metal-contaminated groundwater. The contamination of groundwater by heavy metal originating from natural soil sources or from anthropogenic sources is a matter of immense concern to the public health and hygiene. Groundwater remediation stands in the midst of deep scientific research question and definite vision. Science and engineering of groundwater remediation and bioremediation are vastly intricate and extremely far-reaching. In the developing and the developed world, chemical engineers, environmental scientists, geoscientists, geologists, and physicians are taking every step in eradicating this vexing environmental issue. The scientific success, the human scientific endeavor, and the futuristic vision will veritably lead a long and visionary way in the true emancipation and realization of water technology.[5] Remediation of contaminated groundwater is of highest priority since billions of people all over the world use it for drinking purpose and provision of clean drinking water is a veritable challenge to the vast scientific landscape. Hashim et al.[5] elucidated 35 approaches for groundwater treatment in a well-researched review. The remediation has been classified under three categories, namely chemical, biochemical/biological/biosorption, and physicochemical treatment procedures. Selection of a suitable technology for contaminant remediation is a difficult process due to extremely complex soil chemistry and intricate aquifer characteristics. In the past decade, iron-based techniques, microbial remediation, biological sulfate reduction, and various adsorbents played a pivotal role in the true emancipation and true realization of remediation procedure. Mankind's deep scientific vision, the scientific girth and determination, and the scientific fortitude will all lead a wide and visionary way in the true realization of remediation technology and the holistic world of environmental engineering science.[5,11,12]

Human civilization is moving from one visionary paradigm toward another. Technology validation, scientific truth, and the vast scientific vision are the torchbearers toward a newer visionary eon in chemical engineering. Nanotechnology has tremendous applications in chemical engineering, and groundwater remediation and membrane science are some of the effective scientific endeavor. Groundwater remediation is the utmost need of the hour and the scientific domain is gearing up for newer challenges in this respect. The scientific prowess of mankind is at definite stake as civilization marches forward. In this treatise, the author pointedly focuses on the today's nanotechnology applications of chemical engineering in a visionary attempt toward true realization of environmental engineering science. Chemical process engineering and environmental engineering science are two opposite sides of the visionary coin. This treatise gives wide glimpse on the present and future advances in the field of nanotechnology, chemical engineering, and environmental engineering.[12,13,14]

8.15 FUTURE FRONTIERS AND FUTURE FLOW OF THOUGHTS

Future frontiers of science and technology are ever-growing and far-reaching. Nanotechnology is a revolutionary avenue of scientific research pursuit. Scientific truth, scientific cognizance, and scientific sagacity are the pallbearers toward a greater emancipation of nanotechnology science. The future of environmental engineering science and groundwater remediation are the research questions of chemical engineering. Engineering science has few answers to the intricacies of drinking water treatment and industrial wastewater treatment. The future flow of thoughts should be directed toward the scientific success, vision, and truth behind environmental engineering science. Nanotechnology applications in chemical engineering also involve the environmental engineering research questions. Human civilization and research pursuit need to be reenvisioned and reorganized as engineering and science trudge a weary path toward a newer scientific destiny. Water science and technology is the next step toward scientific emancipation. The challenge, vision, and targets need to be overhauled with the progress of scientific and academic rigor in environmental engineering and nanotechnology.[13,14]

8.16 FUTURE RESEARCH TRENDS

Scientific subtleties and deep profundity are in a state of immense scientific difficulties and rejuvenation. The science of nanotechnology is witnessing immense revamping. Human scientific endeavor and the march of human civilization are challenged and need to be reenvisioned with the progress of scientific and academic rigor in the field of chemical process engineering and environmental engineering. In present times targets of science should be directed toward energy and environmental sustainability and holistic sustainable development. Energy, power, and water are the utmost need of human civilization. In such a difficult context of scientific vision, future research trends should be reenvisioned and restructured toward emancipation of basic human needs. Developed and developing countries throughout the world are in the midst of immense economic as well as social crisis and intricacies of scientific endeavor. Scientific and academic rigor in the field of chemical engineering and nanotechnology is witnessing immense regeneration and revamping. This state of mankind needs to be resolved and reorganized. Science and technology are in the state of a new beginning as environmental engineering emancipation ushers in a new eon in research pursuit. Drinking water treatment, desalination, and wastewater treatment are the utmost need of the hour. Social and economic sustainability are the burning issues facing human civilization today. Thus the need for a holistic beginning toward scientific research trends.

8.17 SUMMARY, CONCLUSION, AND SCIENTIFIC PERSPECTIVES

Human civilization and its scientific perspectives are undergoing drastic challenges and revolutionary changes. Scientific forbearance and scientific fortitude are the need of the hour today. Nanotechnology applications in chemical engineering and CNTs are the wonders of science and technology. The success of scientific endeavor needs to be restructured and revamped with every step of today's scientific challenges and scientific forays. In this treatise, the author pointedly focuses on the visionary achievements in the field of nanotechnology, CNTs, and chemical engineering. Scientific perspectives need to be restructured and revamped with the ever-growing concerns for environmental and energy sustainability. The challenges of science are immense and groundbreaking. In this treatise, the author rigorously points out toward the vast scientific vision, the feasible solutions, and the technological profundity in nanotechnology applications. Sustainability is the issue which needs to be enshrined and envisaged with the passage of scientific history, scientific sagacity, and visionary timeframe. This chapter opens up a new generation of scientific thoughts and scientific cognizance in a deep effort toward today's nanotechnology applications. Perspectives of science and engineering are wide and bright. This chapter redefines and revisits the world of nanotechnology and environmental engineering and opens up a new chapter in the field of scientific emancipation. Technological vision and sagacity stands in the midst of deep introspection and contemplation. Nanotechnology and CNTs are new avenues in scientific introspection. This treatise mainly redefines and reasserts, in a widely watershed text, the scientific success, the scientific brilliance, and the immense world of challenges in tackling nanotechnology issues today.

ACKNOWLEDGMENT

The author with great respect acknowledges the contribution of his late father Shri Subimal Palit, an eminent textile engineer from India who taught the author the rudiments of chemical engineering.

KEYWORDS

- **nanotechnology**
- **carbon**
- **nanotubes**
- **vision**
- **overview**

REFERENCES

1. Ajayan, P. M.; Zhou, O. Z. Application of Carbon Nanotubes, (Chapter) Carbon Nanotubes. In *Topics in Applied Physics;* Dresselhaus, M. S., Dresselhaus, G. Avouris, Ph., Eds.; Springer-Verlag: Berlin, Heidelberg, Germany, 2001; Vol. 80, pp 391–425.
2. Baughman, R. H.; Zakhidov, A. A.; De Heer, W. A. Carbon Nanotubes—The Route Towards Application. *Science* **2002,** *297*, 787–792.
3. Cheryan, M. *Ultrafiltration and Microfiltration Handbook;* Technomic Publishing Company Inc.: USA, 1998.
4. CMP Cientifica, Nanotubes White Paper, 2003.
5. Hashim, M. A.; Mukhopadhyay, S.; Sahu, J. N.; Sengupta, B. Remediation Technologies for Heavy Metal Contaminated Groundwater. *J. Environ. Manage.* **2011,** *92,* 2355–2388.
6. Hussain, C. M.; Carbon Nanomaterials as Adsorbents for Environmental Analysis, Chapter 14. In *Nanomaterials for Environmental Protection;* Kharisov, B. I., Kharissova, O. V., Rashika Dias, H. V., Eds.; John Wiley: USA, 2014; pp 217–236.
7. Ong, Y. T.; Yee, K. F.; Yeang, Q. W.; Zein, S. H. S.; Tan, S. H. Carbon Nanotubes: Next Generation Nanomaterials for Clean Water Technologies, Chapter 8. In *Nanomaterials for Environmental Protection;* Kharisov, B. I., Kharissova, O. V., Rashika Dias, H. V., Eds.; John Wiley: USA, 2014; pp 127–142.
8. Palit, S. Advanced Oxidation Processes, Nanofiltration and Application of Bubble Column Reactor, Chapter 13. In *Nanomaterials for Environmental Protection;* Kharisov, B. I., Kharissova, O. V., Rashika Dias, H. V., Eds.; John Wiley: USA, 2014; pp 207–215.
9. Popov, V. Carbon Nanotubes: Properties and Application. *Mater. Sci. Eng.* **2004,** *R43,* 61–102.
10. Seetharamappa, J.; Yellappa, S.; D'Souza, F. Carbon Nanotubes: Next Generation of Electronic Materials. *Electrochem. Soc. Interface* **2006,** *15*(2), 23–25 and 61.
11. Shannon, M. A.; Bohn, P. W.; Elimelech, M.; Georgiadis, J. G.; Marinas, B. J.; Mayes, A. M. *Science and Technology for Water Purification in the Coming Decades;* Nature Publishing Group, 2008; pp 301–310.
12. Van der Bruggen, B.; Manttari, M.; Nystrom, M. Drawbacks of Applying Nanofiltration and How to Avoid them: A Review. *Sep. Purif. Technol.* **2008,** *63,* 251–263.
13. www.google.com.
14. www.wikipedia.com.

CHAPTER 9

PROGRESS IN POLYMER NANOCOMPOSITES FOR ELECTROMAGNETIC SHIELDING APPLICATION

RAGHVENDRA KUMAR MISHRA[1,2,*], SRAVANTHI LOGANATHAN[3], JISSY JACOB[3], PROSANJIT SAHA[4], and SABU THOMAS[1,3]

[1]*International and Inter University Centre for Nanoscience and Nanotechnology, Mahatma Gandhi University, Kottayam, Kerala, India*
[2]*Indian Institute of Space Science and Technology, ISRO, Thiruvananthapuram, Kerala, India*
[3]*School of Chemical Sciences, Mahatma Gandhi University, Kottayam, Kerala, India*
[4]*Dr. M. N. Dastur School of Materials Science and Engineering, Indian Institute of Engineering Science and Technology, Shibpur, Howrah 711103, India*

[*]*Corresponding author. E-mail: raghvendramishra4489@gmail.com*

CONTENTS

ABSTRACT

Polymer nanocomposites are being extensively used for electromagnetic interference shielding application. The current chapter is focused on investigating the benefits of utilizing foam structure, elastomer, and thermoset polymer matrices for electromagnetic interference shielding purpose. The effect of carbonaceous fillers such as carbon nanotubes and graphene on the electromagnetic interference shielding properties of respective polymer matrices is discussed in detail. The factors that influence on increasing electrical conductivity for polymer nanocomposites are pressed. The structure–property relationship for different types of polymer nanocomposites is given main importance. Apart from carbon-based fillers, the importance of metallic magnetic nanofillers on improving the electromagnetic interference shielding properties of elastomer nanocomposites is also highlighted. Comparative analysis of electromagnetic interference shielding properties of several polymer nanocomposites is carried out.

9.1 INTRODUCTION

Electrically conducting polymer composites are being increasingly used for several applications such as electromagnetic interference shielding, electrostatic dissipation as well as painting, capacitors with great charge storage ability, and integrated circuits. The electrically conducting polymer composites generally consist of nanofillers with electrical conduction feature and offer various advantages that include cheaper cost, low weight, easy processing capability, and conductivities with tuning ability.[1–4]

The utilization of high content of nanofillers is in practice to achieve electrically conducting polymer composites with percolation threshold and excellent electrical conductivity characteristics. However, the presence of high amount of nanofillers in the polymer matrices often leads to tedious processing activities, poor mechanical characteristics, and expensive product cost. In order to overcome high product cost as well as difficulty in processing, it is necessary to reduce the magnitude for percolation threshold which, in turn, becomes a significant step for fabrication of high-quality electrically conducting polymer composites. The most effective way considered to minimize the magnitude of percolation threshold is to form "segregated network" and alignment of nanofillers. The segregated network can be formed by enabling selective dispersion of nanofillers and concentrating the presence of fillers at interfacial domain.[5–12] The drawbacks in connection with the segregated network-based electrically conducting polymer composites include (i) the presence of fillers at the interface prevents diffusion among polymeric molecules and therefore affects the mechanical characteristics and (ii) the occurrence of possibility for damaged segregated structure through post-processing.[13–15] The main aim of this chapter is to focus the electromagnetic shielding performance of conducting polymer composites. The important factors that should be considered to achieve electrically conducting polymer composites with effective electromagnetic magnetic interference shielding capacity are optimized electrical conductivity and dielectric and magnetic properties. In general, it is regarded that electrically conducting polymer composites with electrical conductivity in the range of 1 S/m can exhibit excellent electromagnetic interference shielding property. It is highly difficult to attain the electrically conducting polymer composites with least electrical conductivity, below 1 S/m, which is due to the presence of a relatively greater concentration of nanofillers. In order to make electrically conducting polymer composites useful for commercial application, electromagnetic interference shielding effectiveness value lesser than 20 dB should be achieved.[16–18] The electrically conducting polymer composites containing multiwalled carbon nanotubes (MWCNTs) are reported to exhibit an electromagnetic interference shielding effectiveness value of ~20 dB.[19,20] However, such an effective electromagnetic magnetic interference shielding value is achieved only at higher loadings (~7 wt.%) of MWCNTs in initial attempts. In the later years, polystyrene (PS)-based nanocomposites containing a combination of MWCNTs and graphite nanoplatelets (GNPs) as reinforcement materials are developed through typical method. Through the peculiar method adopted, it became possible for the researchers to fabricate PS-based electrically conducting polymer composites with an effective electromagnetic interference shielding property of ~20.2 dB, which, in turn, contained only 2 wt.% of MWCNTs and ~1.5 wt.% of GNPs.[19,20]

Apart from excellent electromagnetic interference shielding effectiveness offered by electrically conducting polymer composites, it is also recommended to have percolation threshold in the lower range along with easy processing conditions; the conducting polymer composites with higher filler cause the processing of composites. It is quite challenging to fabricate electrically conducting polymer composites with a combination of effective electromagnetic interference shielding feature (reflection or absorption or both reflection and absorption) combined with lower percolation threshold.[21,22] At present, electrically conducting polymer composites with high electromagnetic interference shielding property are also being developed through co-continuous combination method. Through this method, it is possible to obtain electrically conducting polymer composites with a hierarchic configuration which, in turn, is achieved through melt processing of blends.[23–27] The high aspect ratio as well as electrical conductivity properties enabled by carbon-based nanofillers such as SWCNTs or MWCNTs, graphene nanoplatelets, and carbon-based nanofibers make them attractive to be used as reinforcements as well as fillers for fabrication of electrically conducting polymer composites useful for energy and electronic field.[18,28–30]

Several research articles are reported for carbon filler-incorporated polymer nanocomposites for electromagnetic interference shielding application.[30–32] It is noticed that the nanocomposites demonstrated improved electromagnetic interference shielding characteristics with respect to increase in nanofiller content as well as the thickness of the shielding material.[33] Studies correlating structure–property relationship for graphene nanoplatelets-reinforced polymer nanocomposites are also increasingly reported.[34–39] Figure 9.1 shows the various forms of carbon-based structures.

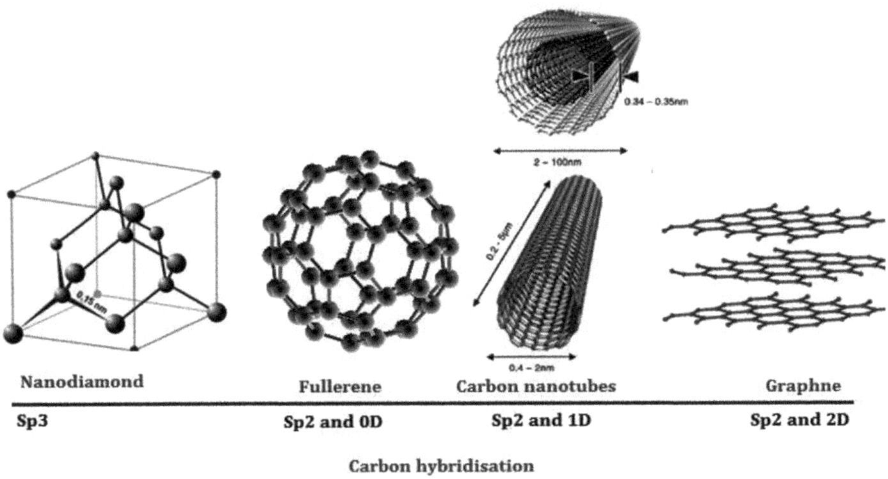

FIGURE 9.1 Various forms of carbon-based structures.

Figure 9.2 shows the scanning electron microscopy (SEM) morphology for CNTs-reinforced epoxy composites obtained in charge contrast mode. While obtaining the SEM images, different parameters such as the mechanism of imaging, penetrating depth and conductivity of the sample, as well as nature of SEM detector were studied. It is declared that there is a possibility to view CNTs even when the operating voltage is below 1 kV. This finding is considered to be highly significant because it is possible to avoid the charging problems associated with the CNTs-based sample when the SEM is operated at high-voltage conditions.[35,40]

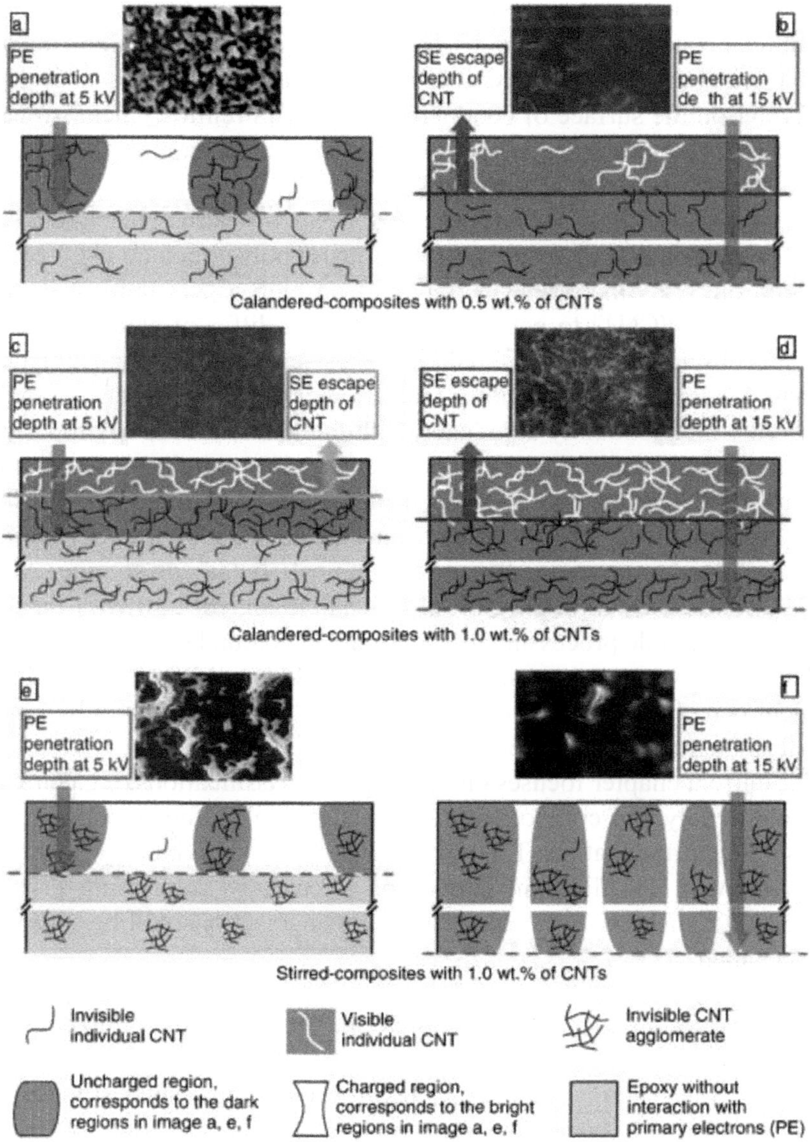

FIGURE 9.2 (See color insert.) Scanning electron microscopy (SEM) images for carbon nanotube (CNT)-reinforced epoxy calendered and stirred nanocomposites obtained using charge contrast mode.

Source: Adapted with permission from ref 35. © 2014 Elsevier.

In order to achieve lower percolation threshold property for electrically conducting polymer composites, attempts were made to localize the conducting reinforcement at the interfacial region. For example, an attempt is made to enable localization of MWCNTs at the interfacial region polyamide (PA)/ethylene–methyl acrylate-based blends obtained through random copolymerization. Interestingly, it is reported that there occurs absorption of polymer on the surface of conductive MWCNTs reinforcement irreversibly. Such an absorption phenomenon is observed to occur mainly due to the thermodynamic events and filler shape.[35]

A comparative investigation of the performance of carbon black (CB) and MWCNT-reinforced poly(styrene acrylonitrile)/polycarbonate (PC) blends is carried out.[41–48] Since MWCNTs possessed high aspect ratio, it is quite easy for the MWCNTs to get transferred among different phases. Interestingly, it is observed that CB offered higher resistance toward transfer among different phases. Hence, confined localization of CB at the interfacial region is noticed which is mainly associated with the low aspect ratio property of CB. The CB being spherical in shape with low aspect ratio exhibited more stability at the interfacial region which can be seen from Figure 9.3. Efforts are also undertaken to selectively distribute CNTs in the interfacial region of polymeric blends comprising PC and acrylonitrile-butadiene-styrene (ABS). This is indeed tried through grafting of maleic anhydride (MA) on ABS followed by suitable processing methodology. It is found that the electrically conducting polymer composites made of PC/MA-g-ABS/CNTs exhibited excellent percolation threshold property which, in turn, is driven by kinetic as well as thermodynamic aspects.[41–48]

The current chapter focuses on the different classifications of electrically conducting polymer composites fabricated for electromagnetic interference shielding application. The structure–property correlation relationship associated with the different classes of electrically conducting polymer composites is discussed in detail. The following sections will be addressed in this chapter.

i. Polymer-based foam structures for electromagnetic interference shielding
ii. Polymer-based nanocomposites for electromagnetic interference shielding
iii. Elastomeric polymer-based nanocomposites for electromagnetic interference shielding
iv. Thermoset polymer-based nanocomposites for electromagnetic interference shielding

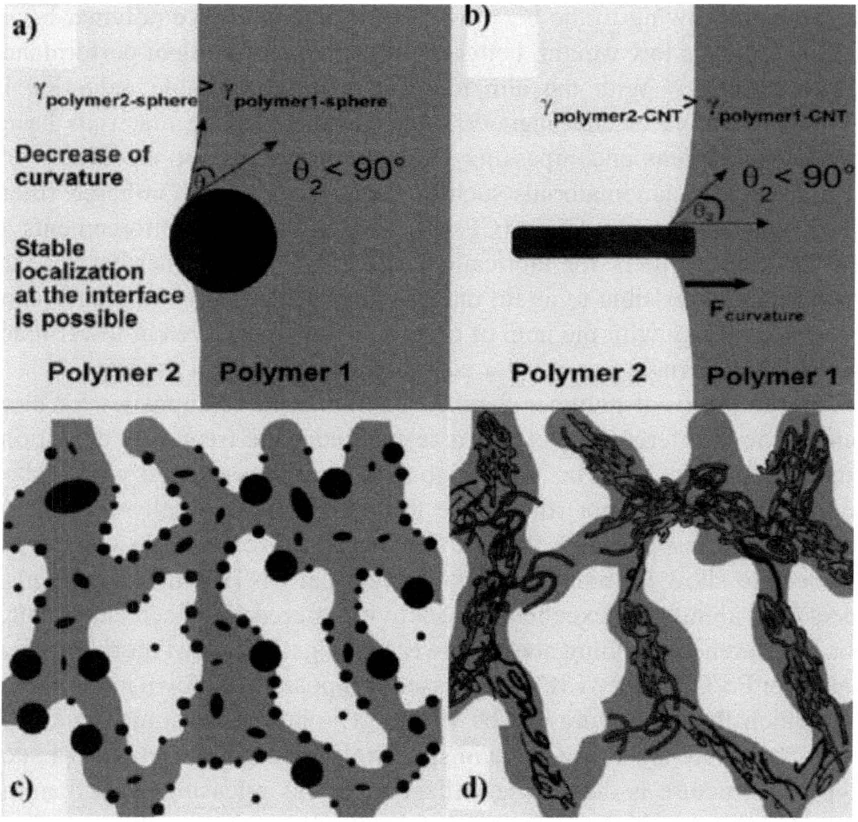

FIGURE 9.3 (a) Carbon black (CB) possessing low aspect ratio located at the interfacial region in polymeric blend after melt processing, (b) CNTs possessing large aspect ratio located at the interfacial region in polymeric blend after melt processing, (c) hypothetical distribution profile for CB with low aspect ratio in the polymeric blend, and (d) hypothetical distribution profile for carbon nanotubes with high aspect ratio in the polymeric blend.

Source: Adapted with permission from ref 35. © 2014 Elsevier.

9.2 POLYMER-BASED FOAM STRUCTURES FOR ELECTROMAGNETIC INTERFERENCE SHIELDING

At present, there is a significant increase in electromagnetic interference shielding issue which is in connection with enhanced growth observed in wireless networks as well as portable digital electrical and electronics system.[49–52] In order to overcome this problem, research activities on the fabrication of conductive polymer-based materials for electromagnetic interference shielding application are being carried out in a rapid phase.[18,53–55]

This is mainly owing to the benefits offered by conductive polymer-based materials such as low weight, better processing, and excellent performance characteristics.[56–59] With the aim to achieve a considerable reduction in weight, innovative electromagnetic interference shielding materials based on polymeric foams encompassing porous morphology are invented.[16,59–64] Several types of nanomaterials such as metals[65,66] and carbon-based fillers [CB, GNPs, and carbon fibers (CFs)][67–70] are utilized as reinforcements as well conducting fillers for fabrication of conductive polymer-based foam structures. It is possible to attain the required electromagnetic interference shielding property with the help of carbon-based fillers even at lower loadings which, in turn, paves way for economically beneficial aspects.[71–73]

The suitability of polymer foam/carbon nanofiber composites for electromagnetic interference shielding is tested under the frequency conditions ranging from 8 to 12 GHz.[20] The high aspect ratio associated with carbon nanofiber played a major role in the formation of electrically conducting network. With respect to increase in frequency and conductivity, skin depth is noticed to show decrement. It is observed that CNTs which exhibit high aspect ratio along with excellent conductivity offered enhanced electromagnetic interference shielding at even lower loadings. The SEM morphological images for PS/CNTs (5 wt.%) foam-based composite are shown in Figure 9.4 from which foam structure can be visible. The bubbles that are developed in foam structure are found to be in spherical shape. The formation of such a type of structure is due to the utilization of gas releasing solid foaming agent. This agent which produces foam is found to present stability at room temperature. However, the integrity of foam structure is lost at high-temperature conditions. The PS-/CNTs-based foam structure containing uniform pore size is developed by using compression molding process.

In general, conventional polymeric foams consist of only two state which includes (i) solid state formed by polymer and (b) gaseous state formed by a solid foaming agent. Unlike conventional polymeric foams, the foam-based composite includes two different solid phases, namely (i) PS and (ii) CNTs. From Figure 9.4b, it can be visualized that CNTs are distributed uniformly in PS matrix and interconnected network emerged out of CNTs. Such an interconnected network made of CNTs forms the conductive route in the PS/CNTs foam composite system which, in turn, acts as electromagnetic interference shielding material.[74] Table 9.1 shows the comparative analysis of electromagnetic interference shielding effectiveness for PS/CNTs and PS/carbon nanofiber foam composites. It can be seen from Table 9.1 that both the composites differ in their electromagnetic interference shielding effectiveness. The difference in effectiveness values exhibited by these composites arose from shorter diameter and high aspect ratio possessed by CNTs as compared to carbon nanofibers.

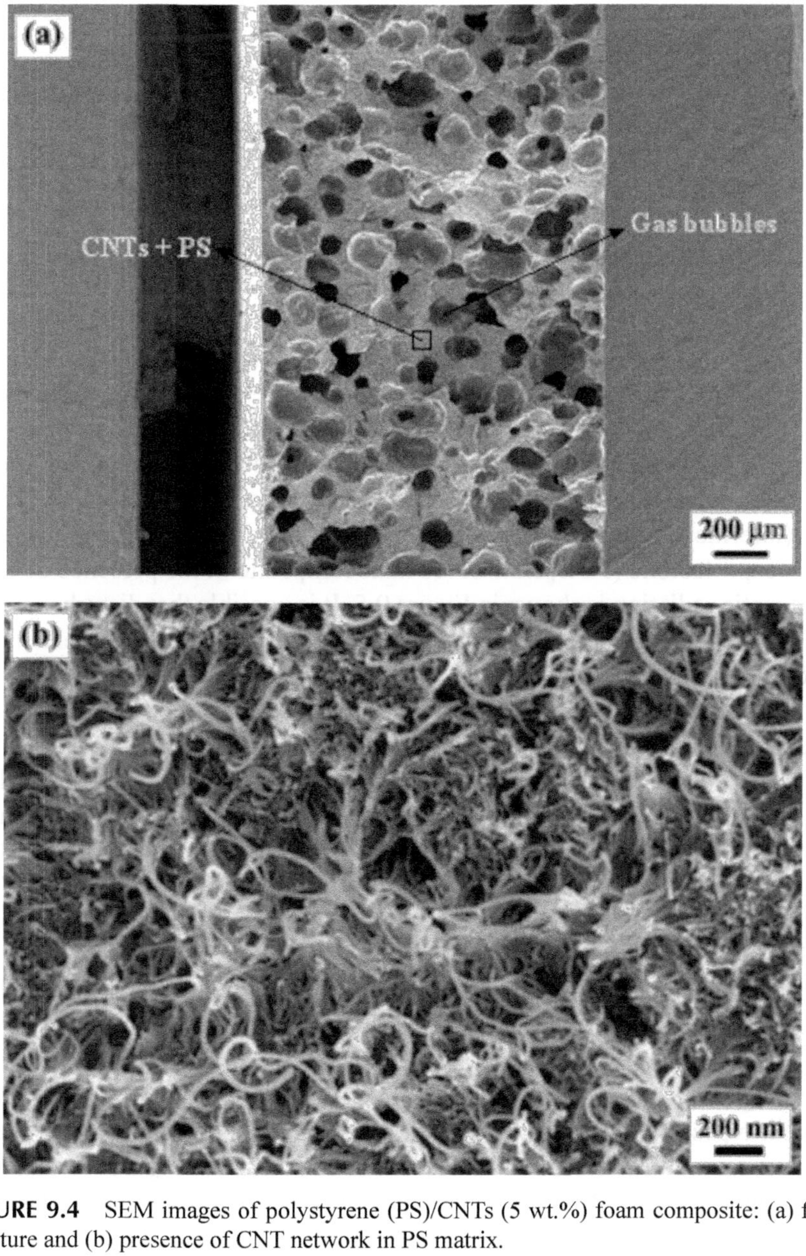

FIGURE 9.4 SEM images of polystyrene (PS)/CNTs (5 wt.%) foam composite: (a) foam structure and (b) presence of CNT network in PS matrix.

Source: Adapted with permission from ref 74. © 2005 American Chemical Society.

TABLE 9.1 Comparative Analysis of Electromagnetic Interference Shielding Effectiveness for Polystyrene (PS)/Carbon Nanotube (CNT) and PS-/Carbon Nanofiber-Based Foam Composites. (Source: Adapted with permission from ref 74. Copyright (2016) American Chemical Society)

Content of CNTs and carbon nanofibers (wt.%)	Electromagnetic interference shielding effectiveness for PS/CNT foam composite (dB)	Electromagnetic interference shielding effectiveness for PS/carbon nanofiber foam composite (dB)
0.5	2.84	0.41
1	5.73	0.73
5	10.30	3.09
7	18.56	8.53

The highly electrically conducting nanofillers used in polymer matrices show a significant drawback by reflecting the incident electromagnetic radiation instead of absorption. In spite, nanofillers do not allow the propagation of wave to cross the nanocomposite, prevalence of electromagnetic waves inside the composite due to numerous reflections that arise from the inner walls (interfacial reflection due to the high surface area). The behavioral action of polycaprolactone (PCL)/MWCNTs nanocomposite in the presence of electromagnetic radiation is shown in Figure 9.5. It is possible that disturbance can occur to certain components such as resistors, transistors present in electronic devices, and signal as a result of electromagnetic interferences. In order to nullify such adverse actions, it is necessary to absorb the electromagnetic waves by the composite material.[73–75] For PCL/MWCNTs nanocomposites, it is observed that the shielding effectiveness is greater than 25 dB which is indicative of higher absorption or lower reflection. The incident power which enters the PCL/MWCNTs nanocomposites is partly dissipated or completely absorbed by the material and hence higher absorption results. The PCL/MWCNTs nanocomposites are thus reported to exhibit dual role by acting as a good absorbent and excellent shielding material against electromagnetic interference. A schematic representation of electromagnetic interference shielding and electromagnetic interference absorber materials is shown in Figure 9.5.

Figure 9.6 presents SEM images for PCL foam-based nanocomposite containing MWCNTs at different weight loadings (0, 0.1, and 0.2 vol.%). Two different types of methods such as melt blending and coprecipitation are used for the preparation of PCL/MWCNTs. It can be seen from Figure 9.6 that neat PCL shows open-cell structure containing pores of size greater than 100 μm. In case of PCL/MWCNTs (0.2 vol.%) foam composites, pore size is comparatively smaller than neat PCL. However, increase in cell density for nanocomposite is observed which, in turn, is due to the increment in internal viscosity. Therefore, MWCNTs play an effective role as nucleating agent and help to attain increased cell density. Upon addition of 0.222 vol.% MWCNTs, the porous morphology is better defined with smaller pores and a higher cell density.[75]

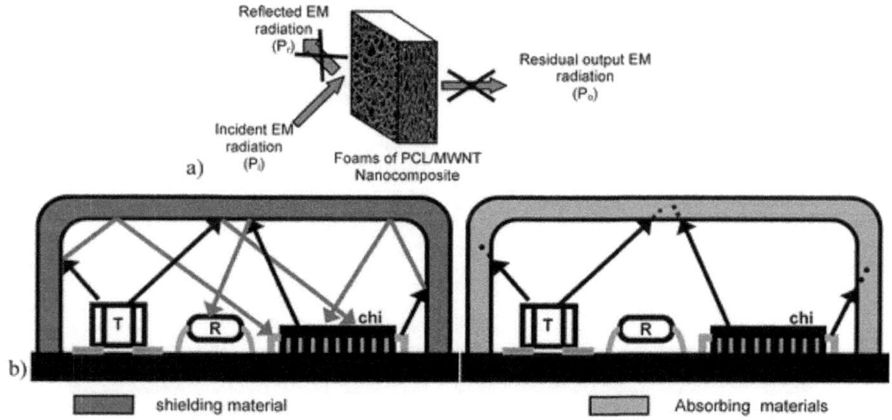

FIGURE 9.5 (a) Behavioral action of polycaprolactone (PCL)/MWCNTs nanocomposite in the presence of electromagnetic radiation and (b) differences among an electromagnetic interference shielding and an electromagnetic interference absorber material.

Source: Adapted with permission from ref 76. © 2008 Royal Society of Chemistry.

The CNTs with lesser size leads to the greater interfacial area and therefore conductivity for the interconnected network formed by CNTs gets improved enormously. In addition to this, the higher aspect ratio possessed by CNTs also improves the conductivity and favors transport of electrons in the polymeric foam-based nanocomposite structure even at lower loadings.[74] In fact, the efficiency of electromagnetic interference shielding for a nanocomposite material is dependent on many factors such as (i) conductivity of filler, (ii) aspect ratio, (iii) smaller diameter, (iv) dielectric constant, (v) magnetic property, and (vi) dispersion and distribution. As CNTs are smaller in diameter and exhibit high aspect ratio along with excellent mechanical strength, they are regarded as potential nanofillers for fabrication of polymer foam-based conductive nancomposites for electromagnetic interference shielding. PS-/MWCNTs-based foam nanocomposites are shown to achieve an electromagnetic shielding efficiency of ~20 dB at 7 wt.% loadings of MWCNTs.[74] In case of poly(methyl methacrylate) (PMMA)-/MWCNTs-based foam nanocomposites, ~27 dB of electromagnetic interference shielding efficiency is achieved at 40 wt.% loadings of MWCNTs.[77]

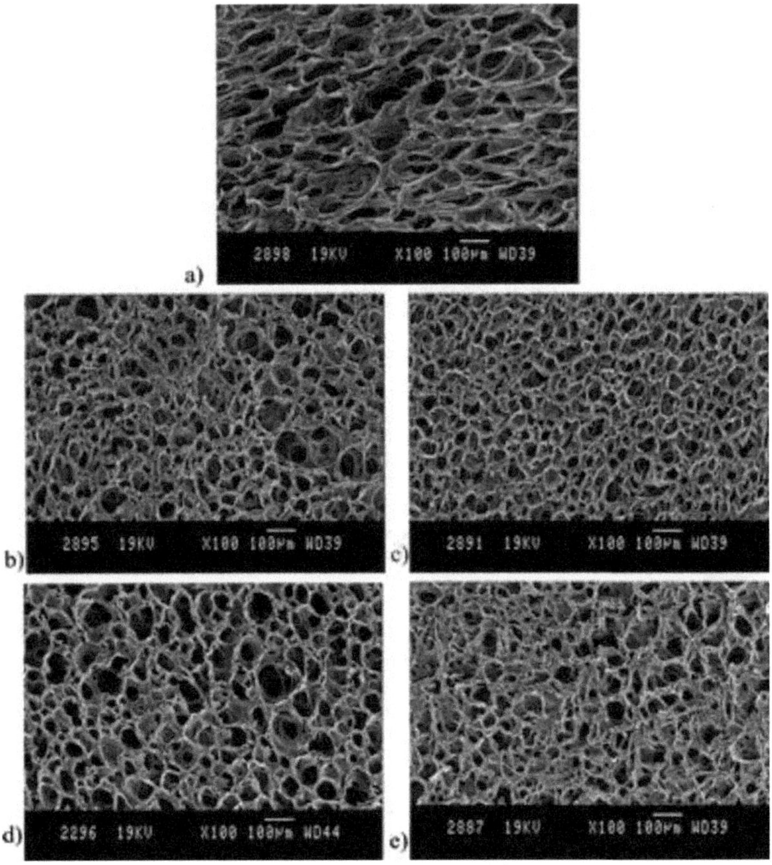

FIGURE 9.6 SEM images for (a) PCL, (b) PCL/multiwalled carbon nanotubes (MWCNTs) (0.1 vol.%) prepared by melt blending, (c) PCL/MWCNTs (0.2 vol.%) prepared by melt blending, (d) PCL/MWCNTs (0.1 vol.%) prepared by co-precipitation, and (e) PCL/MWCNTs (0.2 vol.%) prepared by co-precipitation.

Source: Adapted with permission from ref 76. © 2008 Royal Society of Chemistry.

Apart from CNTs, graphene is regarded as the most advanced nanofiller due to its astonishing electrical, mechanical, as well as thermal characteristics. Such type of special characteristics makes graphene to find its significant position in order to fabricate multifunctional application.[78–80] Particularly, graphene is provided with special consideration to be used as nanofiller for development of polymeric foam-based nanocomposites suitable for electromagnetic interference shielding application.[77,81,82] The graphene-reinforced polymeric foam-based nanocomposites are reported to be electrically conductive and exhibit corrosion resistance as well as tunable conductivity properties.[2,83–86] If the pore size of polymer/graphene foam composites is bigger in size, then there is a great possibility to develop brittleness due to the formation of cracks. The PMMA/graphene microcellular foams are reported to show improved toughness and fatigue life.[60] The morphological images for PMMA/graphene microcellular foams are shown in Figure 9.7. Table 9.2 shows bulk density values for bulk PMMA/graphene and PMMA/graphene foam composites.

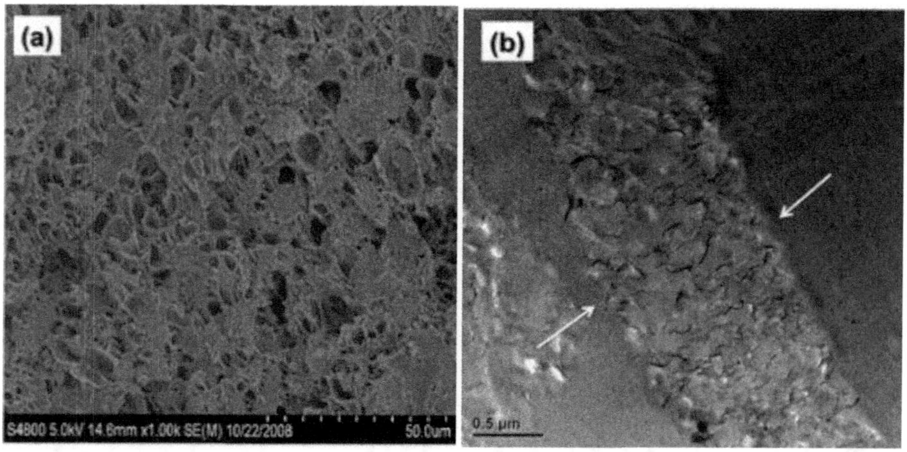

FIGURE 9.7 (a) SEM image for poly(methyl methacrylate) (PMMA)/graphene microcellular foam and (b) transmission electron microscopy (TEM) image for PMMA/graphene microcellular foam. The arrows represent cell wall surfaces.
Source: Adapted with permission from ref 60. © 2011 American Chemical Society.

The polyethylenimine (PEI)/graphene foam-based nanocomposite developed through phase separation process induced by water vapor is reported.[87,88] The schematic representation of water vapor-induced phase-separated PEI/graphene foams is shown in Figure 9.8. Initially, graphene is uniformly dispersed in PEI matrix through both stirring and sonication. Upon exposure of PEI/graphene to air, induction of cell nucleation takes place. The cell nucleation is induced due to phase separation process as dimethylformamide (DMF) and water diffuse into dispersion and air, respectively. Thereafter, the size of nuclei continues to grow with further progress in the diffusion of solvent. When the maximum expansion of nuclei takes place, coalescence among adjacent bubbles may occur. Soon after the DMF is removed from the cells, the formation of PEI/graphene foam-based nanocomposites is observed.

TABLE 9.2 Density Values for Bulk Poly(methyl methacrylate) (PMMA)/Graphene and PMMA/Graphene Foam Composites. (Source: Adapted with permission from ref 60. Copyright (2016) American Chemical Society)

Graphene content in bulk (wt.%)	Graphene content in bulk (vol.%)	Density of bulk (g·cm−3)	Graphene content in foam (vol.%)	Density of foam (g·cm−3)
0.5	0.3	1.19	0.2	0.65
1.0	0.5	1.20	0.3	0.58
1.2	0.6	1.20	0.4	0.67
2.0	1.1	1.20	0.6	0.65

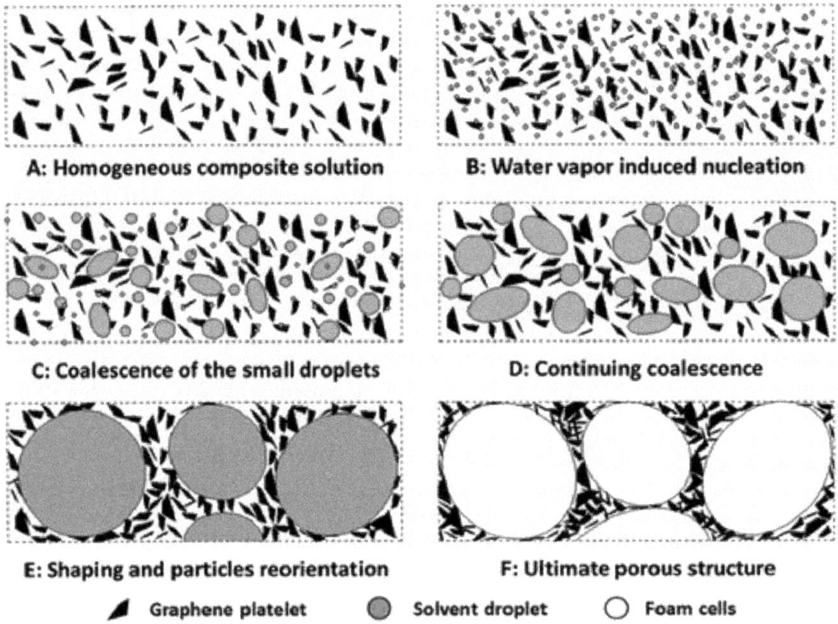

FIGURE 9.8 Schematic representation of polyethylenimine (PEI)/graphene foam-based nanocomposites through water vapor-induced phase separation process.

Source: Adapted with permission from ref 88. © 2013 American Chemical Society.

It is noticed that PEI foam exhibits a density of 0.28 g/cm^3. After addition of 10 wt.% graphene in the PEI matrix, there is no change in the density for PEI/graphene foam-based nanocomposites. In case of PC/nanosilica foam-based composites prepared using a physical blowing agent, it is observed that the density of the composites increased to 0.92 g/cm^3, whereas the density of pure PC is noticed to be 0.75 g/cm^3. It is concluded that addition of 9 wt.% of nanosilica in PC matrix tends to enhance the density of pure PC foam which resulted owing to improved stiffness property of PC matrix.[17] Comparative analysis of both the methodologies used for fabrication of polymer foam-based composites indicates the fact that water vapor-induced phase separation process helps in producing lighter weight foams as compared to the other method. In pure PEI foam, the cell size is evaluated to be 15.3 μm. With the addition of 1 wt.% of graphene, the cell size increased up to 16.6 μm. Upon an increase in loading of graphene further, it is observed that the cell size tends to decrease. In case of 10 wt.% of graphene in PEI matrix, the cell size of ~9 μm is reported. The decrease in cell size at higher loadings of graphene is due to result of enhanced viscosity and poorer dispersion.[87,88]

The SEM images for pure PEI and PEI/graphene foam-based composites are shown in Figure 9.9. In case of 1 wt.% loading of graphene, cell wall appears to be smooth. For 7 wt.% loading of graphene, it is observed that porous structure is present in the nanocomposite foam structure due to the formation of voids. With further increase in graphene loading, it is noticed that gaps are also visualized among graphene and PEI matrix along with voids. The gaps arise owing to the fact that dispersion of graphene is poorer in PEI matrix at higher loadings.[60]

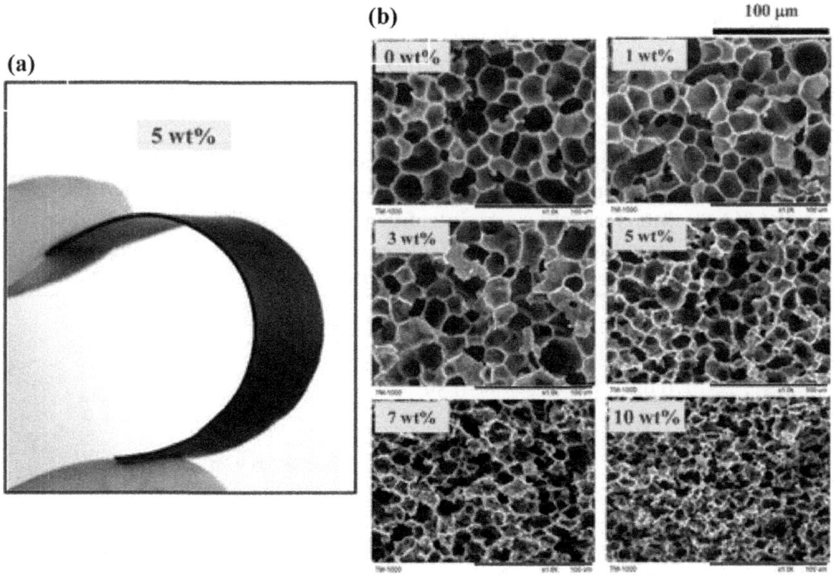

FIGURE 9.9 Optical image for PEI/graphene (5 wt.%) foam-based nanocomposite and (b) SEM images for PEI/graphene foam-based nanocomposites with respect to graphene loadings.

Source: Adapted with permission from ref 88. © 2013 American Chemical Society.

9.3 POLYMER NANOCOMPOSITES FOR ELECTROMAGNETIC SHIELDING

In the recent years, there is considerable utilization of CNTs for imparting electromagnetic interference shielding characteristics for polymers such as PS,[74] PMMA,[77] PC,[89] and PCL.[90,91] Table 9.3 shows the theoretical as well as experimental electromagnetic interference shielding efficiency values for PC/MWCNTs nanocomposites as a function of MWCNTs loading. It can be seen from the table that PC/MWCNTs (0.7 wt.%) nanocomposites exhibit electromagnetic shielding effectiveness value of ~15 dB both theoretically and experimentally at a frequency 0.6 GHz. Figure 9.10 shows the tensile strength (TS) results for PC/MWCNTs with respect to MWCNTs loading. It can be interpreted from the figure that TS increases up to ~10% for 2.5 wt.% loading of MWCNTs. With further increase in CNTs loadings, it is found that TS for PC/MWCNT nanocomposites decreases as compared to neat PC. The ductility decreases and brittleness increases for PC/MWCNT nanocomposites at higher loadings of MWCNTs. When MWCNTs content in PC matrix increases, the mobility of polymer chains and felicity are restricted and hence stiffness for nanocomposites gets increased.[89]

TABLE 9.3 Theoretical and Experimental Electromagnetic Interference Shielding Effectiveness for Polycarbonate (PC)/Mutiwalled Carbon Nanotube (MWCNT) Nanocomposites as a Function of MWCNTs Loading. (Source: Reproduced with permission from ref 89. © 2006 Springer.)

S. No.	MWCNT content (wt.%)	Theoretical electromagnetic interference shielding effectiveness (dB)	Experimental electromagnetic interference shielding effectiveness at 0.6 GHz (dB)	Experimental electromagnetic interference shielding effectiveness at 1.2 GHz (dB)
1	0	0	0	0
2	0.5	2.0×10^{-9}	2.0×10^{-2}	2.1×10^{-1}
3	1.5	0.08	0.59	0.83
4	2.5	1.28	1.94	2.04
5	4.5	4.03	4.13	4.40
6	7.0	15.49	15.82	15.95

In case of polymer nanocomposites, the pattern for glass transition temperature (T_g) can be changed. Certain factors including the size of filler and its content as well as aspect ratio could bring changes in glass transition pattern for polymer nanocomposites[92,93] Several polymer nanocomposites such as poly(vinyl alcohol)/silica, PC/CF, and PMMA/silica systems exhibiting double T_g pattern are been reported in the literature.[92,94] The presence of double T_g for polymer nanocomposites is related to two concurrent facts. The initial T_g recorded depicts the change of polymer into a glassy state. The T_g that could be noticed at higher temperature conditions is associated with the restriction provided by filler for the mobility of polymeric chain.

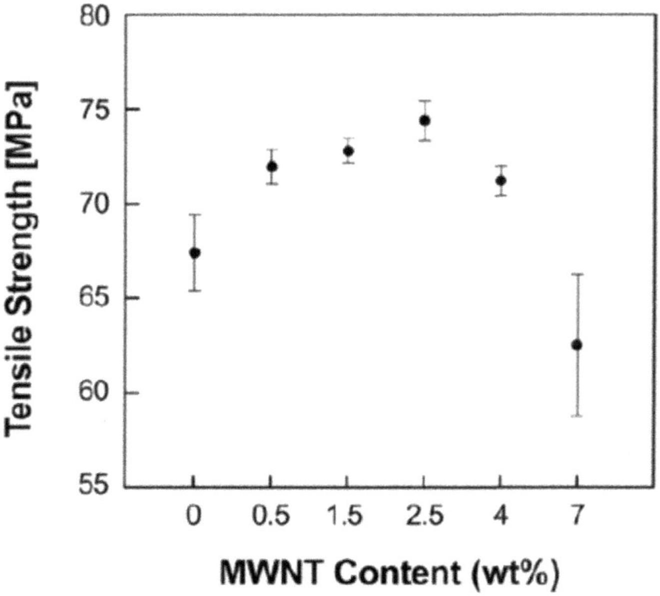

FIGURE 9.10 Tensile strength results for polycarbonate (PC)/MWCNT nanocomposites with respect to MWCNTs content.

Source: Adapted with permission from ref 89. © 2006 Springer.

Figure 9.11 depicts the storage modulus (G') for neat PC- and PC/MWCNT-based nanocomposites. It can be seen from the figure that the storage modulus below T_g (glassy region) of both neat PC and PC/MWCNT nanocomposites upholds plateau and exhibits no significant difference in storage modulus values for all the samples. However, the storage modulus values tend to decrease above T_g (rubbery region) for both PC and PC/MWCNT nanocomposites. It can be noticed that the storage modulus values for nanocomposites are higher as compared to neat PC. Moreover, it is observed that storage modulus for PC nanocomposites increases with respect to MWCNT content.

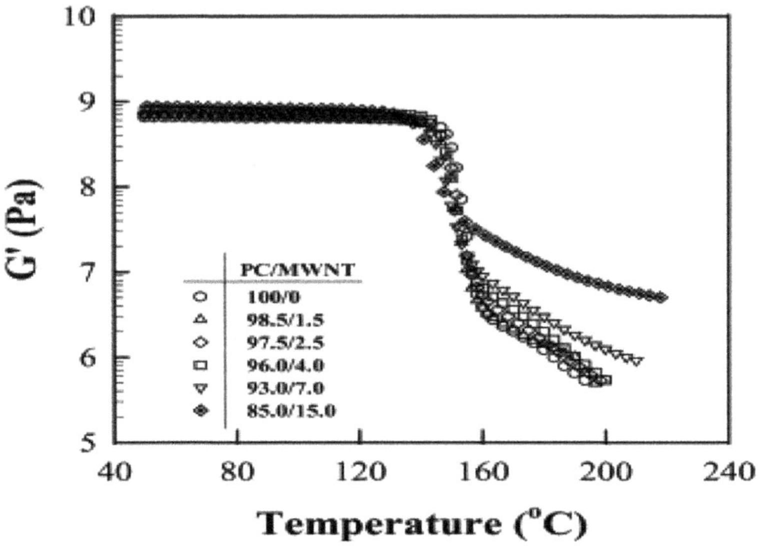

FIGURE 9.11 Storage modulus (G') for PC and PC/MWCNT nanocomposites.
Source: Adapted with permission from ref 109.

Figure 9.12 presents tanδ versus temperature graph for PC and PC/MWCNT nanocomposites. It can be observed from tanδ curve that PC exhibits a single peak at 156.6°C which corresponds to its T_g. In case of PC/MWCNTs, the values of tanδ curves tend to decrease with rise in MWCNT content. Interestingly, PC/MWCNT (93.0/7.0) nanocomposite demonstrates double tanδ behavior. The initial tanδ peak is associated with T_g of PC, whereas tan δ peak observed in later stages is in correspondence to the restriction of polymer mobility by the presence of MWCNTs in the PC matrix. Further, tanδ peak that appears in the later stages in case of PC/MWCNT (85.0/15.0) nanocomposite is found to be broader. The reason behind this fact may be due to the enhanced restriction of polymer chain mobility promoted by increased MWCNT content.

In general, three routes are utilized for incorporation of nanotubes in polymer matrices which include (i) solution casting of dispersed nanotube/polymer suspension, (ii) in situ polymerization of polymer monomer in the presence of nanotubes, and (iii) melt mixing of nanotubes with polymer matrices. Solution casting is the most fundamental method used for investigation of the effect of dispersion as well as the orientation of nanotubes, mechanisms involving deformation, and interfacial bonding on the final properties of polymers used as matrices. Studies related to the investigation on interfacial bonding among nanotube and polymer matrix are carried out using Raman spectroscopy. The degree at which the nanotubes are aligned as well as oriented in the polymer matrix is investigated using X-ray diffraction. In addition to this, reports are available which focus on the mechanical as well as electrical conductivity properties of polymer/nanotube composites.[96–99]

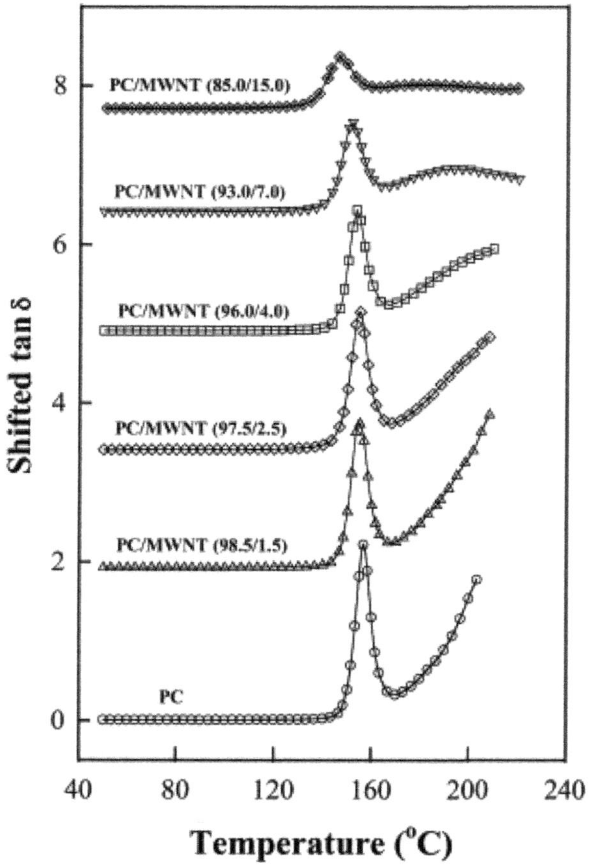

FIGURE 9.12 tanδ versus temperature for neat PC and PC/MWCNT nanocomposites.
Source: Adapted with permission from ref 95.

Figure 9.13 presents the influence of MWCNTs on volume resistivity characteristics. It can be observed from the figure that resistivity tends to decrease with increase in nanotubes content. This is due to the fact that interconnected MWCNTs network formation takes place which, in turn, is regarded as the electrical percolation threshold. When the MWCNTs content is below 2 wt.%, the possibility for the flow of electrons in higher percentage occurs because of formation of interconnected conductive network route. When MWCNTs content exceeds 2 wt.%, volume resistivity values decrease. The resistivity values presented in Figure 9.13 are measured at various applied voltages. It is well known that volume resistivity is always dependent on applied voltage. Hence, the resistivity values that can be interpreted from Figure 9.13 may vary slightly if the applied voltage is held constant. Therefore, the values presented in Figure 9.13 may have been slightly different if the applied voltage was held constant.[100]

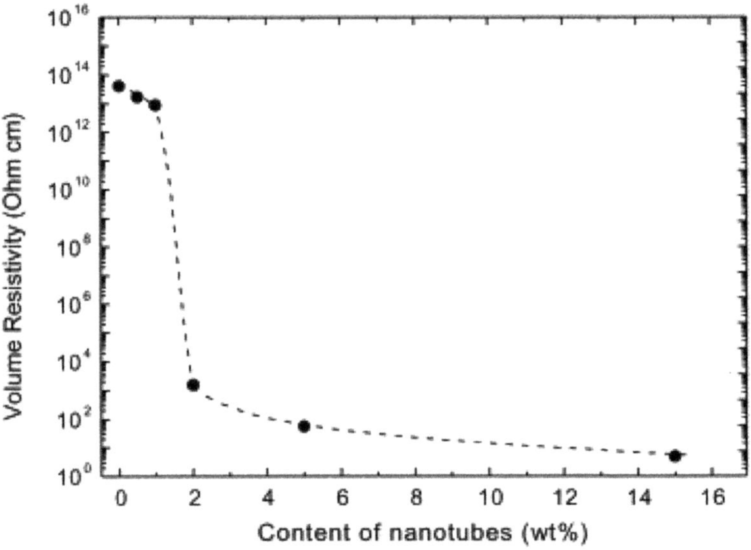

FIGURE 9.13 Influence of MWCNTs content on volume resistivity characteristics of PC/MWCNT composite.

Source: Adapted with permission from ref 100. © 2002 Elsevier.

Figure 9.14 shows the cryofractured SEM image obtained for PC/MWCNTs (5 wt.%) nanocomposites. It can be seen from Figure 9.14 that the propagation of crack occurs through PC matrix without affecting the nanotubes. The nanotubes are seen to be oriented in a random fashion in PC matrix. Interestingly, nanotubes tend to bridge the crack and enhance strength for PC composites by effectively involving in load transfer process.[100]

In general, it is not possible to attain good dispersion of unmodified CNTs in the polylactic acid matrix. In particular, poorer dispersion of unmodified CNTs in polypropylene (PP) matrix is reported and this, in turn, is known to affect electrical as well as mechanical properties of PP/CNT composites.[101,102] Therefore, both covalent as well as non-covalent modification methods using organic functional moieties are practiced in order to enhance dispersion of CNTs in polymer matrices. For achieving covalent bonding among CNTs and polymer, two procedures being followed include (i) grafting from and (ii) grafting to. In case of "grafting from" approach, initiators are initially attached to the CNTs which are then subjected to surface-initiated polymerization process.[103] In the second method, functional polymer moieties are allowed for reaction with unmodified or already modified CNTs.[104–105] In case of non-covalent bonding, polymers with aromatic groups connect with the surface of CNTs through π–π interaction.[106–108] The advantage of non-covalent bonding lies in keeping the structure of CNTs and their electronic structure intact without modification. The utilization of compatibilizer MA is made to enhance the dispersion properties of CNTs in the PP matrix. In spite compatibilization helped in the proper dispersion of CNTs in PP, improvement in terms of electrical conductivity property is not significantly evidenced.[109,110]

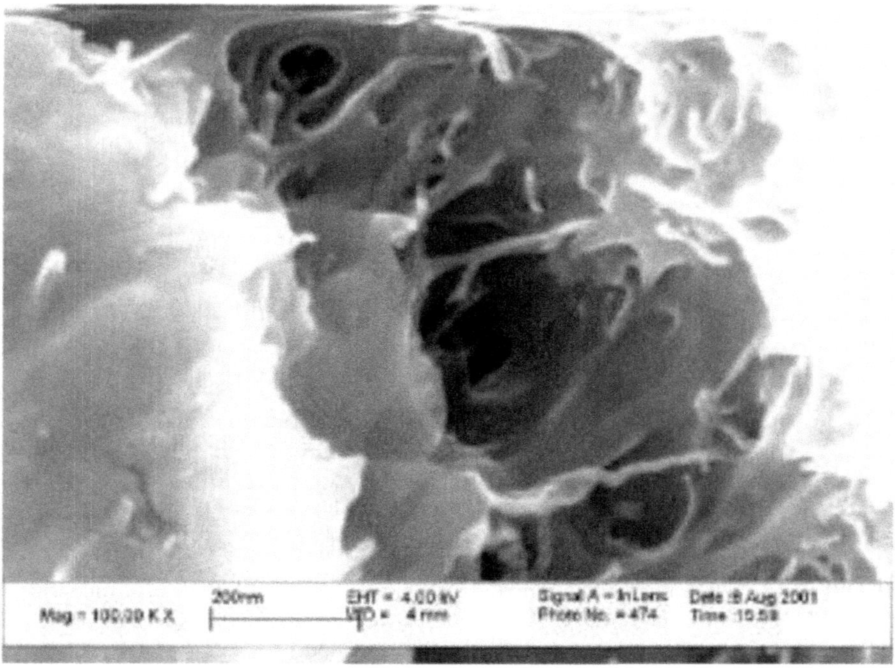

FIGURE 9.14 SEM image for cryo-fractured PC/MWCNTs (5 wt.%) composites.
Source: Adapted with permission from ref 100. © 2002 Elsevier.

Figure 9.15 shows transmission electron microscopy micrographs for thin and thick MWCNTs dispersed in PCL matrix through melt blending and coprecipitation methods. It can be seen from the figure that melt-blended PCL/MWCNTs composites exhibit breakdown for nanotubes, which is more pronounced in case of thicker nanotubes. The MWCNTs with greater breakdown aspect are visualized with enhanced structural defects which, in turn, impart weakness in MWCNTs and affect the targeted properties of PCL matrix. However, breakdown issue is not evidenced in case of coprecipitated PCL/MWCNT composites. The rheological characteristics of PP/MA/CNT (10 wt.%) composites prepared through melt-blending approach are also not improved significantly. This indicates the poorer dispersion of CNTs in the PP matrix when the composites are prepared through melt blending. If coprecipitation method is followed, slight enhancement in viscoelastic behavior is obtained for PP composites.

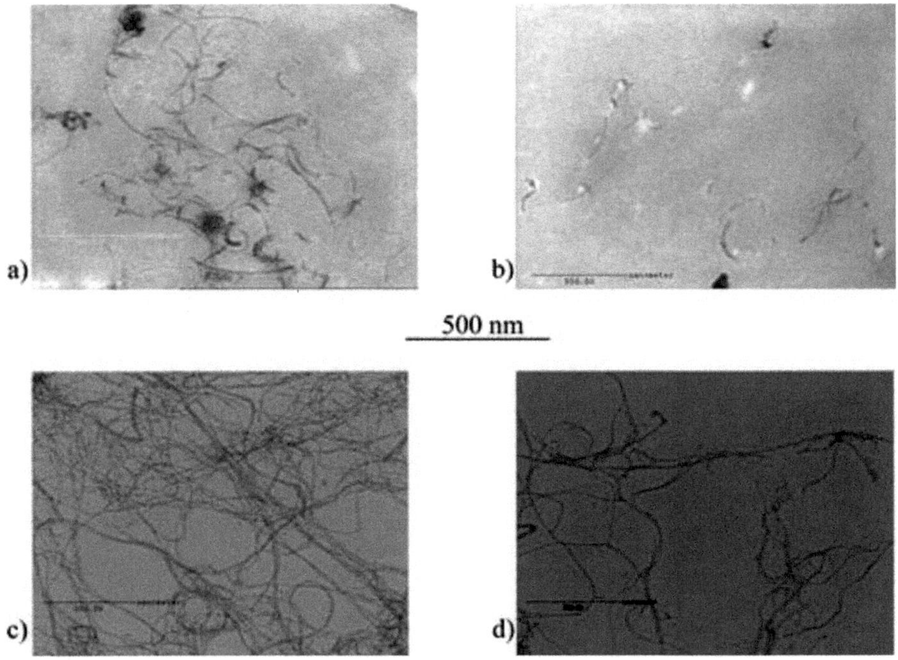

FIGURE 9.15 TEM micrographs for (a) melt-blended PCL/thin MWCNT composites, (b) melt-blended PCL/thick MWCNT composites, (c) coprecipitated PCL/thin MWCNT composites, and (d) coprecipitated PCL/thin MWCNT composites.

Source: Adapted with permission from ref 90. © 2007 American Chemical Society.

Figure 9.16 shows the electromagnetic shielding properties for PP/CNT, PP/PP-g-aminomethyl pyridine (AMP)/CNT, PP/PP-g-pyrene (Py)/CNTs composites. For all the samples, CNTs loading is reported to be ~2 wt. It is found that both AMP and Py acted efficiently as compatibilizers in order to achieve enhanced electromagnetic interference shielding efficiency. However, AMP showed upper hand in terms of shielding effectiveness as compared to Py. It can be seen from the figure that electromagnetic interference shielding efficiency of ~20 dB is achieved with the composites made using compatibilizers, whereas the same efficiency is not reached for PP/CNTs composites.[91]

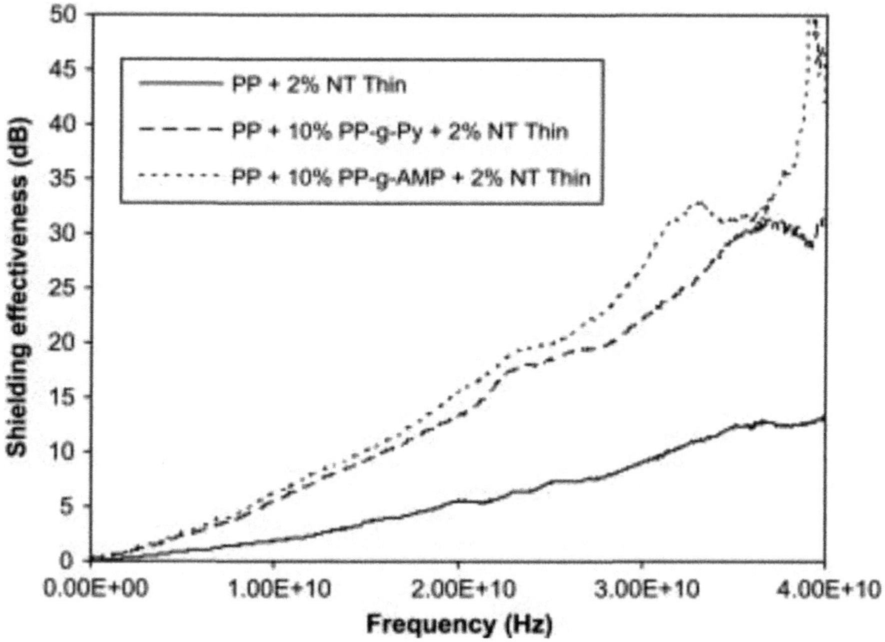

FIGURE 9.16 Electromagnetic interference shielding effectiveness for polypropylene (PP)/CNT, PP/PP-g-Py/CNT, and PP/PP-g-AMP/CNT composites with respect to frequency. **Source:** Adapted with permission from ref 91. © 2010 Elsevier.

The segregated electrically conducting polymer composites which include PS and copper (Cu) nanowires demonstrating excellent shielding effectiveness of ~42 dB at 13 wt.% loading of Cu is reported in the year 2011.[111] The polyethylene (PE)/CNT nanocomposites containing three variant conductive networks are reported.[112] The three different networks consisted of (i) segregated structure (PE/*s*-CNT), (ii) partially segregated structure (PE/*p*-CNT), and (iii) randomly distributed structure (PE/*r*-CNT) as shown in Figure 9.17. All these structures are made with the help of hot compaction followed by mechanical or solution blending.

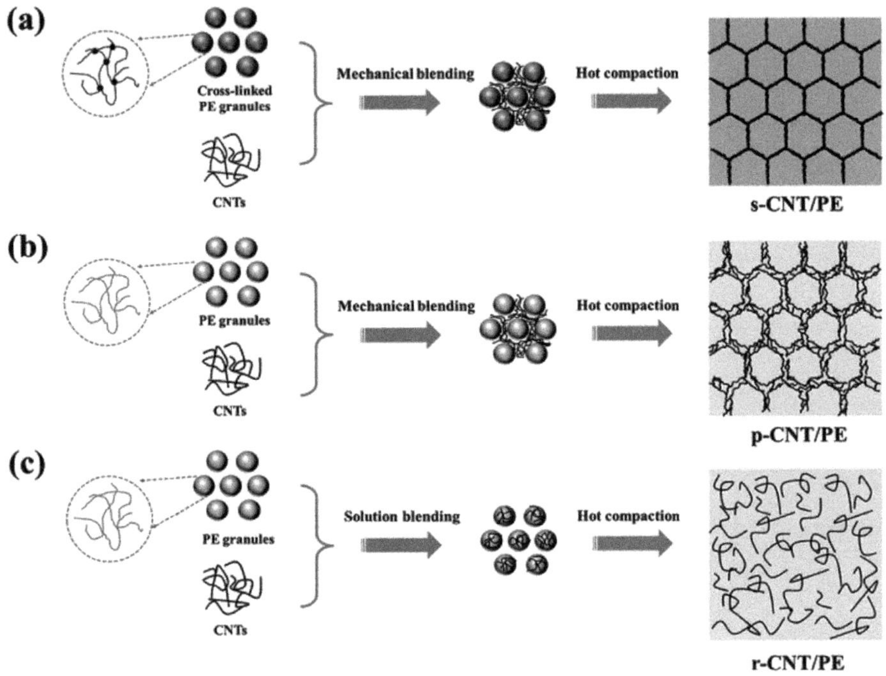

FIGURE 9.17 Steps involved in the fabrication of (a) polyethylene (PE)/*s*-CNT, (b) PE/*p*-CNT, and (c) PE/*r*-CNT composites.

Source: Adapted with permission from ref 112. © 2015 Royal Society of Chemistry.

The electromagnetic interference shielding potential for PE/CNT nanocomposites and the mechanism of shielding involved in nanocomposite samples were examined. The electromagnetic interference shielding effectiveness for different PE/CNT nanocomposites under the frequency regime of 8–12 GHz at a CNT loading of 5.0 wt.% is shown in Figure 9.18a. It can be observed from the figure that PE/*s*-CNT exhibits greater electromagnetic interference shielding effectiveness in comparison with PE/*p*-CNT and PE/*r*-CNT throughout the analyzed frequency regime. The maximum value for electromagnetic interference shielding effectiveness achieved by PE/*s*-CNT nanocomposites is found to be 46 dB. The enormous electromagnetic interference shielding potential offered by PE/*s*-CNT nanocomposites indicates that segregated structure is promising for the said application. Moreover, the content of *s*-CNTs in PE matrix is varied and the electromagnetic interference shielding effectiveness is measured as shown in Figure 9.18b. With respect to frequency, significant improvement in shielding effectiveness is not noticed; however, the effect is found to be more pronounced with respect to CNTs loading and the results are compared with literature data shown in Table 9.4.

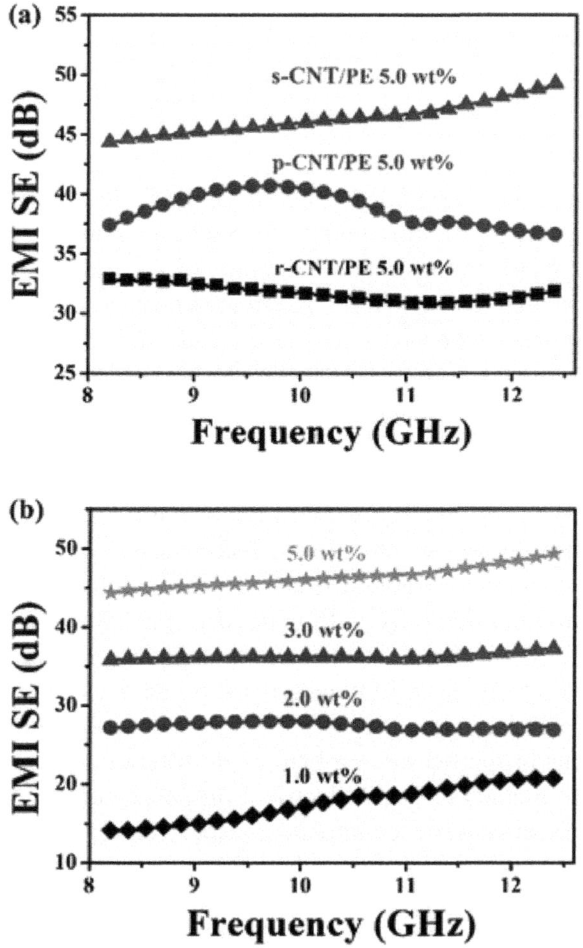

FIGURE 9.18 (a) Electromagnetic interference shielding effectiveness for three different PE/CNT nanocomposites containing 5.0 wt.% CNT loading of 5 wt.% under the frequency regime of 8–12 GHz and (b) electromagnetic interference shielding effectiveness for PE/*s*-CNT nanocomposites with respect to the content of CNT.

Source: Adapted with permission from ref 112. © 2015 Royal Society of Chemistry.

TABLE 9.4 Comparative Analysis of Electromagnetic Interference Shielding Effectiveness for Different Polymer Nanocomposites.

Composite	CNT content (wt.%)	Thickness (mm)	Electromagnetic interference (dB)	References
PC/CNT	10	Not available	27.2	113
PS/CNT	5.0	2.0	20.0–25.0	114
PMMA/CNT	10.0	2.1	40.0	115
Polypropylene (PP)/CNT	10.0	2.8	25.0	116
Polyethylene (PE)/CNT	10.0	3.0	35.0	117
PE/*s*-CNT	5.0	2.1	46.4	112
PE/CNT	10	1.65	22.4	118

Apart from CNTs, expanded graphite and CB have also been extensively investigated as reinforcements in polymer matrices and tested for their electromagnetic interference shielding properties. In spite excellent electromagnetic interference shielding effectiveness is achieved with CNTs-reinforced polymer composites, it is admissible that CNTs are expensive. In case of CB, high amount of filler should be used even for achieving a small improvement in electromagnetic interference shielding effectiveness. The cheaper cost and easy availability of graphite in bulk volume as well as easy fabrication procedure for graphene (Gr) from graphite make it an excellent option for reinforcement in polymer matrices in order to improve electromagnetic interference shielding effectiveness.[56]

Graphene is shown to demonstrate significant characteristics such as excellent thermal conductivity with enhanced mechanical as well as electrical conductivity[119,120] Such intrinsic characteristics exhibited by graphene make it potential to be used in new-generation devices such as solar cells, sensors, and electronic circuits.[121–129] Table 9.5 shows the comparative chart for various properties of graphene with different materials. Tables 9.6 and 9.7 present mechanical and electrical properties of various graphene-/graphite-reinforced polymer composites, respectively.

It is documented that graphene-based composites have demonstrated greater electrical conductivity and hence could help in the easy formation of conductive networks as compared to CNT-based composites.

However, graphene exhibits nonmagnetic nature and poor microwave-absorbing characteristics. In order to overcome this problem, graphene is combined with magnetic compounds. This is because of high saturation magnetization as well as Snoek's limit possessed by magnetic materials. Moreover, skin effect which is usually observed at higher frequency

TABLE 9.5 Comparative Chart for Various Properties of Graphene with Different Materials.

Materials	Tensile strength (TS)	Thermal conductivity (W/mK) at room temperature	Electrical conductivity (S/m)	References
Graphene	130 ± 10 GPa	$(4.84 \pm 0.44) \times 10^3$	7200	129, 130
CNT	60–150 GPa	3500	3000–4000	131
Nano-sized steel	1769 MPa	5–6	1.35×10^6	132
Plastic (high-density polyethylene; HDPE)	18–20 MPa	0.46–0.52	Insulator	133
Rubber (natural rubber)	20–30	0.13–0.142	Insulator	134
Fiber (Kevlar)	3620 MPa	0.04	Insulator	135

Source: Adapted with permission from ref 121. © 2010 Elsevier.

TABLE 9.6 Mechanical Properties of Graphene-/Graphite-Reinforced Polymer Composites.

Matrix	Filler type	Filler loading (wt.%^{a}, vol.%^{b})	Process	% Increase E	% Increase TS	% Increase flexural strength	References
Epoxy	Expanded graphite (EG)	1^a	Sonication	8	−20		136
	EG	1^a	Shear	11	−7		136
	EG	1^a	Sonication and shear	15	−6		136
	EG	0.1^a	Solution			87	137
PMMA	EG	21^a	Solution	21			138
	Graphite nanoplatelet (GNP)	5^a	Solution	133			139
PP	EG	3^b	Melt			8	140
	xGnP-1	3^b	Melt			26	141
	xGnP-15	3^b	Melt			8	141
	Graphite	2.5^b	SSSP		60		142
LLDPE	xGnP	15^a	Solution		200		143
	Paraffin-coated xGnP	30	Solution		22		144
HDPE	EG	3^a	Melt	100	4		145
	UG	3^a	Melt	33			145
PPS	EG	4^a	Melt			−20	146
	S-EG	4^a	Melt			−33	146

TABLE 9.6 *(Continued)*

Matrix	Filler type	Filler loading (wt.%[a], vol.%[b])	Process	% Increase E	% Increase TS	% Increase flexural strength	References
Poly(vinyl alcohol) PVA	Graphite oxide (GO)	0.7[a]	Solution		76		82
	Graphene	1.8[b]	Solution		150		147
TPU	Graphene	5.1[b]	Solution	200			148
	Sulfonated graphene	1[a]	Solution		75		149
PETI	EG	5[a]	In situ	39			150
	EG	10[a]	In situ	42			151

[a]wt% loading

[b]vol% loading

PMMA: poly(methyl methacrylate); PP: polypropylene; LLDPE: linear low-density polyethylene; HDPE: high-density polyethylene; PPS: polyphenylene sulfide; TPU: thermoplastic polyurethane; PETI: poly(ethylene terephthalate-co-isophthalate)

Source: Adapted with permission from ref 116. © 2009 Elsevier.

TABLE 9.7 Electrical Properties for Graphene-/Graphite-Based Polymer Composites

Matrix	Filler	Filler loading (wt.%[a], vol.%[b])	Process	σ (S/m) of matrix	σ (S/m) of composite	Reference
Epoxy	EG	3.00[a]	Sonication	1E−13	1E−4	152
	EG	2.50[b]	Solution	1E−15	1E−2	153
	Graphene	0.52[b]	Solution	1E−10	1E−2	56
PMMA	NanoG	0.68[b]	In situ	1E−13	1E−3	154
	EG	1.00[a]	Solution	1E−15	1E−3	138
	EG	10[a]	In situ	–	77.65	155
PS	NanoG	1.00[a]	In situ	1E−14	1E−4	156
	Graphene	0.10[b]	Solution	1E−16	1E−5	2
	GNS^{C4P}	0.40[b]	Solution	1E−14	1E−5	157
	GNS^{C4P}	0.10[b]	Solution	1E−14	4	157
	GNS^{8B}	0.20[b]	Solution	1E−14	1E−5	157
	GNS^{5D}	0.30[b]	Solution	1E−14	1E−5	157
	Graphene	–	Solution	1E−16	24	158
	Graphene	2.0[a]	In situ	1E−10	1E−2	159
	EG	1.50[b]	In situ	1E−16	1E−4	160
	K-GIC	8.20[a]	Solution	NA	–	161
Nylon-6	EG	1.50[b]	In situ	1E−15	0.1	161
	FG	0.75[b]	In situ	1E−15	1E−5	162
PP	xGnP-1	3.00[b]	Coating	1E−12	0.1	140
	xGnP-1	3.00[b]	Solution	1E−12	1E−2	140
	xGnP-15	7.00[b]	Melt	1E−12	1E−3	140

TABLE 9.7 *(Continued)*

Matrix	Filler	Filler loading (wt.%[a], vol.%[b])	Process	σ (S/m) of matrix	σ (S/m) of composite	Reference
	xGnP-15	5.00[b]	Coating	1E−12	0.1	140
	EG	0.67[b]	Solution	1E−16	0.1	163
HDPE	EG	3.00[a]	Melt	1E−16	1E−8	145
	UG	5.00[a]	Melt	1E−16	1E−10	145
PPS	EG	4.0[a]	Melt	1E−12	1E−3	146
	S-EG	4.0[a]	Melt	1E−12	1E−2	146
PANI	Graphite	1.5[a]	In situ	5.0	3300.3	164
	GO	–	In situ	2.0	1000	165
PVDF	FGS	2.0[a]	Solution	1E−11	1E−2	166
	EG	5.0[a]	Solution	1E−11	1E−3	166
PVA-S	NanoG	0.2[a]	Solution	1E−13	1E−3	167
Polyethylene terephthalate	Graphene	0.47[b]	Melt	1E−14	7.4E−2	168
PC	FGS	2.0[a]	Melt	1E−14	1E−9	169
	Graphite	12	Melt	1E−14	6.6E−11	169

[a]wt% loading

[b]vol% loading

PPS: polyphenylene sulfide; PANI: polyaniline; PVDF: polyvinylidene fluoride; FGS: functionalized graphing sheet

Source: Adapted with permission from ref 96. © 2000 AIP Publishing.

conditions can be avoided by soft materials which, in turn, can assist in effective passage of electromagnetic waves. Amidst of several soft magnetic compounds, much attention has been paid to magnetic iron oxide, maghemite (γ-Fe_2O_3 or magnetite Fe_3O_4) nanoparticles, which is mainly owing to the fact that these materials are biocompatible in nature. It is possible to impart magnetic characteristics to graphene by the decoration of Fe_2O_3 nanoparticles on grapheme.[170–173]

The fabrication steps involved in reduced graphite oxide (RGO)/Fe_3O_4 composite is shown in Figure 9.19. Using Hummer's method, graphite is converted into graphite oxide (GO) which is further reduced to RGO using solvothermal process. For reduction of GO into RGO, ethylene glycol is employed as reducing agent. Moreover, ethylenediamine used in the fabrication of RGO/Fe_3O_4 composite helps in coordination of metal cations in order to aid the formation of stable complex.[173,174]

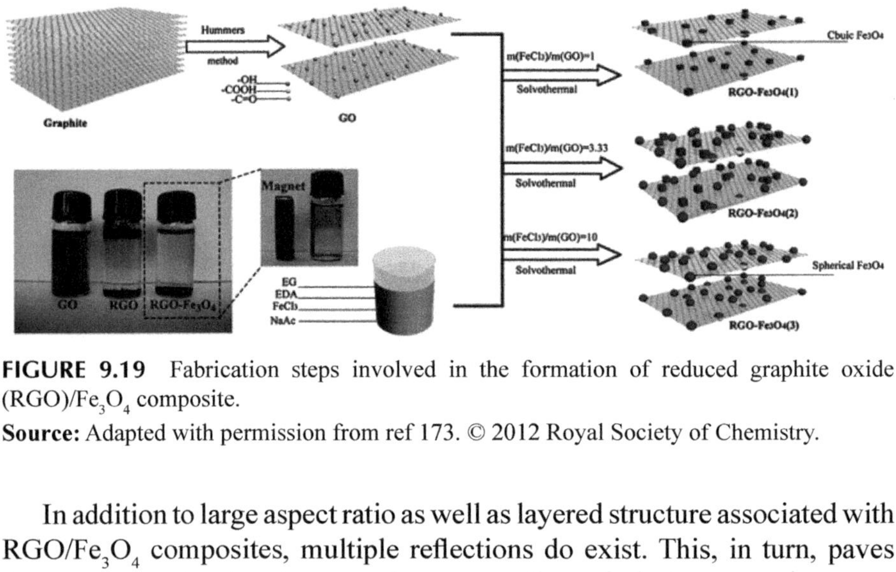

FIGURE 9.19 Fabrication steps involved in the formation of reduced graphite oxide (RGO)/Fe_3O_4 composite.

Source: Adapted with permission from ref 173. © 2012 Royal Society of Chemistry.

In addition to large aspect ratio as well as layered structure associated with RGO/Fe_3O_4 composites, multiple reflections do exist. This, in turn, paves the way for extension in route for propagation of electromagnetic waves through the composites and hence enhancement in absorption capacity can be achieved. Figure 9.20 presents the schematics for electromagnetic waves absorption in RGO/Fe_3O_4 composites. The compensatory characteristics available in RGO/Fe_3O_4 composite make it a potential material for electromagnetic interference shielding application.[173,175,176]

For fabrication of highly transparent graphene-based composites, several methods are adopted which include chemical vapor deposition,[177] spin-coating,[178] and electrophoretic deposition (EPD) process.[179] Among all these methods, EPD is regarded to be promising for the fabrication of thin graphene sheets. This is because the technique offers a greater rate of deposition even on complex substrates with precise control over purity and thickness. In case of EPD technique, thin graphene layers are deposited on the metal substrate through which enhanced electromagnetic interference shielding property and transparency are achieved. It should be also noted that the strong electrophoretic squeezing force, which is usually generated in the EPD process, could facilitate cohesive flocculation of graphene sheets in the in-plane direction, desirably eliminating the use of binder materials.[180]

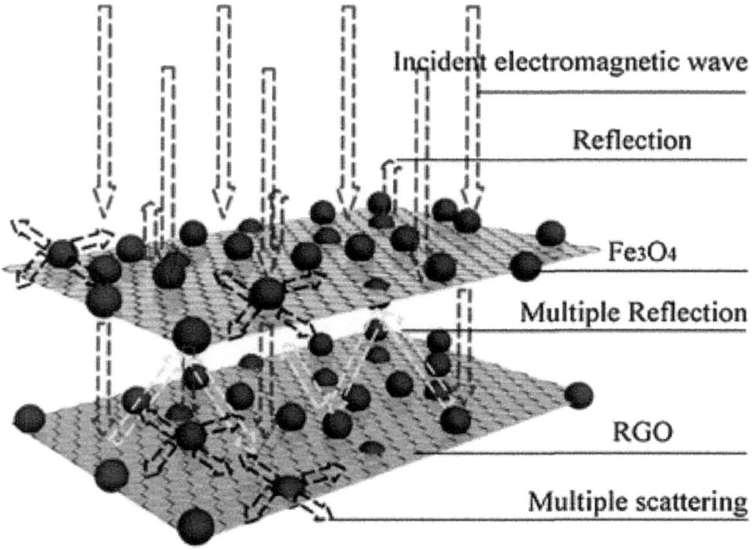

FIGURE 9.20 Schematics for electromagnetic wave absorption mechanism in RGO/Fe_3O_4 composites.

Source: Adapted with permission from ref 173. © 2012 Royal Society of Chemistry.

Figure 9.21a shows basics involved in electromagnetic interference shielding mechanism for RGO. The absorbance as well as reflectance values for single PEI/RGO and double PEI/RGO composites can be seen from Figure 21b–d. It is found that double PEI/RGO composites show greater electromagnetic interference shielding efficiency as compared to single PEI/RGO composite. The electromagnetic interference shielding effectiveness value recorded for double PEI/RGO composite is shown to be two times higher than single PEI/RGO in the frequency regime ranging from 0.5 to 8.5 GHz.[181]

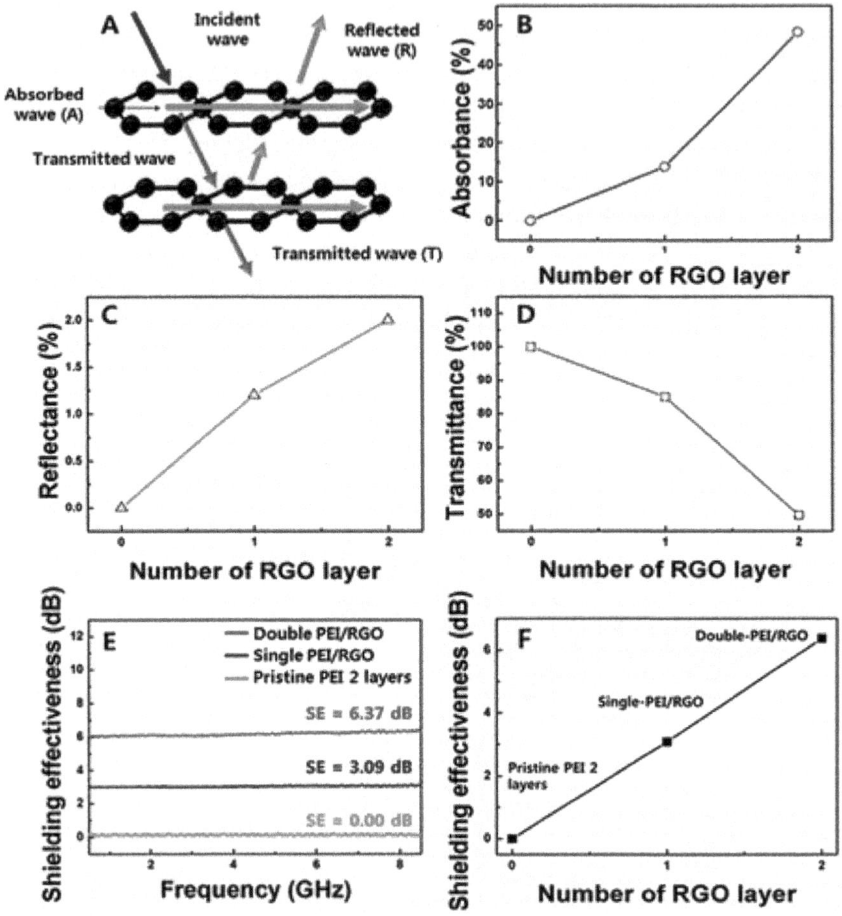

FIGURE 9.21 (See color insert.) (a) Electromagnetic interference shielding mechanism for RGO, (b) absorbance, (c) reflectance, (d) transmittance as a function of RGO layers, and (e) shielding efficiency for PEI/RGO composites.

Source: Adapted with permission from ref 181. © 2014 American Chemical Society.

9.4 ELASTOMERIC NANOCOMPOSITES FOR ELECTROMAGNETIC SHIELDING

Rotorcraft is the main application where elastomeric polymers are extensively used. In order to overcome the primary failure experienced by elastomers in several applications, CNTs are introduced as reinforcement in the elastomer matrices. The utilization of CNTs as reinforcement materials in elastomer matrices actually stemmed while aimed at designing a strain energy accumulator.[182] It is found that both TS and elastic modulus get enhanced as a function of CNTs. In case of the military as well as shipboard rotorcraft purposes, CNTs-reinforced rubber composites with high electromagnetic interference shielding effectiveness are of great interest.

In the recent years, magnetic rubber elastomer (MRE) is extensively tried as vibration absorption and impedance devices.[183] Mainly, the magnetic as well as mechanical characteristics of MRE determine its performance. The magnetic nanoparticles are reinforced in several elastomer matrices such as natural rubber, silicon rubber, polybutadiene, acrylonitrile, and polyurethane.[184,185] The types of matrix used do not usually determine the properties of composites whereas the structure and doped content are reported to have an effect on the targeted characteristics. Therefore, research articles focused on the investigation of doped magnetic content on the magnetic as well as mechanical properties of composites are reported.[185–187] It is found that magnetic properties increase with respect to rise in doped magnetic content, whereas the mechanical characteristics are not improved which is because of the weak force of attraction among magnetic nanoparticles and rubber.[188,189] The excellent adhesion properties are actually determined by the surface energy possessed by the filler. It is expected that good adhesion can be established among filler and the matrix in the case where surface energy of filler incorporated is equal or higher as compared to the polymer matrix. The large size of magnetic nanoparticles is always associated with low surface energy due to which adhesion properties with elastomer matrices is affected.

In order to overcome this drawback, magnetic nanoparticles are subjected to surface modification process. In case of utilization of magnetic oxides such as ferrites, it is necessary to incorporate huge volume of filler in the elastomer matrix in order to achieve necessary magnetization property. This is due to the fact that magnetic oxides are of low density as compared to metals. The physical characteristics of the polymer matrix are affected detrimentally due to the incorporation of huge volume fraction of magnetic oxide filler alone in the polymer matrix. It has been regarded that introduction of metallic magnetic fillers such as iron, cobalt, and nickel can overcome the abovementioned drawback. Among the metallic magnetic fillers, nickel offers more chemical stability and hence considered as a better choice as reinforcement in elastomer matrices in order to achieve improved flexibility as molding characteristics.[190,191] The experimental and theoretically calculated magnetization values for nickel composites are shown in Table 9.8.

TABLE 9.8 Experimental and Theoretically Calculated Magnetization Values for Nickel Composites. (Source: Adapted with permission from ref 183. © 2009 Elsevier.)

Sample (phr) (mass of nickel in 100 g rubber)	Total mass (nickel + rubber + curing agents)	Calculated magnetization (emu/g)	Observed magnetization (emu/g)
20	129.5	7.3	7
40	149.5	12.7	11
60	169.5	16.8	14
80	189.5	20	17.5
100	209.5	22.7	20
20	129.5	7.3	7

Figure 9.22 shows the effect of nickel content on dielectric permittivity (ε') at 40°C and under different frequency conditions. It can be seen from the figure that permittivity tends to increase with respect to nickel content at all the frequency conditions. This is due to the presence of a metallic component in the elastomer matrix which permitted the increase in dielectric permittivity.[183,192,193]

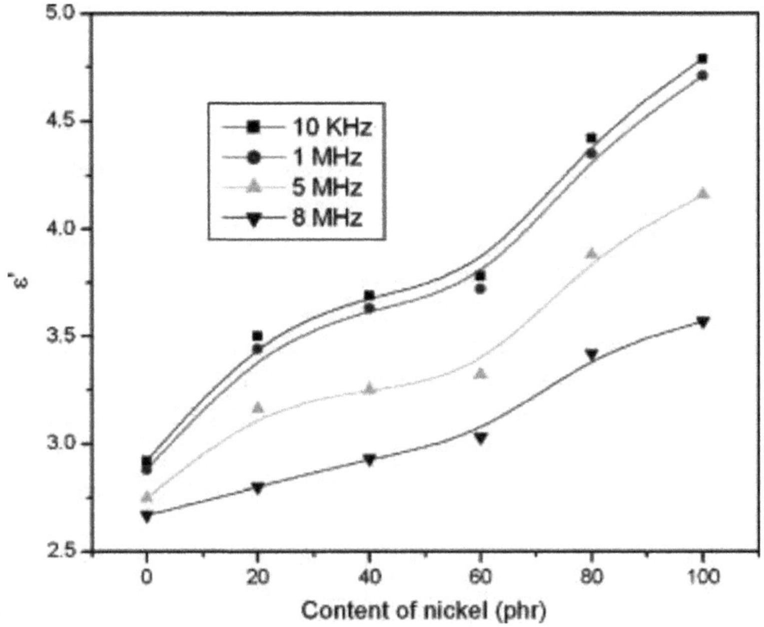

FIGURE 9.22 Effect of nickel content on dielectric permittivity.
(Source: Adapted with permission from ref 183. © 2009 Elsevier.)

Figure 9.23 presents dielectric loss values for rubber/nickel composites as a function of frequency. It can be seen from the figure that rubber/nickel composites do not show dielectric relaxation under the frequency conditions studied which is not the case with neat rubber.

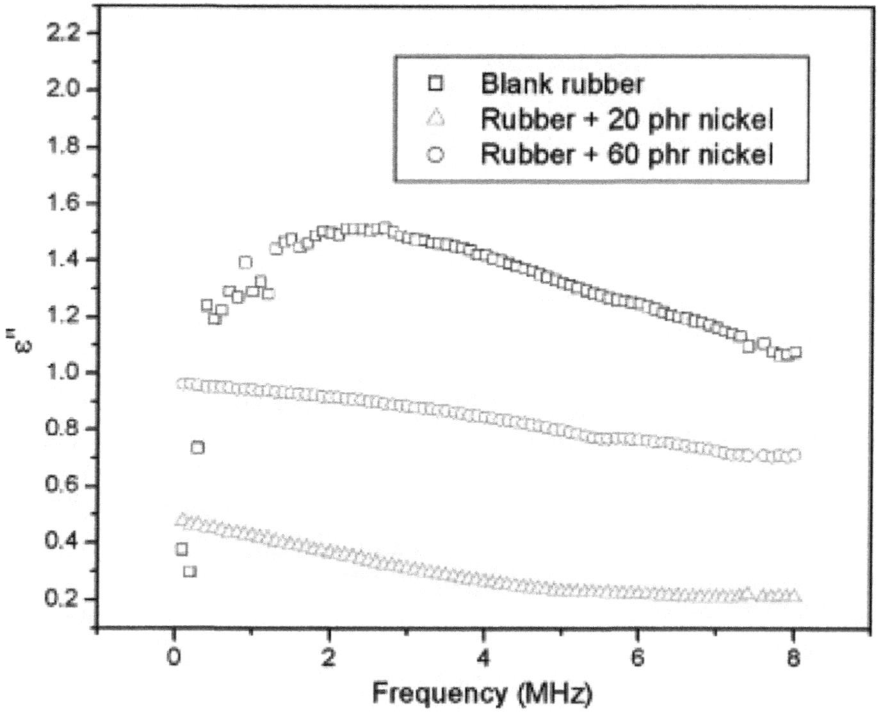

FIGURE 9.23 Dielectric loss versus frequency for rubber/nickel composites as a function of frequency and at 120°C.

(Source: Adapted with permission from ref 183. © 2009 Elsevier.)

Natural rubber is extensively used for fabrication of elastomeric composites doped with metallic magnetic properties due to its adequate availability and cheaper cost. Apart from natural rubber, synthetic elastomers are also used for fabrication of composites imparted with magnetic properties. Figure 9.24 shows the SEM images for natural rubber/nickel and neoprene/nickel composites.

It is evaluated that incorporated magnetic fillers help in imparting reinforcing properties to the composites apart from improving magnetic characteristics. This can be evidenced by evaluating mechanical and magnetic properties of elastomeric composites.

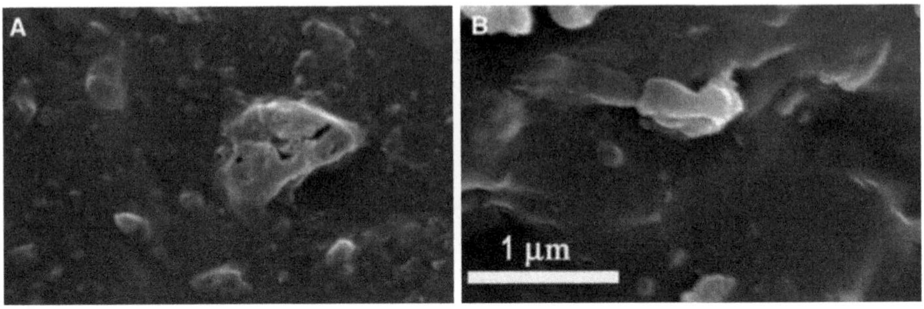

FIGURE 9.24 SEM images for (a) natural rubber/nickel nanocomposites and (b) neoprene/nickel nanocomposites.
Source: Reprinted with permission from ref 184. © Elsevier.

9.5 THERMOSETTING NANOCOMPOSITES FOR ELECTROMAGNETIC SHIELDING

Several types of thermoset-based nanocomposites are also reported for electromagnetic interference shielding application. A study is reported on CNTs/CF-reinforced epoxy matrix for electromagnetic interference shielding purpose.[194,195] Initially, CNTs are grown on the surface of CF and the performance of CNTs/CF is used as reinforcement in an epoxy matrix. The CNTs grown on CF surface can be visualized from SEM image shown in Figure 9.25. It can be seen from inset of Figure 9.25 that plenty of CNTs are grown on CF whose surface is completely covered by CNTs.[194,195]

Figure 9.26a shows the variation of electromagnetic interference shielding effectiveness for different epoxy composites. The pure epoxy/CF composites exhibited an electromagnetic interference shielding effectiveness value of ~29.4 dB. In case of epoxy/CNT/CF composites, it has been noticed that electromagnetic interference shielding effectiveness value increased up to ~51 dB. In the same study, shielding effectiveness due to reflection and absorption is also calculated individually and the values are shown in Figure 9.26b and c. It can be observed that the shielding effectiveness due to absorption is higher as compared to reflection for the composites. In case of lower frequency conditions, the electromagnetic interference shielding effectiveness depends on reflection wastage. On the other hand, absorption wastage shows the effect on electromagnetic interference shielding effectiveness under high-frequency conditions.[196]

The improved absorption phenomenon is associated with the reduced skin depth. It is evident from Figure 9.26d that increase in skin depth leads to

FIGURE 9.25 SEM image for CNTs grown on the surface of carbon fiber.[195]

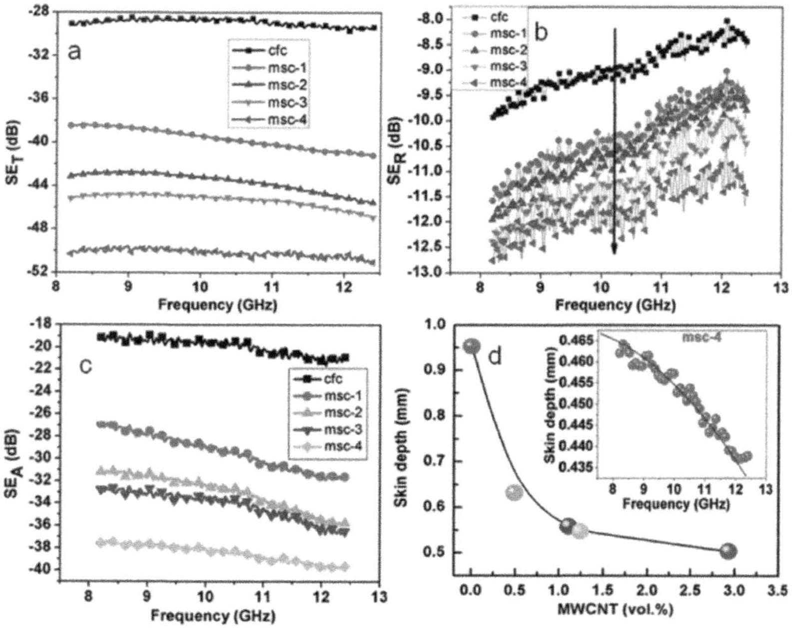

FIGURE 9.26 (See color insert.) (a) Variation of electromagnetic interference shielding effectiveness with respect to frequency, (b) electromagnetic interference shielding effectiveness due to reflection, (c) electromagnetic interference shielding effectiveness due to absorbance for epoxy composites, and (d) variation of skin depth as a function of MWCNT loading.[195]

reduced electromagnetic interference shielding effectiveness.[195] The electromagnetic interference shielding values for different epoxy composites can be seen from Table 9.9. A study reported on epoxy-/nickel-coated polyethylene terephthalate (PET) fiber-based composites exhibited an electromagnetic interference shielding efficiency of ~30 dB in the frequency regime of 3 kHz to 1500 MHz.[197] The effective electromagnetic interference shielding phenomenon is due to the electrical conductivity of epoxy-/nickel-coated PET fiber-based composites.

TABLE 9.9 Electromagnetic interference Shielding Effectiveness Values for Epoxy/Carbon Composites.[195]

S. No.	Type of filler/% filler/thickness	Polymer	Frequency	Shielding effectiveness (dB)	References
1	Single-walled CNT (SWCNT)/ 15 wt.%/2 mm	Epoxy	8.2–12.4 GHz	20–30	198
2	SWCNT/ 20 wt.%/2 mm	Polyurethane	8.2 GHz	17	199
3	SWCNT/ 15 wt.%/1.5 mm	Ethylene vinyl acetate	8.2–12.4 GHz	22–23	200
4	SWCNT/ 4.5 vol.%/2 mm	Reactive ethylene terpolymer	12.4 GHz	30	201
5	Carbon fibers fabric/ 30 vol.%/2 mm	Epoxy	12.4 GHz	29.4	195

9.6 CONCLUSIONS

Various types of polymer-/carbon-based nanocomposites and their electromagnetic interference shielding potential are discussed in detail. CNTs and graphene nanofillers act as effective reinforcements for achieving improvement in terms of electromagnetic interference shielding properties. The high aspect ratio and electrical conductivity characteristics offered by such nanofillers help in the formation of the conductive network. Different fabrication steps for formation of polymer foam-based structures with enhanced electromagnetic interference shielding and mechanical properties are discussed. It is interpreted that factors associated with carbon-based fillers such as aspect ratio, smaller diameter, mechanical strength, electrical conductivity, and dielectric constant show definite influence on electromagnetic interference shielding properties. It is evaluated that magnetic nanofillers help in imparting reinforcing characteristics apart from enhancing magnetic properties. The polymer-/carbon-based nanocomposites hold high potential to be used as electromagnetic interference shielding materials.

KEYWORDS

- **polymer nanocomposites**
- **electromagnetic interference shielding**
- **carbon nanotubes**
- **graphene**
- **conductive polymer**
- **elastomer**
- **thermosets**

REFERENCES

1. Wang, D.; et al. Functionalized Graphene–BaTiO3/Ferroelectric Polymer Nanodielectric Composites with High Permittivity, Low Dielectric Loss, and Low Percolation Threshold. *J. Mater. Chem. A* **2013,** *1*(20), 6162–6168.
2. Stankovich, S.; et al. Graphene-Based Composite Materials. *Nature* **2006,** *442*(7100), 282–286.
3. Klionsky, D. J.; et al. Guidelines for the Use and Interpretation of Assays for Monitoring Autophagy. *Autophagy* **2012,** *8*(4), 445–544.
4. Villmow, T.; et al. Liquid Sensing: Smart Polymer/CNT Composites. *Mater. Today* **2011,** *14*(7), 340–345.
5. Huang, J. -C. Carbon Black Filled Conducting Polymers and Polymer Blends. *Adv. Polym. Tech.* **2002,** *21*(4), 299–313.
6. Tchoudakov, R.; et al. Conductive Polymer Blends with Low Carbon Black Loading: Polypropylene/Polyamide. *Polym. Eng. Sci.* **1996,** *36*(10), 1336–1346.
7. King, J. A.; et al. Factorial Design Approach Applied to Electrically and Thermally Conductive Nylon 6,6. *Polymer Compos.* **2001,** *22*(1), 142–154.
8. Pang, H.; et al. Conductive Polymer Composites with Segregated Structures. *Prog. Polym. Sci.* **2014,** *39*(11), 1908–1933.
9. Göldel, A.; et al. The Kinetics of CNT Transfer between Immiscible Blend Phases during Melt Mixing. *Polymer* **2012,** *53*(2), 411–421.
10. Yan D. -X.; et al. Structured Reduced Graphene Oxide/Polymer Composites for Ultra-Efficient Electromagnetic Interference Shielding. *Adv. Funct. Mater.* **2015,** *25*(4), 559–566.

11. Pang, H.; et al. Super-Tough Conducting Carbon Nanotube/Ultrahigh-Molecular-Weight Polyethylene Composites with Segregated and Double-Percolated Structure. *J. Mater. Chem.* **2012,** *22*(44), 23568–23575.
12. Pang, H.; et al. Double-Segregated Carbon Nanotube–Polymer Conductive Composites as Candidates for Liquid Sensing Materials. *J. Mater. Chem. A* **2013,** *1*(13); 4177–4181.
13. Grunlan, J. C.; et al. Water-Based Single-Walled-Nanotube-Filled Polymer Composite with an Exceptionally Low Percolation Threshold. *Adv. Mater.* **2004,** *16*(2), 150–153.
14. Regev, O.; et al. Preparation of Conductive Nanotube–Polymer Composites Using Latex Technology. *Adv. Mater.* **2004,** *16*(3), 248–251.
15. Grossiord, N.; Loos, J.; Koning, C. E. Strategies for Dispersing Carbon Nanotubes in Highly Viscous Polymers. *J. Mater. Chem.* **2005,** *15*(24), 2349–2352.
16. Yan, D. -X.; et al. Efficient Electromagnetic Interference Shielding of Lightweight Graphene/Polystyrene Composite. *J. Mater. Chem.* **2012,** *22*(36), 18772–18774.
17. Hsiao, S. -T.; et al. Using a Non-Covalent Modification to Prepare a High Electromagnetic Interference Shielding Performance Graphene Nanosheet/Water-Borne Polyurethane Composite. *Carbon* **2013,** *60,* 57–66.
18. Cao, M. -S.; et al. Ultrathin Graphene: Electrical Properties and Highly Efficient Electromagnetic Interference Shielding. *J. Mater. Chem. C* **2015,** *3*(26), 6589–6599.
19. Li, N.; et al. Electromagnetic Interference (EMI) Shielding of Single-Walled Carbon Nanotube Epoxy Composites. *Nano Lett.* **2006,** *6*(6), 1141–1145.
20. Yang, Y.; et al. Conductive Carbon Nanofiber–Polymer Foam Structures. *Adv. Mater.* **2005,** *17*(16), 1999–2003.
21. Maiti, S.; et al. Polystyrene/MWCNT/Graphite Nanoplate Nanocomposites: Efficient Electromagnetic Interference Shielding Material through Graphite Nanoplate–MWCNT–Graphite Nanoplate Networking. ACS Appl. Mater. *Interfaces* **2013,** *5*(11), 4712–4724.
22. Ayana, B.; Suin, S.; Khatua, B. B. Highly Exfoliated Eco-Friendly Thermoplastic Starch (TPS)/Poly (Lactic Acid) (PLA)/Clay Nanocomposites Using Unmodified Nanoclay. *Carbohydr. Polym.* **2014,** *110,* 430–439.
23. Li, L. -P.; et al. Characterization of PA6/EPDM-g-MA/HDPE Ternary Blends: The Role of Core-Shell Structure. *Polymer* **2012,** *53*(14), 3043–3051.
24. Dou, R.; et al. Structuring Tri-Continuous Structure Multiphase Composites with Ultralow Conductive Percolation Threshold and Excellent Electromagnetic Shielding Effectiveness Using Simple Melt Mixing. *Polymer* **2016,** *83,* 34–39.
25. Dou, R.; et al. Insight into the Formation of a Continuous Sheath Structure for the PS Phase in Tri-Continuous PVDF/PS/HDPE Blends. *RSC Adv.* **2016,** *6*(1), 439–447.
26. Bai, Z.; Guo, H. Interfacial Properties and Phase Transitions in Ternary Symmetric Homopolymer–Copolymer Blends: A Dissipative Particle Dynamics Study. *Polymer* **2013,** *54*(8), 2146–2157.
27. Li, S. -L.; et al. Effect of the Mutiwalled Carbon Nanotubes Selective Localization on the Dielectric Properties for PVDF/PS/HDPE Ternary Blends with in situ Formed Core–Shell Structure. *RSC Adv.* **2016,** *6*(63), 58493–58500.
28. Mittal, G.; et al. A Review on Carbon Nanotubes and Graphene as Fillers in Reinforced Polymer Nanocomposites. *J. Ind. Eng. Chem.* **2015,** *21,* 11–25.
29. Kingston, C.; et al. Release Characteristics of Selected Carbon Nanotube Polymer Composites. *Carbon* **2014,** *68,* 33–57.
30. Al-Saleh, M. H.; Sundararaj, U. A Review of Vapor Grown Carbon Nanofiber/Polymer Conductive Composites. *Carbon* **2009,** *47*(1), 2–22.

31. Remya, V. R.; et al. Biobased Materials for Polyurethane Dispersions. *Chem. Int.* **2016,** *2*(3), 158–167.
32. Chen, Z.; et al. Lightweight and Flexible Graphene Foam Composites for High-Performance Electromagnetic Interference Shielding. *Adv. Mater.* **2013,** *25*(9), 1296–1300.
33. Al-Saleh, M. H.; Sundararaj, U. Electromagnetic Interference Shielding Mechanisms of CNT/Polymer Composites. *Carbon* **2009,** *47*(7), 1738–1746.
34. Basuli, U.; et al. Electrical Properties and Electromagnetic Interference Shielding Effectiveness of Multiwalled Carbon Nanotubes-Reinforced EMA Nanocomposites. *Polym. Compos.* **2012,** *33*(6), 897–903.
35. Deng, H.; et al. Progress on the Morphological Control of Conductive Network in Conductive Polymer Composites and the Use as Electroactive Multifunctional Materials. *Prog. Polym. Sci.* **2014,** *39*(4), 627–655.
36. Sengupta, R.; et al. A Review on the Mechanical and Electrical Properties of Graphite and Modified Graphite Reinforced Polymer Composites. *Prog. Polym. Sci.* **2011,** *36*(5), 638–670.
37. Du, J.; Cheng, H. M. The Fabrication, Properties, and Uses of Graphene/Polymer Composites. *Macromol. Chem. Phys.* **2012,** *213*(10–11), 1060–1077.
38. Inuwa, I.M.; et al. Characterization and Mechanical Properties of Exfoliated Graphite Nanoplatelets Reinforced Polyethylene Terephthalate/Polypropylene Composites. *J. Appl. Polym. Sci.* **2014,** *131*, 15.
39. Al-Saleh, M. H. Electrical and Electromagnetic Interference Shielding Characteristics of GNP/UHMWPE Composites. *J. Phys. D: Appl. Phys.* **2016,** *49*(19), 195302.
40. Li, W.; et al. The Imaging Mechanism, Imaging Depth, and Parameters Influencing the Visibility of Carbon Nanotubes in a Polymer Matrix Using an SEM. *Carbon* **2011,** *49*(6), 1955–1964.
41. Göldel, A.; et al. Shape-Dependent Localization of Carbon Nanotubes and Carbon Black in an Immiscible Polymer Blend during Melt Mixing. *Macromolecules* **2011,** *44*(15), 6094–6102.
42. Zhang, S. M.; et al. Synergistic Effect in Conductive Networks Constructed with Carbon Nanofillers in Different Dimensions. *Express Polym. Lett.* **2012,** *6*(2), 159–168.
43. Sumfleth, J.; Cordobes Adroher, X.; Schulte, K. Synergistic Effects in Network Formation and Electrical Properties of Hybrid Epoxy Nanocomposites Containing Multi-Wall Carbon Nanotubes and Carbon Black. *J. Mater. Sci.* **2009,** *44*(12), 3241–3247.
44. Wen, M.; et al. The Electrical Conductivity of Carbon Nanotube/Carbon Black/Polypropylene Composites Prepared Through Multistage Stretching Extrusion. *Polymer* **2012,** *53*(7), 1602–1610.
45. Yin C. -L.; et al. Effect of Compounding Procedure on Morphology and Crystallization Behavior of Isotactic Polypropylene/Carbon Black Ternary Composites. *Polym. Adv. Technol.* **2012,** *23*(7), 1112–1120.
46. Shen, L.; et al. Thermodynamically Induced Self-Assembled Electrically Conductive Networks in Carbon-Black-Filled Ternary Polymer Blends. *Polym. Int.* **2012,** *61*(2), 163–168.
47. Baudouin, A. -C.; et al. Polymer Blend Emulsion Stabilization Using Carbon Nanotubes Interfacial Confinement. *Polymer* **2011,** *52*(1), 149–156.
48. Chen, J.; et al. A Simple Strategy to Achieve Very Low Percolation Threshold via the Selective Distribution of Carbon Nanotubes at the Interface of Polymer Blends. *J. Mater. Chem.* **2012,** *22*(42), 22398–22404.

49. Shen B.; Zhai, W.; Zheng, W. Ultrathin Flexible Graphene Film: an Excellent Thermal Conducting Material with Efficient EMI Shielding. *Adv. Funct. Mater.* **2014,** *24*(28), 4542–4548.
50. Shen, B.; et al. Compressible Graphene-Coated Polymer Foams with Ultralow Density for Adjustable Electromagnetic Interference (EMI) Shielding. *ACS Appl. Mater. Int.* **2016,** *8*(12), 8050–8057.
51. Chung, D. D. L. Electromagnetic Interference Shielding Effectiveness of Carbon Materials. *Carbon* **2001,** *39*(2), 279–285.
52. Umrao, S.; et al. Microwave-Assisted Synthesis of Boron and Nitrogen Co-Doped Reduced Graphene Oxide for the Protection of Electromagnetic Radiation In Ku-Band. *ACS Appl. Mater. Int.* **2015,** *7*(35), 19831–19842.
53. Huang, X.; et al. Graphene-Based Composites. *Chem. Soc. Rev.* **2012,** *41*(2), 666–686.
54. Zhu, Y.; et al. Graphene and Graphene Oxide: Synthesis, Properties, and Applications. *Adv. Mater.* **2010,** *22*(35), 3906–3924.
55. Cao, W. -Q.; et al. Temperature Dependent Microwave Absorption of Ultrathin Graphene Composites. *J. Mater. Chem. C* **2015,** *3*(38), 10017–10022.
56. Liang, J.; et al. Electromagnetic Interference Shielding of Graphene/Epoxy Composites. *Carbon* **2009,** *47*(3), 922–925.
57. Li, C.; et al. The Preparation and Properties of Polystyrene/Functionalized Graphene Nanocomposite Foams Using Supercritical Carbon Dioxide. *Polym. Int.* **2013,** *62*(7), 1077–1084.
58. Zhang, C. -S.; et al. Electromagnetic Interference Shielding Effect of Nanocomposites with Carbon Nanotube and Shape Memory Polymer. *Compos. Sci. Technol.* **2007,** *67*(14), 2973–2980.
59. Yang, Y.; Gupta, M. C.; Dudley, K. L. Towards Cost-Efficient EMI Shielding Materials Using Carbon Nanostructure-Based Nanocomposites. *Nanotechnology* **2007,** *18*(34), 345701.
60. Zhang, H. -B.; et al. Tough Graphene–Polymer Microcellular Foams For Electromagnetic Interference Shielding. *ACS Appl. Mater. Interfaces* **2011,** *3*(3), 918–924.
61. Eswaraiah, V.; Sankaranarayanan, V.; Ramaprabhu, S. Functionalized Graphene–PVDF Foam Composites for EMI Shielding. *Macromol. Mater. Eng.* **2011,** *296*(10), 894–898.
62. Ameli, A.; Jung, P. U.; Park, C. B. Electrical Properties and Electromagnetic Interference Shielding Effectiveness of Polypropylene/Carbon Fiber Composite Foams. *Carbon* **2013,** *60,* 379–391.
63. Thomassin, J. -M.; et al. Polymer/Carbon Based Composites as Electromagnetic Interference (EMI) Shielding Materials. *Mater. Sci. Eng. R* **2013,** *74*(7), 211–232.
64. Chung, D. D. L. Carbon Materials for Structural Self-Sensing, Electromagnetic Shielding and Thermal Interfacing. *Carbon* **2012,** *50*(9), 3342–3353.
65. Joo, J.; Lee, C. Y. High Frequency Electromagnetic Interference Shielding Response of Mixtures and Multilayer Films Based on Conducting Polymers. *J. Appl. Phys.* **2000,** *88*(1), 513–518.
66. Shui, X.; Chung, D. D. L. Submicron Diameter Nickel Filaments and Their Polymer-Matrix Composites. *J. Mater. Sci.* **2000,** *35*(7), 1773–1785.
67. Wu, J.; Chung, D. D. L. Improving Colloidal Graphite for Electromagnetic Interference Shielding Using 0.1 μm Diameter Carbon Filaments. *Carbon* **2003,** *41*(6), 1313–1315.
68. Tripathy, O. K.; Mallick, A. K.; Annadura, P. Studies on Microwave Shielding Materials Based on Ferrite and Carbon Black Filled EPDM Rubber in the X-Band Frequency. *Proceedings of Recent Advances in Polymers and Composites*. Allied Publishers, 2000.

69. Farid, E. -T.; Kamada, K.; Ohnabe, H. On the 'Curiosity' of Electrical Self-Heating, Static Charge and Electromagnetic Shielding Effectiveness from Carbon Black/Aluminium Flakes Reinforced Epoxy-Resin Composites. *Polym. Int.* **2002,** *51*(7), 635–646.
70. Paligova, M.; et al. Electromagnetic Shielding of Epoxy Resin Composites Containing Carbon Fibers Coated with Polyaniline Base. *Phys. A* **2004,** *335*(3), 421–429.
71. Ramasubramaniam, R.; Chen, J.; Liu, H. Homogeneous Carbon Nanotube/Polymer Composites for Electrical Applications. *Appl. Phys. Lett.* **2003,** *83*(14), 2928–2930.
72. Yang, Y.; et al. The Fabrication and Electrical Properties of Carbon Nanofibre–Polystyrene Composites. *Nanotechnology* **2004,** *15*(11), 1545.
73. Yang, Y.; et al. A Comparative Study of EMI Shielding Properties of Carbon Nanofiber and Multi-Walled Carbon Nanotube Filled Polymer Composites. *J. Nanosci. Nanotechnol.* **2005,** *5*(6), 927–931.
74. Yang, Y.; et al. Novel Carbon Nanotube-Polystyrene Foam Composites for Electromagnetic Interference Shielding. *Nano Lett.* **2005,** *5*(11), 2131–2134.
75. Cao, X.; et al. Polyurethane/Clay Nanocomposites Foams: Processing, Structure and Properties. *Polymer* **2005,** *46*(3), 775–783.
76. Thomassin, J. -M.; et al. Foams of Polycaprolactone/MWNT Nanocomposites for Efficient EMI. *J. Mater. Chem.* **2008,** *18*(7), 792–796.
77. Kim, H. M.; et al. Electrical Conductivity and Electromagnetic Interference Shielding of Multiwalled Carbon Nanotube Composites Containing Fe Catalyst. *Appl. Phys. Lett.* **2004,** *84*(4), 589–591.
78. Zhang, J. Recent Advances in Graphene Based Polymer Composites [J]. *New Chem. Mater.* **2012,** *8,* 004.
79. Loh, K. P.; et al. The Chemistry of Graphene. *J. Mater. Chem.* **2010,** *20*(12), 2277–2289.
80. Geim, A. K. Graphene: Status and Prospects. *Science* **2009,** *324*(5934), 1530–1534.
81. Fugetsu, B.; et al. Electrical Conductivity and Electromagnetic Interference Shielding Efficiency of Carbon Nanotube/Cellulose Composite Paper. *Carbon* **2008,** *46*(9), 1256–1258.
82. Liang, J.; et al. Molecular-Level Dispersion of Graphene into Poly (vinyl alcohol) and Effective Reinforcement of Their Nanocomposites. *Adv. Funct. Mater.* **2009,** *19*(14), 2297–2302.
83. Park, S. H.; et al. The Influence of Coiled Nanostructure on the Enhancement of Dielectric Constants and Electromagnetic Shielding Efficiency in Polymer Composites. *Appl. Phys. Lett.* **2010,** *96*(4), 043115.
84. Li, D.; et al. Processable Aqueous Dispersions of Graphene Nanosheets. *Nature Nanotechnol.* **2008,** *3*(2), 101–105.
85. Lee, L. J.; et al. Polymer Nanocomposite Foams. *Compos. Sci. Technol.* **2005,** *65*(15), 2344–2363.
86. Zeng, C.; et al. Polymer–Clay Nanocomposite Foams Prepared Using Carbon Dioxide. *Adv. Mater.* **2003,** *15*(20), 1743–1747.
87. Park, H. C.; et al. Membrane Formation by Water Vapor Induced Phase Inversion. *J. Membr. Sci.* **1999,** *156*(2), 169–178.
88. Ling, J.; et al. Facile Preparation of Lightweight Microcellular Polyetherimide/Graphene Composite Foams for Electromagnetic Interference Shielding. *ACS Appl. Mater. Interfaces.* **2013,** *5*(7), 2677–2684.
89. Kum, C. K.; et al. Effects of Morphology on the Electrical and Mechanical Properties of the Polycarbonate/Multi-Walled Carbon Nanotube Composites. *Macromol. Res.* **2006,** *14*(4), 456–460.

90. Thomassin, J. -M.; et al. Multiwalled Carbon Nanotube/Poly (ε-caprolactone) Nanocomposites with Exceptional Electromagnetic Interference Shielding Properties. *J. Phys. Chem. C* **2007,** *111*(30), 11186–11192.
91. Thomassin, J. -M. et al. Functionalized Polypropylenes as Efficient Dispersing Agents for Carbon Nanotubes in a Polypropylene Matrix; Application to Electromagnetic Interference (EMI) Absorber Materials. *Polymer* **2010,** *51*(1), 115–121.
92. Landry, C. J. T.; et al. Poly (vinyl acetate)/Silica-Filled Materials: Material Properties of in situ vs. Fumed Silica Particles. *Macromolecules* **1993,** *26*(14), 3702–3712.
93. Li, G. Z.; et al. Viscoelastic and Mechanical Properties of Epoxy/Multifunctional Polyhedral Oligomeric Silsesquioxane Nanocomposites and Epoxy/Ladderlike Polyphenylsilsesquioxane Blends. *Macromolecules* **2001,** *34*(25), 8686–8693.
94. Tsagaropoulos, G.; Eisenberg, A. Dynamic Mechanical Study of the Factors Affecting the Two Glass Transition Behavior of Filled Polymers. Similarities and Differences with Random Ionomers. *Macromolecules* **1995,** *28*(18), 6067–6077.
95. Sung, Y. T.; et al. Dynamic Mechanical and Morphological Properties of Polycarbonate/Multi-Walled Carbon Nanotube Composites. *Polymer* **2005,** *46*(15), 5656–5661.
96. Qian, D.; et al. Load Transfer and Deformation Mechanisms in Carbon Nanotube-Polystyrene Composites. *Appl. Phys. Lett.* **2000,** *76*(20), 2868–2870.
97. Stephan, C.; Nguyen, T. P.; de la Lamy, C. M.; Lefrant, S.; Journet, C.; Bernier, P. *Synth. Met.* **2000,** *108,* 139.
98. Shaffer, M. S. P.; Windle, A. H. Fabrication and Characterization of Carbon Nanotube/Poly (vinyl alcohol) Composites. *Adv. Mater.* **1999,** *11*(11), 937–941.
99. Cheng, Y.; Zhou, O. Electron Field Emission from Carbon Nanotubes. *C. R. Phys.* **2003,** *4*(9), 1021–1033.
100. Pötschke, P.; Fornes, T. D.; Paul, D. R. Rheological Behavior of Multiwalled Carbon Nanotube/Polycarbonate Composites. *Polymer* **2002,** *43*(11), 3247–3255.
101. Manchado, M. A. L.; et al. Thermal and Mechanical Properties of Single-Walled Carbon Nanotubes–Polypropylene Composites Prepared by Melt Processing. *Carbon* **2005,** *43*(7), 1499–1505.
102. Mičušík, M.; et al. A Comparative Study on the Electrical and Mechanical Behaviour of Multi-Walled Carbon Nanotube Composites Prepared by Diluting a Masterbatch with Various Types of Polypropylenes. *J. Appl. Polym. Sci.* **2009,** *113*(4), 2536–2551.
103. Kong, H.; Gao, C.; Yan, D. Controlled Functionalization of Multiwalled Carbon Nanotubes by In Situ Atom Transfer Radical Polymerization. *J. Am. Chem. Soc.* **2004,** *126*(2), 412–413.
104. Zhang, Y.; He, H.; Gao, C. Clickable Macroinitiator Strategy to Build Amphiphilic Polymer Brushes on Carbon Nanotubes. *Macromolecules* **2008,** *41*(24), 9581–9594.
105. Zhang, Y.; et al. Covalent Layer-by-Layer Functionalization of Multiwalled Carbon Nanotubes by Click Chemistry. *Langmuir* **2009,** *25*(10), 5814–5824.
106. Chen, J.; et al. Noncovalent Engineering of Carbon Nanotube Surfaces by Rigid, Functional Conjugated Polymers. *J. Am. Chem. Soc.* **2002,** *124*(31), 9034–9035.
107. Chen, R. J.; et al. Noncovalent Sidewall Functionalization of Single-Walled Carbon Nanotubes for Protein Immobilization. *J. Am. Chem. Soc.* **2001,** *123*(16), 3838–3839.
108. Yuan, W. Z.; et al. Wrapping Carbon Nanotubes in Pyrene-Containing Poly (phenylacetylene) Chains: Solubility, Stability, Light Emission, and Surface Photovoltaic Properties. *Macromolecules* **2006,** *39*(23), 8011–8020.

109. Lee, S. H.; et al. Rheological and Electrical Properties of Polypropylene/MWCNT Composites Prepared with MWCNT Masterbatch Chips. *Eur. Polym. J.* **2008,** *44*(6), 1620–1630.
110. Koval'chuk, A. A.; et al. Effect of Carbon Nanotube Functionalization on the Structural and Mechanical Properties of Polypropylene/MWCNT Composites. *Macromolecules* **2008,** *41*(20), 7536–7542.
111. Gelves, G. A.; Al-Saleh, M. H.; Sundararaj, U. Highly Electrically Conductive and High Performance EMI Shielding Nanowire/Polymer Nanocomposites by Miscible Mixing and Precipitation. *J. Mater. Chem.* **2011,** *21*(3), 829–836.
112. Jia, L-C; Yan, D-X; Cui, C-h; Jiang, X; Li, Z-M. Electrically Conductive and electromagnetic Interference Shielding of Polyethylene Composites with Devisable Carbon Nanotube Networks. *J. Mater. Chem. C* **2015,** *3,* 9369–9378.
113. Babal, A. S.; Gupta R; Singh, B. P.; Singh, V. N.; Dhakate, S. R.; Mathur, R. B. *RSC Adv.* **2014,** *4,* 64649.
114. Arjmand, M.; et al. Comparative Study of Electromagnetic Interference Shielding Properties of Injection Molded Versus Compression Molded Multi-Walled Carbon Nanotube/Polystyrene Composites. *Carbon* **2012,** *50*(14), 5126–5134.
115. Pande, S.; et al. Improved Electromagnetic Interference Shielding Properties of MWCNT–PMMA Composites Using Layered Structures. *Nanoscale Res. Lett.* **2009,** *4*(4), 327.
116. Al-Saleh, M. H.; Sundararaj, U. *Carbon* **2009,** *47,* 1738.
117. Yim, Y. -J.; Park, S. -J. Electromagnetic Interference Shielding Effectiveness of High-Density Polyethylene Composites Reinforced with Multi-Walled Carbon Nanotubes. *J. Ind. Eng. Chem.* **2015,** *21,* 155–157.
118. Singh, B. P.; et al. Designing of Multiwalled Carbon Nanotubes Reinforced Low Density Polyethylene Nanocomposites for Suppression of Electromagnetic Radiation. *J. Nanopart. Res.* **2011,** *13*(12), 7065–7074.
119. Wang, G.; et al. Facile Synthesis and Characterization of Graphene Nanosheets. *J. Phys. Chem. C* **2008,** *112*(22), 8192–8195.
120. Blake, P.; et al. Graphene-Based Liquid Crystal Device. *Nano Lett.* **2008,** *8*(6), 1704–1708.
121. Kuilla, T.; et al. Recent Advances in Graphene Based Polymer Composites. *Prog. Polym. Sci.* **2010,** *35*(11), 1350–1375.
122. Sundaram, R. S. et al. Electrochemical Modification of Graphene. *Adv. Mater.* **2008,** *20*(16), 3050–3053.
123. Matsuo, Y.; Hatase, K.; Sugie, Y. Selective Intercalation of Aromatic Molecules into Alkyltrimethylammonium Ion-Intercalated Graphite Oxide. *Chem. Lett.* **1999,** *10,* 1109–1110.
124. Hirata, M.; Gotou, T.; Ohba, M. Thin-Film Particles of Graphite Oxide. 2: Preliminary Studies for Internal Micro Fabrication of Single Particle and Carbonaceous Electronic Circuits. *Carbon* **2005,** *43*(3), 503–510.
125. Becerril, H. A.; et al. Evaluation of Solution-Processed Reduced Graphene Oxide Films as Transparent Conductors. *ACS Nano* **2008,** *2*(3), 463–470.
126. Wang, X.; Zhi, L.; Müllen, K. Transparent, Conductive Graphene Electrodes for Dye-Sensitized Solar Cells. *Nano Lett.* **2008,** *8*(1), 323–327.
127. Li, X.; et al. Chemically Derived, Ultrasmooth Graphene Nanoribbon Semiconductors. *Science* **2008,** *319*(5867), 1229–1232.

128. Miranda, R.; Amadeo, L.; Vázquez de Parga. Graphene: Surfing ripples towards new devices. *Nat. Nanotechnol.* **2009,** *4*(9), 549–550.
129. Van Lier, G.; et al. Ab Initio Study of the Elastic Properties of Single-Walled Carbon Nanotubes and Graphene. *Chem. Phys. Lett.* **2000,** *326*(1), 181–185.
130. Robertson, D. H.; Brenner, D. W.; Mintmire, J. W. Energetics of Nanoscale Graphitic Tubules. *Phys. Rev. B* **1992,** *45*(21), 12592.
131. Li, Y.; et al. Tensile Properties of Long Aligned Double-Walled Carbon Nanotube Strands. *Carbon* **2005,** *43*(1), 31–35.
132. Shin, D. -I., Gitzhofer, F.; Moreau, C. Thermal Property Evolution of Metal-Based Thermal Barrier Coatings with Heat Treatments. *J. Mater. Sci.* (Full Set) **2007,** *42*(15), 5915–5923.
133. Woo, M. W.; et al. Melting Behavior and Thermal Properties of High Density Polythylene. *Polym. Eng. Sci.* **1995,** *35*(2), 151–156.
134. Sun, Y.; Luo, Y. Jia, D. Preparation and Properties of Natural Rubber Nanocomposites with Solid-State Organomodified Montmorillonite. *J. Appl. Polym. Sci.* **2008,** *107*(5), 2786–2792.
135. Ventura, G.; Martelli, V. Thermal Conductivity of Kevlar 49 between 7 and 290 K. *Cryogenics* **2009,** *49*(12), 735–737.
136. Yasmin, A.; Luo, J. -J. Daniel, I. M. Processing of Expanded Graphite Reinforced Polymer Nanocomposites. *Compos. Sci. Technol.* **2006,** *66*(9), 1182–1189.
137. Debelak, B.; Lafdi, K. Use of Exfoliated Graphite Filler to Enhance Polymer Physical Properties. *Carbon* **2007,** *45*(9), 1727–1734.
138. Zheng, W., Wong, S. -C. Electrical Conductivity and Dielectric Properties of PMMA/Expanded Graphite Composites. *Compos. Sci. Technol.* **2003,** *63*(2), 225–235.
139. Ramanathan, T.; et al. Graphitic Nanofillers in PMMA Nanocomposites—An Investigation of Particle Size and Dispersion and Their Influence on Nanocomposite Properties. *J. Polym. Sci. Part B Polym. Phys.* **2007,** *45*(15), 2097–2112.
140. Kalaitzidou, K.; Fukushima, H.; Drzal, L. T. A New Compounding Method for Exfoliated Graphite–Polypropylene Nanocomposites with Enhanced Flexural Properties and Lower Percolation Threshold. *Compos. Sci. Technol.* **2007,** *67*(10), 2045–2051.
141. Kalaitzidou, K.; Fukushima, H.; Drzal, L. T. Mechanical Properties and Morphological Characterization of Exfoliated Graphite–Polypropylene Nanocomposites. *Compos. Part A Appl. Sci. Manuf.* **2007,** *38*(7), 1675–1682.
142. Wakabayashi, K.; et al. Polymer-Graphite Nanocomposites: Effective Dispersion and Major Property Enhancement via Solid-State Shear Pulverization. *Macromolecules* **2008,** *41*(6), 1905–1908.
143. Kim, S.; Do, I.; Drzal, L. T. Multifunctional xGnP/LLDPE Nanocomposites Prepared by Solution Compounding Using Various Screw Rotating Systems. *Macromol. Mater. Eng.* **2009,** *294*(3), 196–205.
144. Kim, S.; Seo, J.; Drzal, L. T. Improvement of Electric Conductivity of LLDPE Based Nanocomposite by Paraffin Coating on Exfoliated Graphite Nanoplatelets. *Compos. Part A Appl. Sci. Manuf.* **2010,** *41*(5), 581–587.
145. Zheng, W.; Lu, X.; Wong, S. -C. Electrical and Mechanical Properties of Expanded Graphite-Reinforced High-Density Polyethylene. *J. Appl. Polym. Sci.* **2004,** *91*(5), 2781–2788.
146. Zhao, Y. F.; et al. Preparation and Properties of Electrically Conductive PPS/Expanded Graphite Nanocomposites. *Compos. Sci. Technol.* **2007,** *67*(11), 2528–2534.

147. Zhao, X.; et al. Enhanced Mechanical Properties of Graphene-Based Poly (vinyl alcohol) Composites. *Macromolecules* **2010,** *43*(5), 2357–2363.
148. Lee, Y. R.; et al. Properties of Waterborne Polyurethane/Functionalized Graphene Sheet Nanocomposites Prepared by an in situ Method. *Macromol. Chem. Phys.* **2009,** *210*(15), 1247–1254.
149. Liang, J.; et al. Infrared-Triggered Actuators from Graphene-Based Nanocomposites. *J. Phys. Chem. C* **2009,** *113*(22), 9921–9927.
150. Du, X. S.; Xiao, M.; Meng, Y. Z. Synthesis and Characterization of Polyaniline/Graphite Conducting Nanocomposites. *J. Polym. Sci. Part B Polym. Phys.* **2004,** *42*(10), 1972–1978.
151. Cho, D.; et al. Dynamic Mechanical and Thermal Properties of Phenylethynyl-Terminated Polyimide Composites Reinforced With Expanded Graphite Nanoplatelets. *Macromol. Mater. Eng.* **2005,** *290*(3), 179–187.
152. Jović, N.; et al. Temperature Dependence of the Electrical Conductivity of Epoxy/Expanded Graphite Nanosheet Composites. *Scr. Mater.* **2008,** *58*(10), 846–849.
153. Celzard, A.; et al. Composites Based on Micron-Sized Exfoliated Graphite Particles: Electrical Conduction, Critical Exponents and Anisotropy. *J. Phys. Chem. Solids* **1996,** *57*(6), 715–718.
154. Chen, G.; et al. PMMA/Graphite Nanosheets Composite and its Conducting Properties. *Eur. Polym. J.* **2003,** *39*(12), 2329–2335.
155. Wang, W.; et al. Synthesis and Characteristics of Poly (Methyl Methacrylate)/Expanded Graphite Nanocomposites. *J. Appl. Polym. Sci.* **2006,** *100*(2), 1427–1431.
156. Chen, G.; et al. Preparation of Polystyrene/Graphite Nanosheet Composite. *Polymer* **2003,** *44*(6), 1781–1784.
157. Liu, N.; et al. One-Step Ionic-Liquid-Assisted Electrochemical Synthesis of Ionic-Liquid-Functionalized Graphene Sheets Directly from Graphite. *Adv. Funct. Mater.* **2008,** *18*(10), 1518–1525.
158. Eda, G.; Chhowalla, M. Graphene-Based Composite Thin Films for Electronics. *Nano Lett.* **2009,** *9*(2), 814–818.
159. Hu, H.; et al. Preparation and Properties of Graphene Nanosheets–Polystyrene Nanocomposites via in situ Emulsion Polymerization. *Chem. Phys. Lett.* **2010,** *484*(4), 247–253.
160. Zou, J. -F.; et al. Conductive Mechanism of Polymer/Graphite Conducting Composites with Low Percolation Threshold. *J. Polym. Sci. Part B Polym. Phys.* **2002,** *40*(10), 954–963.
161. Xiao, M.; et al. Synthesis and Properties of Polystyrene/Graphite Nanocomposites. *Polymer* **2002,** *43*(8), 2245–2248.
162. Weng, W.; et al. Fabrication and Characterization of Nylon 6/Foliated Graphite Electrically Conducting Nanocomposite. *J. Polym. Sci. Part B Polym. Phys.* **2004,** *42*(15), 2844–2856.
163. Chen, X. -M.; Shen, J. -W.; Huang, W. -Y. Novel Electrically Conductive Polypropylene/Graphite Nanocomposites. *J. Mater. Sci. Lett.* **2002,** *21*(3), 213–214.
164. Du, X. S.; Xiao, M.; Meng, Y. Z. Facile Synthesis of Highly Conductive Polyaniline/Graphite Nanocomposites. *Eur. Polym. J.* **2004,** *40*(7), 1489–1493.
165. Yu, C.; Li, B. Morphology and Properties of Conducting Polyvinyl Alcohol Hydrosulfate/Graphite Nanosheet Composites. *J. Compos. Mater.* **2008,** *42*(15), 1491–1504.
166. Ansari, S.; Giannelis, E. P. Functionalized Graphene Sheet—Poly (Vinylidene Fluoride) Conductive Nanocomposites. *J. Polym. Sci. Part B Polym. Phys.* **2009,** *47*(9), 888–897.

167. Wang, H.; et al. Graphene Oxide Doped Polyaniline for Supercapacitors. *Electrochem. Commun.* **2009,** *11*(6), 1158–1161.
168. Zhang, H.-B.; et al. Electrically Conductive Polyethylene Terephthalate/Graphene Nanocomposites Prepared by Melt Compounding. *Polymer* **2010,** *51*(5), 1191–1196.
169. Kim, H.; Macosko, C. W. Processing-Property Relationships of Polycarbonate/Graphene Composites. *Polymer* **2009,** *50*(15), 3797–3809.
170. Sun, G.; et al. Hierarchical Dendrite-Like Magnetic Materials of Fe_3O_4, γ-Fe_2O_3, and Fe with High Performance of Microwave Absorption. *Chem. Mater.* **2011,** *23*(6), 1587–1593.
171. Lu, A. -H.; Salabas, E. L.; Schüth, F. Magnetische Nanopartikel: Synthese, Stabilisierung, Funktionalisierung und Anwendung. *Angew. Chem. Int. E* **2007,** *119*(8), 1242–1266.
172. Frey, N. A.; et al. Magnetic Nanoparticles: Synthesis, Functionalization, and Applications in Bioimaging and Magnetic Energy Storage. *Chem. Soc. Rev.* **2009,** *38*(9), 2532–2542.
173. Sun, X.; et al. Laminated Magnetic Graphene with Enhanced Electromagnetic Wave Absorption Properties. *J. Mater. Chem. C* **2013,** *1*(4), 765–777.
174. Deng, H.; et al. Monodisperse Magnetic Single-Crystal Ferrite Microspheres. *Angew. Chem. interference Int. E* **2005,** *117*(18), 2842–2845.
175. Rutter, G. M.; et al. Scattering and Interference in Epitaxial Graphene. *Science* **2007,** *317*(5835), 219–222.
176. Sun, S.; et al. Gradient-Index Meta-Surfaces as a Bridge Linking Propagating Waves and Surface Waves. *Nature Mater.* **2012,** *11*(5), 426–431.
177. Sukang, B.; et al. Roll-to-Roll Production of 30-inch Graphene Films for Transparent Electrodes. *Nature Nanotechnol.* **2010,** *5*(8), 574–578.
178. Hwang, E. H.; Sarma, S. D. Acoustic Phonon Scattering Limited Carrier Mobility in Two-Dimensional Extrinsic Graphene. *Phys. Rev. B* **2008,** *77*(11), 115449.
179. Neirinck, B.; et al. Aqueous Electrophoretic Deposition in Asymmetric AC Electric Fields (AC–EPD). *Electrochem. Commun.* **2009,** *11*(1), 57–60.
180. Besra, L.; Liu, M. A Review on Fundamentals and Applications of Electrophoretic Deposition (EPD). *Prog. Mater. Sci.* **2007,** *52*(1), 1–61.
181. Kim, S.; et al. Electromagnetic Interference (EMI) Transparent Shielding of Reduced Graphene Oxide (RGO) Interleaved Structure Fabricated by Electrophoretic Deposition. *ACS Appl. Mater. Interfaces* **2014,** *6*(20), 17647–17653.
182. Cummins, J. J.; et al. Advanced Strain Energy Accumulator: Materials, Modeling and Manufacturing. *ASME/BATH 2014 Symposium on Fluid Power and Motion Control,* American Society of Mechanical Engineers, 2014.
183. Jamal, E. M. A.; et al. Synthesis of Nickel–Rubber Nanocomposites and Evaluation of Their Dielectric Properties. *Mater. Sci. Eng. B* **2009,** *156*(1), 24–31.
184. Jun, J. -B.; et al. Synthesis and Characterization of Monodisperse Magnetic Composite Particles for Magnetorheological Fluid Materials. *Colloids Surf. A Physicochem. Eng. Aspects* **2005,** *260*(1), 157–164.
185. Youyi, S.; et al. Effect of Magnetic Nanoparticles on the Properties of Magnetic Rubber. *Mater. Res. Bull.* **2010,** *45*(7), 878–881.
186. Makled, M. H.; et al. Magnetic and Dynamic Mechanical Properties of Barium Ferrite–Natural Rubber Composites. *J. Mater. Process. Technol.* **2005,** *160*(2), 229–233.
187. Malini, K. A.; Kurian, P.; Anantharaman, M. R. Loading Dependence Similarities on the Cure Time and Mechanical Properties of Rubber Ferrite Composites Containing Nickel Zinc Ferrite. *Mater. Lett.* **2003,** *57*(22), 3381–3386.

188. Brzozowska, M.; Krysinski, P. Synthesis and Functionalization of Magnetic Nanoparticles with Covalently Bound Electroactive Compound Doxorubicin. *Electrochim. Acta* **2009,** *54*(22), 5065–5070.
189. Iwamoto, T.; Yoshitaka K.; Toshima, N. Anomalous Magnetic Behavior in FePt Nanoparticles Chemically Synthesized with Polymer Protective Agent. *Phys. B Condens. Matter* **2009,** *404*(14), 2080–2085.
190. Zhang, B.; et al. Microwave-Absorbing Properties of De-aggregated Flake-Shaped Carbonyl-Iron Particle Composites at 2–18 GHz. *IEEE Trans. Magn.* **2006,** *42*(7), 1778–1781.
191. Sindhu, S.; et al. Evaluation of ac Conductivity of Rubber Ferrite Composites from Dielectric Measurements. *Bull. Mater. Sci.* **2002,** *25*(7), 599–607.
192. Boris, T. M. *Physics of Dielectric Materials*; Mir publishers, 1975.
193. Musameh, S. M.; et al. Some Electrical Properties of Aluminum-Epoxy Composite. *Mater. Sci. Eng. B* **1991,** *10*(1), 29–33.
194. Mathur, R. B.; Chatterjee, S.; Singh, B. P. Growth of Carbon Nanotubes on Carbon Fibre Substrates to Produce Hybrid/Phenolic Composites with Improved Mechanical Properties. *Compos. Sci. Technol.* **2008,** *68*(7), 1608–1615.
195. Singh, B. P.; et al. Designing of Epoxy Composites Reinforced with Carbon Nanotubes Grown Carbon Fiber Fabric for Improved Electromagnetic Interference Shielding. *AIP Adv.* **2012,** *2*(2), 022151.
196. Mishra, R. K.; Thomas, S.; Karikal, N. *Micro and Nano Fibrillar Composites (MFCs and NFCs) from Polymer Blends*, 1st ed., 2017.
197. Zeng, W; Tan, S. T. Preparation and Electromagnetic Interference Shielding Properties of Nickel-Coated PET Fiber Filled Epoxy Composites. *Polym. Compos.* **2006,** *27*(1), 24–29.
198. Huang, Y.; et al. The Influence of Single-Walled Carbon Nanotube Structure on the Electromagnetic Interference Shielding Efficiency of its Epoxy Composites. *Carbon* **2007,** *45*(8), 1614–1621.
199. Liu, Z.; et al. Reflection and Absorption Contributions to the Electromagnetic Interference Shielding of Single-Walled Carbon Nanotube/Polyurethane Composites. *Carbon* **2007,** *45*(4), 821–827.
200. Das, N. C.; Maiti, S. Electromagnetic Interference Shielding of Carbon Nanotube/Ethylene Vinyl Acetate Composites. *J. Mater. Sci.* **2008,** *43*(6), 1920–1925.

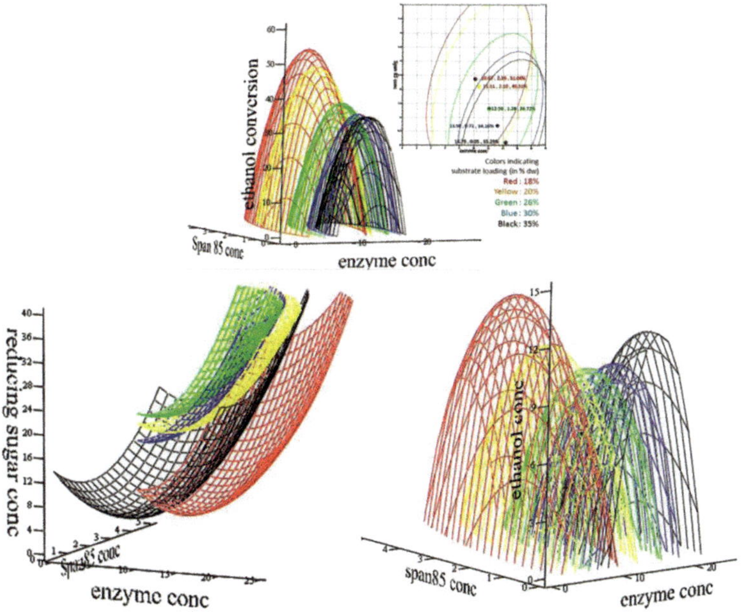

FIGURE 2.2 Profile of cellulose conversion, reducing sugar concentration and ethanol concentration that predicted by the obtained mathematical model.

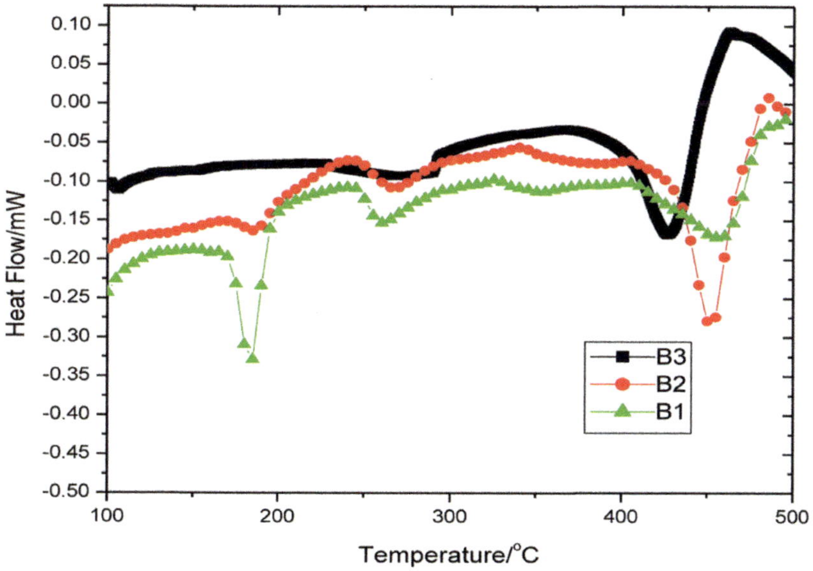

FIGURE 6.5 Differential scanning calorimetry thermogram of different compositions B_1, B_2, and B_3 at a heating rate of 10 K/min.

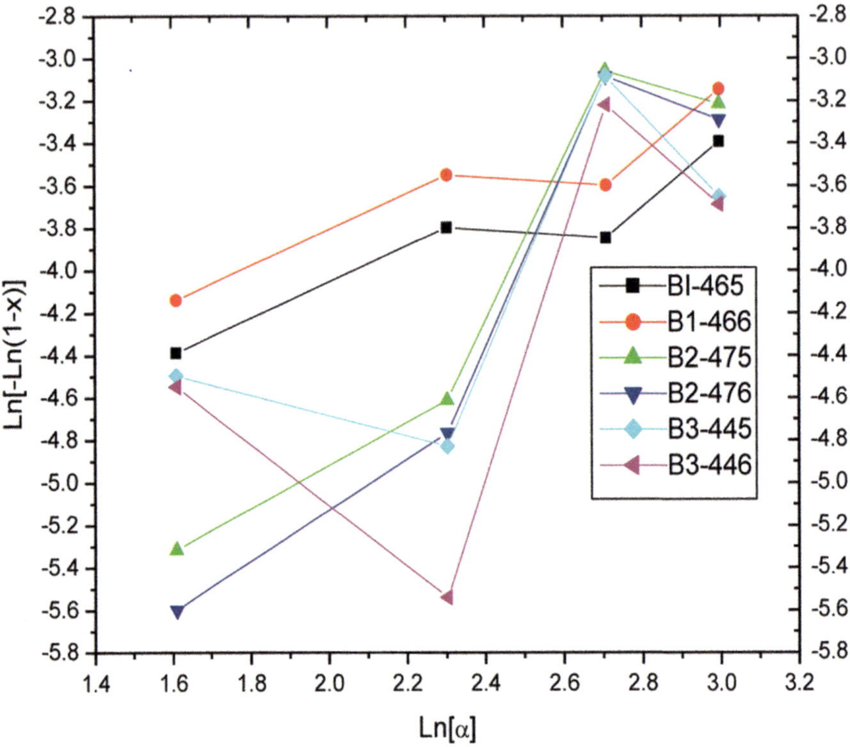

FIGURE 6.6 Evaluation of Avrami index for all the samples.

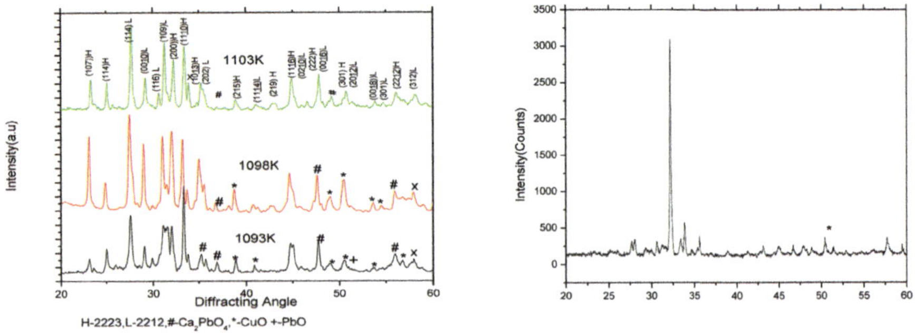

(a) (b)

FIGURE 6.7 Diffraction pattern of the composition ($n=4$) sintered at different temperatures (a) 1093, 1098, and 1103 and (b) 1113 K.

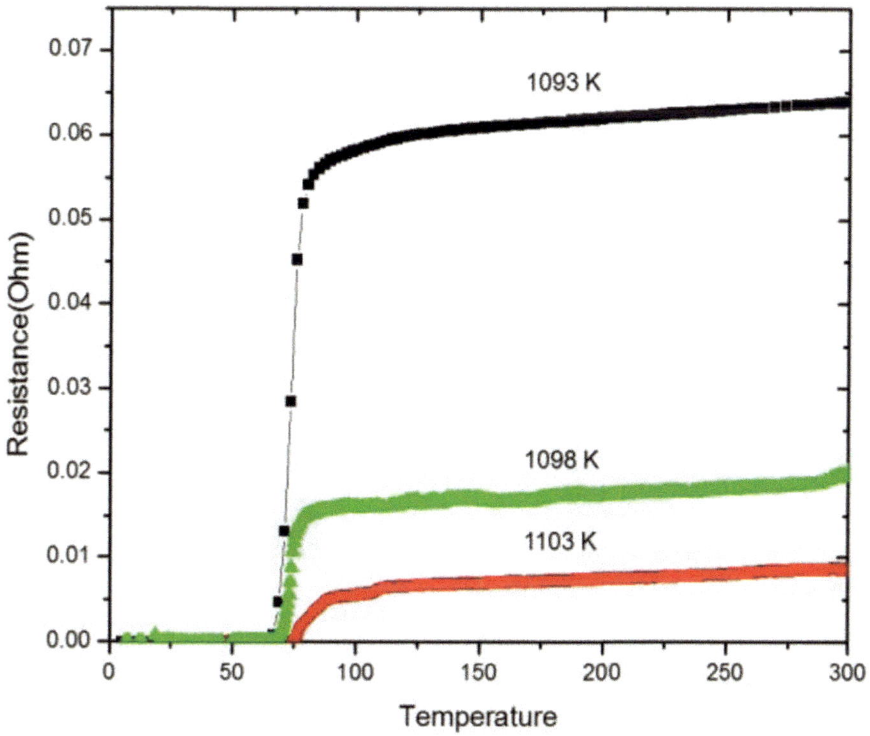

FIGURE 6.8 Variation of resistance with temperature composition ($n=4$) sintered for 24 h at different temperatures.

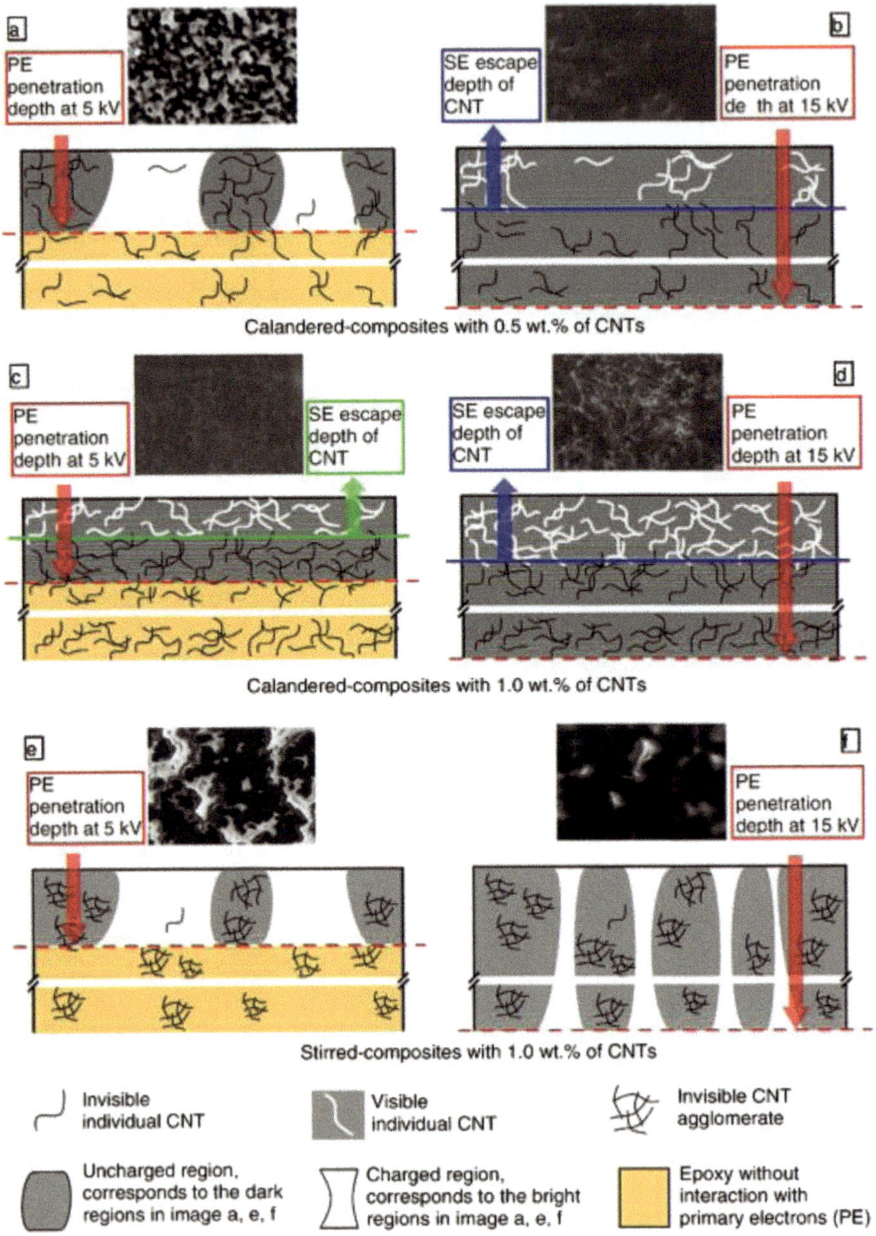

FIGURE 9.2 Scanning electron microscopy (SEM) images for carbon nanotube (CNT)-reinforced epoxy calendered and stirred nanocomposites obtained using charge contrast mode.

Source: Adapted with permission from ref 35.

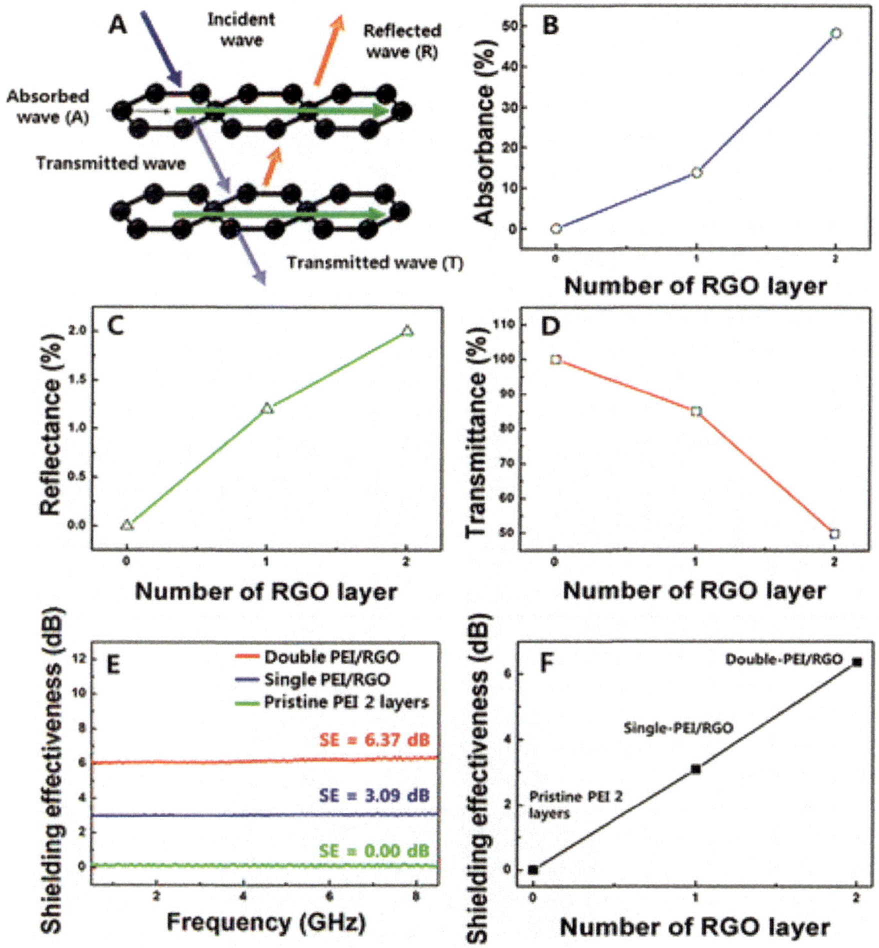

FIGURE 9.21 (a) Electromagnetic interference shielding mechanism for RGO, (b) absorbance, (c) reflectance, (d) transmittance as a function of RGO layers, and (e) shielding efficiency for PEI/RGO composites.

Source: Adapted with permission from ref 181. Copyright (2016) American Chemical Society.

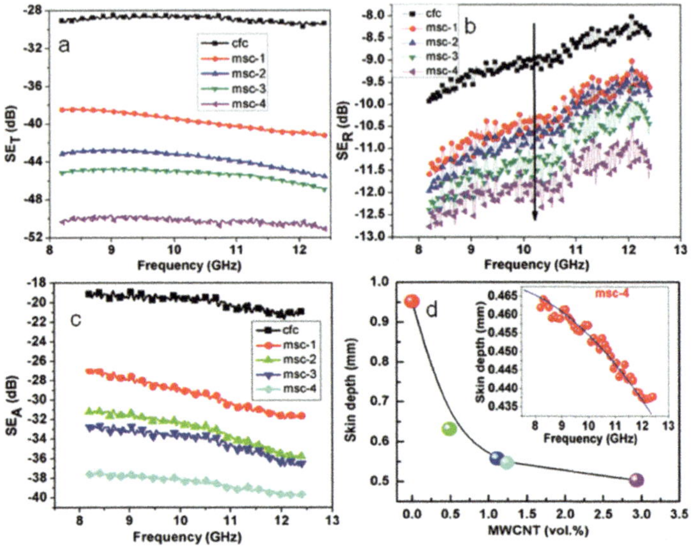

FIGURE 9.26 (a) Variation of electromagnetic interference shielding effectiveness with respect to frequency, (b) electromagnetic interference shielding effectiveness due to reflection, (c) electromagnetic interference shielding effectiveness due to absorbance for epoxy composites, and (d) variation of skin depth as a function of MWCNT loading.[195]

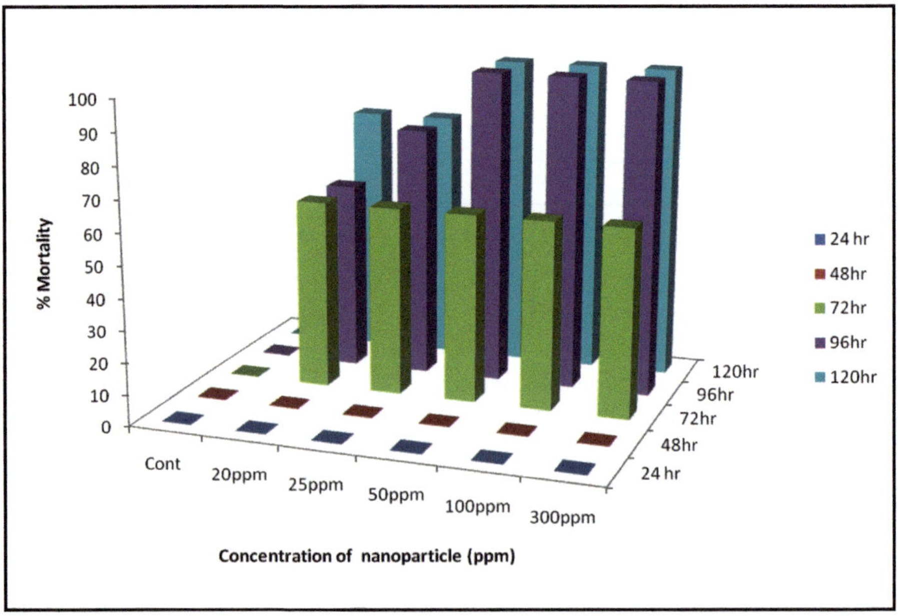

FIGURE 11.1 Mortality rate (in percentage) measurement.

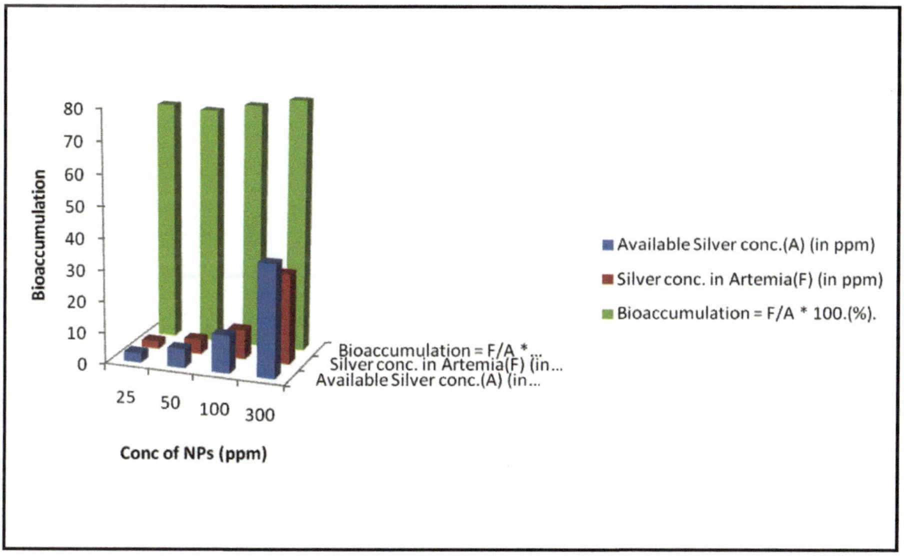

FIGURE 11.3 Inductively coupled plasma optical emission spectrometry–bioaccumulation of silver nanoparticles (NPs).

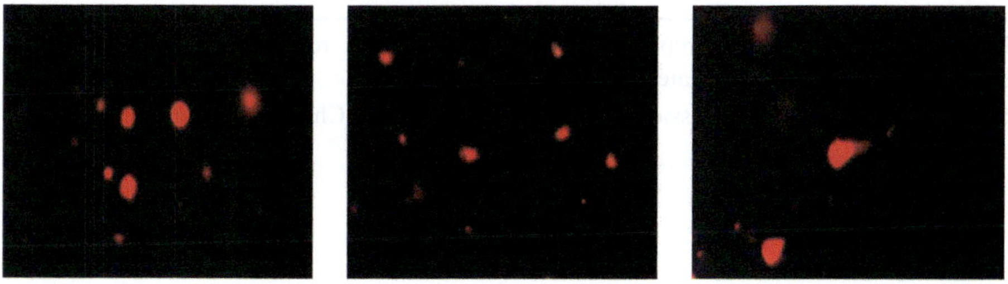

FIGURE 11.5 (a) Comet at control, (b) comet at 20 ppm, (c) comet at 300 ppm.

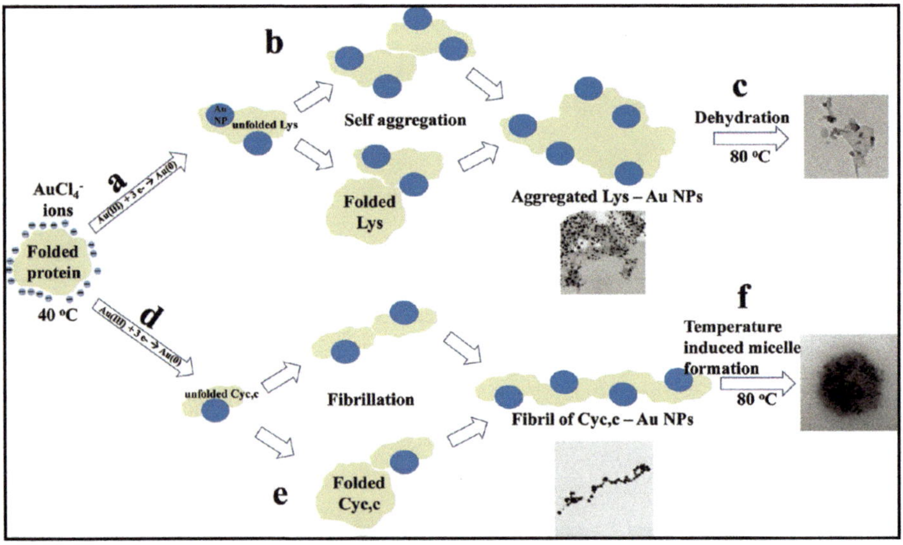

FIGURE 12.1 Schematic representation of the proposed reaction mechanism for the synthesis of Lys/cyt. c-conjugated AuNPs (nanoparticles).

Source: Reprinted with permission ref 39. © 2010. American Chemical Society.

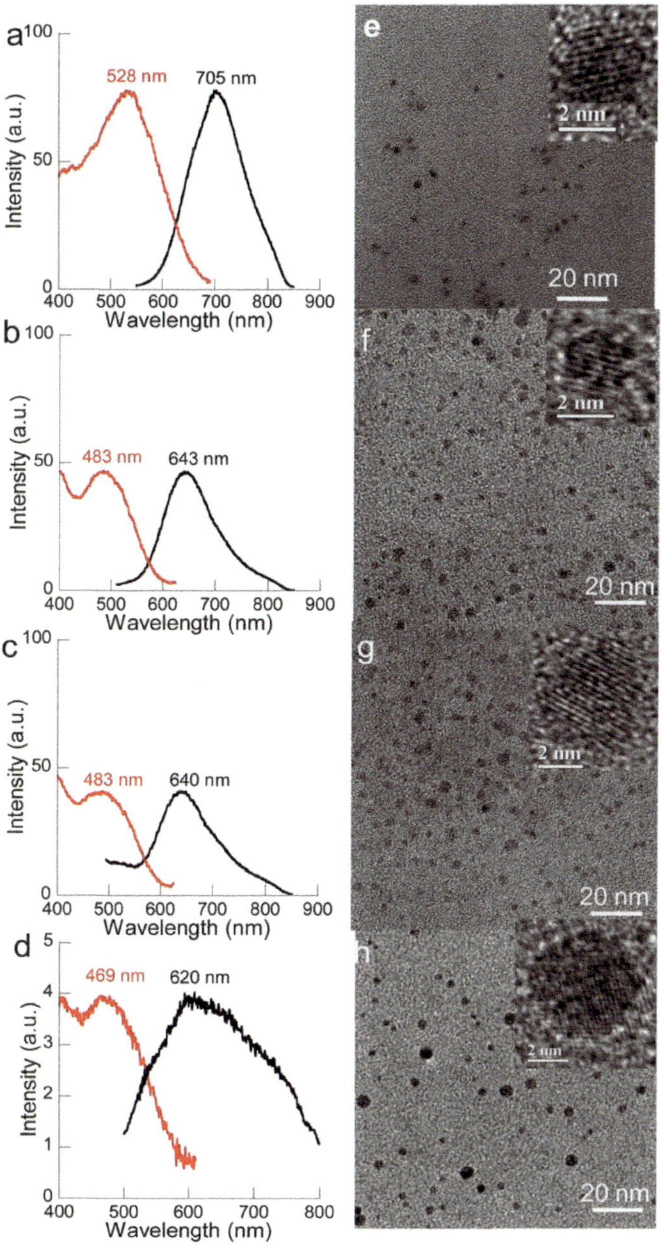

FIGURE 12.2 Protein-encapsulated Au nanoclusters (NCs): fluorescent emission (black)/excitation (red) spectra and transmission electron microscopy (TEM) images of AuNCs generated from BSA (a and e), trypsin (b and f), lysozyme (c and g), and pepsin (d and h).

Source: Reprinted with permission from Royal Society of Chemistry, ref 51.

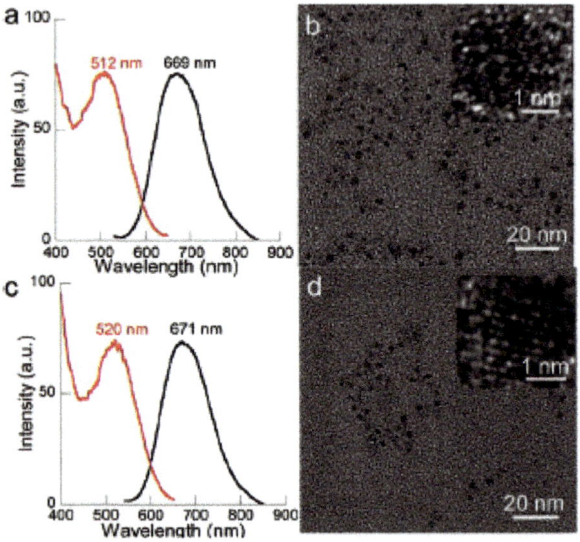

FIGURE 12.3 Fluorescent emission (black)/excitation (red) spectra and TEM images of AuNCs generated from excess trypsin (a and b) and lysozyme (c and d).

Source: Reprinted with permission from Royal Society of Chemistry, ref 51.

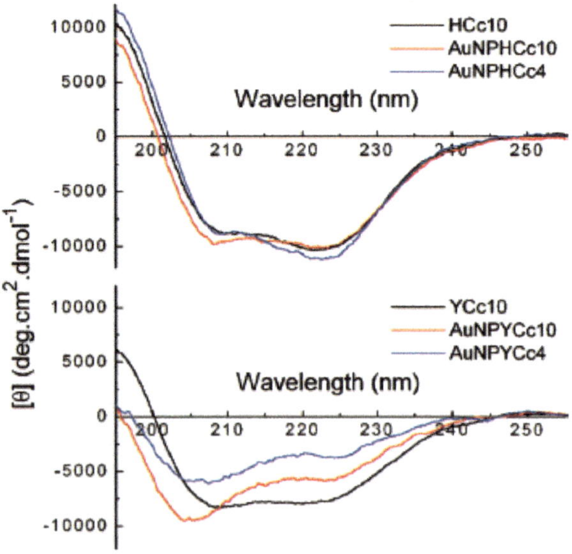

FIGURE 12.4 Far-ultraviolet (UV) circular dichroism (CD) spectra for cyt c alone at pH 10 (black traces), in the bionanoconjugates at pH 10 (red traces), or at pH 4 where aggregation occurs (blue traces). The top spectra correspond to HCc and HCc–AuNP and the bottom spectra are for YCc and YCc–AuNP.

Source: Reprinted with permission from American Chemical Society, ref 59.

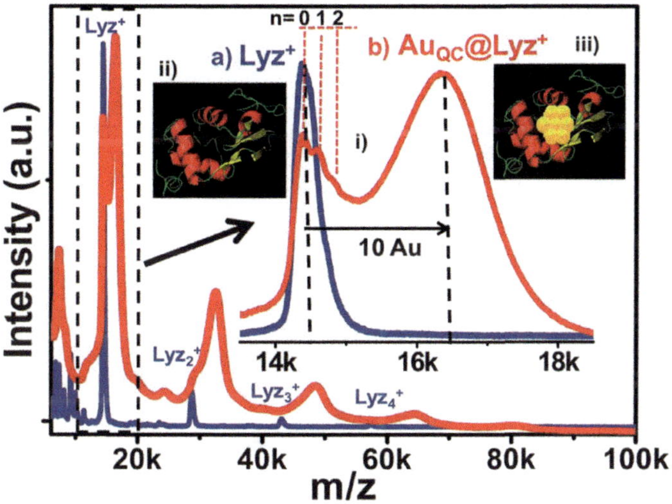

FIGURE 12.7 Positive-ion matrix-assisted laser desorption/ionization (MALDI) mass spectroscopy (MS) of Lyz at pH 12 in linear mode (a) and Au_{QC}–Lyz after 24 h of incubation. (b) All the spectra were measured in the linear positive mode over the m/z range of 20,000–100,000. Both Lyz and Au_{QC}–Lyz showed aggregate formation. The expanded monomer region in inset (i) clearly shows a separation of 10 Au atoms from the parent protein. In the dimer, trimer, tetramer, and pentamer regions, the separations are of 20, 30, 40, and 50 Au atoms, respectively. In insets (ii) and (iii), schematic representations of Lyz and Au_{QC}–Lyz, respectively, are shown.

Source: Reprinted with permission from Royal Society of Chemistry, ref 74.

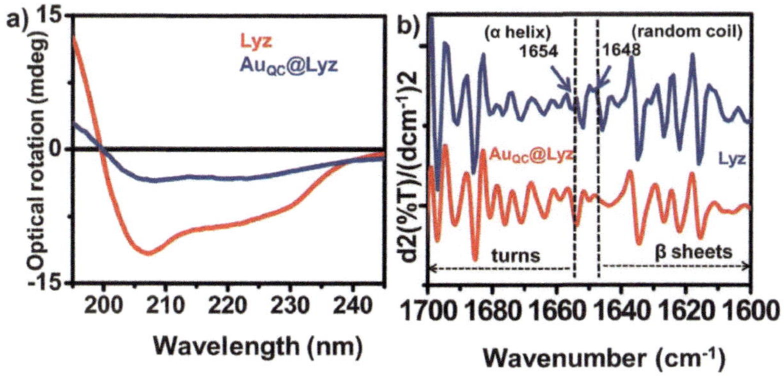

FIGURE 12.8 (a) CD spectra of Lyz and as prepared Au_{QC}–Lyz showing a clear change in ellipticity of the spectra, which indicates a huge change in the α-helical structure. (b) Double derivative of the infrared spectra shows the disappearance of the peak at 1654 cm^{-1} in the case of Au_{QC}–Lyz.

Source: Reprinted with permission from Royal Society of Chemistry, ref 74.

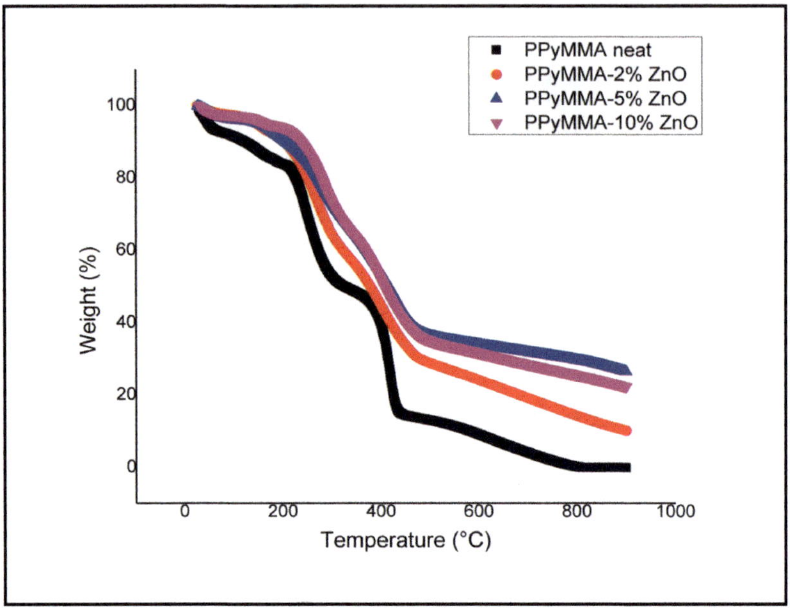

FIGURE 14.6 Thermogravimetric analysis of PPyMMA and PPyMMA/ZnO with different loadings of OA-ZnO.

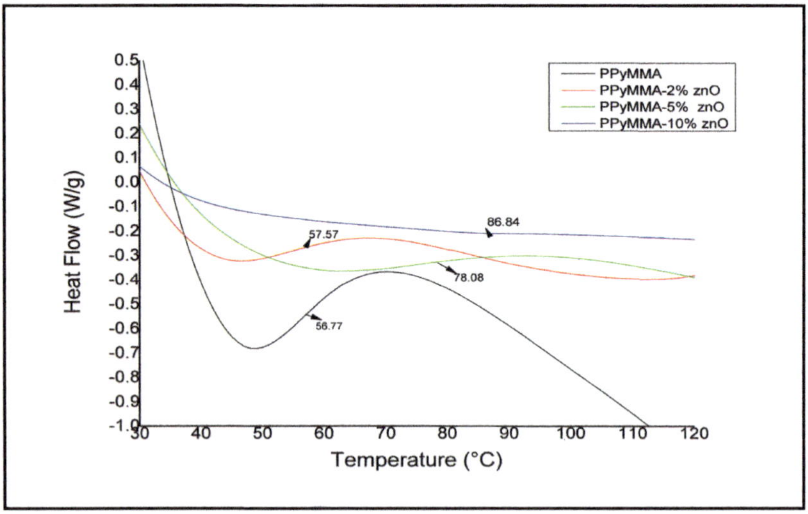

FIGURE 14.7 Differential scanning calorimetry spectra of PPyMMA and PPyMMA/ZnO with different loadings of OA-ZnO.

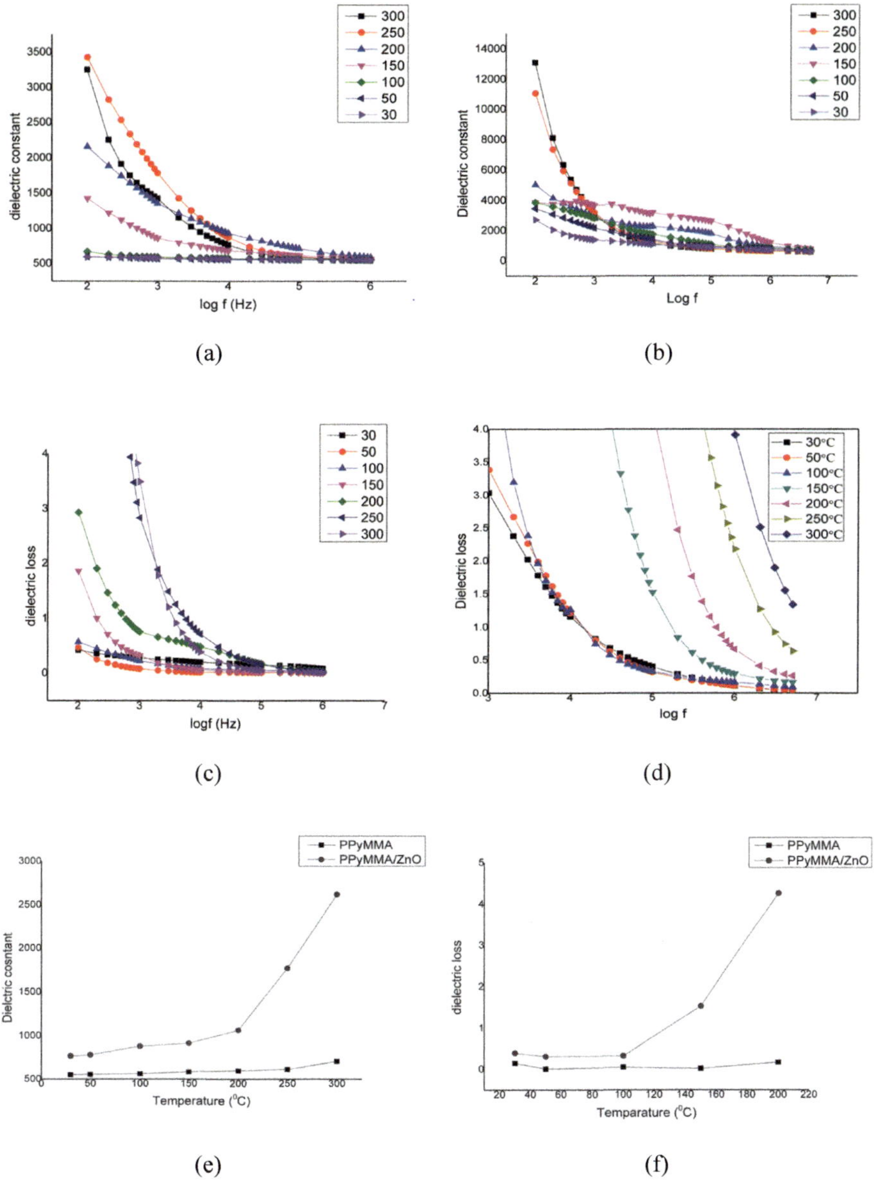

FIGURE 14.9 (a) Dielectric constant of PPyMMA, (b) dielectric constant for PPyMMA/ZnO (2 wt.%), (c) dielectric loss for PPyMMA, (d) dielectric loss for PPyMMA/ZnO (2 wt.%), (e) variations of the dielectric constant with the temperature at 100 kHz, and (f) variations of dielectric loss with the temperature at 100 kHz.

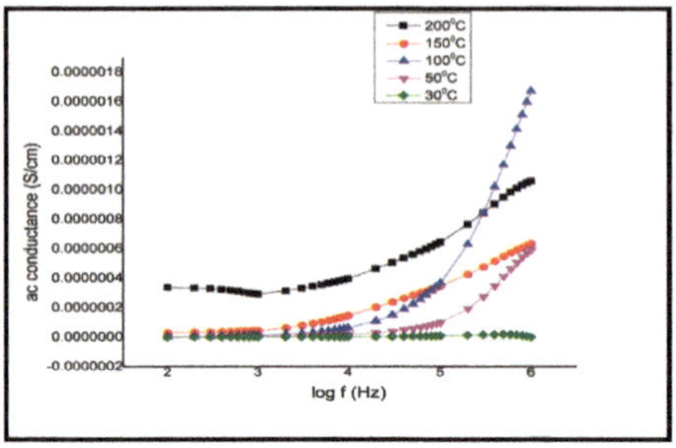

(a)

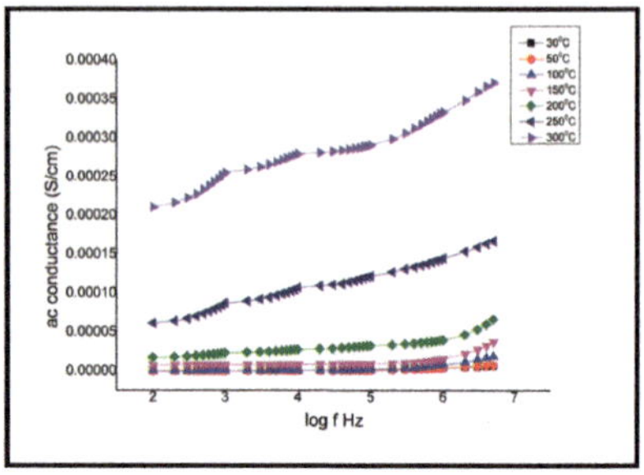

(b)

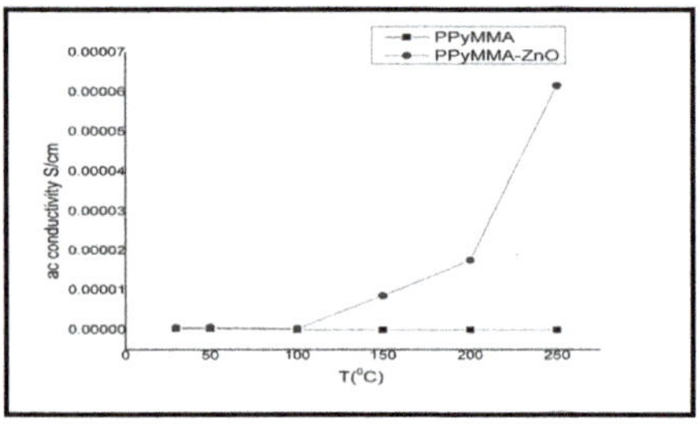

(c)

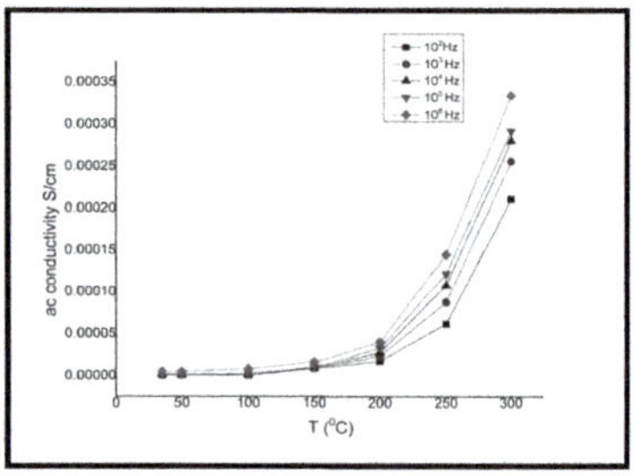

(d)

FIGURE 14.10 (a) Alternating current (AC) conductance for PPyMMA, (b) AC conductance for PPyMMA/ZnO (2 wt.%), (c) AC conductivity of PPyMMA and PPyMMA/ZnO at 105 Hz, and (d) AC conductivity of PPyMMA/ZnO at different temperatures.

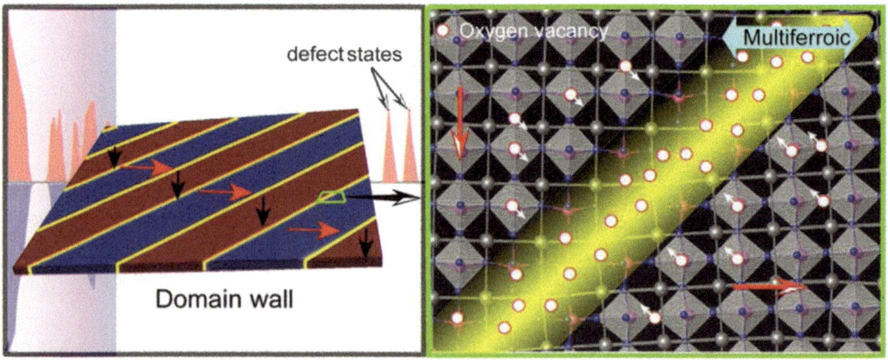

FIGURE 17.2 Multiferroic domain walls.

Source: Reprinted (adapted) with permission from ref 58. © 2016 American Chemical Society.

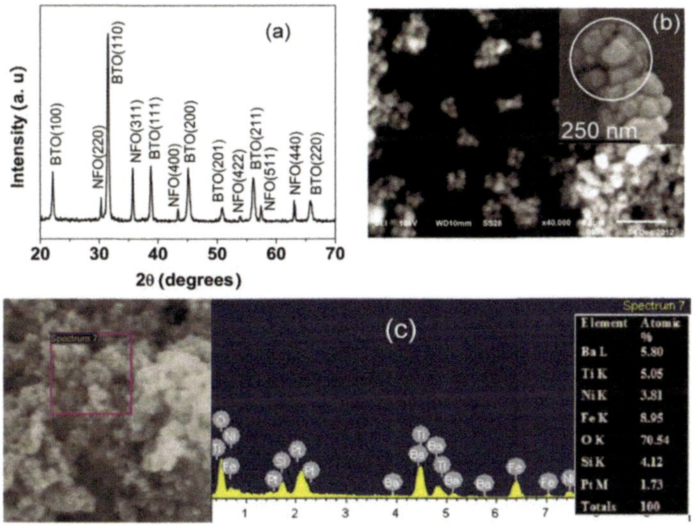

FIGURE 17.3 (a) X-ray diffraction (XRD) data for as-assembled core–shell particles with 100 nm NFO core and 50 nm BTO shell (sample-A). (b) SEM micrograph of clusters of sample-A. (c) energy-dispersive X-ray spectroscopy data for sample-A.

Source: Reprinted with permission from ref 47. © 2014 AIP Publishing LLC.

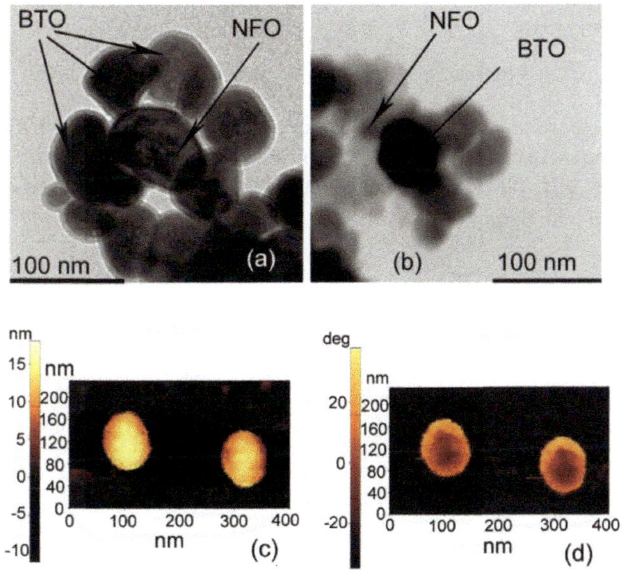

FIGURE 17.4 (a) TEM micrograph showing core–shell structures for sample-A. (b) Similar TEM micrograph for clusters of 50 nm BTO core and 10-nm NFO shell (sample-B). MFM (c) amplitude and (d) phase image of sample-B.

Source: Reprinted with permission from ref 47. © 2014 AIP Publishing LLC.

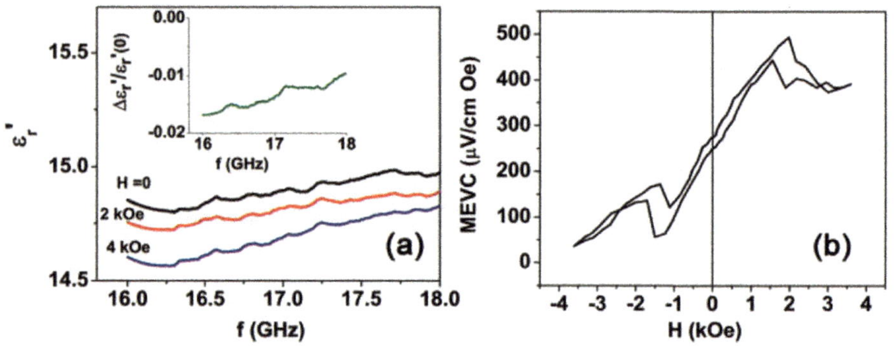

FIGURE 17.5 (a) Real part of the relative permittivity ε_r' versus frequency for an as-assembled sample-A and the relative variation in ε_r' in H i4 kOe estimated from the data. (b) Low-frequency ME voltage coefficient versus bias field H data for a film of sample-A.

Source: Reprinted with permission from ref 47. © 2014 AIP Publishing LLC.

CHAPTER 10

BREAKTHROUGHS IN NANOFIBROUS MEMBRANES FOR INDUSTRIAL WASTEWATER TREATMENT

PREMLATA AMBRE[1], JOGINDER SINGH PANEYSAR[1], EVANS COUTINHO[1], SUKHWINDER KAUR BHULLAR[2,*]

[1]*Department of Pharmaceutical Chemistry, Bombay College of Pharmacy, Mumbai 400098, India*

[2]*Department of Mechanical Engineering, Osmangazi Campus, Gaziakdemir Mah., Mudanya Cad. No. 4/10, 16190 Osmangazi, Bursa Technical University, Bursa, Turkey*

Corresponding author. E-mail: sbhullar@uvic.ca

CONTENTS

ABSTRACT

Traditionally industries are employing pH neutralization, coagulation and sedimentation, pressure floatation, bio-oxidation, filtration and absorption techniques for the wastewater treatment before channelizing it to industrial effluent treatment (IET), however the most suitable, recently used technology of nanofibers have gained lot of importance due to their nonwoven structure, porosity, surface area, chemical properties, scalability, and economic viability for purification of industrial waste water. Nanofibers can be fabricated using suitable chemistry to synthesize stronger and more stable materials, modify surfaces to produce fluid channels with proper dimensions that will ensure rapid throughput on an industrial scale.

This chapter throws light on the plenteous of applicability for nanofibers over conventional and revolutionary methods of industrial waste water purification. It also accentuates the preparation and characterization of nanofibres using sophisticated technologies, their types and array of applications in water treatment. The later section highlights the future directions and advantages of chemical modification from a various polymeric material to a nonwoven nanofibrous membrane for elimination of organic as well as inorganic impurities from wastewater. Thus, projecting the superiority of nanofibres over the existing nanosystems for water purification due to their ease of handling, reusability and recylability.

10.1 INTRODUCTION

Industrial activities are responsible for a significant proportion of toxic waste poured into the environment. This toxic waste is defined as an "industrial effluent." The industrial effluent in the form of liquid or wastewater is generated from manufacturing processes and/or from any activity occurring in an industrial premise. Water and sewerage authorities usually set and define the limits of pollutants in an effluent that can be discharged into the sewers. The specifications typically include concentration (and possible load), limits for chemical oxygen demand (COD), and suspended solids (SSs). Often discharged effluent needs to be treated to meet the standards set by the controlling authorities to minimize pollutants that can be discharged into the sewers. Irrespective of the size of the company, it is a legal requirement to comply with trade effluents as specified by the permitted limits.

Currently, industrial effluent treatment (IET) technology employs comprehensive processes such as pH neutralization, coagulation and sedimentation, pressure floatation, bio-oxidation, filtration, and absorption. Heavy metal ions such as Pb^{+2}, Cr^{+6}, As^{+5}, Cu^{+2}, Zn^{+2}, Cd^{+2}, and Hg^{+2} or organic compounds such as formaldehyde, phenols, chlorides, sulfides, oils/grease, colorants, and ammoniacal nitrogen or oils and grease in water are a serious concern worldwide because of their high toxicity both to human beings and the ecology. Various methods have been employed for the purification of wastewater by industries before channelizing to IET, namely screening, evaporation, flocculation, electrolysis, electrodialysis, chemical redox followed by precipitation, physical and biological adsorption, ion exchange, solvent extraction, reverse osmosis (RO), and membrane separation.[1] A recently discovered inexpensive and scalable technology to clean wastewater uses nanofibers. These next generations of membranes or nonwoven filter media made from polymeric nanofibers have unique physicochemical properties. Polymeric nano-/microfibers are ubiquitous in many spheres of human life as shown in Figure 10.1. Clothing, apparel, cosmetics, cigarette filters, air-conditioning filters, fishing nets, composites, surgical masks, extracorporeal devices, vascular grafts, heart valves are few examples. Fibers used in these applications are typically 5–50 μm in diameter and are made from a variety of polymers, both synthetic and natural origin.

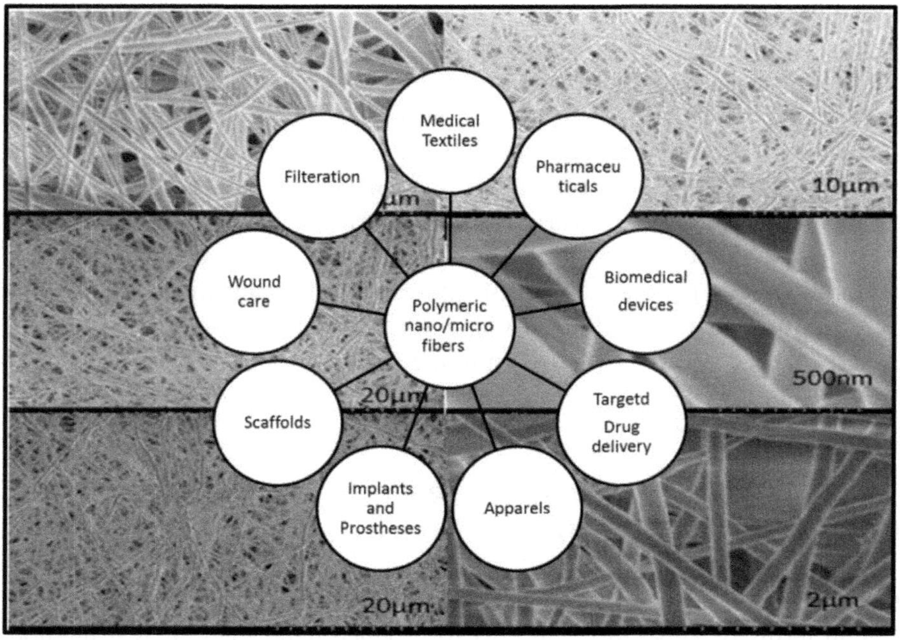

FIGURE 10.1 Applications of nano-/microfibers.

According to one estimate, the world fiber consumption touches 46 million t/year. Polymer fibers with a thousand times smaller diameter are called nanofibers. These help to increase both the surface-area-to-weight ratio and surface-area-to-volume ratio significantly. Increased surface area means a feasibility to enhance the functionality of fibers. This could lead to a highly sensitive filter that is able to selectively capture matter small in size.

In this chapter, we present the current research on nanofibrous membranes and membranes designed for the removal of contaminations from effluents. This will first summarize the novel technologies used for the preparation of nanofibers; subsequently, the removal of major impurities by these nanofibers from effluents will be discussed. Finally, we present techniques for the design of newer adsorbents that will have the potential to extract simultaneously a large number of effluents from wastewater. The chapter also focuses on recent developments and future directions in this domain and the ability to enhance the filtration performance in the presence of support materials.

10.1.1 STATE OF SCIENCE AND TECHNOLOGY: METHODS TO FABRICATE POLYMER NANOFIBERS, ALIGNMENTS OF NANOFIBERS, AND APPLICATION IN WASTEWATER TREATMENT

Nanofibers have extremely high surface-area-to-weight ratio as compared to conventional fibers. The diameter, surface area, size, and pore volume influence the diffusion process and thus are considered important factors for filtration. The diameter is generally less than 100 nm. All these parameters can be tailored by altering various factors during fabrication to make them usable for intended applications. For this reason, polymeric nanofibrous assemblies have generated a lot of interest in pharmaceuticals and in the field of treatment of effluents. Generally, the filter efficiency increases linearly with decreasing filter membrane thickness and increasing applied pressure.

10.1.2 TECHNIQUES USED FOR FABRICATION OF POLYMERIC NANOFIBERS

Electrospinning process is perhaps the most promising of all nanotechnologies in terms of versatility and cost to produce nanofibers. It is a method which produces strongly elongated polymer or ceramic structures with submillimeter diameter by means of applied electric fields. The first step of the process generally leads to a pendant droplet at the end of the fluidic stream. An electric (positive or negative) field (2–100 kV) is applied between the solution and the (often grounded) collecting surface by connecting electrodes to a high-voltage generator. The voltage applied between the spinneret and collector may be either positive or negative. Upon increasing the applied voltage, electric forces ultimately prevail and an electrically charged solution jet is extruded. During this process, the droplet at the spinneret progressively becomes more and more elongated. The next step in the process is solvent evaporation (drying), which controls the nanofiber porosity. The type of fiber, that is, solid or porous is obtained when the ratio of tPS (phase separation time)/tD (drying time) is greater than or less than 1, respectively; on the other hand, when the phase separation phenomena is absent or very slow, the fiber interior becomes solid. With electrospinning, nanofibrous assemblies can be prepared from various materials such as polymers, ceramics, or metals with full control of the fiber fineness, surface morphology, orientation, and cross-sectional configuration. This gives the method a distinct advantage over other processes. However, electrospinning has disadvantages of solvent recovery, jet instability, and low productivity; these limitations can be overcome by the relatively newer technique of force spinning.[2–4]

10.1.3 ELECTROSPINNING TECHNIQUES FOR NANOFABRICATION

A variety of techniques of electrospinning, namely direct-dispersed electrospinning, gas–solid reaction, in situ photoreduction, solgel processes, emulsions, co-evaporation, and coaxial electrospinning have been reported by a number of investigators in the literature. Different electrospinning techniques are illustrated in Figure 10.2.

a) *Layer-by-layer electrospinning* is a complementary way to generate multicomponent samples, which consists of sequentially deposited layers of nanofibers of different polymer solutions onto the same collector. The result is a multilayered sample with optical, electrical, biochemical, or mechanical anisotropy.[5]
b) *Electro-blowing* is a method which involves a gas distributor. It consists of conduits surrounding the spinneret that produce a controlled flow of gas all around the jet. Electro-blowing method

can be used for producing nanofibers from high-viscosity materials such as hyaluronic acid.[6]

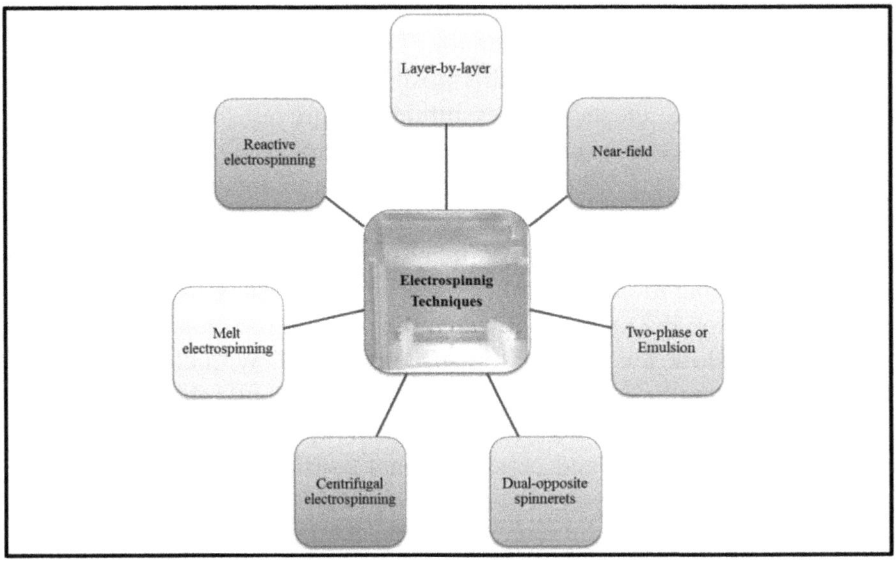

FIGURE 10.2 Different electrospinning techniques.

c) *Reactive electrospinning* involves a reaction with a polymer and a cross-linking agent along the jet. This process produces a new material from two or more chemically different solutions that are injected into the spinneret. This method is usually applicable to polymers whose solutions are too difficult to electrospin directly. It also includes techniques which enhance the cross-linking while the jet travels toward the collector, by means of an external polymerization agent such as ultarviolet (UV) or laser light that drives the polymerization reaction.[7–10]

d) *Melt electrospinning* produces fibers by driving a polymer through a region where heaters are present that heat the material above its T_g; this reduces the viscosity to a value low enough to allow spinning. Melt electrospinning is useful for those thermoplastic compounds that are difficult to dissolve in organic solvents at room temperature. This method results in fibers with a slightly larger (< 1 mm) diameter than produced by other electrospinning methods described above. This is because of the much larger number of molecular entanglements present per unit volume compared to solutions. The process can be complicated and expensive as the electric field to be applied must be higher than in standard electrospinning, besides it also requires a vacuum for preparation.[11,12]

e) *Centrifugal electrospinning or double-spinning or force spinning* is a safe, simple, and more versatile method for making nanofibers. The method can produce a new class of nanofibers that are not feasible by electrospinning. It involves a combination of centrifugal forces and conventional electrospinning techniques. It produces a good alignment of the nanofibers along with extra degrees of freedom, which is controlled by means of the rotational speed of the spinner. Two processes such as force spinning and rotary jet spinning can deliver nanofibers using just centrifugal forces without the application of electric field.[13–15]
f) *Dual-opposite-spinnerets electrospinning* method consists of two counterfacing spinnerets with opposite electric potentials provided by a high-voltage power supply. It produces a cluster of nanofibers due to the attraction and collision of the two complex helical whipping jets. It reduces the whipping and bending instability of the electrified jets by highly concentrated polymer solutions.
g) *Two-phase or emulsion electrospinning* involves two solutions from a single spinneret, namely a polymer solution in an organic solvent and a relatively immiscible aqueous solution containing a water-soluble doping agent such as a protein or a hydrophilic drug.[16] This method is widely used to tailor the fiber morphology, especially for phase-separating polymeric blends.
h) *Near-field electrospinning technique* aims to improve the degree of control in positioning single fiber during the deposition stage by decreasing the distance between the spinneret and the collector (down to the range of 0.5–5 mm). It helps to eliminate the jet instability issue and hence enables a much more accurate positioning of the fibers than in conventional electrospinning.[17–23]

10.1.4 OTHER APPROACHES FOR FABRICATION OF NANOFIBERS

10.1.4.1 SELF-ASSEMBLY AND POLYMERIZATION METHOD

This method creates a spontaneous arrangement of physical systems that produces a stable organized self-assembled polymeric structure. The arrangement of inter-component interactions is driven by relatively weak (i.e., non-covalent) intermolecular interactions (van der Waals, electrostatic, dipolar, hydrogen bonds, and so forth). The resultant assembly is thermodynamically more stable than the original state of the composites.

This method delivers a degree of adjustability to the overall supramolecular system that helps adjacent spatial and energetic configurations to find the most favorable, eventually ordered arrangements.

Fabrication of self-assembly nanofibers using marginal solvents: Self-assembled nanofibers can be prepared by precipitation with solvents. These solvents have a poor ability to dissolve a given polymer species with the solubility strongly dependent on temperature (these are so-called marginal solvents). In this case, the solubility is relatively good only at a high temperature. Consequently, cooling the solution under controlled conditions leads to the precipitation of solid nanofibers, which often exhibit enhanced microcrystalline order. However, the weakness of solvent-based assembly is the lack of an ability to gain spatial control, accurate positioning, or mutual alignment of the synthesized nanostructures.

10.1.4.2 CASTING SELF-ASSEMBLY METHOD

This is favored by the evaporation of a solution containing a solid substrate that has interacted with the solvent. Nanofibrous structures can then be assembled simply as a consequence of spin-coating processes, where the rapid solvent evaporation assists the formation and supramolecular organization of the polymer molecules.

10.1.4.3 POLYMERIZATION METHOD

Many light-emitting polymers have demonstrated conductive properties because of conjugation properties of the materials. The nanostructures of such conducting polymers (i.e., nanorods, -tubes, -wires, and -fibers) have the advantage of exhibiting organic conduction with low dimensionality and therefore create interesting physicochemical properties and potentially useful applications. The introduction of a "structural directing agent" such as surfactants,[24–26] liquid crystals,[27] polyelectrolytes,[28] nanowire seeds,[29] oligomers of the desired polymer,[30] or bulky organic dopants can guide the formation of desired nanostructures. It is believed that such functional molecules can either directly act as templates (e.g., polyelectrolytes) or promote the self-assembly of ordered "soft templates" (e.g., micelles, emulsions). The template is a chemical which strongly influences the polymerization reaction dynamics for monomers in solution or in vapor phase. These templates are also known as "molecular anchors." For example, anionic sites in polyheterocyclic compounds containing acidic groups bind the newly generated polymer nanostructure. The structure of the nanofibers depends on the polymerization time. Longer the polymerization time, thicker would be the wall of nanostructures resulting in solid nanofibers. However, by limiting the growth time of polymerization with molecular anchors, one can achieve hollow nanocylinders with very thin walls.

The synthetic process is quite robust and can be implemented using camphor, sulfonic acid, hydrochloric acid (HCl), perchloric acid ($HClO_4$), methanesulfonic acid (CH_3SO_3H), hydrogen-sulfated fullerenol with $-(O)SO_3H$ groups as the doping agents.[31] Recently, methods based on sodium chlorite ($NaClO_2$) and potassium biiodate $[KH(IO_3)_2]$[32a,32b] as oxidants allow nanofibers to be produced with especially good conduction properties.[33]

Bui et al.[34] prepared a nanomembrane by in situ polymerization of polyamide as the top coating layer over polyethersulfone-electrospun nanofibers. The group prepared several thin-layer nanofiber composite (TNFC) membranes with different polymer/solvent compositions and ratios for tailored osmosis. They observed 2–5 times higher water flux and 100 times salt rejection compared with a standard forward osmosis commercial membrane.

10.1.4.4 NANOFLUIDICS

It is defined as the study and application of fluid flow in and around nano-sized objects like nanochannels. The penetration of a polymer solution or melt within nanochannels can be significantly affected by non-covalent interactions at the interface with the surrounding walls of the capillary. Nanofluidics uses the filling of capillaries with size below 100 nm by the polymer in order to template the formation of nanofibers. The motion of the solution or of the melt in the nanochannels can either be driven by spontaneous capillarity or be assisted by external pumping. After capillary filling, several processes such as the evaporation of the residual solvent from the possibly open terminals of the nanochannel, diffusion of the solvent molecules through the material constituting the walls of the capillary, and the solidification of the polymer melt upon cooling below T_g finally lead to the formation of solid nanostructures.

10.1.4.5 INTERFACIAL POLYMERIZATION

It is a template-free polymerization method that leads to high-throughput nanofibers. The morphology and functional properties of nanofibers depend on the monomer structure, dopants, concentration of the reagents, and process temperature. In case of interfacial polymerization, one can induce polymerization at the interface of aqueous and organic phases. An example of oxidative polymerization of aniline in water using interfacial polymerization technique gives a polymer with the perfect nanofibrous morphology. In this process, an aqueous solution comprising the acid and the oxidant component is brought in contact with a solution of aniline in an organic solvent. The contact is maintained for a few seconds when the aniline monomer polymerizes at the interface forming nanofibers, and with the passage of time, the newly formed nanostructures gradually diffuse into the aqueous solution.[35] The length of contact is critical and rapid removal of the nanofibers from the reactive interface would avoid formation of secondary growth processes, which can transform fibers in more complex agglomerates as in conventional oxidative polymerization method.

The interfacial method can be realized by camphor, sulfonic acid,[36,37] hydrochloric, sulfuric, or nitric acid as the acid dopants.[38] In particular, the average diameter of the fibers shows an interesting dependence on the doping acid; this varies from 30 nm with use of hydrochloric acid to 120 nm with perchloric acid.

10.1.5 ALIGNMENT OF NANOFIBERS

Nanofibers are generally collected in the form of many layers in which fibers are deposited sequentially without forming mutual entanglements or kinks and are highly disordered. This disordered nature of the arrangement is termed as nonwoven fibers, which imply a random distribution of fiber orientations in the plane of deposition. The nonwoven structure comes about when the fibers are stacked layer by layer in the direction normal to the collector surface. In principle, if the electrified jet is stably produced and it follows its own trajectory from the spinneret to the collector without interruptions, an electrospun nonwoven structure can be constituted as a single, extremely long nanofiber, being deposited with many bends that form many overlapping layers of nanostructures. These mats are useful for many bulky applications, for example, catalysis, electrode materials, sensors, nanocomposites, medicine, and cosmetics.

Nonwoven filter media (microfilters) have certain limitations. The nonwoven media can remove particles between 10 and 200 μm in diameter due to the considerably larger average pore size generated by the micron-size fibers. Apart from this, particles are easily trapped and get lodged within the complex path of the nonwoven media. Hence, they are not easily cleaned and difficult to reuse. In addition, a thicker fibrous layer is required to reduce the overall average pore size of the media to separate particles of less than 10 μm in diameter. Nevertheless, nonwoven media are mechanically robust and hence are widely used as the base to support porous membranes. The limitations of nonwoven filter media discussed above can be reduced using electrospun nanofibrous membranes (ENMs). ENMs have a large dirt-loading capacity due to their large internal surface area. The current state of the art that uses prefilters to take the load off the downstream membranes can be replaced by ENMs. The porosity of ENMs is 10–20% greater than conventional phase-inverted membranes.[39] This higher porosity is highly beneficial as it provides higher fluxes, which are made possible of prefilters prepared from ENMs.[40,41]

10.2 CHARACTERIZATION OF NANOFIBERS

Nanofibers are characterized by their geometry, chemistry, surface properties, and performance.

10.2.1 GEOMETRICAL CHARACTERIZATION

The geometry of the ENMs includes characteristics such as fiber size, pore size, pore diameter, and porosity. Scanning electron microscopy (SEM), transmission electron microscopy (TEM), and atomic force microscopy (AFM) are used to characterize fiber size and pore diameter. Some SEM images for fiber size and diameter are illustrated in Figure 10.3.

TEM has the advantage of producing images under dry or wet stages. AFM produces two-dimensional (2D) images of fibers but it cannot record smaller objects accurately due to the size of the cantilever.

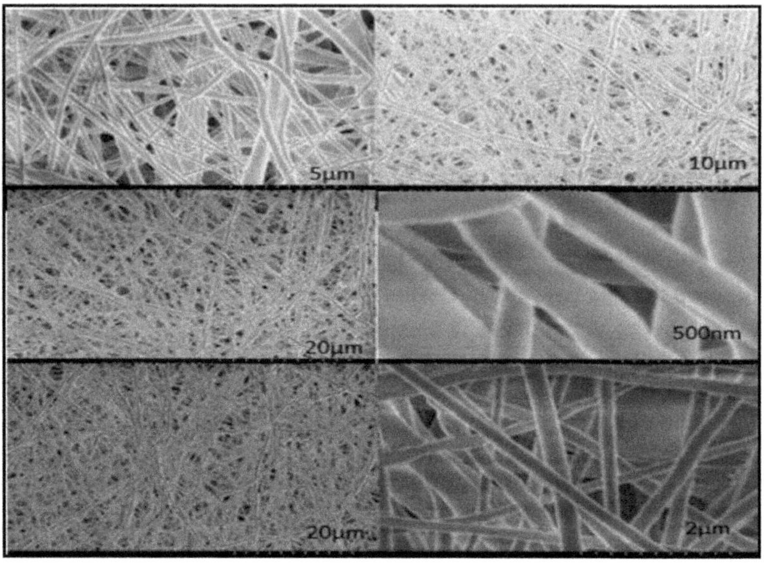

FIGURE 10.3 Scanning electron microscopy images of nanofibers.

10.2.2 PORE CHARACTERISTICS

This is measured using mercury intrusion techniques such as mercury intrusion porosimetry and liquid extrusion porosimetry.[42] Mercury is non-wetting to nanofibrous membranes and does not spontaneously penetrate the pores. In mercury intrusion technique, pressure is used to force mercury into the pores. The pore structure is determined through the intrusion pressure and volume. The disadvantage of this technique is the need of a high pressure and it is possible that the pore structure may get distorted by the pressure exerted in the experiment.

In liquid extrusion technique, the membrane is soaked in a liquid that spontaneously wets the membrane. Such a liquid would ideally form a zero contact angle with the membrane. Then, a relatively small pressure is used to force the liquid out of the membrane pores. This technique effectively measures the diameter of the pores.[43]

10.2.3 CHEMICAL CHARACTERIZATION OF NANOFIBERS

Fourier-transform infrared (FTIR) and nuclear magnetic resonance spectra can be used to determine chemical structure and the interaction between polymer molecules. X-ray photoelectron spectroscopy and attenuated total reflectance Fourier-transform infrared (ATR-FTIR) can be used for characterization of the surface chemistry of ENMs.

Surface properties: Surface properties of nanofiber membranes include roughness, surface charge, and hydrophobicity/hydrophilicity.

10.2.4 HYDROPHOBIC AND HYDROPHILIC PROPERTIES

Hydrophobicity and hydrophilicity are important parameters for the application of nanofibers. In the field of membrane distillation, the process needs hydrophobic membranes as it is a thermally driven process, where separation occurs due to phase change. Membrane distillation is a thermally driven separation program in which separation occurs due to phase change. A hydrophobic membrane presents a barrier for the liquid phase but allows the vapor phase (e.g., water vapor) to pass through the membrane's pores. The process is driven by a partial vapor pressure difference that is commonly triggered by a temperature difference. A hydrophobic membrane displays a barrier for the liquid phase, allowing the vapor phase (e.g., water vapor) to pass through the membrane's pores. On the other hand, in the case of separation of protein solutions, hydrophilic membranes are preferred, if possible, to minimize protein adsorption. Hydrophilic membranes will generally show optimum wettability and form filters with low pressure because of the lower water contact angle. It will significantly reduce the capillary pressure of the filtration medium and increase the flow rate of the liquid and the separation ability of the suspended particles. In addition, it will greatly eliminate the biofouling and clogging of the filter media during the filtration process. Currently, researchers are improving the hydrophilicity/hydrophobicity of the membrane by surface modification techniques.[39]

10.2.5 PERFORMANCE

Efficiency of nanofibers for water treatment has been evaluated by three methods: (a) pure water permeation, (b) molecular weight cutoff (MWCO), and (c) liquid entry pressure of water.

10.2.6 PURE WATER PERMEATION TEST

Water is pushed through the membrane at a certain pressure and under steady state. The flux is calculated by the ratio of water permeation rate and the area of the membrane.

10.2.7 MWCO

It is defined as the lowest molecular weight (in daltons) at which greater than 90% of a solute with a known molecular weight is retained by the membrane. In this process, water containing solutes such as dextran, polyethylene glycol, and proteins of various molecular weights of varied molecular weight are passed through the membrane and the solute removal efficiency of the membrane is checked. At a certain molecular weight, no more solute passes through the membrane and this point is referred to as the MWCO.

10.2.8 LIQUID ENTRY PRESSURE OF WATER (LEPW)

Pressure is measured when water penetrates into the membrane. This method is frequently used in membrane distillation process. To determine *LEPw*, a dry membrane is placed in a separation cell. Both feed and permeate sides of the membrane cell are filled with pure water. The transmembrane pressure is increased in small increments (e.g., 0.1 bar) until a steady flow of water begins. This pressure is recorded as the *LEPw*.[44]

10.3 PURIFICATION PROCESS USED FOR EFFLUENT WATER

10.3.1 CONVENTIONAL METHODS

a) *Preliminary effluent treatment*: This step mainly involves removal of grit from water. Grit is a mixture of heavy inorganic particles such as sand, gravel, and other heavy particulate matter, which are normally removed by settling in grit channels. After use, the grit is separated and can be used as filling material or for landfill.
b) *Primary effluent treatment*: Primary treatment significantly removes a large amount of harmful substances and harmful pollutants from wastewater.[45]

c) In this process, the effluent is made to flow through sedimentation tanks, where majority of the solid material settles at the bottom of the tank and is called sludge. The effluent stays inside the tanks for several hours; by this time, the sludge is formed while a scum is formed on the surface; this scum is skimmed off and the sludge is removed from the bottom. This partially treated wastewater moves on to the secondary treatment stage. The primary treatment generally removes about 40% of the biochemical oxygen demand (BOD; these are substances that utilize the oxygen in the water), around 80–90% of SSs, and up to 55% of fecal coliforms.

   Following the above primary treatment, inorganic-based effluents are removed by techniques such as neutralization, oxidation, reduction, coagulation and sedimentation, pressure floatation, filtration, adhesion, and dehydration.[46]

d) *Secondary treatment*: To remove organic matter and any residual SSs, the waste goes through a secondary treatment. This involves biological processes that can further reduce the BOD level and remove most of the organic matter, as well as a small amount of SS material. In secondary treatment, the sewage undergoes strong aeration that encourages growth of aerobic bacteria and other microorganisms that oxidize the dissolved organic matter to methane, carbon dioxide, and water. The methane gas produced during the process can be used as a fuel for heating the digesters, as well as for running other power equipment in the plant. The treated sludge is then used for landfilling or sent to solid-waste-handling plants. To remove nutrients such as nitrogen and phosphorus present in the sewage, nitrification and luxury cell uptake processes (biological processes that remove phosphorous and nitrogen by exposing the biomass to alternating anaerobic and aerobic conditions; this results in high uptake of phosphorus and nitrogen by the cells) have been used. The main reason for nutrient removal is to avoid eutrophication and depletion of the oxygen level of the receiving water body into which the treated sewage is discharged.[47]

   The sludge generated from secondary treatment mainly contains unreleased inorganic as well as organic compounds. These impurities are mainly removed using filter presses, screw presses, and vacuum dehydrators to match the amount and condition of the sludge as per the recommended standards. Sometimes, dehydration agents are added as preprocessing step to improve the purification process. Organic compounds in effluents are removed using screens, oil removers, coagulating sedimentation, pressure floatation, standard activated sludge, methane fermentation, nitrification, filtering, adhesion, dehydration, and so forth; these techniques are either used independently or in combination.[48]

d) *Tertiary treatment*: This process is also called a disinfection process. Disinfection of the clarified treated effluent has become an increasingly frequent requirement to obtain potable water standards. Either calcium or sodium hypochlorite is often used as the disinfectant. Disinfectant is provided in the form of a powder or as a liquid. The water after tertiary treatment can meet the standards of drinking water. However, the process of tertiary treatment is extremely costly and seldom adopted by industries.[49]

Following parameters must be considered while designing a process for treating effluent wastewater:

a) Suspended solids (SSs)
b) Temperature
c) Oil and grease
d) Organic content in terms of BOD or COD
e) pH
f) Specific metals and/or specific organic compounds
g) Nitrogen and/or phosphorus
h) Indicator microorganisms (e.g., *Escherichia coli*) or specific microorganisms

Table 10.1 and 10.2 summarize the limits to control the impurities from the industrial wastewater plants.

10.4 REVOLUTIONARY METHODS FOR WASTEWATER TREATMENT

Conventional membranes have been widely used in water treatment decades back, but they suffer from intrinsic limitations such as low flux and susceptibility to fouling. Furthermore, formation of pinholes during preparation of the membrane is common, which contributes to inefficiency in the separation process. Usually, conventional membranes consist of a selective layer supported by a thick nonselective and highly permeable substrate. The pore size of the selective layer is controlled by different techniques that are able

TABLE 10.1 Acceptable conditions for discharge of industrial effluent or mixed effluent of standards A and B: (i) Standard A, as shown in the third column any inland waters within the catchment areas as specified in the sixth schedule or (ii) Standard B, as shown in the fourth column of any other inland waters (Environmental Quality Industrial Effluent Regulation 2009).

Parameter	Unit	Standard A	Standard B
1. Temperature	°C	40	40
2. pH value		6.0–9.0	5.5–9.0
3. BOD_5 at 20°C	mg/l	20	50
4. Suspended solids	mg/l	50	100
5. Mercury	mg/l	0.005	0.05
6. Cadmium	mg/l	0.01	0.02
7. Chromium, hexavalent	mg/l	0.05	0.05
8. Chromium, trivalent	mg/l	0.20	1.0
9. Arsenic	mg/l	0.05	0.10
10. Cyanide	mg/l	0.05	0.10
11. Lead	mg/l	0.10	0.5
12. Copper	mg/l	0.20	1.0
13. Manganese	mg/l	0.20	1.0
14. Nickel	mg/l	0.20	1.0
15. Tin	mg/l	0.20	1.0
16. Zinc	mg/l	2.0	2.0
17. Boron	mg/l	1.0	4.0
18. Iron (Fe)	mg/l	1.0	5.0
19. Silver	mg/l	0.1	1.0
20. Aluminum	mg/l	10	15
21. Selenium	mg/l	0.02	0.5
22. Barium	mg/l	1.0	2.0
23. Fluoride	mg/l	2.0	5.0
24. Formaldehyde	mg/l	1.0	2.0
25. Phenol	mg/l	0.001	1.0
26. Free chlorine	mg/l	1.0	2.0
27. Sulfide	mg/l	0.50	0.50
28. Oil and grease	mg/l	1.0	10
29. Ammoniacal nitrogen	mg/l	10	20
30. Color	ADMI	100	200

ADMI—American Dye Manufacturers Institute.

TABLE 10.2 Acceptable Conditions for Discharge of Industrial Effluent Containing Chemical Oxygen Demand for Specific Trade or Industry Sector.

Trade/industry	Unit	Standard A	Standard B
1. Pulp mill	mg/l	80	350
2. Paper mill (recycled)	mg/l	80	250
3. Pulp and paper mill	mg/l	80	300
4. Textile industry	mg/l	80	250
5. Fermentation and distillery industry	mg/l	400	400
6. Other industries	mg/l	80	200
7. Mixed effluents	mg/l	80	200

to produce membranes capable of microfiltration (MF), ultrafiltration (UF), nanofiltration (NF), and RO.[44] MF, UF, and NF membranes typically have pore sizes that allow them to effectively remove micron-sized (10^{-6} m) substances from water. UF membranes remove substances that are 10^{-9} m (nanometers) and larger. RO is also referred to as "hyperfiltration," and it effectively removes ion-sized substances such as sodium, calcium, sulfate, chloride, and nonpolar organic molecules. RO is capable of providing the purest water.

Nanofilters are usually rated by the smallest molecular weight substance that is effectively removed. These small substances are generally organic molecules. Molecular shape, polarity, and molecular size influence the effectiveness of the removal of organic impurities. Of all techniques, an NF membrane is considered novel, and has the ability to quickly and economically remove total dissolved solids from surface and groundwater, pathogens (bacteria, virus, molds, and fungus), monovalent and multivalent anions and cations (water-softening agents), salts, minerals, and other suspended nanoparticles.[50,51] The NF membrane has extensive applications in several industries such as beverage, food, textile, oil, chemical, and others. The pore size of the NF membrane can be as small as 1 nm in order to selectively remove larger molecules from smaller molecules and is able to size the nanomaterials into different fractions.[52–55]

In conventional membranes, the interconnectivity of the pores eliminates the inefficiency of the dead-end pores, thereby significantly increasing flux, enhancing the operational pressure and energy loss and thus increasing the overall cost of water purification. There has been significant research and development in recent times to produce membranes that resist fouling of any type. The newer nanofibrous membranes have a nonwoven assembly, which helps to enhance flux rate, thereby lowering the operational pressure bringing about energy savings. This development has enabled nanomembranes to be used for waste treatment at the source, as part of an overall pollution prevention program. In some cases, it may also be feasible to replace clarifiers within wastewater treatment systems with one or more membrane technology systems, thus greatly simplifying separation of solids. The final water may be suitable for reuse in one or more processes, and the solids may be suitable for reclamation.

Ramakrishna and Shirazi[56] have reported that ENMs or polymeric webs will gain worldwide attention in the coming days due to their wide application in water treatment. Thin-film composite membranes comprise three fundamental layers: the top ultrathin selective layer, a middle porous support layer, and the bottom nonwoven fabric layer. An ENM is generally used as the middle support layer.

Materials used for preparation of membranes: Membranes are made from organic polymers such as cellulose acetate, polysulfone, polycarbonate, polyamide, and inorganics such as aluminum oxide (ceramic) and porous carbon. The preparation of a given polymer can be varied to produce membranes of different porosity or MWCO. These membranes can be configured in stacks of plates, spiral-wound modules, or bundles of hollow fibers.

10.5 TYPES OF NANOFIBROUS MEMBRANES AND THEIR APPLICATION IN WATER TREATMENT

10.5.1 ADSORBENTS

The phenomenon of adsorption is considered one of the most suitable water treatment methods due to its ease of operation and the availability of a wide range of adsorbents. The principles of adsorption can be applied for the removal of both soluble and insoluble organic, inorganic, and biological pollutants. Additionally, adsorption can also be used to remove impurities from industrial wastewater, making it potable and usable for a variety of other purposes. However, adsorption has certain limitations and has not been able to achieve a good status commercially.[57] This is probably due to lack of suitable adsorbents with high adsorption capacity that can be scaled commercially. Moreover, it is difficult to find a single adsorbent that can be used for all kind of pollutants; consequently, different adsorbents have to be used for different pollutants. In the past two decades, nanosize adsorbents have been prepared and used for the removal of water pollutants and some of these are discussed below (Table 10.3).

A nanosize adsorbent made from electrospun fiber membrane impregnated with boehmite nanoparticles has been fabricated with sorptive characteristics intended for the removal of heavy metals. Two polymers, the hydrophobic/PCL and hydrophilic/Nylon-6, were chosen to serve as the support for the boehmite. Using atomic absorption spectroscopy, the authors have shown sorption of Cd(II) by electrospun membrane impregnated with boehmite with a removal capacity of 0.20 mg/g.[58]

Aliabadi et al.[59] have reported adsorption of nickel (Ni), cadmium (Cd), lead (Pb), and copper (Cu) from aqueous solution using reusable polyethylene oxide/chitosan nanofiber membranes prepared by the electrospinning technique. The sorption selectivity of lead, copper, cadmium, and nickel onto the membrane was in order of Pb(II) < Cd(II) < Cu(II) < Ni(II). Sang et al.[91] using chlorinated polyvinylchloride nanofibrous membrane has observed the following order of adsorption Cu(II) < Pd(II) < Cd(II).

TABLE 10.3 Nanofibers Used for Removal of Various Effluents Present in Water

Sr. No.	Name of the material used for nanofiber	Method of synthesis	Organic substances	Inorganic substances	Bacteria	Average diameter (D) and thickness (T)	References
1	PCL/Nylon-6 Impregnated with boehmite nanoparticles	Electrospinning		Cadmium		400–850 nm	58
2	Polyethylene oxide/chitosan solution	Electrospinning		Nickel, cadmium, copper, lead		98 nm	59
3	Chloridized polyvinyl chloride	Electrospinning		Cu(II), Pb(II), Cd(II)		1000 nm	60
4	Carbonized polyacrylonitrile (PAN)	Electrospinning	Disinfectants—chloroform and monochloroacetic acid			206–490 nm	61
5	Polyvinylpyrrolidone/SiO_2 composite	Electrospinning		Cr(III)		100–700 nm	62
6	Polyvinylpyrrolidone TiO_2/CuO composites	Electrospinning	AO7 dye			100 nm	63
7	Cellulose acetate	Electrospinning	Bovine serum albumin (BSA)			200 nm to 1 μm	64
8	Iron ferrocyanide immobilized PAN	Electrospinning		Cesium		220–411 nm	65
9	Surface-modified PAN nanofibers with surface modification using polyethylene diamine tetraacetic acid (EDTA) and ethylenediamine as cross-linker	Electrospinning		Cd(II) and Cr(VI)			66

TABLE 10.3 *(Continued)*

Sr. No.	Name of the material used for nanofiber	Method of synthesis	Organic substances	Inorganic substances	Bacteria	Average diameter (D) and thickness (T)	References
10	Glutaraldehyde-cross-linked chitosan	Electrospinning	Peptone	Ca, Fe, Mg, Mn, Zn, K, NH_4		D:150–450 nm Pore size: 0.1–0.5 μm	67
11	PAN scaffolds with chitosan coating	Electrospinning	Vegetable oil			D:124–720 nm T: 1 μm	68
12	Polyvinylidene fluoride	Electrospinning	Polystyrene beads	NaCl (seawater desalination)		500 nm	69
13	PAN modified with EDTA	Electrospinning		Cd(II), Cr(VI)		250 nm	66
14	PAN reinforced with halloysite nanotubes	Electrospinning	Bilge water	Cu(II)		Pore size—30–80 nm	70
15	Aminated PAN	Electrospinning		Cu(II), Ag(I), Pb(II), Fe(II)		150–250 nm	71
16	Cellulose acetate/silica composite	Electrospinning		Cr(VI)		100–500 nm	72
17	Polyvinyl alcohol/nanozeolite composite	Electrospinning		Cd(II) and Ni(II)		~170 nm	73
18	PAN with jute cellulose nanowhiskers	Electrospinning	Oil			173 nm	74
19	Nylon-6	Electrospinning	Particulate matter—0.5–10 μm (polystyrene particles)			30–110 nm	75

TABLE 10.3 *(Continued)*

Sr. No.	Name of the material used for nanofiber	Method of synthesis	Organic substances	Inorganic substances	Bacteria	Average diameter (D) and thickness (T)	References
20	Poly(methyl methacrylate) blended with β-cyclodextrin (β-CD)	Electrospinning	Phenolphthalein			900 nm	76
21	Poly(methacrylic acid) modified by grafting	Graft and Cross-linking	Bisphenol A, salicylic acid, and ibuprofen				77
22	Polyvinylidene fluoride with surface-modified chitosan	Electrospinning	BSA			170 nm	78
23	Polyurethane urea	Electrospinning	Sulfides			Average diameter: 220 nm	
24	Polyvinylidene fluoride copolymerized with acrylic and methacrylic acid	Electrospinning	Polyethylene oxide			100–600 nm	79
25	PAN functionalized with β-CD	Electrospinning	Reactive blue 13			200–700 nm	80
26	Polyimide	Electrospinning	Lignocellulosic hydrolysate		Yeast cells		81
27	Surface-modified polyethylene terephthalate	Electrospinning	Particulate matter (latex beads)		*Escherichia coli, Staphylococcus aureus*	100 nm	82
28	Silica–polyamide composite	Electrospinning	Chlorobenzenes			<170 nm	83
29	Polyacrylic acid/silica composites	Electrospinning	Malachite green			300–700 nm	84

TABLE 10.3 *(Continued)*

Sr. No.	Name of the material used for nanofiber	Method of synthesis	Organic substances	Inorganic substances	Bacteria	Average diameter (D) and thickness (T)	References
30	Polycellulose acetate/polyvinylchloride with silver metal	Electrospinning			*E. coli* and *Pseudomonas aeruginosa*	D: 200–588 nm T: 46–91 nm	85
31	PAN with silver nanoparticles	Electrospinning			*S. aureus* and *E. coli*	450 nm	86
32	Chitosan/poly(vinyl alcohol) with silver nanoparticles	Electrospinning			*E. coli*	58–109 nm	87
33	Polyurethane	Electrospinning			Pathogen	D: 147 nm T: 2.5 g/m^2	88
34	Polyvinylidene fluoride	Electrospinning		Salts—sodium chloride and magnesium sulfate		D: 180–500 nm T: 120 nm	89
35	Polyethersulfone	Electrospinning				550–1300 nm	90

A study on the removal of by-products produced by the use of disinfectants was carried out by Singh et al.[61] Chloroform and monochloroacetic acid were used as model disinfection by-products. The ability of carbonized polyacrylonitrile (PAN) membrane to adsorb these two molecules was evaluated; the results indicate efficient adsorption of chloroform due to the hydrophobic nature of the membrane; nevertheless, a significant adsorption of monochloroacetic acid was also observed.

10.5.1.1 CARBON NANOFIBER (CNF) MEMBRANES

The use of carbon as an adsorbent for the removal of metal ions from effluent water has been explored and proven to be highly effective by many researchers. A comparative study of CNFs supported on activated carbon and commercial CNFs has been carried out for the removal of iron, manganese, and boron from industrial effluent water by Abdullah et al.[92] claiming that the pH of the effluent and type of adsorbents play important roles for the adsorption of metal ions. Further to improve the efficiency of the carbon nanotubes (CNTs), Das et al.[93] suggested few strategies of functionalization of CNT membranes to improve biofouling and self-cleaning by conjugating the membranes with nanoparticles that possess antibacterial activity such as silver and TiO_2. The resultant CNT membranes have a small size and soft condensed architecture and are able to auto-arrange without any external input. Such assemblies can be used for water desalination, a process which demands selective salt rejection at high water flux. Another group of scientists, Ahmed et al.[94] have synthesized and characterized the CNF using the field emission scanning electron microscopy technique. These CNFs have diameter ranging from 100 to 160 nm and have been explored for adsorption of lead from industrial wastewater. The highest sorption capacity and lead removal efficiency was 77 mg Pb^{2+}/g of synthesized CNF. Further when the synthesized CNFs were treated with powdered activated carbon, the fibers could eliminate nearly 67% of Pb^{2+} from wastewater collected from a semiconductor electronic industry.

10.5.1.2 BIOLOGICALLY MODIFIED ZEOLITE MEMBRANES

Zeolites are hydrated aluminosilicates obtained from natural resources and are used for adsorption of heavy metal ions. The filtration efficiency of porous zeolite is high due to a substantially larger surface area than the other natural adsorbents like sand. Thus, the increased surface area improves adsorption of SSs, microorganisms, and other materials present in solution. Zeolite has the ability to remove ammonium ions from water which serve as a nutrient for microorganisms (nitrifying bacteria). Nitrifying bacteria convert the ammonium ions (NH_4^+) to nitrite ions (NO_3^-), which can be efficiently removed from water. Thus, water purification can be achieved by a combination of ion-exchange, filtration, and/or microbiological processes using beds of zeolite-rich materials or zeolite nanomembranes.[95–98] Ouki and Kavannagh have studied the adsorption of lead and cadmium using two natural zeolites such as chabazite and clinoptilolite. The two natural zeolites when pretreated with NaOH have demonstrated very high adsorption capacity for lead (Pb^{+2}) and cadmium (Cd^{+2}), with removal efficiency of more than 99%. The high porosity of zeolite leads to significant adsorption capacity and the photocatalytic reduction ability of zeolite aids to reduce metal ions in a higher oxidation state to a lower valence state, thus decreasing the toxicity of metals. Nanozeolites are most frequently used as a dopant in thin-film nanocomposite (TFN) membranes and have shown a potential to enhance membrane permeability. It was hypothesized that the small, hydrophilic pores of nanozeolites create preferential paths for water. However, water permeability has been found to be efficient even with pore-filled zeolites, albeit lower than the pore-open ones; this could only be attributed to defects at the zeolite polymer interface. Nanozeolites were also tested as carriers for antimicrobial agents such as Ag^+ by Lind et al.[99] who have also reported an anti-fouling property of the membrane. The zeolite TFN technology has reached the stage of commercialization and QuantumFlux, a TFN RO membrane for seawater, is now commercially available (www.nanoH$_2$O.com).

10.5.1.3 METAL OXIDE MEMBRANES

A composite nanofiber membrane composed of polyvinylpyrolidone and silicon dioxide (PVP/SiO_2) with a mesostructure functionalized with amino groups was designed. These membranes exhibit high adsorptions properties for Cr^{3+} ions. In addition, the mesoporous nanofiber membranes can be conveniently fixed and reclaimed. These composite materials have a great potential for application in water treatment for adsorption of heavy metal ions, dyes, and other pollutants from aqueous solutions.[100]

Similarly, electrospun TiO_2/CuO and polyvinylpyrolidone (PVP) composite nanofibers were designed by Lee et al.[63] for generation of both energy and clean water from wastewater containing dyestuffs by photocatalysis. The energy generated is in the form of hydrogen and this is the first study wherein composite nanofibers were used for simultaneous photocatalytic organic pollutant degradation and hydrogen production from wastewater generated by the dye industry. AO7 was used as the model dye wastewater since azo dyes represent the major portion of impurities in wastewater generated by the industry. It was observed that bare TiO_2 nanofibers were only useful for removal of organic material while TiO_2/CuO composite nanofibers exhibit multifunctional ability to evolve H_2 gas concomitantly with degradation of the dyes.

10.5.1.4 AFFINITY MEMBRANES

Affinity membranes were developed to purify water based on physical/chemical properties or biological functions rather than on molecular weight/size. Rather than operate purely on the sieving mechanism, affinity membranes are able to "capture" molecules, that is, immobilize specific ligands on the membrane surface. Affinity membranes are a result of the technological advances in both fixed-bed liquid chromatography and membrane filtration, and combine both the outstanding selectivity of the chromatography resins while eliminating the reduced pressure drop associated with filtration membranes.[101] Application of such a technology was shown by Ma et al.[64] who fabricated a cellulose acetate nanofiber membrane and showed a significant capture capacity for bovine serum albumin and bilirubin. In another study, polyacrylonitrile PAN nanofiber membranes were used as a support for iron ferrocyanide. Iron ferrocyanide can selectively capture radioactive cesium ion. This is normally done by direct addition of iron ferrocyanide to the contaminated water that forms a coagulant with cesium, which ultimately sediments and is separated. However, because this method involves solid–liquid separation, handling becomes quite complex. Therefore, it is desirable from the standpoint of ease in handling to immobilize iron ferrocyanide[65] through use of nanofibers.

10.5.1.5 SURFACE-MODIFIED POLYMERS AS NANOFIBROUS ADSORBENTS

An application of surface-modified nanofibers in adsorption of metal ions such as chromium (Cr^{+6}) and cadmium (Cd^{+2}) was demonstrated by Chaúque et al.[66] They constructed PAN nanofibers with the surface modified by cross-linking with ethylenediamine (EDA) to polyethylene diamine tetraacetic acid (EDTA). The EDTA–EDA–PAN nanofibers showed effective sorption affinity for both Cd(II) and Cr(VI), achieving maximum adsorption capacities of 32.68 and 66.24 mg/g.

10.5.2 ULTRAVIOLET MEMBRANES

A bilayered hybrid biofoam composed of bacterial nanocellulose (BNC) and reduced graphene oxide for the generation of steam using solar heat localized at the evaporation surface has been proposed by Jiang et al.[102] BNC is composed of highly pure cellulose nanofibrils. This is produced from dextrose by bacteria through a series of biochemical steps; the secreted cellulose fibrils then self-assemble in the culture medium. It is a highly attractive material for the fabrication of functional foams due to its large specific surface area, open microporous structure, excellent mechanical properties, and a synthesis that is facile and scalable. The bilayer structure of the functional foam has been tailored to produce high absorption of light resulting in photothermal conversion, localization of heat, and transport of water to the evaporation surface; these intrinsic properties render a material that is highly efficient for solar steam generation. The bilayer structure exhibits excellent stability even under vigorous mechanical agitation and harsh chemical conditions, which is quite remarkable considering the simplicity of the two-step fabrication approach. The novel bilayered structure exhibits a remarkably high solar thermal efficiency. The material can be used for generation of potable drinking water from seawater.

10.5.3 ULTRAFILTRATION MEMBRANES

In the nonwoven fiber industry, one of the fastest growing applications is filtration. Traditionally, wet-laid, melt-blown, and spun-bonded nonwoven articles, containing micron-size fibers have been popular for this application because fibers prepared in this way are inexpensive, easy to produce, and have good filtration efficiency. In case of liquid filtration, nonwoven articles have been mostly used as substrates to support porous nanofilters, which can exclude particles larger than a few nanometers.[103] The pores in a nonwoven structure (i.e., the empty space) are highly interconnected and pores that have become partially blocked by particles would necessitate hydraulic pressure to aid fluid filtration. For liquid filtration, porous polymeric membranes manufactured by conventional methods have their intrinsic limitations, for example, low flux and easily fouled, due to the geometrical structure of the pores and the corresponding pore size distribution. On the other hand, electrospun fibers, with diameter 10–100 times smaller than those of nonwoven articles, have higher effective porosity and larger surface-area-to-volume ratio, making them apposite for filtration applications. One unique feature in electrospinning is its capability to control the fiber diameter (from tens of nanometers to a few micrometers) by varying the process variable(s), such as solution concentration, applied voltage, fluid flow rate, surface tension, and so forth. The ability to change the fiber diameter provides an opportunity to fine-tune the membrane porosity that also depends on the membrane thickness. It thus appears that the nanofibrous membranes produced by electrospinning can overcome some of these limitations associated with nonwoven membranes.[68]

The use of this technology was demonstrated by Yoon et al.[68] who constructed a new type of high-flux UF or NF composite membranes containing three layers—a thin layer of hydrophilic chitosan coating, a mid-layer support made of asymmetric electrospun PAN nanofibrous and lastly a nonwoven PET substrate. This composite membrane exhibited flux rates of an order of magnitude higher than the commercial NF filter (e.g., NF 270 made by Dow Chemical Company) after 24 h of operation. It was able to maintain a good filtration efficiency with rejection ratios better than 99.9% (i.e., less than 1 ppm in the permeate). Gopal et al.[69] proposed the use of electrospun nanomembranes (ENMs) made from polyvinylidene fluoride nanofibers in water prefiltration. This membrane can replace the three-step process of coagulation, flocculation, and sedimentation followed in conventional water treatment. An efficient application of NF in producing drinking water and purifying industrial wastewater was explored by Cao et al.[74] They made electrospun composite PAN with selectively oxidized jute cellulose nanowhisker nanofibers. The versatile PAN/cellulose ES composite membranes exhibited not only superior mechanical properties but also a high filtration efficiency of particles 7–40 nm in size. Aussawasathien et al.[75] have demonstrated that membranes made from nylon-6 electrospun nanofibers when used as prefilters prior to UF or NF are able to increase the filtration efficiency and prolong the life of downstream membranes.

10.6 SURFACE MODIFICATION

10.6.1 ELECTROSPUN NANOFIBERS FOR HEAVY METAL IMPURITIES

A nanofiber membrane of chlorinated polyvinyl chloride was fabricated by using high-voltage electrospinning process for the removal of divalent metal cations (Cu^{2+}, Cd^{2+}, and Pb^{2+}) from simulated groundwater.[91,104] With this membrane, the group investigated several methods to obtain the maximum removal of heavy metals; some of these methods were static adsorption, direct filtration, soil-aided filtration, diatomaceous earth-aided filtration, and micellar-enhanced filtration (MEF). The experimental results revealed that the removal of copper in the simulated groundwater by MEF can reach more than 73%; the removal of lead more than 82%; and the removal of cadmium more than 91%. It was also indicated that the nanofibers membranes can be used for the treatment of groundwater containing Cu^{2+}, Pb^{2+}, and Cd^{2+} with a high removal efficiency.

Zander et al.[82] fabricated surface-functionalized nanofibers from recycled PET as liquid filtration membranes. On successfully making fibers with a diameter of 100 nm, it was found that mats with such fibers were not suitable for UF being ineffective for the removal of particles below 500 nm. However, the mats were found to be suitable in MF applications when used as a prefilter in a wastewater treatment system. The problem of biofouling was remarkably reduced by surface modification of the fibers with cationic biocides. In particular, covalent attachment of the Vantocil biguanide led to a 7-log reduction in both gram-positive and gram-negative bacteria.

An attempt was made by Kaur et al.[76] to incorporate β-cyclodextrin (β-CD) onto the surface of the nanofiber to treat organic waste. Phenylcarbomylated and azidophenylcarbomylated β-CDs were synthesized and successfully blended with poly(methyl methacrylate) and the blend electrospun into nanofibrous membranes with an approximate diameter of 900 nm. These membranes exhibited promising results for capturing of small molecules like phenolphthalein. This suggests that nanofibers blended with functionalized β-CD have the potential to capture similar small organic molecules in organic waste. To improve the removal of certain pollutants, nanofibrous membranes need to be modified by various methods such as graft polymerization and cross-linking. One such modification of a raw nanofibrous membrane involved adding a thin polyamide skin and a thick polysulfone support layer. The raw membrane (polyamide skin and thick polysulfone) was modified by graft polymerization with methacrylic acid (MA) for the rejection of organic micropollutants such as endocrine-disrupting chemicals (bisphenol A [BPA]) and pharmaceutically active ingredients (ibuprofen and salicylic acid). Graft polymerization of MA on the raw NF membrane increased the hydrophilicity and negative surface charge of the membrane in proportion to the amount of carboxylic acid in the grafted polymer chains. This increased steric hindrance and negative surface charge of the MA-membrane improved rejection of target pollutants: BPA, ibuprofen, and salicylic acid. To further improve rejection of target pollutants, the grafted polymer chains on the MA-membrane surface were cross-linked with ethylene diamine.[77] Although rejection of uncharged solutes (e.g., BPA) improved relative to that of the MA-membrane, rejection of negatively charged solutes decreased due to loss of negative surface charge on the MA-membrane upon cross-linking. However, the water flux with the cross-linked membrane decreased severely due to the increased hydraulic resistance compared to the MA-membrane.

10.6.2 ANTIMICROBIAL FUNCTIONALITY

Antimicrobial polymers are used in several areas such as medical devices, health care, hygienic application, water purification systems, hospitals, dental surgery equipment, textiles, food packaging, and storage.[105] The contamination of polymer surfaces by microorganisms such as pathogenic bacteria, bacteria generating odor, molds, fungi, and viruses is likely to occur and is of a great concern. As mentioned earlier, polymer fibers prepared by electrospinning have unique properties such as high porosity, high surface-area-to-volume and length-to-diameter ratios. Lala et al.[85] have fabricated nanofibers containing silver nanoparticles from three model polymers and tested them against two different strains of gram-negative bacteria, namely *E. coli* and *Pseudomonas aeruginosa*. The polymers for the preparation of the nanofibers were cellulose acetate, PAN, and polyvinylchloride. Polymers in various combinations were blended with silver metal. Of all the combinations, PAN in DMF (as solvent) containing 5%wt. of $AgNO_3$ and irradiated by UV for 30 min was found to be the most effective. A convenient and cost-effective approach to developing antimicrobial nanofibrous membranes that are particularly useful for filtration of water was proposed by Zhang et al.[86] They described a method where PAN was bonded to the silver ion (as nanoparticles) by a coordinate bond. These membranes were capable of killing microorganisms such as *Staphylococcus aureus and E. coli* in 30 min. A method of incorporation of silver nanoparticles in chitosan/

poly(vinyl alcohol) nanofibers from the removal of *E. coli* during filtration was described by Adibzadeh et al.[87]

10.7 FUTURE DIRECTIONS OF NANOFIBROUS MEMBRANES IN WASTEWATER MANAGEMENT

The current methods for wastewater treatment are attempts to remove both organic as well as inorganic impurities from water; no single method is truly able to eliminate completely both organic as well as inorganic impurities from industrial effluent water. Therefore, these methods have limited utility and are also uneconomical because they are unable to completely purify water and have limitations on the reuse of filtration materials. Among the current methods, nanotechnology focuses on use of nanofibers as filtration media to improve traditional methods by increasing efficiency of the process and reusability of the filter media. This should bring about a savings in the cost of operation of the plant or processes. This is possible because nanofibers have unique properties such as high surface-to-volume ratio and high reactivity and sensitivity; attributes that make them capable of self-assembling onto substrates to form fibers with high adsorption power. All these characteristics make them ideal for the treatment of wastewater. They are effective against various organic and inorganic pollutants, heavy metals, as well as against several harmful microbes present in contaminated water when suitably linked with antibacterial agents such as silver nanoparticles.

Currently, researchers are exploring combination of functionally modified or advanced nanofibers with conventional technologies to usher in a new family of nanotechnology that will enable multifunctional water treatment, that is, creating a device with a capacity to perform multiple tasks. Such multifunctional systems will be able to enhance the overall performance and avoid excessive redundancy while miniaturizing the footprint. Owing to the high efficiency of nanofibers, it is envisaged that nanofibers with different functionalities can be assembled as layers in a cartridge or as modules arranged in series, which will allow optimization/regeneration of each functionality separately.[106] The capacity and functionality of such nanotechnology enabled systems can be easily manipulated by plugging in or pulling out respective modules.

We hypothesize that nanofibrous technology when integrated with natural adsorbents and smart polymeric materials such as thermoresponsive or pH-responsive polymers will efficiently extract organic as well as inorganic impurities simultaneously.[107] Recently, we have published work on thermoresponsive graft assemblies with a unique and exclusive property of adsorption, which begins at the lower critical solution temperature that is around room temperature. The graft assemblies have been made by hybridization of thermoresponsive synthetic polymers with the naturally occurring adsorbent through graft polymerization. These synergistic properties of the two polymers coupled with their reproducibility and elimination spectrum can be used for water purification. However, further investigation will focus on nanofabrication of such polymeric assemblies.

Another method to decontaminate water in a cost-effective manner is to harvest solar energy, which is freely available. The process involves photocatalysis by visible light using TiO_2 or fullerenes attached to surface-functionalized advanced polymers followed by nanosizing. Nanofibrous materials will become an essential component of industrial and wastewater treatment systems in the near future as progress is being made in terms of fabrication of economically efficient and eco-friendly multifunctional devices that will be able to eliminate all impurities simultaneously.[45] The success will depend on how issues like reusablility and reproducibility of nanofibrous membranes are addressed, as these are key factors for the economical commercialization of fabricated devices for water treatment. The long-term efficacy of these nanotechnologies will also determine their success, at present very little is known, as most laboratory studies have been conducted for relatively short periods of time. Thus, research that addresses the long-term performance of nanotechnologies in wastewater treatment is the need of the hour. To be truly successful, these materials must be able to resist fouling easily to ensure efficient and reproducible filtration. Additionally, these materials must possess high water permeability that will ensure energy savings, and also have a broad spectrum for organic, inorganic, viral, microbial, and gaseous impurities.

10.8 CONCLUDING REMARKS

The transformation of polymer nanofibrous membranes from lab scale to commercialization scale is indeed a herculean task for researchers due to involvement of assorted technologies. However, one can achieve this goal using appropriate chemistry to synthesize stronger and more stable materials, modify surfaces to produce fluid channels with proper dimensions that will ensure rapid throughput, alter materials to adsorb and exclude impurities by introduction of desired functional groups, understand the appropriate physics to accurately characterize the structure and the morphology, engineer processes to fabricate nanomaterial on an industrial scale, and finally understand the biology of interaction with living systems. To sum up, if nanofibrous devices are to become successful, they must be produced in a cost-effective manner, be durable, be reusable, and be scalable for use in effluent water treatment.

KEYWORDS

- **electrospinning**
- **industrial effluent treatment**
- **nanofibrous membranes**
- **nanofiltration**

REFERENCES

1. Gupta, V. K.; Rastogi, A. Biosorption of Lead (II) from Aqueous Solutions by Non-Living Algal Biomass *Oedogonium sp.* and *Nostoc sp.*—A Comparative Study. *Colloids Surf. B* **2008,** *64*(2), 170–178.
2. (a) Reneker, D. H.; Fong, H. *Polymeric Nanofibers;* ACS Publications: Washington D. C., 2006; (b) Ramakrishna, S.; Fujihara, K.; Teo, W.-E.; Lim, T.-C.; Ma, Z. *An Introduction to Electrospinning and Nanofibers;* World Scientific: Singapore, 2005; Vol. 90.
3. (a) Wendorff, J. H.; Agarwal, S.; Greiner, A. *Electrospinning: Materials, Processing, and Applications;* John Wiley and Sons: United Kingdom, 2012; (b) He, J.-H.; Liu, Y.; Mo, L.-F.; Wan, Y.-Q.; Xu, L. *Electrospun Nanofibres and Their Applications;* iSmithers: Shawbury, Shrewsbury, Shropshire, UK, 2008.
4. Li, D.; Xia, Y. Electrospinning of Nanofibers: Reinventing the Wheel? *Adv. Mater.* **2004,** *16*(14), 1151–1170.
5. Kidoaki, S.; Kwon, I. K.; Matsuda, T. Mesoscopic Spatial Designs of Nano- and Microfiber Meshes for Tissue-Engineering Matrix and Scaffold Based on Newly Devised Multilayering and Mixing Electrospinning Techniques. *Biomaterials* **2005,** *26*(1), 37–46.
6. Um, I. C.; Fang, D.; Hsiao, B. S.; Okamoto, A.; Chu, B. Electro-Spinning and Electro-Blowing of Hyaluronic Acid. *Biomacromolecules* **2004,** *5*(4), 1428–1436.
7. Ji, Y.; Ghosh, K.; Li, B.; Sokolov, J. C.; Clark, R. A.; Rafailovich, M. H. Dual-Syringe Reactive Electrospinning of Cross-Linked Hyaluronic Acid Hydrogel Nanofibers for Tissue Engineering Applications. *Macromol. Biosci.* **2006,** *6*(10), 811–817.
8. Kim, S. H.; Kim, S.-H.; Nair, S.; Moore, E. Reactive Electrospinning of Cross-Linked Poly(2-Hydroxyethyl Methacrylate) Nanofibers and Elastic Properties of Individual Hydrogel Nanofibers in Aqueous Solutions. *Macromolecules* **2005,** *38*(9), 3719–3723.

9. Gupta, P.; Trenor, S. R.; Long, T. E.; Wilkes, G. L. In Situ Photo-Cross-Linking of Cinnamate Functionalized Poly(methyl Methacrylate-Co-2-Hydroxyethyl Acrylate) Fibers During Electrospinning. *Macromolecules* **2004,** *37*(24), 9211–9218.
10. Penchev, H.; Paneva, D.; Manolova, N.; Rashkov, I. Electrospun Hybrid Nanofibers Based on Chitosan or N-Carboxyethylchitosan and Silver Nanoparticles. *Macromol. Biosci.* **2009,** *9*(9), 884–894.
11. Larrondo, L.; St John Manley, R. Electrostatic Fiber Spinning from Polymer Melts. I. Experimental Observations on Fiber Formation and Properties. *J. Polym. Sci. Polym. Phys. Ed.* **1981,** *19*(6), 909–920.
12. Larrondo, L.; St John Manley, R., Electrostatic Fiber Spinning from Polymer Melts. III. Electrostatic Deformation of a Pendant Drop of Polymer Melt. *J. Polym. Sci. Polym. Phys. Ed.* **1981,** *19*(6), 933–940.
13. Sarkar, K.; Gomez, C.; Zambrano, S.; Ramirez, M.; de Hoyos, E.; Vasquez, H.; Lozano, K. Electrospinning to Forcespinning™. *Mater. Today* **2010,** *13*(11), 12–14.
14. McEachin, Z.; Lozano, K. Production and Characterization of Polycaprolactone Nanofibers via Forcespinning™ Technology. *J. Appl. Polym. Sci.* **2012,** *126*(2), 473–479.
15. Badrossamay, M. R.; McIlwee, H. A.; Goss, J. A.; Parker, K. K. Nanofiber Assembly by Rotary Jet-Spinning. *Nano Lett.* **2010,** *10*(6), 2257–2261.
16. Sanders, E. H.; Kloefkorn, R.; Bowlin, G. L.; Simpson, D. G.; Wnek, G. E. Two-Phase Electrospinning from a Single Electrified Jet: Microencapsulation of Aqueous Reservoirs in Poly(ethylene-co-vinyl Acetate) Fibers. *Macromolecules* **2003,** *36*(11), 3803–3805.
17. (a) Rinaldi, M.; Ruggieri, F.; Lozzi, L.; Santucci, S. Well-Aligned TiO_2 Nanofibers Grown by Near-Field-Electrospinning. *J. Vac. Sci. Technol. B* **2009,** *27*(4), 1829–1833; (b) Zheng, G.; Li, W.; Wang, X.; Wu, D.; Sun, D.; Lin, L. Precision Deposition of a Nanofibre by Near-Field Electrospinning. *J. Phys. D: Appl. Phys.* **2010,** *43*(41), 415–501.
18. Kim, H.-Y.; Lee, M.; Park, K. J.; Kim, S.; Mahadevan, L. Nanopottery: Coiling of Electrospun Polymer Nanofibers. *Nano Lett.* **2010,** *10*(6), 2138–2140.
19. (a) Chang, C.; Tran, V. H.; Wang, J.; Fuh, Y.-K.; Lin, L. Direct-Write Piezoelectric Polymeric Nanogenerator with High Energy Conversion Efficiency. *Nano Lett.* **2010,** *10*(2), 726–731; (b) Bisht, G. S.; Canton, G.; Mirsepassi, A.; Kulinsky, L.; Oh, S.; Dunn-Rankin, D.; Madou, M. J. Controlled Continuous Patterning of Polymeric Nanofibers on Three-Dimensional Substrates Using Low-Voltage Near-Field Electrospinning. *Nano Lett.* **2011,** *11*(4), 1831–1837.
20. Zhou, F.-L.; Hubbard, P. L.; Eichhorn, S. J.; Parker, G. J. Jet Deposition in Near-Field Electrospinning of Patterned Polycaprolactone and Sugar-Polycaprolactone Core–Shell Fibres. *Polymer* **2011,** *52*(16), 3603–3610.
21. Chen, D.; Lei, S.; Chen, Y. A Single Polyaniline Nanofiber Field Effect Transistor and its Gas Sensing Mechanisms. *Sensors* **2011,** *11*(7), 6509–6516.
22. Hellmann, C.; Belardi, J.; Dersch, R.; Greiner, A.; Wendorff, J.; Bahnmueller, S. High Precision Deposition Electrospinning of Nanofibers and Nanofiber Nonwovens. *Polymer* **2009,** *50*(5), 1197–1205.
23. Wang, H.; Zheng, G.; Li, W.; Wang, X.; Sun, D. Direct-Writing Organic Three-Dimensional Nanofibrous Structure. *Appl. Phys. A* **2011,** *102*(2), 457–461.
24. Zhang, X.; Manohar, S. K. Polyaniline Nanofibers: Chemical Synthesis Using Surfactants. *Chem. Commun.* **2004,** (20), 2360–2361.

25. Yu, L.; Lee, J. I.; Shin, K. W.; Park, C. E.; Holze, R. Preparation of Aqueous Polyaniline Dispersions by Micellar-Aided Polymerization. *J. Appl. Polym. Sci.* **2003,** *88*(6), 1550–1555.
26. Michaelson, J. C.; McEvoy, A. Interfacial Polymerization of Aniline. *J. Chem. Soc. Chem. Commun.* **1994,** (1), 79–80.
27. Huang, L.; Wang, Z.; Wang, H.; Cheng, X.; Mitra, A.; Yan, Y. Polyaniline Nanowires by Electropolymerization from Liquid Crystalline Phases. *J. Mater. Chem.* **2002,** *12*(2), 388–391.
28. Liu, J.-M.; Yang, S. C. Novel Colloidal Polyaniline Fibrils Made by Template Guided Chemical Polymerization. *J. Chem. Soc. Chem. Commun.* **1991,** (21), 1529–1531.
29. Zhang, X.; Goux, W. J.; Manohar, S. K. Synthesis of Polyaniline Nanofibers by "Nanofiber Seeding". *J. Am. Chem. Soc.* **2004,** *126*(14), 4502–4503.
30. Li, W.; Wang, H.-L. Oligomer-Assisted Synthesis of Chiral Polyaniline Nanofibers. *J. Am. Chem. Soc.* **2004,** *126*(8), 2278–2279.
31. Qiu, H.; Wan, M.; Matthews, B.; Dai, L. Conducting Polyaniline Nanotubes by Template-Free Polymerization. *Macromolecules* **2001,** *34*(4), 675–677.
32. (a) Li, X.; Li, X. Oxidative Polymerization of Aniline Using $NaClO_2$ as an Oxidant. *Mater. Lett.* **2007,** *61*(10), 2011–2014; (b) Rahy, A.; Yang, D. J. Synthesis of Highly Conductive Polyaniline Nanofibers. *Mater. Lett.* **2008,** *62*(28), 4311–4314.
33. Huang, J. Syntheses and Applications of Conducting Polymer Polyaniline Nanofibers. *Pure Appl. Chem.* **2006,** *78*(1), 15–27.
34. Bui, N.-N.; Lind, M. L.; Hoek, E. M.; McCutcheon, J. R. Electrospun Nanofiber Supported Thin Film Composite Membranes for Engineered Osmosis. *J. Membr. Sci.* **2011,** *385*, 10–19.
35. Huang, J.; Virji, S.; Weiller, B. H.; Kaner, R. B. Nanostructured Polyaniline Sensors. *Chem. A Eur. J.* **2004,** *10*(6), 1314–1319.
36. Zhang, X.; Chan-Yu-King, R.; Jose, A.; Manohar, S. K. Nanofibers of Polyaniline Synthesized by Interfacial Polymerization. *Synth. Met.* **2004,** *145*(1), 23–29.
37. Huang, J.; Virji, S.; Weiller, B. H.; Kaner, R. B. Polyaniline Nanofibers: Facile Synthesis and Chemical Sensors. *J. Am. Chem. Soc.* **2003,** *125*(2), 314–315.
38. (a) Huang, J.; Kaner, R. B. A General Chemical Route to Polyaniline Nanofibers. *J. Am. Chem. Soc.* **2004,** *126*(3), 851–855; (b) Huang, J.; Kaner, R. B. The Intrinsic Nanofibrillar Morphology of Polyaniline. *Chem. Commun.* **2006,** (4), 367–376.
39. Kaur, S.; Ma, Z.; Gopal, R.; Singh, G.; Ramakrishna, S.; Matsuura, T. Plasma-Induced Graft Copolymerization of Poly(methacrylic Acid) on Electrospun Poly(vinylidene Fluoride) Nanofiber Membrane. *Langmuir* **2007,** *23*(26), 13085–13092.
40. (a) Kaur, S.; Rana, D.; Matsuura, T.; Sundarrajan, S.; Ramakrishna, S. Preparation and Characterization of Surface Modified Electrospun Membranes for Higher Filtration Flux. *J. Membr. Sci.* **2012,** *390*, 235–242; (b) Kaur, S.; Gopal, R.; Ng, W. J.; Ramakrishna, S.; Matsuura, T. Next-Generation Fibrous Media for Water Treatment. *MRS Bull.* **2008,** *33*(01), 21–26.
41. Subramanian, S.; Seeram, R. New Directions in Nanofiltration Applications—Are Nanofibers the Right Materials as Membranes in Desalination? *Desalination* **2013,** *308*, 198–208.
42. Manickam, S. S.; McCutcheon, J. R. Characterization of Polymeric Nonwovens Using Porosimetry, Porometry and X-Ray Computed Tomography. *J. Membr. Sci.* **2012,** *407*, 108–115.

43. Jena, A.; Gupta, K. Pore Structure Characterization Techniques: Four Widely Used Techniques Are Described. *Am. Ceram. Soc. Bull.* **2005,** *84*(3), 28–30.
44. Tabe, S. Electrospun Nanofiber Membranes and their Applications in Water and Wastewater Treatment. In *Nanotechnology for Water Treatment and Purification;* Hu, A., Abblett, A., Eds.; Springer: Switzerland, 2014; pp 111–143.
45. Bora, T.; Dutta, J. Applications of Nanotechnology in Wastewater Treatment—A Review. *J. Nanosci. Nanotechnol.* **2014,** *14*(1), 613–626.
46. Cloete, T. E. *Nanotechnology in Water Treatment Applications;* Horizon Scientific Press: United Kingdom, 2010.
47. Templeton, M. R.; Butler, D. *Introduction to Wastewater Treatment;* Bookboon: United Kingdom, 2011.
48. Einschlag, F. S. G. *Waste Water: Evaluation and Management;* InTech: Croatia, 2011.
49. Tchobanoglous, G.; Burton, F.; Stensel, H. *Metcalf and Eddy Wastewater Engineering: Treatment and Reuse;* McGraw Hill: New York, NY, 2003; p 384.
50. Muppalla, H. Highly Hydrophilic Electrospun Fibers for the Filtration of Micro and Nanosize Particles Treated with Coagulants, Wichita State University, 2011.
51. Faccini, M.; Borja, G.; Boerrigter, M.; Martín, D. M.; Crespiera, S. M.; Vázquez-Campos, S.; Aubouy, L.; Amantia, D. Electrospun Carbon Nanofiber Membranes for Filtration of Nanoparticles from Water. *J. Nanomater.* **2015,** *2015*, 2.
52. Alharbi, A. R.; Alarifi, I. M.; Khan, W. S.; Asmatulu, R. Highly Hydrophilic Electrospun Polyacrylonitrile/Polyvinypyrrolidone Nanofibers Incorporated with Gentamicin as Filter Medium for Dam Water and Wastewater Treatment. *J. Membr. Sep. Technol.* **2016,** *5*(2), 38–56.
53. Mittal, K. L.; Lee, K.-W. *Polymer Surfaces and Interfaces: Characterization, Modification and Application.* VSP: The Netherlands, 1997.
54. Kim, H.; Abdala, A. A.; Macosko, C. W. Graphene/Polymer Nanocomposites. *Macromolecules* **2010,** *43*(16), 6515–6530.
55. (a) Kaur, S. Surface Modification of Electrospun Poly(vinylidene Fluoride) Nanofibrous Microfiltration Membrane. 2007; (b) Khan, W. S.; Asmatulu, R.; Eltabey, M. M. Electrical and Thermal Characterization of Electrospun PVP Nanocomposite Fibers. *J. Nanomater.* **2013,** *2013.*
56. Ramakrishna, S.; Shirazi, M. M. A. Electrospun Membranes: Next Generation Membranes for Desalination and Water/Wastewater Treatment. *Desalination* **2013,** *308*, 198–208.
57. Ali, I. New Generation Adsorbents for Water Treatment. *Chem. Rev.* **2012,** *112*(10), 5073–5091.
58. Hota, G.; Kumar, B. R.; Ng, W.; Ramakrishna, S. Fabrication and Characterization of a Boehmite Nanoparticle Impregnated Electrospun Fiber Membrane for Removal of Metal Ions. *J. Mater. Sci.* **2008,** *43*(1), 212–217.
59. Aliabadi, M.; Irani, M.; Ismaeili, J.; Piri, H.; Parnian, M. J. Electrospun Nanofiber Membrane of Peo/Chitosan for the Adsorption of Nickel, Cadmium, Lead and Copper Ions from Aqueous Solution. *Chem. Eng. J.* **2013,** *220*, 237–243.
60. Sang, Y.; Li, F.; Gu, Q.; Liang, C.; Chen, J. Heavy Metal-Contaminated Groundwater Treatment by a Novel Nanofiber Membrane. *Desalination* **2008,** *223*(1–3), 349–360.
61. Singh, G.; Rana, D.; Matsuura, T.; Ramakrishna, S.; Narbaitz, R. M.; Tabe, S. Removal of Disinfection by Products from Water by Carbonized Electrospun Nanofibrous Membranes. *Sep. Purif. Technol.* **2010,** *74*(2), 202–212.

62. Taha, A. A.; Qiao, J.; Li, F.; Zhang, B. Preparation and Application of Amino Functionalized Mesoporous Nanofiber Membrane via Electrospinning for Adsorption of Cr^{3+} from Aqueous Solution. *J. Environ. Sci.* **2012,** *24*(4), 610–616.
63. Lee, S. S.; Bai, H.; Liu, Z.; Sun, D. D. Novel-Structured Electrospun TiO_2/CuO Composite Nanofibers for High Efficient Photocatalytic Cogeneration of Clean Water and Energy from Dye Wastewater. *Water Res.* **2013,** *47*(12), 4059–4073.
64. Ma, Z.; Kotaki, M.; Ramakrishna, S. Electrospun Cellulose Nanofiber as Affinity Membrane. *J. Membr. Sci.* **2005,** *265*(1), 115–123.
65. Mukai, Y.; Mizuno, A. Preparation of Iron Ferrocyanide-Supported Nanofiber Membrane for Purification of Cesium-Contaminated Water. *J. Water Resour. Prot.* **2014,** *2014*.
66. Chaúque, E. F.; Dlamini, L. N.; Adelodun, A. A.; Greyling, C. J.; Ngila, J. C. Modification of Electrospun Polyacrylonitrile Nanofibers with EDTA for the Removal of Cd and Cr Ions from Water Effluents. *Appl. Surf. Sci.* **2016,** *369*, 19–28.
67. Qiu, S. Nanofiber as Flocculant or Modifier in Membrane Bioreactors for Wastewater Treatment, University of Akron, 2005.
68. Yoon, K.; Kim, K.; Wang, X.; Fang, D.; Hsiao, B. S.; Chu, B. High Flux Ultrafiltration Membranes Based on Electrospun Nanofibrous PAN Scaffolds and Chitosan Coating. *Polymer* **2006,** *47*(7), 2434–2441.
69. Gopal, R.; Kaur, S.; Ma, Z.; Chan, C.; Ramakrishna, S.; Matsuura, T. Electrospun Nanofibrous Filtration Membrane. *J. Membr. Sci.* **2006,** *281*(1), 581–586.
70. Makaremi, M.; De Silva, R. T.; Pasbakhsh, P. Electrospun Nanofibrous Membranes of Polyacrylonitrile/Halloysite with Superior Water Filtration Ability. *J. Phys. Chem. C* **2015,** *119*(14), 7949–7958.
71. Kampalanonwat, P.; Supaphol, P. Preparation and Adsorption Behavior of Aminated Electrospun Polyacrylonitrile Nanofiber Mats for Heavy Metal Ion Removal. *ACS Appl. Mater. Interfaces* **2010,** *2*(12), 3619–3627.
72. Taha, A. A.; Wu, Y.-N.; Wang, H.; Li, F., Preparation and Application of Functionalized Cellulose Acetate/Silica Composite Nanofibrous Membrane via Electrospinning for Cr(VI) Ion Removal from Aqueous Solution. *J. Environ. Manage.* **2012,** *112*, 10–16.
73. Rad, L. R.; Momeni, A.; Ghazani, B. F.; Irani, M.; Mahmoudi, M.; Noghreh, B. Removal of Ni^{2+} and Cd^{2+} Ions from Aqueous Solutions Using Electrospun PVA/Zeolite Nanofibrous Adsorbent. *Chem. Eng. J.* **2014,** *256*, 119–127.
74. Cao, X.; Huang, M.; Ding, B.; Yu, J.; Sun, G. Robust Polyacrylonitrile Nanofibrous Membrane Reinforced with Jute Cellulose Nanowhiskers for Water Purification. *Desalination* **2013,** *316*, 120–126.
75. Aussawasathien, D.; Teerawattananon, C.; Vongachariya, A. Separation of Micron to Sub-micron Particles from Water: Electrospun Nylon-6 Nanofibrous Membranes as Pre-filters. *J. Membr. Sci.* **2008,** *315*(1), 11–19.
76. Kaur, S.; Kotaki, M.; Ma, Z.; Gopal, R.; Ramakrishna, S.; Ng, S. Oligosaccharide Functionalized Nanofibrous Membrane. *Int. J. Nanosci.* **2006,** *5*(01), 1–11.
77. Kim, J.-H.; Park, P.-K.; Lee, C.-H.; Kwon, H.-H. Surface Modification of Nanofiltration Membranes to Improve the Removal of Organic Micro-Pollutants (EDCs and PhACs) in Drinking Water Treatment: Graft Polymerization and Cross-Linking Followed by Functional Group Substitution. *J. Membr. Sci.* **2008,** *321*(2), 190–198.
78. Zhao, Z.; Zheng, J.; Wang, M.; Zhang, H.; Han, C. C. High Performance Ultrafiltration Membrane Based on Modified Chitosan Coating and Electrospun Nanofibrous PVDF Scaffolds. *J. Membr. Sci.* **2012,** *394*, 209–217.

79. Savoji, H.; Rana, D.; Matsuura, T.; Tabe, S.; Feng, C. Development of Plasma and/or Chemically Induced Graft Co-polymerized Electrospun Poly(vinylidene Fluoride) Membranes for Solute Separation. *Sep. Purif. Technol.* **2013,** *108*, 196–204.
80. Borhani, S. Removal of Reactive Dyes from Wastewater Using Cyclodextrin Functionalized Polyacrylonitrile Nanofibrous Membranes. *J. Text. Polym.* **2016,** *4*(1), 45–52.
81. Gautam, A. K.; Lai, C.; Fong, H.; Menkhaus, T. J. Electrospun Polyimide Nanofiber Membranes for High Flux and Low Fouling Microfiltration Applications. *J. Membr. Sci.* **2014,** *466*, 142–150.
82. Zander, N. E.; Gillan, M.; Sweetser, D. Recycled PET Nanofibers for Water Filtration Applications. *Materials* **2016,** *9*(4), 247.
83. Bagheri, H.; Roostaie, A. Electrospun Modified Silica-Polyamide Nanocomposite as a Novel Fiber Coating. *J. Chromatogr. A* **2014,** *1324*, 11–20.
84. Xu, R.; Jia, M.; Zhang, Y.; Li, F. Sorption of Malachite Green on Vinyl-Modified Mesoporous Poly(acrylic Acid)/SiO_2 Composite Nanofiber Membranes. *Microporous Mesoporous Mater.* **2012,** *149*(1), 111–118.
85. Lala, N. L.; Ramaseshan, R.; Bojun, L.; Sundarrajan, S.; Barhate, R.; Ying-jun, L.; Ramakrishna, S. Fabrication of Nanofibers with Antimicrobial Functionality Used as Filters: Protection Against Bacterial Contaminants. *Biotechnol. Bioeng.* **2007,** *97*(6), 1357–1365.
86. Zhang, L.; Luo, J.; Menkhaus, T. J.; Varadaraju, H.; Sun, Y.; Fong, H. Antimicrobial Nano-Fibrous Membranes Developed from Electrospun Polyacrylonitrile Nanofibers. *J. Membr. Sci.* **2011,** *369*(1), 499–505.
87. Adibzadeh, S.; Bazgir, S.; Katbab, A. A. Fabrication and Characterization of Chitosan/Poly(vinyl Alcohol) Electrospun Nanofibrous Membranes Containing Silver Nanoparticles for Antibacterial Water Filtration. *Iran. Polym. J.* **2014,** *23*(8), 645–654.
88. Lev, J.; Holba, M.; Kalhotka, L.; Szostková, M.; Kimmer, D. In *Application of the Electrospun Nanofibers in Wastewater Treatment*, Proceedings of the International Conference NANOCON, 2011; pp 21–23.
89. Kaur, S.; Sundarrajan, S.; Gopal, R.; Ramakrishna, S. Formation and Characterization of Polyamide Composite Electrospun Nanofibrous Membranes for Salt Separation. *J. Appl. Polym. Sci.* **2012,** *124*(S1).
90. Yoon, K.; Hsiao, B. S.; Chu, B. Formation of Functional Polyethersulfone Electrospun Membrane for Water Purification by Mixed Solvent and Oxidation Processes. *Polymer* **2009,** *50*(13), 2893–2899.
91. Sang, Y.; Li, F.; Gu, Q.; Liang, C.; Chen, J. Heavy Metal-Contaminated Groundwater Treatment by a Novel Nanofiber Membrane. *Desalination* **2008,** *223*(1), 349–360.
92. Abdullah, N.; Mohamad, I. S.; Abd Hamid, S. B. Removal of Iron, Manganese and Boron from Industrial Effluent Water Using Carbon Nanofibers. *Adv. Mater. Res. Trans. Tech. Publ.* **2015,** *1109*, 158–162.
93. Das, R.; Ali, M. E.; Hamid, S. B. A.; Ramakrishna, S.; Chowdhury, Z. Z. Carbon Nanotube Membranes for Water Purification: A Bright Future in Water Desalination. *Desalination* **2014,** *336*, 97–109.
94. Ahmed, Y. M.; Al-Mamun, A.; Al Khatib, M. a. F. R.; Jameel, A. T.; AlSaadi, M. A. H. A. R. Efficient Lead Sorption from Wastewater by Carbon Nanofibers. *Environ. Chem. Lett.* **2015,** *13*(3), 341–346.
95. Zamzow, M.; Eichbaum, B.; Sandgren, K.; Shanks, D. Removal of Heavy Metals and Other Cations from Wastewater Using Zeolites. *Sep. Sci. Technol.* **1990,** *25*(13–15), 1555–1569.

96. (a) Maliou, E.; Malamis, M.; Sakellarides, P. Lead and Cadmium Removal by Ion Exchange. *Water Sci. Technol.* **1992,** *25*(1), 133–138; (b) Ouki, S. K.; Kavannagh, M. Performance of Natural Zeolites for the Treatment of Mixed Metal-Contaminated Effluents. *Waste Manage. Res.* **1997,** *15*(4), 383–394.
97. Ibrahim, K.; Khoury, H. Use of Natural Chabazite–Phillipsite Tuff in Wastewater Treatment from Electroplating Factories in Jordan. *Environ. Geol.* **2002,** *41*(5), 547–551.
98. Pansini, M.; Colella, C.; De Gennaro, M. Chromium Removal from Water by Ion Exchange Using Zeolite. *Desalination* **1991,** *83*(1), 145–157.
99. Lind, M. L.; Jeong, B.-H.; Subramani, A.; Huang, X.; Hoek, E. M. Effect of Mobile Cation on Zeolite-Polyamide Thin Film Nanocomposite Membranes. *J. Mater. Res.* **2009,** *24*(05), 1624–1631.
100. Taha, A. A.; Qiao, J.; Li, F.; Zhang, B. Preparation and Application of Amino Functionalized Mesoporous Nanofiber Membrane via Electrospinning for Adsorption of Cr^{3+} from Aqueous Solution. *J. Environ. Sci.* **2012,** *24*(4), 610–616.
101. Klein, E. Affinity Membranes: A 10-Year Review. *J. Membr. Sci.* **2000,** *179*(1), 1–27.
102. Jiang, Q.; Tian, L.; Liu, K. K.; Tadepalli, S.; Raliya, R.; Biswas, P.; Naik, R. R.; Singamaneni, S. Bilayered Biofoam for Highly Efficient Solar Steam Generation. *Adv. Mater.* **2016,** *28*(42), 9400–9407.
103. Zeman, L. J.; Zydney, A. *Microfiltration and Ultrafiltration: Principles and Applications;* Marcel Dekker, Inc.: New York, NY, USA, 1996.
104. Feng, C.; Khulbe, K. C.; Matsuura, T.; Tabe, S.; Ismail, A. F. Preparation and Characterization of Electro-Spun Nanofiber Membranes and Their Possible Applications in Water Treatment. *Sep. Purif. Technol.* **2013,** *102*, 118–135.
105. Gottenbos, B.; van der Mei, H. C.; Klatter, F.; Nieuwenhuis, P.; Busscher, H. J. In Vitro and In Vivo Antimicrobial Activity of Covalently Coupled Quaternary Ammonium Silane Coatings on Silicone Rubber. *Biomaterials* **2002,** *23*(6), 1417–1423.
106. Qu, X.; Alvarez, P. J.; Li, Q. Applications of Nanotechnology in Water and Wastewater Treatment. *Water Res.* **2013,** *47*(12), 3931–3946.
107. Paneysar, J. S.; Barton, S.; Chandra, S.; Ambre, P.; Coutinho, E. Novel Thermoresponsive Assemblies of Co-grafted Natural and Synthetic Polymers for Water Purification. *Water Sci. Technol.* **2017,** *75*(5), 1084–1097.

CHAPTER 11

GENOTOXIC EFFECTS OF SILVER NANOPARTICLES ON MARINE INVERTEBRATE

S. VIJAYAKUMAR, S. THANIGAIVEL, AMITAVA MUKHERJEE, NATARAJAN CHANDRASEKARAN, and JOHN THOMAS*

Centre for Nanobiotechnology, VIT University, Vellore, Tamil Nadu India

**Corresponding author. E-mail: john.thomas@vit.ac.in, th_john28@yahoo.co.in*

CONTENTS

ABSTRACT

The main role of this study was to investigate the effect of silver nanoparticles (NPs) on *Donax faba.* The animals were exposed to different concentrations (ranging from 20 to 300 ppm) of silver NPs, and the LC_{50} value was calculated at 96 h. Mortality was highest after 72 h of exposure to different concentrations. The tissues of the *D. faba* were examined for histopathological analysis, and the results indicated that an increase in the concentration of silver NPs showed complete damage. Biochemical parameters (protein estimation, catalase, glutathione s-transferase, and superoxide dismutase) were employed to study the effect of silver NPs on the exposed tissues. These antioxidant parameters showed an increase in their activity at the higher concentration. The tissues were further assessed using inductively coupled plasma optical emission spectrometry to check the actual amount of silver present in the animal. It showed that with an increase in the NP concentration, there was close to 80% of bioaccumulation of silver in the animal. In this study, the DNA damages were observed by comet assay carried out in the tissue samples of *D. faba* exposed to silver NP at 20 and 300 ppm concentrations. However, the control tissue did not show any damage. The length of comet tail was used to determine the DNA damage. The results revealed that the length of the comet tail increased with increasing concentration of silver NPs.

11.1 INTRODUCTION

Nanotechnology is one of the most advanced and emerging fields of this century as its application in every sector is enormous. It has various applications in the fields of engineering, agriculture, medical research, and so forth.[30] Nanotechnology is expected to reach new highs, and the industry is expected to be a billion-dollar industry by 2025. Nanoparticles (NPs) are the foundation or basis on which the field of nanotechnology is based. It is expected that nanotechnology will be used or exploited in several areas.[48] Nanotoxicology is the study of the toxic properties of nanomaterials or nanoproducts. It deals with the toxic effects of certain NPs on several vertebrates, invertebrates, fishes, and so forth. The impact of NPs on human beings can be disastrous and can lead to several disorders of the brain, kidney, heart, and other vital organs of the body.[34]

Silver NPs have antimicrobial activity. The NPs used in the manufacturing of a wide range of products such as toothpastes, vacuum cleaners, socks, and washing machines owing to its size range of 1–100 nm.[26] The NPs of silver tend to occur naturally and are also being processed industrially. Owing to this, silver has a potential risk of getting into the aquatic systems. In these aquatic systems, they may cause an environmental risk to the aquatic life due to its toxicity. The use of silver in nanotechnology raises many concerns about its toxic effects on the environment.[18] Silver NPs cause toxicity to an organism at the cellular level through may mechanisms. They include adhesion of NPs to the cell membrane, thereby causing a change in its properties. This, in turn, affects processes such as respiration and cell permeability. Silver NPs also cause DNA damage and are toxic to the genetic structure of an organism.[26]

Donax faba is an invertebrate animal that is a tiny, edible saltwater bivalve mollusk having various local names across the globe. This animal is found on temperate and tropical coasts worldwide. If a wave washes this clam out of the sand, it can dig back quite rapidly. On most of the beaches in Mandapam, nature of the bottom is not conducive for young or seed clams to thrive. Therefore, large clams only occur at the bottom. This is the general distribution pattern. However, because of the wave action during high tide, sometimes the young and the seed clams appear near the low water mark, probably working their way up to the fine sands. The clams have not been obtained from below low water mark.[1]

This animal is used for toxicology studies since it has a unique "filter-feeding" mechanism that helps the animal to retain certain particulate matter and hence can absorb silver. It can be used to monitor environmental pollution in aquatic habitats. It also accumulates toxic substances and easily acclimatizes to laboratory conditions. Another advantage of using this animal for research is that it is found in abundance throughout the year in shallow areas of beaches, and hence samples can be collected easily. In addition, this animal is a good indicator of environmental pollution.[24]

In the recent years, the use of in vitro systems for toxicological studies has been rapidly growing field.[42] The in vivo experiments allow exploring the entire effect of toxicant without excluding any biochemical pathway,[8] and at the same time, in vitro experiments provide basic information on the nature of the tested agents and/or the cellular response.[7] In the present study, total protein content, superoxide dismutase (SOD) activity, as well as glutathione s-transferase (GSH) released have been performed.

Comet assay is one of the ways to evaluate the genotoxic potential of physical, chemical, and biological substances. Small DNA damages such as single-strand breaks, double-strand breaks, and alkali-labile sites of individual cells were detected by using comet assay.[44] Comet assay was first applied to ecotoxicology and it has become one of the popular tests for detecting strand breaks in aquatic animals. Comet assay is considered sensitive compared to others. It is cheap and fast, and it needs a few cells to perform the assay.[32,43,27] In erythrocytes of many species such as fish and mollusks, this comet assay was applied with great success.[33] Nowadays, routinely used methods in the laboratory are the fluorescent, ethidium bromide staining methodology, and nonfluorescent and silver staining technique.[44,12]

11.2 MATERIALS AND METHODS

11.2.1 COLLECTION AND MAINTENANCE OF DONAX FABA

Adult *D. faba* of uniform size (1 ± 0.5 cm length) were collected by hand from Thoppuvalasai, Palk Bay Coast, and the southeast coast of India. They were transported to the lab immediately in plastic buckets with natural seawater and mud and acclimated to room temperature in square aquaria containing filtered seawater. Seawater was brought from the Chennai coast to maintain the animals in aquarium tanks. Animals were acclimated for 1 week in an ambient room condition. Phytoplankton was given as food for *D. faba* and water was aerated gently during the acclimation period. Physicochemical parameters of the filtered seawater were estimated before and during the experiment using standard protocol[3] and they were listed in Table 11.1.

TABLE 11.1 Physicochemical Parameters of Water.

Parameters	Before experiment	During experiment
T (°C)	29.6	29.8
pH	7.5	7.6
Hardness ($mg \cdot l^{-1}$)	187	189
Dissolved oxygen ($mg \cdot l^{-1}$)	7.8	7.7
O_2 (% saturation)	61.50%	61.02%
Total solids ($g \cdot l^{-1}$)	0.006	0.007

11.2.2 PREPARATION OF SILVER NANOPARTICLE (NP) SOLUTION

The silver NP solution was prepared by Creighton's method. About 30 ml of 2 mM ice-cold sodium borohydride solution was mixed with 2 ml of

1% trisodium citrate with stirring for 5 min. About 10 ml of 1 mM silver nitrate solution was added drop by drop under dark conditions without stirring. The appearance of pale yellow color marked the end of the reaction and confirmed the formation of the silver NPs. Then, the NPs were sonicated using ultrasonification and the DLS size was found to be 95 nm. A preliminary range-finding bioassay was performed initially. Then, the concentration gradient series was prepared from the stock solution based on the results.[24] Test concentration and, also, seawater control replicates were maintained. Dead animals were removed immediately during observation. When the bivalves were fully opened and settled down at the bottom, death was confirmed. Tests were conducted in an ambient room temperature. Experiments were carried out at 29±2°C, salinity 30% (ppt) under a 12:12 light:dark regime without aeration.

11.2.3 DETERMINATION OF LETHAL CONCENTRATIONS

The lethal concentration (LC_{50}) value was determined at 96 h of exposure to the silver NPs. Continuous aeration was provided. The control (seawater) and the test animals were maintained in glass aquarium tanks. The experiment was repeated in triplicates to obtain the lethal concentration value. For the experiment, varying concentrations of silver NPs were used: 20, 25, 50, 100, and 300 ppm. The percentage of mortality is shown in Figure 11.1. The test animals were not fed during the course of the experiment.[39] The LC_{50}–96-h values were calculated using the Probit analysis method. The dead animals were removed from the tank. The toxicity experiment was carried out for 120 h at different concentrations of the test chemicals. Mortality was determined. The lethal concentration was also determined.

11.2.4 HISTOPATHOLOGICAL ANALYSIS

The tissue samples after being exposed to silver NPs at different concentrations were processed for histopathological analysis. For histological investigation, organs were dissected out from the body tissue of *Donax,* which was exposed to silver NPs. They were examined and compared with that of the normal body tissue. The organs were fixed in Davidson's fixative and then stored in screw cap bottles containing the fixative.

11.2.5 BIOACCUMULATION STUDY

At the end of the 96-h exposure of the animals to silver NPs, they were taken for bioaccumulation study. It was done to estimate the amount of silver, which accumulated in the animal after being exposed to silver NPs. The tissues were acid digested with concentrated nitric acid and then diluted and analyzed for bioaccumulation of NPs. Water samples were also collected at different intervals ranging from 24 to 120 h to determine the actual available content of silver present.

11.2.6 BIOCHEMICAL STUDIES

Subsequent to the exposure to silver NPs for 96 h, the total protein content and antioxidant and non-antioxidant enzymes were estimated in the body tissues of the exposed *D. faba.* It was done to study the extent of damage caused by the animals due to the toxicity of silver NPs.

11.2.6.1 TOTAL PROTEIN CONTENT

Total protein was ascertained spectrophotometrically at 640 nm based on the method of Binelli et al.[8] Solution A (1% copper sulfate), solution B (2% sodium potassium tartrate), and solution C (2% sodium carbonate in 0.1 N NaOH) were prepared. The solutions were mixed in the ratio of C:B:A (98:1:1). About 4.5 ml of this mix was pipetted into clean, marked glass tubes. The standard bovine serum albumin was added to the tubes at concentrations of 10, 20, 40, 80, and 100 μg. About 10 μl of homogenized samples was added to each marked tube, and then, the tubes were incubated at room temperature for 10 min. Along with that, 0.5 ml of Folin's reagent was added. The test tubes were again incubated in the dark condition for 30 min, and the optical density was read at 640 nm. To know the values of the test samples, a standard graph was plotted from the readings of protein standards.

11.2.6.2 CATALASE ASSAY

Catalase assay was carried out according to the method of Sinha et al.[45] For this assay, phosphate buffer (0.01 M, pH 7.0), 0.2 M solution of hydrogen peroxide, and dichromate acetic acid were the reagents used. Phosphate buffer (0.01 M, pH 7.0) was used to homogenate the tissue. Phosphate buffer (1 ml) was added to 1.0 ml of homogenate tissue. Then, 0.4 ml of hydrogen peroxide was added to that. After that, 2.0 ml dichromate acetic acid reagent was added into the reaction; because of this, the reaction was arrested after 30 and 60 s. Simultaneously, a control reaction was also carried out. The tubes containing solutions were heated in a boiling water bath for 10 min. Then, it was cooled, and measured absorbance was read at 620 nm. The enzyme activity was denoted as units$\cdot$min$^{-1}\cdot$mg$^{-1}\cdot$protein^{-1}.

11.2.6.3 GLUTATHIONE S-TRANSFERASE (GST) ASSAY

Glutathione S-transferase activity was determined by using the standard kit method (HiMedia).[31] About 0.3 M phosphate buffer was prepared, and the pH was adjusted at 6.5. After that, 1-chloro-2,4-dinitrobenzene (CDNB: 30 mM) and reduced glutathione (GSH: 30 mM) were prepared. About 1.0 ml of buffer, 0.1 ml of CDNB, and 0.1 ml of tissue homogenate and remaining water was added to 2.5 ml of the reaction mixture. Then, this mixture was preincubated at 37°C for 5 min. Finally, 0.1 ml of GSH was added, and the absorbance change was measured at 340 nm with 30-s intervals for 3 min.

11.2.6.4 SUPEROXIDE DISMUTASE (SOD) ASSAY

SOD activity was determined according to the modified method of Beauchamp and Fridovich (1995). The relative enzyme activity also called as U$\cdot$mg^{-1} protein. For the control, 100 µl of the homogenized samples were placed under fluorescent light for 2 min at the absorbance of A_{560} optimal density values reached at 0.2–0.25. For this control preparation, nitro blue tetrazolium, 20 µM reaction mixture (0.1 mM EDTA, 13 µM methionine, 0.75 mM nitro blue tetrazolium), and 20 µM riboflavin in phosphate buffer (50 mM, pH 7.8) reagents were prepared and added.

11.2.7 SAMPLE PREPARATION FOR THE COMET ASSAY

After 96-h exposure, gill, foot, and body tissues of the *D. faba* control, as well as treated organs, were removed using scissors. Then, the tissues were cut into small pieces individually and homogenized to obtain a single cell suspension. After that, about 10 mg of tissue was washed three times with chilled phosphate buffered saline. Then, it was transferred to ice-cold homogenization buffer (1X Hank's balanced salt solution, 20 mM EDTA, 10% dimethyl sulfoxide [DMSO], [pH 7–7.5]).[24] Trypan blue exclusion assay was used to differentiate between DNA damage due to cytotoxicity or genotoxicity to check the cell viability. The final suspension showed >80% cell viability, so it was processed for the comet assay. All fish should have the cell viability of the cell suspension between 88 and 96%. When the control cells have below 70–80% cell viability, it may be considered that the cell suspension is excessive.[46] The viability of cell in the suspensions used in this study is acceptable because its values are above the previously mentioned values.[2] To get pellet homogenized tissue samples were centrifuged at 3000 rpm at 4°C for 5 min and the supernatant was discarded.

11.2.7.1 COMET ASSAY

Comet assay was performed according to the modified method of Tice et al.[46] and Arutyunyan et al.[4] About 1% high melting agarose was dissolved in 1X PBS and poured on the slide. Then, 25 μl of homogenized tissue sample was added to 75 μl of 0.5% low melting agar. This was poured on to the existing slide. Coverslip was then placed on the agar surface. After solidified coverslip was removed, 100 μl of 1% agarose dissolved in PBS (100 mM phosphate) was added on to the second layer. It was allowed to solidify. After that, the coverslip was removed, and the slides were kept in 100 ml of lysis buffer for 1 h. Then, the slides were placed in an electrophoresis chamber containing alkaline buffer (1 mM Na_2EDTA, 300 mM NaOH, pH 13.1, 4°C) without running condition for DNA unwinding for 20 min. After 20 min, electrophoresis was carried out at 1.25 $V \cdot cm^{-1}$, 300 mA for 25 min. The slides were removed from the electrophoresis chamber and washed for 10 min with neutralization buffer (0.4 M Tris HCl, pH 7.5, 0.08 M Tris base, pH 7.2). The experiments were repeated thrice. After electrophoresis, the slides were kept in 100% ethanol to fix the cells for 10 min and then left to dry at room temperature. Finally, ethidium bromide (0.02 $mg \cdot mL^{-1}$) was used to stain the slides. Then, slides were observed under Leica DMLB fluorescence microscope.

11.3 RESULTS

11.3.1 DETERMINATION OF LETHAL CONCENTRATIONS

Once the study was over, the mortality was calculated for each concentration, in terms of percentage. It was showed in Figure 11.1. The LC_{50} value was determined at 96 h using Probit analysis method and was found to be 18.525 ppm.

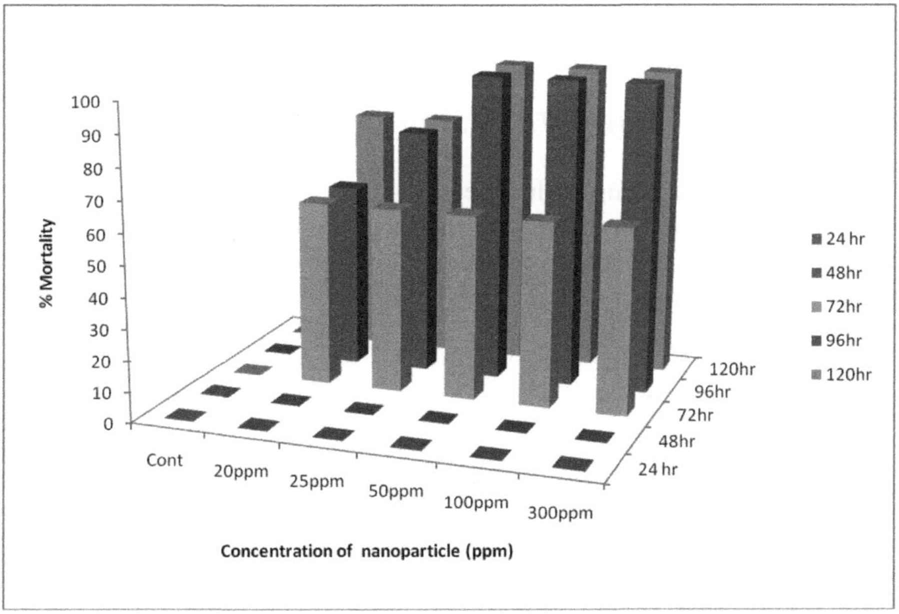

FIGURE 11.1 **(See color insert.)** Mortality rate (in percentage) measurement.

11.3.2 HISTOPATHOLOGICAL ANALYSIS

The histopathological analysis confirmed that there was damage to the *D. faba* tissues after being exposed to silver NP concentrations. In control tissue of *D. faba,* there was no significant damage, whereas the exposure of *D. faba* to 20 ppm of AgNPs caused slight damages in tissues, and at 300 ppm, complete damage to the tissues was observed as shown in Figures 11.2a,b,c. The result proved that exposure to silver NPs was toxic to the tissue. The images of the control and damaged tissues were obtained using a Leica DM2500 fluorescent microscope.

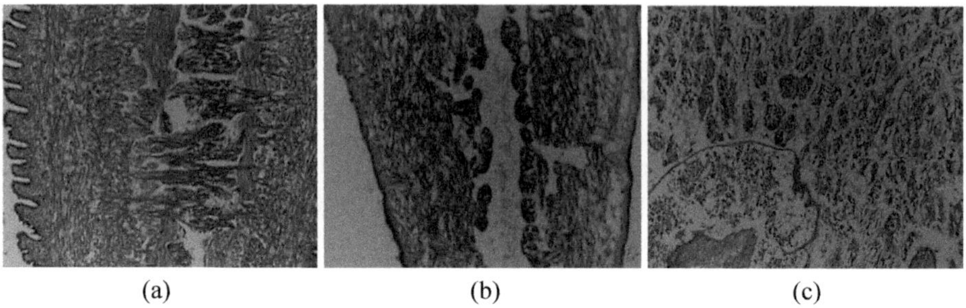

FIGURE 11.2 Histopathological investigation of (a) control tissue, (b) tissue at 20 ppm, (c) tissue at 300 ppm.

11.3.3 BIOACCUMULATION STUDIES

The analysis done using inductively coupled plasma optical emission spectrometry showed an increase in the bioaccumulation of silver with an increasing concentration of silver NPs, which is showed in Figure 11.3. This proves that when concentration of silver NPs is increased, accumulation is more.

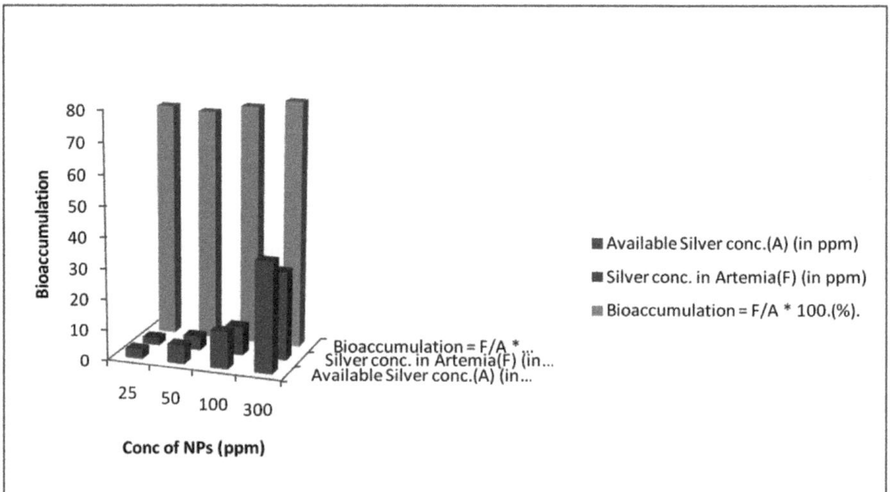

FIGURE 11.3 (See color insert.) Inductively coupled plasma optical emission spectrometry–bioaccumulation of silver nanoparticles (NPs).

11.3.4 BIOCHEMICAL STUDIES

11.3.4.1 TOTAL PROTEIN CONTENT

The toxic effect of silver (Ag) NPs on biochemical parameters of *D. faba* was presented in Figure 11.4a–d. Reductions in the protein concentration in the tissues of *D. faba* exposed to silver (Ag) NPs were statistically significant ($p<0.05$) when compared to the protein concentration in the control animal. Total protein content was decreased, when the tissues were exposed to toxicants (Fig. 11.4a).

11.3.4.2 CATALASE ASSAY

Figure 11.4b represents catalase activity of *D. faba* to toxicity effect of silver (Ag) NPs. When the concentration of the silver NPs was increased, there was a gradual and sporadic increase in the catalase activity, and it showed significant difference ($p<0.05$) between 20 and 300 ppm compared to control.

11.3.4.3 GLUTATHIONE S-TRANSFERASE

The comparisons among GST enzyme activity in muscle tissues of control and silver (Ag) NPs treated (20 and 300 ppm) by one-way analysis of variance revealed statistically significant reduction of GST enzyme in muscle tissues of NP-treated *D. faba* ($p<0.05$) was explained in Figure 11.4c.

11.3.4.4 SOD ACTIVITY

The SOD enzyme activity levels in the tissues of *D. faba* treated with 20 and 300 ppm were found to be increasing significantly ($p < 0.05$) when compared to those in the control (Fig. 11.4d). SOD shows better response, compared to the all other antioxidant enzymes.

11.3.5 COMET ASSAY

The control tissue of *D. faba* (Fig. 11.5a) showed normal structure, whereas the NPs exposed tissues showed some alterations in the comet tail. The

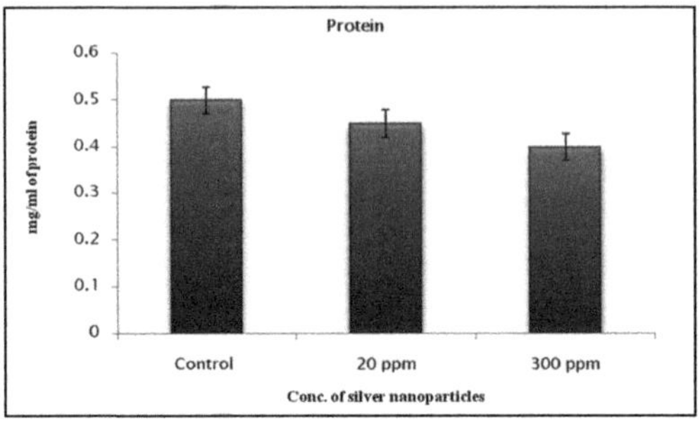

(a)

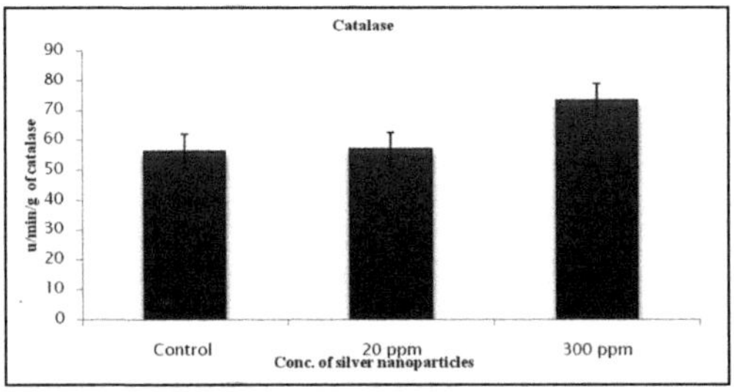

(b)

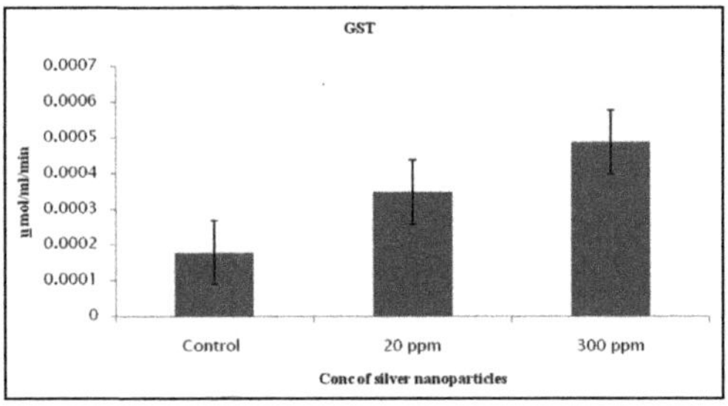

(c)

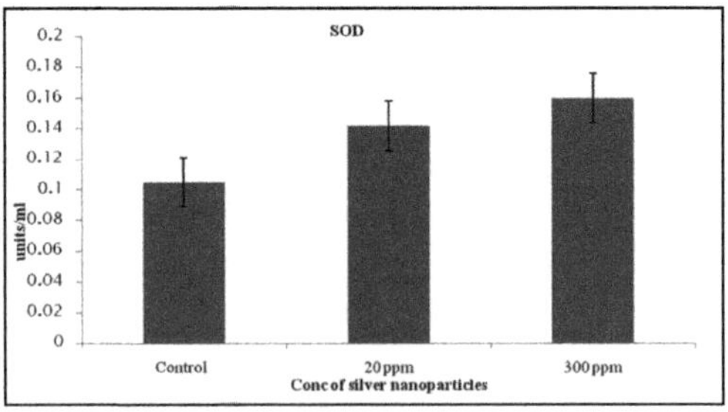

(d)

FIGURE 11.4 (a) Total protein content to determine the protein level in *Donax faba* tissue, (b) Catalase activity to determine the oxidative damage, (c) glutathione S-transferase (GST) activity to determine the reduction of GST enzyme, (d) superoxide dismutase (SOD) activity to determine the SOD enzyme activity.

comet tail length was more in 300 ppm when compared to 20 ppm as shown in Figure 11.5b,c. Figure 11.6 showed that the length of the comet tail increased with the increasing concentration of silver NPs. Tissues exposed to silver NPs for 20 ppm showed significant increases in the score of DNA damage in relation to respective control ($p<0.001$). It showed a significant difference between the comet tail length of control and 20 ppm of tissues treated with silver NPs. The cells exposed to 300 ppm showed damage significantly higher than 20 ppm of treated silver NPs ($p<0.001$).

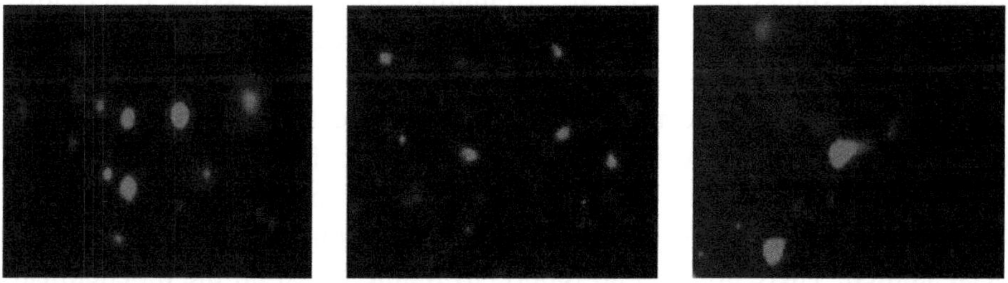

FIGURE 11.5 **(See color insert.)** (a) Comet at control, (b) comet at 20 ppm, (c) comet at 300 ppm.

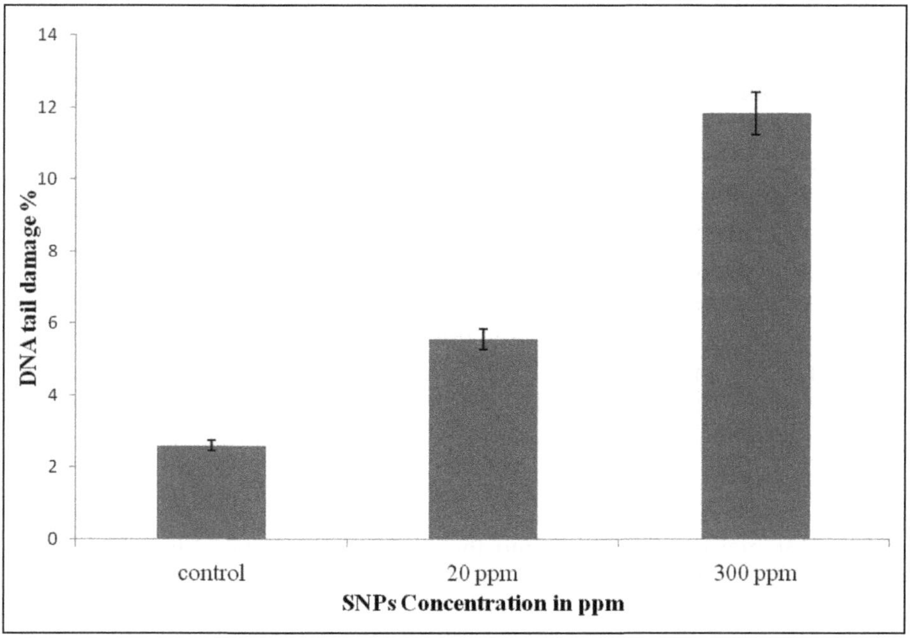

FIGURE 11.6 Comet assay to determine the tail damage of DNA.

11.4 DISCUSSION

Heavy metals and their NPs are majorly responsible for pollution in the aquatic environment. They received considerable attention because of their ability to accumulate in animal tissues, and they can be toxic to the animals at very low levels.[16,10]

Silver was not detected in natural seawater; hence, there was no accumulation of silver in control animal. When exposed to silver NPs, the accumulation of silver NPs increased with increasing concentrations in *D. faba*. The present study shows that silver NPs accumulate in animal tissues at very low concentration. Therefore, the accumulation of toxicant was concentration dependent, that is, higher the concentration, higher the accumulation.

In this study, we were able to understand the mechanism by which silver NPs caused oxidative stress, and by histopathological analysis, we were able to observe the damage to the tissue. There are no reports on the histopathological analysis of the tissue of *D. faba.* Their reproductive cycles alone have been studied. There are also no reports available based on the biochemical aspects of *D. faba.* This work also helped in finding out the mortality rates of the species. It showed that, at the time of 72 h, almost 80% of the animals died and hence LC_{50} was determined. While testing and carrying out experiments, a majority of animals died after being exposed to the particles for 96 h and hence the lethal concentration at 96 h was calculated.

Aquatic animals, usually invertebrates, have always often been used to monitor the quality of surface waters and various other effluents.[9] *D. faba* is edible bivalves and is consumed as seafood. Heavy metals can cause environmental damage under some particular environmental conditions. Bivalves are filter-feeder invertebrates. They can accumulate heavy metals from food, water, and also from the ingestion of inorganic particulate matter. Benthic organisms are playing an important role in maintaining ecological balance. In addition, they act as a major food source for many fish species. Even humans consume some of the large-sized crustaceans and bivalves. Heavy metals such as cadmium, zinc, and so forth, pollutants are conservative and often highly toxic to biota, particularly organisms such as *D. faba*.[51]

Marine organisms such as *D. faba* and other invertebrates exhibit a wide range of bioactivity. Pharmaceutical and pesticide industries are nowadays looking for marine organisms and their secondary metabolites. Along the Indian coast, there have been no reports of the toxicity and hemolytic activities of marine organisms till date. An attempt was made to screen as many organisms as possible for their toxicity and hemolytic activities.[38] In general, bivalves contained metal concentrations that was below the maximum permissible limits and should pose no toxicological risk to consumers. However, nowadays, the toxic level is increasing in bivalves.

According to the Department of Fisheries, Malaysia (2005), in 2005–2006, the Malaysia Fisheries Directory documented 27 species of bivalves in Malaysian coastal areas. According to Goldberg et al.[22] and Phillips,[35] among the common approaches used to survey environmental contamination, the use of bioindicator species has proven to be a valuable and informative tool. Bivalves accumulate heavy metals from food, water, and also from the ingestion of inorganic particulate materials, hence fulfilling the criteria as good bioindicators. The organisms that are sedentary, widespread, and have a long life span are well established as bioindicators for monitoring the concentrations of heavy metals in coastal areas. Biotic and abiotic environmental factors may contribute to the wide variability observed in the heavy metal concentrations in the total soft tissues during the spawning season of the bivalves.

The tissues of the bivalves such as the tissues, gills, foot, and mantles could accumulate heavy metals. The ionic differences between the metals and their affinities to the binding sites of the metallothioneins in the different soft tissues could affect the different metal levels found in the bivalves.[19] Bivalves and snails are marine mollusks and are used as test species for conducting toxicity experiments. They are mostly benthic in habitat and are filter feeders, while some are found along the rocky shores.[37] The sperm cell fertilization methods were achieved using sea urchin which was used to determine the toxicity of heavy metals.[49] There are certain factors that influence the bioconcentration of heavy metals in molluskan tissues. It was reported that the concentrations of heavy metals accumulated by marine organisms depend not only on the quality of water but also on temperature, salinity, seasonal factors, diet or food intake, individual variation, and spawning variation.

Aquatic animals, especially invertebrates, are mostly used in bioassays to monitor water quality of surface waters and effluents.[9] *D. faba* is one of the well-known filter feeders. They can have the capacity for assimilating particulates and associated contaminants directly from sediment and water. One of the best applications of *D. faba* is that it can accumulate toxic substances very well and respond to minute concentration of mutagens. *D. faba* is available throughout the year. It can easily acclimate to lab conditions because of its semi-pelagic nature. It can be good candidate for early detection of aquatic environmental problems because of its wide distribution in shallow waters. LC_{50} provides fundamental data for a safe level or tolerance level of a pollutant, so the determination of the LC_{50} value is important.

In marine field, sublethal effects of NPs have not been studied, so this additional information is critical for the complete understanding of the risk of NPs to invertebrates.

This study showed that the exposure of *D. faba* to the toxicity of silver NPs is both concentration and time dependent. *D. faba* is found to be less resistant than the Amazonian invertebrates, with LC_{50} values of 16 $mg \cdot L^{-1}$ [40] and mysid shrimp showing 0.098 $mg \cdot L^{-1}$. When comparing freshwater invertebrates with marine water invertebrates, organophosphate and carbamide are more toxic to marine water invertebrates.[40] Through several pathways, most of the agrochemicals and NPs have the ability to induce oxidative stress. They inhibit the electron-transfer chain reaction. Oxygen-derived free radicals are reacting with DNA, proteins, lipids, and polysaccharides of biological molecules.[29] As suggested by Girotti,[21] degradation of molecules which is called lipid peroxidation is the major cause of cell damage that is enhanced by free radicals, which react with lipids. Toxicants-caused genotoxicity can directly interact with the nuclear material. Outcome data of genotoxins are insufficiently available for marine invertebrates to be detectable at the population of the genetic level. Only a few generations have passed for the effects to propagate in the population of the genetic level.[50] In filter-feeding bivalve populations, contaminants in the marine environment can have a serious impact on DNA.[23]

The genetic damage is the important assay in the identification of pollution hazards in the marine environment.[47,17] To measure the genotoxicity in marine invertebrates, various techniques are available.[17] The comet assay has been performed in a variety of aquatic organisms, including bivalve mollusks, and has been increasingly used in ecotoxicology.[15,28] Alkaline comet assay has recently improved because it can have the sensitivity to detect low-level DNA damage.[20] Oxidative stress was induced by reducing the GSH content and the formation of GSH conjugates. This could also be another mechanism of DNA damage.[13,25] It is observed that carbendazim possesses a high permeability across the lipid bilayer membrane. This characteristic explains that the toxins directly enter into the cells and cause damage to animals.

Contamination and pollution of aquatic resources are one of the most worrying problems of mankind. The primary responsible contaminants of the aquatic environment were NPs from laboratory and industrial effluents.[14] Commonly used aquatic ecosystem bioindicators are vertebrates and invertebrates because they play a vital role in the food chain. They can take up the toxic substances directly and indirectly by the consumption of both compounds dissolved in water and previously contaminated organism and showed bioaccumulation.[11,6] *D. faba* is an excellent model for the detection of mutagenicity or carcinogenicity.

A whole range of health problems, including diet, disease, and exposure to occupational or environmental toxins is because of DNA damage within the cells. To measure this damage, comet assay is a simple and inexpensive method. Compared to other methods, comet assay technique is more advanced method because it enables DNA damage from disease, exposure to toxins or diet to be identified significantly earlier than before.[5,41] The working principle of comet assay is to determine the number of breaks in the strands of DNA within the cell. A microscope slide which contains gel embedded with cells is washed to remove the soluble cell contents, cell membrane, and the histones from the nucleus of the cells. To attract the remaining clumps of "supercoiled" DNA to the anode, an electric field is activated. There is little movement if the DNA is intact. If loops of DNA are pulled toward the anode, it means that there are breaks in the DNA strands. The technique of the fluorescence microscopy then shows the image. The loops of undamaged DNA (the head) clumps are pulled away, forming a tail. Either visually or by computerized image analysis, the proportion of DNA in the tail enables the number of strand breaks to be determined.[36]

11.5 CONCLUSION

Important to watch in toxicity test is the survival after exposure to contaminated and uncontaminated (control) waters. Holding facilities and handling techniques used in this study were acceptable for conducting such tests because none of the control animals died. The standard protocol represents that survival percentage should be around 90% in control study. No mortality was noticed in animal exposed to seawater, the only medium in the present study. Marine mollusks, including bivalves, have been used as the indicator organisms for marine pollution. Safe concentration of metals and their NPs that should be nontoxic to snails, bivalves, and other marine mollusks.

Our results clearly showed that exposure of bivalve (*D. faba*) to silver NPs (*D. faba*) caused several biochemical changes and were highly toxic to it. The evidence proved that there is an increase in the activity of antioxidants when treated with different doses of the silver NPs. Remarkable histopathological changes were seen in the tissues of the animal. The bioaccumulation studies such as AAS and ICP-OES showed the percentage of silver uptake by the organism on exposure to NPs with their increased concentration. The effects of metals and NPs on aquatic organisms can be confirmed by the detailed studies of enzymes having different end points, which help to identify the oxidative stress and genotoxic status. Our findings support complete understanding of the risk of NPs to invertebrates because effects of NPs have not been completely reported for marine organisms. Powerful indicators of pollution in general across marine sites will be marine mollusks and other long-living sedentary organisms. Our methodologies such as biomarker enzyme and DNA damage assays are flexible and capable of forming successful results with various other methods, which makes it a good tool for investigations in ecotoxicology, ecology, and marine biology.

In conclusion, this study shows that *D. faba* can be used as an ideal organism to monitor environmental pollution. When the silver NPs are released into the aquatic ecosystem, it may lead to disastrous physical, biological, and chemical alterations as well as DNA damage and posing a threat to the life of aquatic animals.

ACKNOWLEDGMENT

The authors acknowledge the support of VIT University, Vellore for providing the facility to carry out the work.

KEYWORDS

- **nanotechnology**
- ***Donax faba***
- **silver nanoparticles**
- **toxicity**
- **lethal concentration**
- **comet assay**

REFERENCES

1. Alagarswami, K. Studies on Some Aspects of Biology of the Wedge-Clam, Donaxfaba Gmelin from Mandapam Coast in the Gulf of Mannar. *J. Mar. Biol. Assoc. India* **1966,** *8*(1), 56–75.

2. Alink, G. M.; Quik, J. T. K.; Penders, E. J. M.; Spenkelink, A.; Rotteveel, S. G. P.; Maas, J. L.; Hoogenboezem, W. Genotoxic Effects in the Eastern Mudminnow (*Umbra pygmaea* L.) after Exposure to Rhine Water, as Assessed by Use of the SCE and Comet Assays: A Comparison Between 1978 and 2005. *Mutat. Res.* **2007,** *631,* 93–100.
3. APHA-AWWAA-WPCF. *Standard Methods for the Examination of Water and Wastewater,* 19th ed.; American Public Health Association: 1015, Eighteenth street NW, Washington, DC, 1995.
4. Arutyunyan, R.; Gebhart, E.; Hovhannisyan, G.; Greulich, K. O.; Rapp, A. Com-FISH Using Peptide Nucleic Acid Probes Detects Telomeric Repeats in DNA Damaged by Bleomycin and Mitomycin C Proportional to General DNA Damage. *Mutagenesis* **2004,** *19*(5), 403–408.
5. Belpaeme, K.; Deldere, K.; Zhu, L.; Kirsch-Volders, M. Cytogenetic Studies of PCB77 on Brown Trout (Salmotruttafario) using Micronucleus Test and the Alkaline Comet Assay. *Mutagenesis* **1996,** *11,* 485–492.
6. Biagini, F. R.; David, J. A. O.; Fontanetti, C. S. The use of Histological, Histochemical and Ultramorphological Techniques to Detect Gill Alterations in Oreochromis Niloticus Reared in Treated Polluted Waters. *Micron* **2009,** *40,* 839–844. DOI: 10.1016/j.micron.2009.10.009.
7. Binelli, A.; Cogni, D.; Parolini, M.; Riva, C.; Provini, A. Cytotoxic and Genotoxic Effects of in Vitro Exposures to Triclosan and Trimethoprimon on Zebra Mussel (D. polymorpha) Hemocytes. *Comp. Biochem. Physiol.* **2009a,** *150C,* 50–56.
8. Binelli, A.; Cogni, D.; Parolini, M.; Riva, C.; Provini, A. In Vivo Experiments for the Evaluation of Genotoxic and Cytotoxic Effects of Triclosan in Zebra Mussel Hemocytes. *Aquat. Toxicol.* **2009b,** *91,* 238–244.
9. Brungs, W. A.; Mount, D. I. Introduction to a Discussion of the use of Aquatic Toxicity Tests for Evaluation of the Effects of Toxic Substances. In Cairns Jr., J., Dickson, K. L., Maki, A. W. Eds.; Estimating the Hazard of Chemical Substances to Aquatic Life. *Am. Soc. Test. Mater.* STP, **1978,** *657,* 1–15.
10. Bryan, G. W. Pollution due to Heavy Metals and their Compounds. In *Vol. V: Marine Ecology, Chap. 3: Ocean Management;* Kinne, O., Ed.; Wiley: Chichester, 1984; pp 1289–1431.
11. Cavas, T.; Ergene-Gozukara, S. Induction of Micronuclei and Nuclear Abnormalities in Oreochromisniloticus Following Exposure to Petroleum Refinery and Chromium Processing Plant Effluents. *Aquat. Toxicol.* **2005,** *74,* 264–271. DOI: 10.106/j.aquatox.2005.06.001.
12. Cerad, H.; Delincee, H.; Haine, H.; Rupp, H. The DNA Comet Assay as a Rapid Screening Technique to Control Irradiated Food. *Mutat. Res.* **1997,** *375,* 167–181.
13. Chakraborty, P.; Ugir Hossain, S. K.; Murmu, N.; Das, J. K.; Pal, S.; Bhattacharya, S. Modulation of Cyclophosphamide-Induced Cellular Toxicity by Diphenylmethyl Selenocyanate In Vivo, an Enzymatic Study. *J. Cancer Mol.* **2009,** *4*(6), 183–189.
14. Claxton, L. D.; Houk, V. S.; Hughes, T. J. Genotoxicity of Industrial Wastes and Effluents. *Mutat. Res.* **1998,** *410,* 237–243. DOI: 10.1016/S1383–5742(98)00008–8.
15. Coughlan, B. M.; Hartl, M. G.; O'Reilly, S. J.; Sheehan, D.; Morthersill, C.; van Pelt, F. N.; O'Halloran, J.; O'Brien, N. M. Detecting Genotoxicity using the Comet Assay Following Chronic Exposure of Manila Clam Tapes Semi Decussates to Polluted Estuarine Sediments. *Mar. Pollut. Bull.* **2002,** *44,* 1359–1365.

16. Depledge, M. H.; Weeks, J. M.; Bjerregaard, P. B. Heavy Metals. In *Handbook of Ecotoxicology;* Calow, P., Ed.; Blackwell Scientific Publications: Cambridge, 1994; pp 79–105.
17. Dixon, D. R.; Pruski, A. M.; Dixon, L. R. J.; Jha, A. N. Marine Invertebrate Ecogenotoxicology: A Methodological Overview. *Mutagen* **2002,** *17,* 495–507.
18. Elzey, S.; Grassian, V. H. Agglomeration, Isolation and Dissolution of Commercially Manufactured Silver Nanoparticles in Aqueous Environments. *J. Nanopart. Res.* **2009,** *12*(5), 1945–1958.
19. Franklin, B. E.; Chee, K. Y.; Ismail, A.; Soon, G. T. Interspecific Variation of Heavy Metal Concentrations in the Different Parts of Tropical Intertidal Bivalves. *Water Air Soil Pollut.* **2009,** *196,* 297–309.
20. Gedik, C.; Ewen, S.; Collins, A. Single-Cell Gel Electrophoresis Applied to the Analysis of UV-C Damage and its Repair in Human Cells. *Int. J. Radiat. Biol.* **1992,** *62,* 313–320.
21. Girotti, A. W. Mechanisms of Lipid Peroxidation. *J. Free Radical Biol. Med.* **1985,** *1,* 87–95.
22. Goldberg, E. D.; Koide, M.; Hodge, V.; Flegal, A. R.; Martin, J. H. U.S. Mussel Watch: 1977–1978 Results on Trace Metals and Radionuclides. *Estuarine Coastal Shelf Sci.* **1983,** *16,* 69–93.
23. Hamoutene, D.; Payne, J. F.; Rahimtula, A.; Lee, K. Use of the Comet Assay to Assess DNA Damage in Hemocytes and Digestive Gland Cells of Mussels and Clams Exposed to Water Contaminated with Petroleum Hydrocarbons. *Mar. Environ. Res.* **2002,** *54,* 471–474.
24. Janaki Devi, V.; Nagarani, N.; Yokesh Babu, M.; Kumaraguru, A. K.; Ramakritinan, C. M. A Study of Proteotoxicity and Genotoxicity Induced by the Pesticide and Fungicide on Marine Invertebrate (*Donax faba*). *Chemosphere* **2012,** *90*(3), 1158–1166.
25. Jia, L.; Garza, M.; Wong, H.; Reimer, D.; Redelmeier, T.; Camden, J. B.; Weitman, S. D. Pharmacokinetic Comparison of Intravenous Carbendazim and Remote Loaded Carbendazim Liposomes in Nude Mice. *J. Pharma. Biomed. Anal.* **2002,** *28,* 65–72.
26. Klaine, S. J.; Alvarez, P. J.; Batley, G. E.; Fernandes, T. F.; Handy, R. D.; Lyon, D.; Mahendra, G. E.; McLaughlin, M. J.; Lead, J. R. Nanomaterials in the Environment: Behaviour, Fate Bioavailability and Effects. *Environ. Toxicol. Chem.* **2008,** *27*(9), 1825–1857.
27. Kosz-Vnenchak, M.; Rokosz, K. The Comet Assay for Detection of Potential Genotoxicity of Polluted Water. *Folia. Biol.* **1997,** *45,* 153–239.
28. Lee, R. F.; Steinert, S. Use of the Single Cell Gel Electrophoresis/Comet Assay for Detecting DNA Damage in Aquatic (Marine and Freshwater) Animals. *Mutat. Res.* **2003,** *544,* 43–64.
29. Lemiere, S.; CossuLeguille, C.; Bispo, A.; Jourdain, M. J.; Lanhers, M. C.; Burnel, D.; Vasseur, P. DNA Damage Measured by the Single-Cell Gel Electrophoresis (Comet) Assay in Mammals Fed with Mussels Contaminated by the 'Erika' Oil–Spill. *Mutat. Res.* **2005,** *581,* 11–21.
30. Liu, Z.; Kiessling, F.; Gatjens, J. Advanced Nanomaterials in Multimodal Imaging: Design, Functionalization, and Biomedical Applications. *J. Nanomater.* **2010,** *2010,* 1–15.
31. Lowry, O. H.; Rosebrough, N. J.; Farr, A. L.; Randall, R. J. Protein Measurement with the Folin Phenol Reagent. *J. Biol. Chem.* **1951,** *193*(1), 265–275.
32. Mitchelmore, C. L.; Chipman, J. K. DNA Strand Breakage in Aquatic Organisms and the Potential Value of the Comet Assay in Environmental Monitoring. *Mutat. Res.* **1998,** *399,* 135–147.

33. Nacci, D. E.; Cayula, S.; Jackmin, E. Detection of DNA Damage in Individual Cells from Marine Organisms using Single Cell Gel Assay. *Aquat. Toxicol.* **1996,** *35,* 197–210. DOI: 10.1016/0166–445X(96)00016–1.
34. Nel, A.; Xia, T.; Mädler, L.; Li, N. Toxic Potential of Materials at the Nanolevel. *Science* **2006,** *311*(5761), 622–627.
35. Phillips, D. J. H. Use of Macroalgae and nvertebrates as Monitors of Metal Levels in Estuaries and Coastal Waters. In *Heavy Metals in the Marine Environment;* Furness, R. W., Rainbow, P. S., Eds.; 1990; pp 81–99.
36. Rajaguru, P.; Suba, S.; Elayaraja, T.; Palanivel, M.; Kalaiselvi, K. Genotoxicity of a Polluted River System Measured using the Alkaline Comet Assay on Fish and Earthworm Tissues. *Environ. Mol. Mutagen.* **2003,** *41*(2), 85–91.
37. Ramakritinan, C. M. Chandruvelan, R.; Kumaraguru, A. K. Acute Toxicity of Metals: Cu, Cd, Hg and Zn on Marine Molluscs. *Indian J. Geo-Mar. Sci.* **2012,** *41*(2).
38. Rao, D. S.; James, D. B.; Pillai, C. G.; Thomas, P. A.; Appukuttan, K. K.; Girijavallabhan, K. G.; Najmuddin, M. *Biotoxicity in Marine Organisms* 1991.
39. Reish, D. L.; Oshida, P. S. Manual of Methods in Aquatic Environment Research. FAO Fisheries Technical paper 1987, 247, 1–62.
40. Rico, A.; Waichman, A. V.; Geber-Correa, R.; van den Brink, P. J. Effects of Malathion and Carbendazim on Amazonian Freshwater Organisms: Comparison of Tropical and Temperate Species Sensitivity Distributions. *Ecotoxicology* **2011,** *20,* 625–634.
41. Rigonato, J.; Mantovani, M. S.; Jordao, B. Q. Comet Assay Comparison of Different Corbicula Fluminea (Mollusca) Tissues for Detection of Genotoxicity. *Genet. Mol. Biol.* **2005,** *28,* 464–468. DOI: 10.1590/S1415–47572005000300023.
42. Sandrini, J. Z.; Bianchini, A.; Trindade, G. S.; Nery, L. E.; Marins, L. F. Reactive Oxygen Species Generation and Expression of DNA Repair-Related Genes after Copper Exposure in Zebrafish (*Danio rerio*) ZFL cells. *Aquat. Toxicol.* **2009,** *95,* 285–291.
43. Sasaki, Y. F.; Izumiyama, F.; Nishidate, E.; Ishibashi, S.; Tsuda, S.; Matsusaka, N.; Asano, N.; Saotome, K.; Sofuni, T.; Hayashi, M. Detection of Genotoxicity of Polluted Sea Water using Shell Fish and Alkaline Single Cell Gel Electrophoresis (SCE) Assay: A Preliminary Study. *Mutat. Res.* **1997,** *393,* 133–139. DOI: 10.1016/S1393–5718(97)00098–3.
44. Singh, N. P.; Mccoy, M. T.; Tice, R. R.; Schnider, E. L. A Simple Technique for Quantitation of Low Levels of DNA Damage in Individual Cells. *Exp. Cell. Res.* **1988,** *175,* 184–191.
45. Sinha, A. K. Colorimetric Assay of Catalase. *Anal. Biochem.* **1972,** *47,* 389–394.
46. Tice, R. R. E. A.; Anderson, D.; Burlinson, B.; Hartmann, A.; Kobayashi, H.; Miyamae, Y.; Rojas, E.; Ryu, J. C.; Sasaki, Y. F. Single Cell Gel/Comet Assay: Guidelines for in Vitro an in Vivo Genetic Toxicology Testing. *Environ. Mol. Mutagen.* **2000,** *35,* 206–221.
47. Tice, R. R.; Agurell, E.; Anderson, D.; Burlinson, B.; Hartmann, A.; Kobayashi, H.; Miyamae, Y.; Rojas, E.; Ryu, J. C.; Sasaki, Y. F. The Single Cell Gel/Comet Assay: Guidelines for in Vitro and in Vivo Genetic Toxicology Testing. *Environ. Mol. Mutagen.* **2000,** *35,* 206–221.
48. Umayaparvathi, S.; Arumugam, M.; Meenakshi, S.; Balasubramanian, T. Biosynthesis of Silver Nanoparticles using Oyster *Saccostrea cucullata*; Study of the in-Vitro Antimicrobial Activity. *Int. J. Sci. Nat.* **2013,** *4*(1), 199–203.
49. Waewtaa Thongra, A. R. Toxicity of Cadmium, Zinc and Copper on Sperm Cell Fertilization of Sea Urchin, *Diadema Setosum*. *J. Sci. Soc. Thailand* **1997,** *23*, 297–306.

50. Whitehead, A.; Kuivila, K. M.; Orlando, J. L.; Kotelevtsev, S.; Anderson, S. L. Genotoxicity in Native Fish Associated with Agricultural Runoff Events. *Environ. Toxicol. Chem.* **2004,** *23*(12), 2868–2877.
51. Yambem, T. S.; Machina, K.; Seetharamaiah, T. Status of Heavy Metals in Tissues of Wedge Clam, *Donax faba* (Bivalvia: Donacidae) Collected from the Panambur Beach Near Industrial Areas. *Recent Res. Sci. Technol.* **2012,** *4*(5), 30–35.

CHAPTER 12

GREEN SYNTHETIC ROUTES FOR SYNTHESIS OF GOLD NANOPARTICLES

DIVYA MANDIAL, RAJPREET KAUR, LAVNAYA TANDON, and POONAM KHULLAR*

Department of Chemistry, BBK DAV College for Women, Amritsar, Punjab 143005, India

**Corresponding author. E-mail: virgo16sep2005@gmail.com*

CONTENTS

ABSTRACT

Protein-templated synthesis of gold nanoparticles (NPs) offers an alternative eco-friendly, greener synthetic route that is free from the toxic chemicals. When proteins are used in the process of biomineralization, any kind of conformational changes along with folding and unfolding mechanism of protein strongly influence the overall reaction. Various important parameters of proteins that strongly influence are, namely, self-aggregation, folding and unfolding of the protein tertiary structure, predominant hydrophobic or hydrophilic nature of the proteins, and so forth. These bioconjugated gold NPs have great potential to be used in various biomedical applications.

12.1 INTRODUCTION

The development of nanotechnology has revolutionized almost every aspect of human life. Coinage metals such as Au, Ag, and Cu have been important materials throughout the history.[1–3] In ancient times, these metals were admired mainly for their ability to reflect light; at present, their applications have become far more sophisticated with our understanding and control of the atomic world. From last few years, the fabrication of the nanomaterials and exploration of their properties have attracted the attention of all the branches of science, that is, Physics, Chemistry, Biology, and Engineering.[4–7] Two types of approaches are normally employed for the fabrication of nanomaterials, that is, top-down and bottom-up approach.

In the top-down approach, metallic nanoparticles (NPs) are fabricated using a number of physical and chemical methodologies for the reduction of suitable bulk material to nanosize, but this approach suffers from a major drawback of imperfection of the surface structure of the metallic NPs, which exerts a significant effect on the physical and surface chemistry properties. Another approach, that is, bottom-up approach also known as self-assembly approach involves the assembly of the atoms/molecules/clusters using a variety of biological and chemical methodologies. However, the major advantages of using the approach is less surface imperfection and cheaper one. In the bottom-up approach, a large number of toxic chemicals, nonpolar, organic solvent, stabilizing agents, capping agent are used, which limits the applications of synthesized nanomaterial in the clinical and biomedical domain.[8–10] Hence, the replacement of these toxic synthesis methods by the nontoxic, biocompatible eco-friendly green routes is the need of the hour.[11] From last few years, noble metal nanomaterials have been studied extensively by researchers for a wide range of application due to their prominent physical, chemical, and optical properties.[12-15] Of all the noble metal nanomaterials, gold nanomaterials have been studied a lot to be used for biological and biomedical applications because of their photostability, good biocompatibility, and cellular uptake.[16-20] The performance of these nanomaterials is highly dependent on the shape, size, and also the capping agent used.[21–24] This chapter accounts for the green synthetic routes for the synthesis of gold nanomaterials using various model proteins such as bovine serum albumin (BSA), cyt C, zein, and so forth, as an alternative route, avoiding the use of hazardous chemicals.

Metal NPs with the size of less than 2 nm have wide applications in many areas such as bioconjugation, catalyst, nanodevices, bioimaging, biosensing, and cancer therapy. As compared to other methods, the protein-directed approach produces metal nanoclusters (NCs) with good water solubility, high stability, strong fluorescence, biocompatibility, and biofunctionality. This approach is not only environment-friendly but also the NCs formed are ready for further bioconjugation. Biomineralization refers to the synthesis of inorganic materials such as silica, and metal alloys by the biological molecules, where biomolecules are directly involved in the synthesis of different materials.[25-29] It is the inherent complexity of the biomolecules that makes even simple reactions difficult to understand. From past million years, nature has evolved mechanisms to produce such nanomaterial for a wide variety of purpose.

When proteins are used in the process of biomineralization, any kind of conformational changes along with folding and unfolding mechanism of protein strongly influence the overall reaction. Various important parameters of proteins that strongly influence are, namely, self-aggregation, folding and unfolding of the protein tertiary structure, and predominant hydrophobic or hydrophilic nature of the proteins.[30]

Lysozyme, a low-molecular-weight protein, contains 129 amino acids arranged in a single polypeptide chain with an average formula weight of 14,300. It contains eight cysteine residues, which form four disulfide bonds located between the positions, 6–127, 30–115, 64–80, and 76–94 and is available in abundance in egg whites, human tears, saliva, and endocrine glands. Owing to the presence of four disulfide bonds, it is highly surface active.[31–34] Amino acid such as cysteine and glutathione act as reducing agents and their location significantly influences the synthesis of nanomaterial. The synthesis of AuNPs using model proteins is characterized by the presence of absorption bands at 285 and 400 nm, respectively, due to tryptophan and tyrosine residues of lysozyme and π–π^* transition of the porphyrin rings of cyt c and surface plasmon resonance band, which is strongly related to the shape and structure of NPs as well as their mode of aggregation.[35–37] In case of proteins, the reduction is primarily achieved by the cysteine residue and as soon as the nucleating center are created, the simultaneously associate with protein. Lysozyme has eight cysteine residues, forming four disulfide bonds and one surface-exposed histidine residue. This enzyme enables studies of its biological activity after forming nanomaterial and the preparation of functional material. Moreover, it is stable over a wide range of pH and temperature that makes it suitable for a variety of synthetic approaches.

Lysozyme has been used for the synthesis of highly fluorescent AuNCs in basic aqueous solution, where lysozyme serves dual purpose of reduction as well as stabilization. The tyrosine and tryptophan residues of lysozyme cause the reduction of $HAuCl_4$. It is also observed that lysozyme from different vendors produced AuNCs with different quantum yields, suggesting that salt and other molecular contamination may affect the synthesis.[38] It is observed that when lysozyme and $HAuCl_4$ are mixed at low pH, Au^{3+} binds lysozyme by coordinating with the O and N of the amino acids.

At 400°C (see Fig. 12.1), Lys exists mainly in the folded state; however, the protein–NP association further induces unfolding by the partial breakdown of some of the disulfide bonds (step a Fig. 12.1), the unfolded protein then interacts with another protein (step b) to induce self-aggregation, which result in the self-assembled state of protein b in case of lysozyme.[39] However, in case of other proteins like cyt c, it shows totally different behavior, which results in the formation of fibril. (step d Fig. 12.1).[40] Fibril is a self-assembled state of protein, in which polypeptide strands composed of antiparallel β-sheets in cross β arrangement exist. It is observed that the native folds of highly α-helical proteins are destabilized as a result of adsorption of proteins on AuNP surface (step e) and subsequently interact with the other free conjugated proteins. At higher temperature, that is, 800°C, extensive dehydration (step c) turns the self-assembled Lys–AuNPs into large liquid crystalline flakes because of its relatively less hydrophilic nature compared to that of cyt c, whereas in case of cyt c, with the increase of temperature, long strands (step f) get converted into spherical micelles. It is due to the presence of both cationic and anionic residues, and with the increase of temperature, the unfolding of cyt c protein results in the greater exposure and contact of hydrophobic residue with the aqueous phase, which results in the formation of roughly spherical micelle. However, at low temperature, that is, 40°C, cyt c remains in the folded state, and therefore, NP micelles are observed.[41]

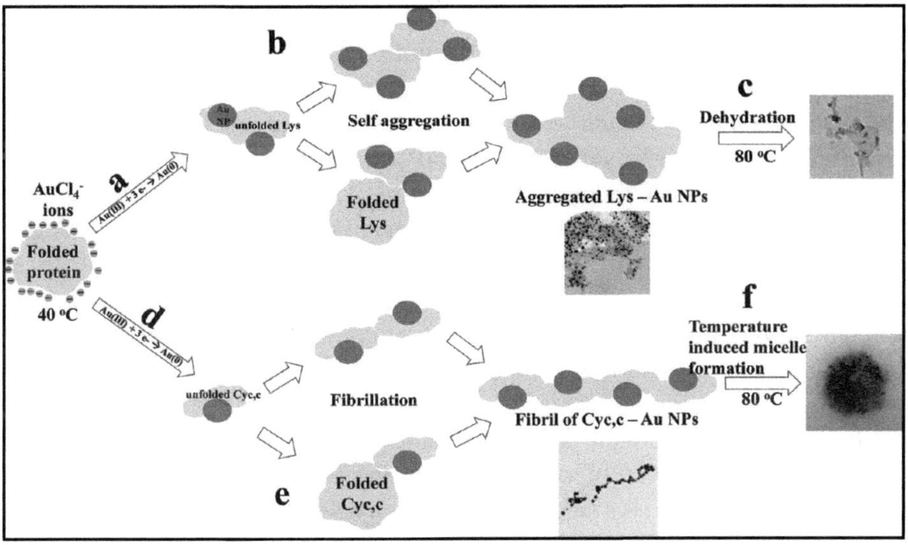

FIGURE 12.1 (See color insert.) Schematic representation of the proposed reaction mechanism for the synthesis of Lys/cyt. c-conjugated AuNPs (nanoparticles).

Source: Reprinted with permission ref 39. © 2010. American Chemical Society.

Cyt c consists of polypeptide chain of 104 amino acid residues that are covalently attached to a heme group with an average molecular weight of 12,400. It serves as a model protein because of its important biological function to act as an electron carrier in the respiratory chain of an aerobic organism and it also plays an important role in the programmed cell death.[42,43] Soret band and Q band in cyt c result from the π–π^* transition of the porphyrin ring and the intensity of the Soret band and Q band are affected by the additive effects of the transition dipole moments between two orbital excitations, that is, a_{2u}–b_{1u} and a_{2u}–e_g. Their intensities are highly affected by the changes in the symmetry of the porphyrin ring. The "C" denomination implies that the heme group is covalently bonded to two cysteines of the amino acids sequence in opposition to the hemes of the "p" type as they are found in hemoglobin and myoglobin in which the heme is noncovalently attached to the protein. Cyt c undergoes several structural changes with the change in pH as revealed from the ultraviolet (UV) studies that show four different pKAs.[44]

Considering the diversity in the amino acid composition of proteins, and hence their reducing ability, a range of proteins such as BSA, fibrinogen, α-lactoalbumin, lysozyme, cyt c, myoglobin, β-lactoglobulin, and α-chymotrypsin has been studied for the synthesis of metal NCs. Fluorescent gold (Au) NCs have attracted much attention due to their emerging photophysical properties and potential applications in biolabeling and sensing.[45–48] Metal NCs are usually fluorescent and do not exhibit localized surface plasmon resonance. When external reducing agent such as $NaBH_4$ or ascorbic acid is used to synthesize AuNPs in the presence of protein, it is referred as an extrinsic method. The addition of an extrinsic reducing agent, that is, extrinsic protocols offers many advantages such as short reaction times, negligible competition from amino acids, and hence, leading to a well-defined reduction and growth mechanism. Of all the proteins, BSA is a much-studied protein as it is an important constituent of blood and it is a carrier protein responsible for the transport of thyroid. It is also involved in other physiological functions such as control of serum osmotic pressure and pH buffering.[49] Fibrinogen is an important protein and it is converted into fibrin in the last step of blood clotting by the enzyme thrombin.[50] Bovine α-lactoalbumin (BLA) is a small protein which resembles with lysozyme with respect to its tertiary structure. cyt c shuttles between the two membrane associates protein complexes, cyt c reductase and cyt c oxidase, inside the inner mitochondrial membrane and participates in the mitochondrial production of ATP. Myoglobin is a monomeric heme-binding protein which is involved in oxygen carrying and it is found in muscle and blood cells. β-lactoglobulin (BLG) is a globular milk protein and binds with small ligands including palmitic and oleic acid.[51,52] The proteolytic enzyme, α-chymotrypsin (CTR), participates in the breakdown of proteins in the digestive system of mammals and other organisms.[50] It cleaves peptide bond specifically on the carboxyl terminal side of large hydrophobic amino acids such as tryptophan, lysine, phenylalanine, and methionine. The photophysical properties of gold NCs prepared using the eight model proteins, BSA, Fib, BLA, Lys, cyt c, Mb, BLG, and CTR, revealed that the capacity for forming large AuNCs can be listed in the increasing order as BLG $\ll$ Lys $\ll$ BLA $\leq$ CTR $\ll$ BSA.[53]

Tryptophan is an intrinsic fluorophore frequently used in the monitoring the changes in the local protein environment such as unfolding.[53] In some studies of biomineralization, it is observed that the tyrosine, in case of BSA, and histidine (His), in case of apoferritin, are responsible for the reduction and nucleation of Au^{3+}. Cysteine is thought to be responsible for the stabilization of the gold NC due to Au–S covalent bonds.[48] However, the comparison of these eight different proteins revealed that the photophysical properties of AuNCs are highly protein dependent. There is no correlation between the number and percentage of Tyr/His residue. Even in case of myoglobin, which is completely devoid of cysteine residue, it is found to be capable of formation and stabilization of small AuNCs. This study indicates that

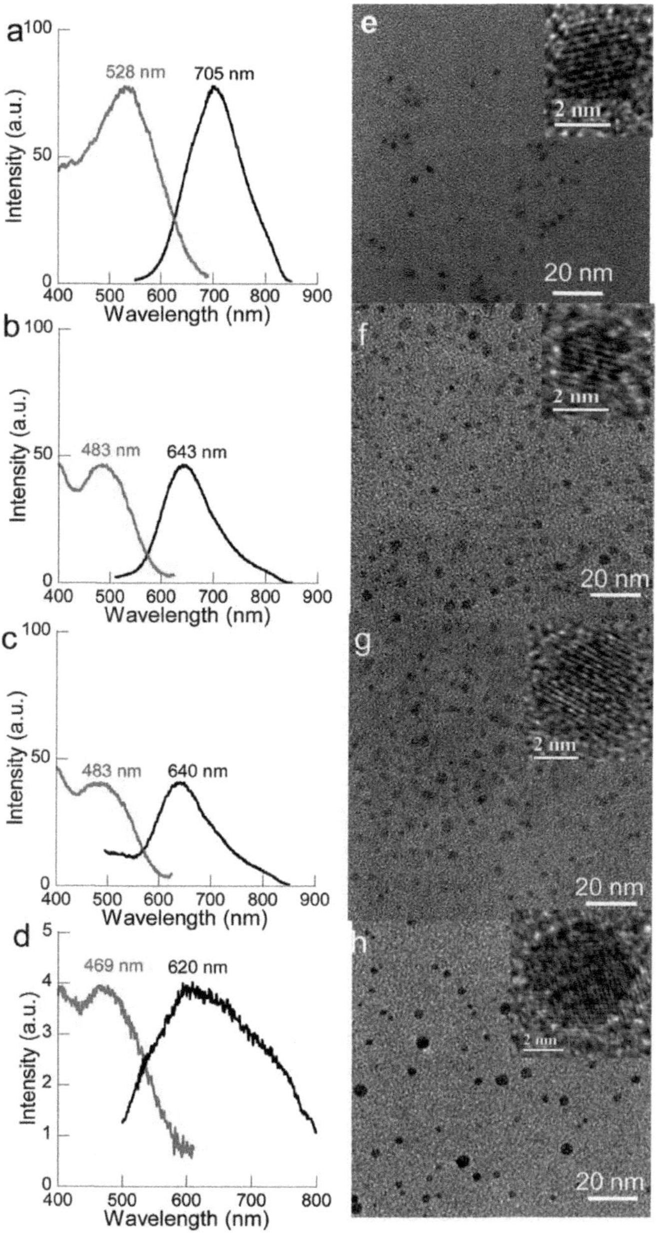

FIGURE 12.2 (See color insert.) Protein-encapsulated Au nanoclusters (NCs): fluorescent emission (black)/excitation (red) spectra and transmission electron microscopy (TEM) images of AuNCs generated from BSA (a and e), trypsin (b and f), lysozyme (c and g), and pepsin (d and h).

Source: Reprinted with permission from Royal Society of Chemistry, ref 51.

cysteine residue is not necessary for the stabilization of the gold NCs. The general protein template gold NC nucleation and growth can be explained by a general mechanism based on nucleation, growth, termination, and stabilization/solubilization events.

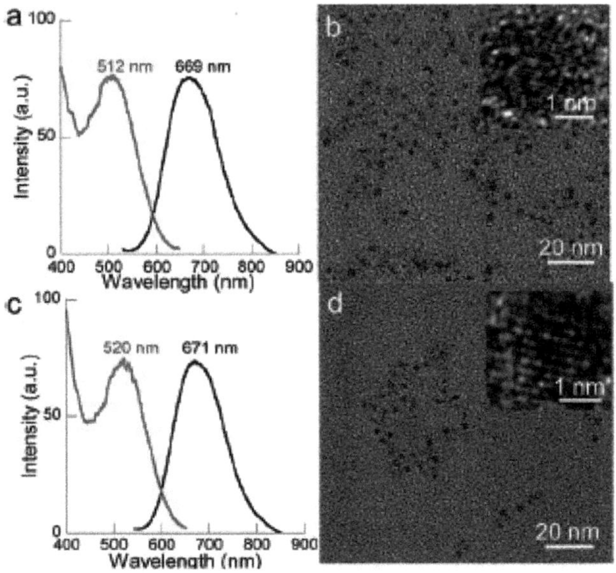

FIGURE 12.3 **(See color insert.)** Fluorescent emission (black)/excitation (red) spectra and TEM images of AuNCs generated from excess trypsin (a and b) and lysozyme (c and d). **Source:** Reprinted with permission from Royal Society of Chemistry, ref 51.

For bioconjugation, mainly two types of method have been used: first, using the functional NPs to covalently link biological molecules[54] and second includes the electrostatic interaction.[55-57] Once the protein gets adsorbed on the surface of AuNPs, the key parameter upon which the activity of the adsorbed protein depends is the conformation. The conformational change of proteins depends on the size of the AuNPs. As in cyt c, the Q band of cyt c shows a redshift of 4–5 nm with a decrease in the absorption intensity and an appearance of shoulder peak at 548 nm when cyt c is adsorbed on 16 nm AuNPs. This difference arises because of the different size of AuNPs, which clearly reflect the different changes of heme environment of the adsorbed cyt c.[58–60]

Circular dichroism (CD) spectroscopy in the UV region is commonly used for the determination of secondary protein structures in solution. On comparison of cyt c from two different sources, that is, cyt c from yeast (YCc) and horse heart (HCc) (Fig. 12.4), it is observed that, in case of bionanoconjugate of HCc–AuNPs, no appreciable variation is observed; however, in contrast, YCc suffers significant alteration in the secondary structure both by interaction with AuNP and pH-induced aggregation. It clearly indicates the loss of secondary structure, occurring at high pH and further decreasing upon aggregation at low pH. The increase in β-sheet content that occurs only upon YCc–AuNP aggregation at low pH is a clear indication of protein aggregation.[61]

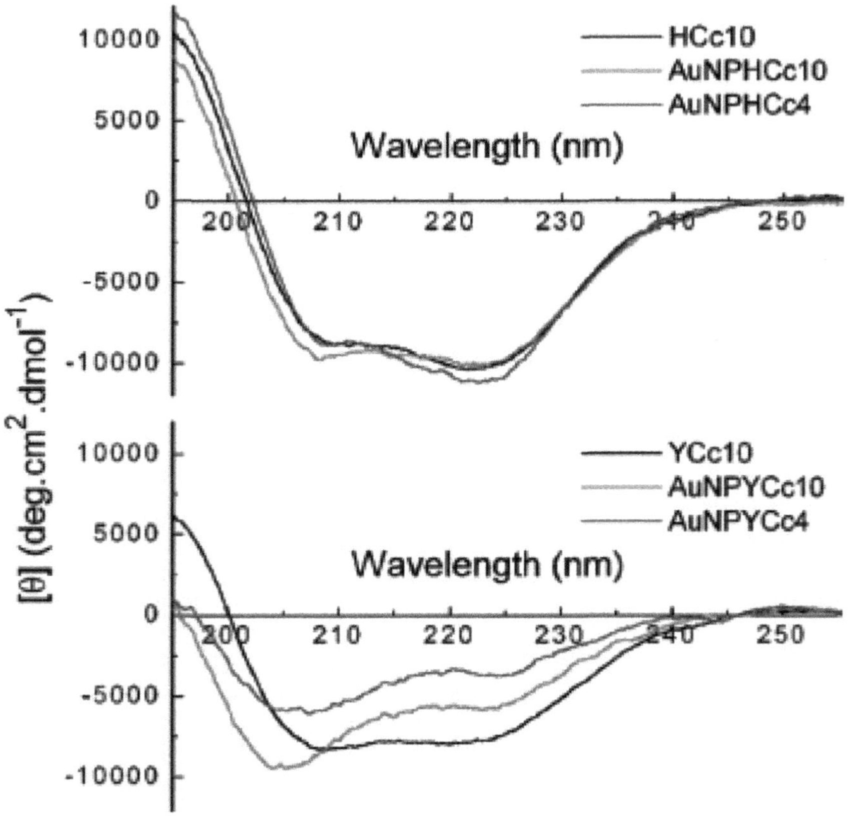

FIGURE 12.4 (See color insert.) Far-ultraviolet (UV) circular dichroism (CD) spectra for cyt c alone at pH 10 (black traces), in the bionanoconjugates at pH 10 (red traces), or at pH 4 where aggregation occurs (blue traces). The top spectra correspond to HCc and HCc–AuNP and the bottom spectra are for YCc and YCc–AuNP.

Source: Reprinted with permission from American Chemical Society, ref 59.

Protein-coated AuNPs can be synthesized using wet chemical method with or without the help of an external reducing agent. Surface-enhanced Raman scattering (SERS) studies describe the mechanism of protein-capped AuNPs.[62] The characteristic peaks at 1484, 1545, and 1583 cm^{-1} correspond to phenylalanine, tyrosine, tryptophan, and histidine residues clearly suggesting the role of these amino acids as stabilizing agents.

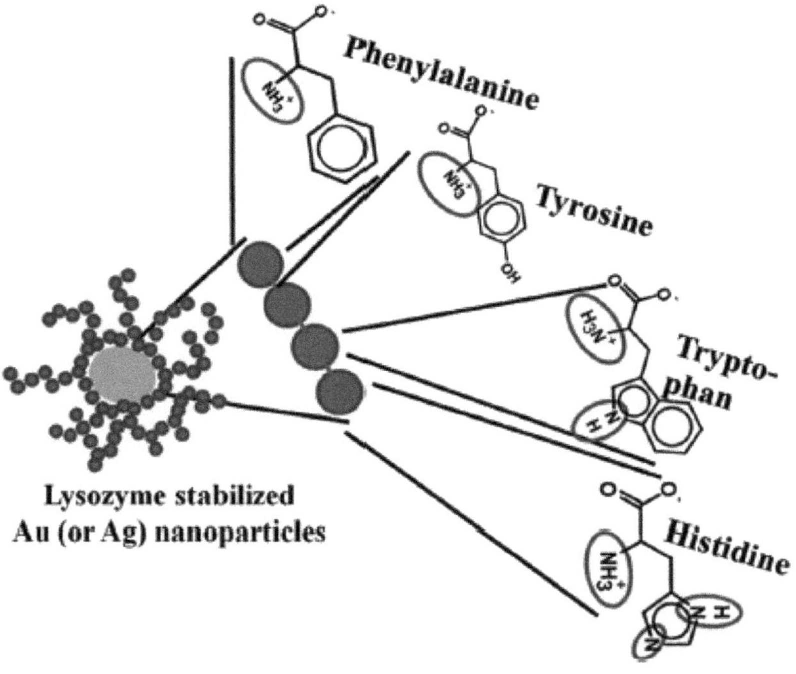

FIGURE 12.5 Surface-enhanced Raman scattering was adopted to elucidate the mechanism of lysozyme-capped Au and Ag nanoparticles (NPs). Certain amino acids played critical roles in stabilizing the NPs.

Source: Reprinted with permission from American Chemical Society, ref 60.

Tuning the size of AuNCs can be achieved using model protein such as BSA. Two types of methods can be employed; first involves the use of an external reducing agent such as ascorbic acid. It is also observed that ascorbic acid plays a minor role in the formation of BSA–AuNCs; however, it yields larger BSA–AuNPs. Tetrachloroauric acid (TCAA) concentration plays a vital role in tuning the size of BSA–AuNPs. At lower TCAA concentration, visible and near-infrared (NIR) AuNCs coexist in the protein backbone and the ratio is tunable. The synthesis of AuNPs using BSA results in the formation of different-sized AuNCs that show size-dependent fluorescence. BSA due to its large value of reduction potential is strong enough to synthesize AuNPs, and also, BSA-capped AuNPs are more stable in biological buffers as compared with citrate-capped AuNPs. The amount of atoms in AuNCs follows a shell model named Jellium model with the corresponding magic numbers 2, 8, 10, 20, 26, 34, and 40. This model resembles the electronic structure of atoms in which each shell is stable with an optimal amount of electrons. Proteins offer a plethora of energy-donating atoms such as sulfur, nitrogen, and oxygen; therefore, it results in the enhancement of fluorescence intensity through surface interactions when they are used as capping agents. In BSA, it is the tyrosine amino acid which is thought to reduce TCAA to NC and NP. The synthesis of AuNPs must be performed only at high pH because it increases the reducing capacity due to the fact that phenol group of tyrosine has a pKa of 10.1.[63] High pH results in the formation of the large gold NCs.

The mechanism at high pH is supposed to be self-assembly mechanism. Initially, at high pH, BSA protein gets negatively charged (step 1, Fig. 12.6).[64] Then, there occurs electrostatic attraction between positively charged gold ions and negatively charged BSA protein, particularly in areas rich in tyrosine and histidine, which are the important reducing amino acids. Nucleation occurs (step 2) when the local concentration of gold ions increases and further growth of AuNCs occurs due to coalescence (step 3). When the concentration of TCAA is high, AUNP synthesis occurs due to coalescence of several proteins surrounding an AuNP core (step 4).[65]

Protein-encapsulated gold clusters have also been synthesized using a small protein lysozyme (Lyz) and characterization of these clusters is done using mass spectrometric and other spectroscopic investigation. It was observed that maximum of 12 Au^0 species could be bound to a single lysozyme molecule irrespective of the molar ratio of Lyz: Au^{3+} used for the cluster growth as compared to the bigger proteins such as BSA or Lf. It was observed that the growth mechanism is highly dependent on the disulfide bond or where cysteine residue can form Au–S bond and the gold core is stabilized by the thiolate linkages. The luminescence of these proteins protected clusters has been used for biolabeling.[66–70] However, large organic molecules such as dendrimers containing multiple thiol groups can completely wrap around Au cluster through gold–thiol interaction. Lysozyme has been used for the synthesis of gold cluster and it is established that the cluster growth happens under the strong influence of cysteine. A comparison of the cluster-encapsulated protein with that of free protein revealed that both form similar types of aggregates while using a large protein such as BSA or Lf, having a large number of thiol groups (34 and 36 cysteine residues) and bulky nature may facilitate the formation of the cluster within a single protein molecule.[71–77]

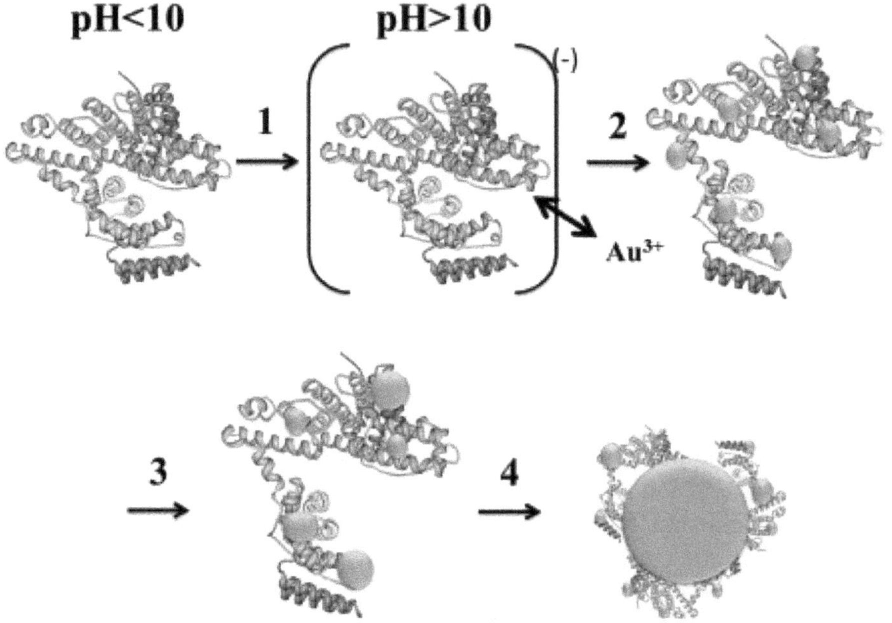

FIGURE 12.6 Proposed mechanism for growth of AuNPs stabilized by BSA. (1) Increasing the pH above 10 causes BSA to be negatively charged which induces electrostatic interactions with the gold ions. (2) Visible (Vis) AuNCs that emit in the visible region of the electromagnetic spectrum start to form, (3) and depending on time and concentration of TCAA, larger near-infrared (NIR) AuNCs, and (4) AuNPs are formed through a shuffling of NIR AuNCs. It should be noted that the structure of the denatured protein is used only as illustration, and not as an accurate description of the conformation.

Source: Reprinted with permission from American Chemical Society, ref 63.

In the matrix-assisted laser desorption/ionization (MALDI) mass spectroscopy (MS) spectrum, the parent Lyz shows a peak at m/z 14,300 and its aggregates as dimer, trimer, and tetramer at m/z 28,800, 42,900, and 57,200 respectively. The total number of gold atoms bound, divided by the number of protein molecules in the aggregate confirmed the distribution of 10 Au atoms per proton in each case. It also indicates that each protein entity contains 10 strongly bound atoms gold nanostructures. It is also observed that binding of gold is totally different from the binding of alkali metal ion. While Au binding is strongly influenced by cysteine residue, alkali metals prefer to bind with the carboxyl and hydroxyl groups of different amino acid residues.[78]

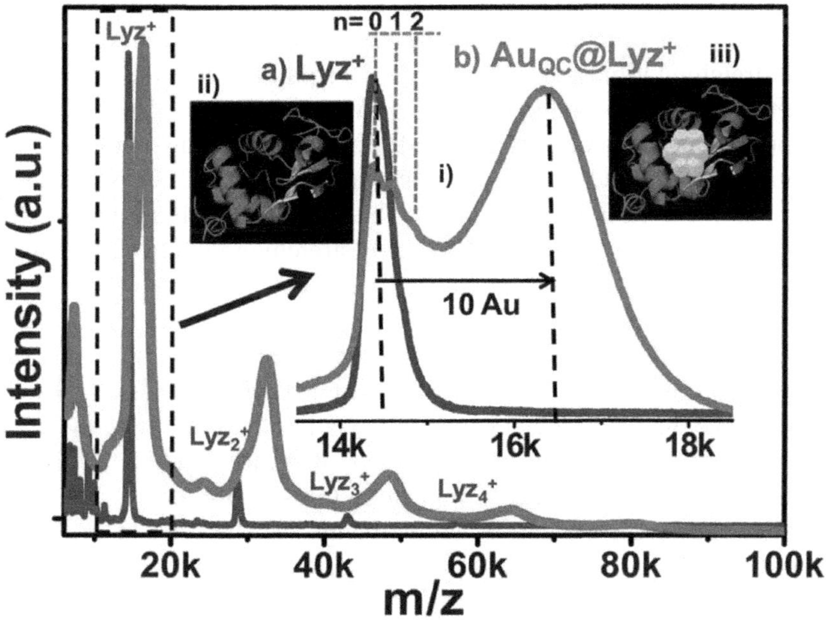

FIGURE 12.7 (See color insert.) Positive-ion matrix-assisted laser desorption/ionization (MALDI) mass spectroscopy (MS) of Lyz at pH 12 in linear mode (a) and Au_{QC}–Lyz after 24 h of incubation. (b) All the spectra were measured in the linear positive mode over the m/z range of 20,000–100,000. Both Lyz and Au_{QC}–Lyz showed aggregate formation. The expanded monomer region in inset (i) clearly shows a separation of 10 Au atoms from the parent protein. In the dimer, trimer, tetramer, and pentamer regions, the separations are of 20, 30, 40, and 50 Au atoms, respectively. In insets (ii) and (iii), schematic representations of Lyz and Au_{QC}–Lyz, respectively, are shown.

Source: Reprinted with permission from Royal Society of Chemistry, ref 74.

The formation of metal nanocrystals greatly affects the secondary structure of proteins. Large changes were observed in the fraction of α-helices after cluster formation as a loss of 28% was observed in CD spectra. It can be attributed to the structure of Lyz where cysteine residues are present (at 6,30,64,80,115,127 amino acid position) and four of which are localized in the vicinity of the α-helices.[79] As the disulfides were broken and used to stabilize the cluster core, drastic changes in the total helix content was observed. Computational studies showed that the disulfide bond can break upon the addition of Au^{3+} to Lyz, as one disulfide bond breaks to give two sulfur ends and two electrons are donated to form the cluster core upon the addition of Au^{3+} alone.

Fourier-transform infrared studies further supported the changes in the secondary structure of proteins. Amides I, II, III are the characteristics of the secondary protein structures.[77,80] Bonds near 1650 cm^{-1} arise mainly because of C=O stretching vibration with a minor contribution from the out-of-plane CN stretching—this is called amide I. Another band near 1550 cm^{-1} is due to an out-of-phase combination of NH in-plane bonding with a smaller contribution from C=O in-plane bending as well as C–C, C–N stretching. The region 1400–1200 cm^{-1} is due to the amide III; vibrational mode in 3300–3000 cm^{-1} range is due to N–H, O–H stretching vibration. This region is a mixture of amide I and amide II. Band near 700 cm^{-1} is due to NH_2 and NH wagging bands at 2950 cm^{-1} due to $-CH_3$, $-CH_2$, and CH stretching vibrations and –OH stretching vibration is observable as a broad peak around 3500 cm^{-1}. Second derivative IR (in the region 1600–1700 cm^{-1}), which is more sensitive, revealed the changes in the amide region due to the cluster formation. Among α-helix (1651–1650 cm^{-1}), β-sheets (1618–1642 cm^{-1}), random coils (1640–1650 cm^{-1}), and turns (1666–1688 cm^{-1}), the α-helix region showed larger change. A large change can be seen in the α-helix feature at 1654 cm^{-1}, which is completely absent in case of cluster because of huge perturbation of the α-helical region, which may be due to the breakage of disulfide bonds for the cluster formation.

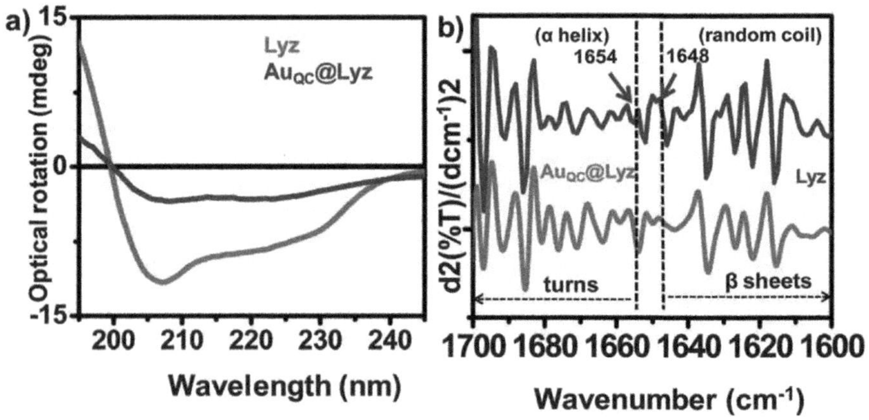

FIGURE 12.8 **(See color insert.)** (a) CD spectra of Lyz and as prepared Au_{QC}–Lyz showing a clear change in ellipticity of the spectra, which indicates a huge change in the α-helical structure. (b) Double derivative of the infrared spectra shows the disappearance of the peak at 1654 cm^{-1} in the case of Au_{QC}–Lyz.

Source: Reprinted with permission from Royal Society of Chemistry, ref 74.

Zein is an alcohol-soluble corn storage protein which find numerous applications in food industry. It consists of a highly robust structure that is made up of 9 homologous repeat units arranged in an antiparallel distorted cylindrical form and is stabilized by hydrogen bonds.[81] The unfolding behavior of zein protein can be understood only by making it aqueous soluble, which can be done using specific surfactants. Cysteine moieties present in the protein cause reduction of Au(III) to Au(0).[82] The reduction potential of protein is directly proportional to its unfolding. Unfolded zein in the aqueous phase exposes hydrophobic domains as a result of which it is expected to act as a stabilizing agent as well as shape-directing agent.[83–85] It is due to the fact that the molecules get adsorbed on some particular crystal plane and hence the crystal growth is directed to unpassivated crystal planes. Completely unfolded protein is always a better shape-directing agent than partially unfolded protein. In the UV–visible spectra (not shown), the aqueous zein gives a clear absorption around 280 nm mainly due to nonpolar amino acid residues such as tyrosine (Tyr) and phenylalanine (Phe) in comparison to tryptophan, which is present in a negligible amount. The solubilization of hydrophobic zein takes place when hydrocarbon tail of SDS molecules is incorporated into the following hydrophobic domains of zein as shown in Figure 12.9 while leaving anionic head groups in aqueous phase as a result of which zein–SDS complex acquires a charge, and therefore it becoming solubilized in water.[86] With the increase in temperature, the secondary structure of proteins gets affected due to the breaking of hydrogen bonds operating among different cylindrical structures and then the breaking of the disulfide bonds to expose the cysteine residues in aqueous phase to initiate reduction.[87,88]

The best way to understand and monitor the surface adsorption and zein and its subsequent unfolding is to follow the change in intensity of adsorption of nonpolar residues such as Tyr and Phe around 280 nm. Since they are deeply buried in the hydrophobic domains, unfolding exposes them to the aqueous phase and enhance their UV absorbance as more and more such residues are aqueously exposed. Globular proteins are found to be less surface active than fibrous proteins. Zein when gets aqueous soluble in the presence of surfactant behaves like a fibrous protein and this feature can be used to form viscoelastic film on the NP surface.

The synthesis of AuNPs when done in the presence of the variety of surfactants such as SDS (C12), SDeS (C10), STS (C14), and SPFO using zein gives a very interesting fact. SDS (C12), SDeS (C10), and STS (C14) show almost similar effect. However, strongly hydrophobic fluorocarbon molecules show dramatic different behaviour. The use of such hydrophobic surfactants (e.g., SPFO) cause significant unfolding even at low temperature due to much different[92] capping and stabilizing behavior in comparison to that in the presence of SDS/STS. Such zein-coated NPs have less toxic and hemolytic effects, which make them a suitable candidate for biomedical applications.

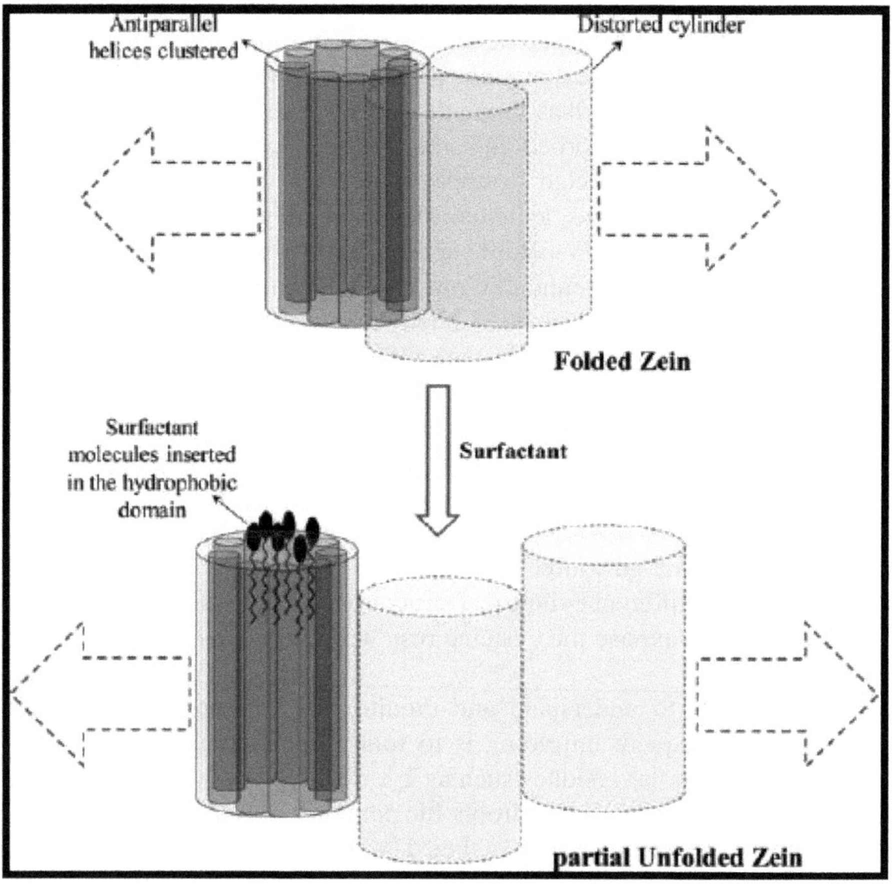

FIGURE 12.9 Schematic representation of the cylindrical structure of folded zein. The addition of surfactant induces unfolding by introducing its hydrocarbon tail into the predominantly hydrophobic cylinder.

Source: Reprinted with permission from American Chemical Society, ref 87.

When zein is used for in vitro synthesis of AuNPs along with BSA and cyt c, it is observed that zein shows strong interaction with BSA throughout the mole fraction range. However, such interactions are

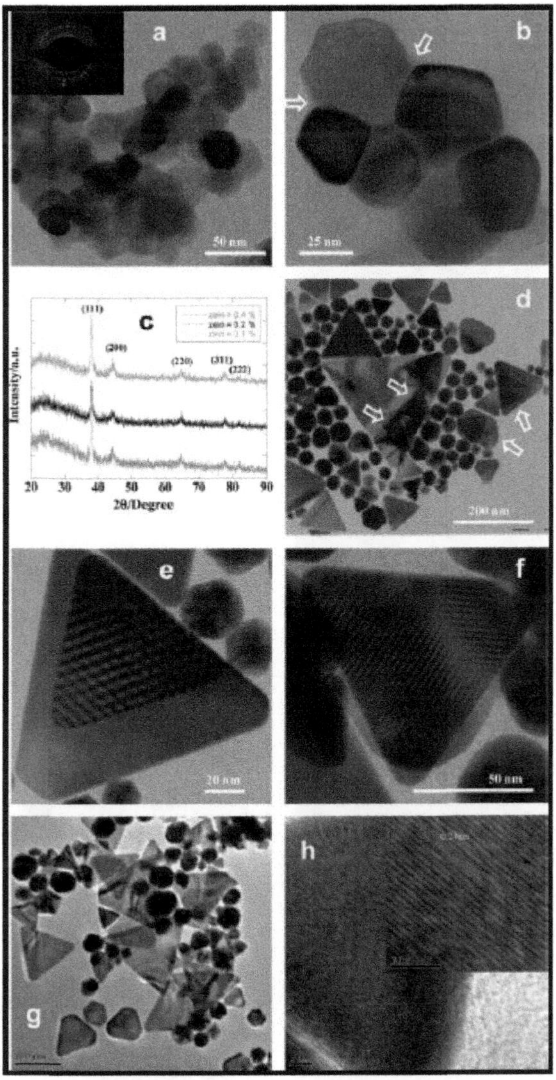

FIGURE 12.10 (a) TEM micrographs of small AuNPs prepared with a reaction of 0.1% zein (in aqueous 24 mM SDS) and 0.25 mM HAuCl4. Note the highly aggregated NPs due to the fusogenic behavior of unfolded zein. (b) Close-up view of faceted NPs in an aggregated state. (c) X-ray diffraction patterns showing the face-centered cubic crystal structure with predominant growth on {111} crystal planes of AuNPs prepared with different amounts of zein. (d) TEM image showing several thin triangular nanoplates prepared with 0.5 mM HAuCl4. (e) and (f) Close-up view of two thin plates lying one above the other. (g) and (h) TEM micrographs of AuNPs and lattice fringes of a nanoplate, respectively, prepared with the same reaction but in aqueous 24 mM STS instead of SDS.

Source: Reprinted with permission from American Chemical Society, ref 87.

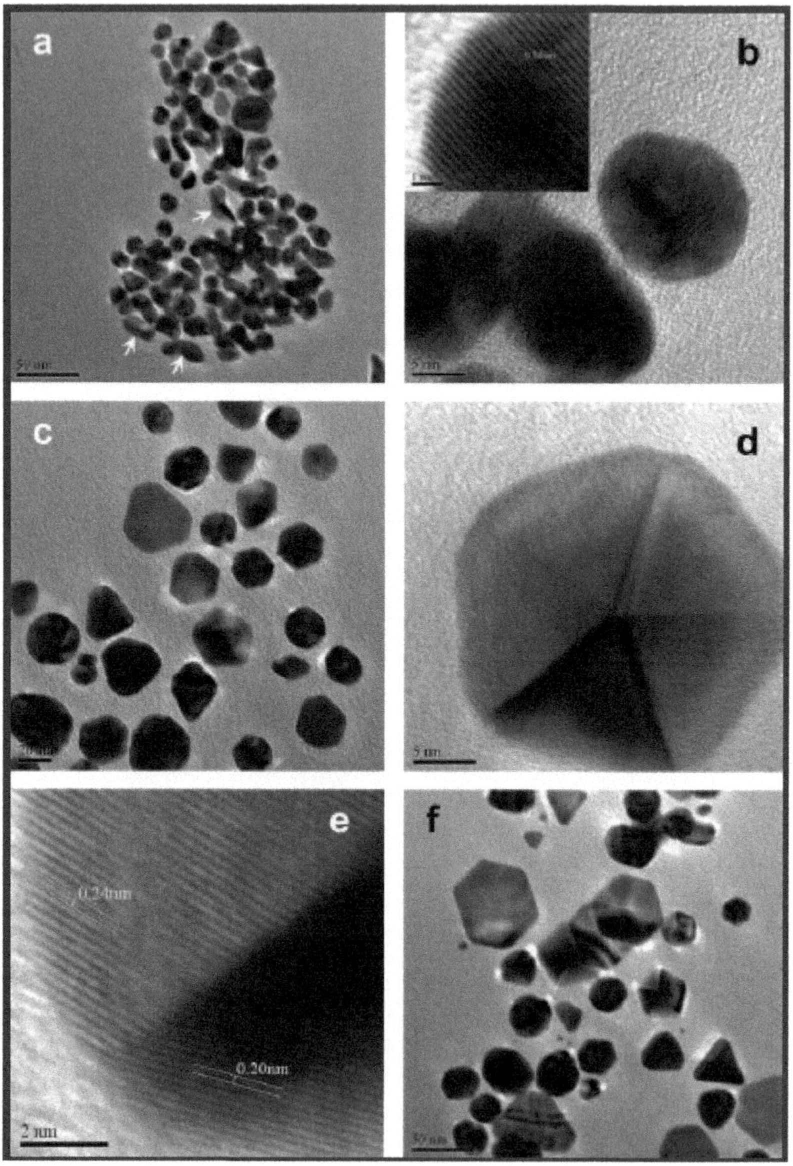

FIGURE 12.11 (a) TEM micrographs of small AuNPs prepared with a reaction of 0.1% zein (in aqueous 24 mM SPFO) and 0.25 mM $HAuCl_4$. Note the relatively unaggregated polyhedral NPs due to the weak fusogenic behavior of unfolded zein in the presence of SPFO. (b) Close-up view of a few NPs with lattice fringes. (c), (d), and (e) TEM images of well-defined NPs prepared with 0.5 mM $HAuCl_4$. (f) TEM image of NPs along with some nanoplates prepared with 1 mM $HAuCl_4$.

Source: Reprinted with permission from American Chemical Society, ref 87.

limited to the zein-rich region. When zein + cyt c mixtures are studied, it is attributed to the electrostatic as well as hydrophobic interaction in case of zein + BSA, whereas solely to the electrostatic interactions in case of zein + cyt c (Fig. 12.12).

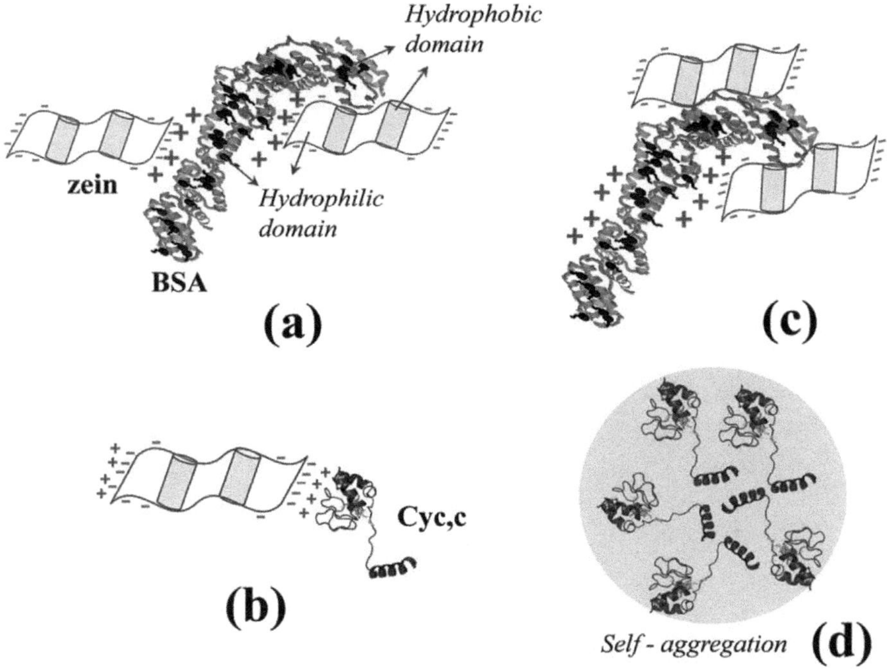

FIGURE 12.12 Schematic representation of possible complex formation in zein + BSA (a) and zein + cyt c (b) mixtures based on the electrostatic interactions between the hydrophilic domains of respective components. (c) The complex formation in zein + BSA driven by the hydrophobic interactions operating between the hydrophobic domains. (d) Self-aggregation in cyt c.

Source: Reprinted with permission from Royal Society of Chemistry, ref 88.

Both the protein mixtures get simultaneously adsorbed on the growing AuNPs and hence act as the shape-directing agent with the result that mostly spherical NPs of the size 20–30 nm are observed in case of zein + cyt c and highly anisotropic morphologies emerge. This difference clearly indicates that cyt c is not involved in the NP stabilization and shape-controlling agents, whereas in case of zein + BSA, there is active participation of protein mixtures in stabilization as well as in shape-control effects.[93]

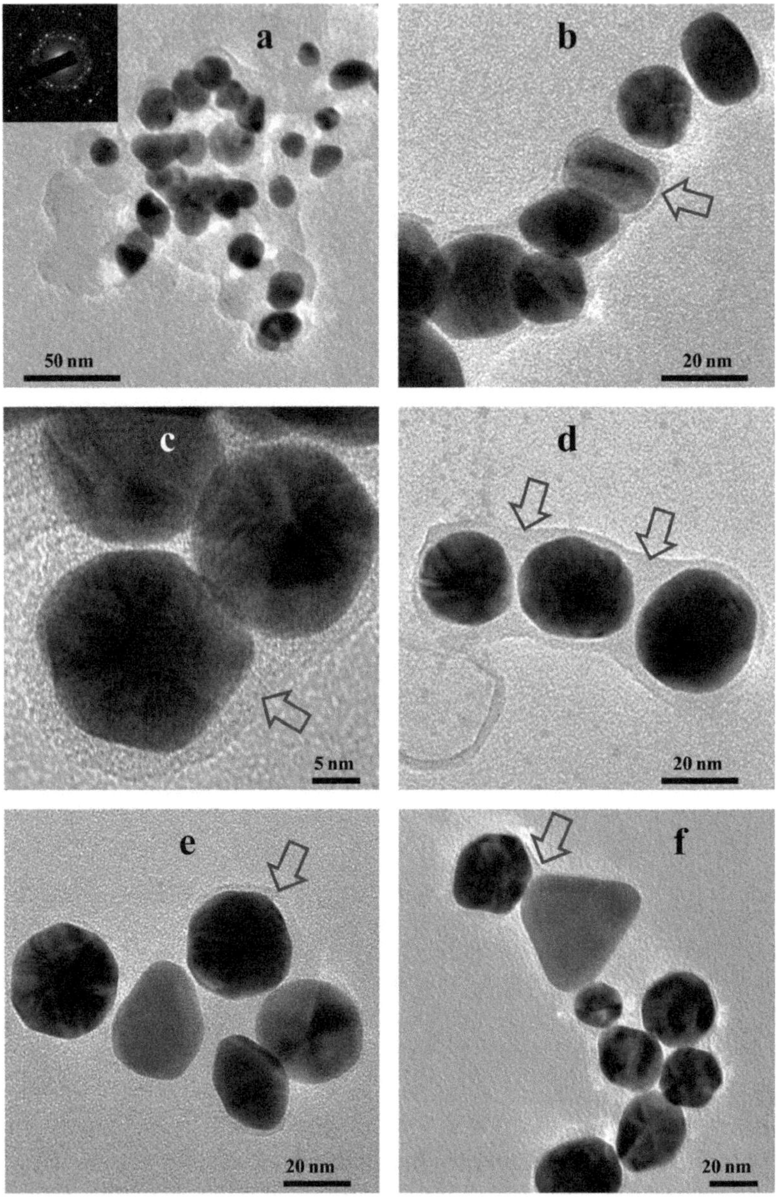

FIGURE 12.13 (a) TEM images of the purified AuNPs coated with zein prepared from aqueous zein+$HAuCl_4$=1 mM reaction. (b) Image of protein-coated NPs prepared from aqueous zein+BSA (X_{BSA}=0.1) with $HAuCl_4$=1 mM reaction. (c) and (d) TEM images of NPs prepared with mole fraction X_{BSA}=0.24, and (e) and (f) of X_{BSA}=0.56. Block arrows indicate thick protein coating on the NPs.

Source: Reprinted with permission from Royal Society of Chemistry, ref 88.

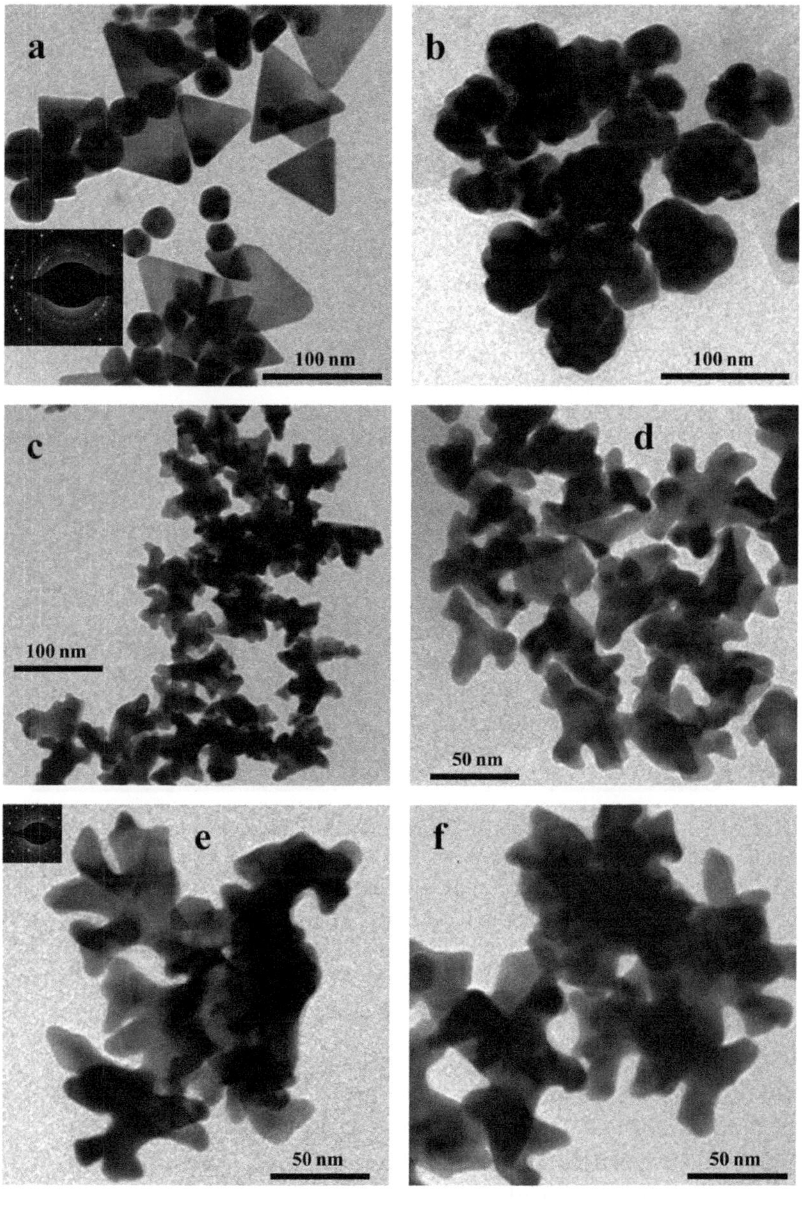

FIGURE 12.14 (a) TEM images of AuNPs prepared from aqueous zein + cyt c ($X_{cyt\ c}$ = 0.16) with $HAuCl_4$ = 0.16) with $HAuCl_4$ = 1 mM reaction. (b) NPs prepared with mole fraction $X_{cyt\ c}$ = 0.53. (c), (d) Images of mole fraction $X_{cyt\ c}$ = 0.72, and (e), (f) $X_{cyt\ c}$ = 0.87. The dendritic nature of the NPs increases with the increase in the amount of cyt c in the zein + cyt c mixtures due to the poor shape-control effects.

Source: Reprinted with permission from Royal Society of Chemistry, ref 88.

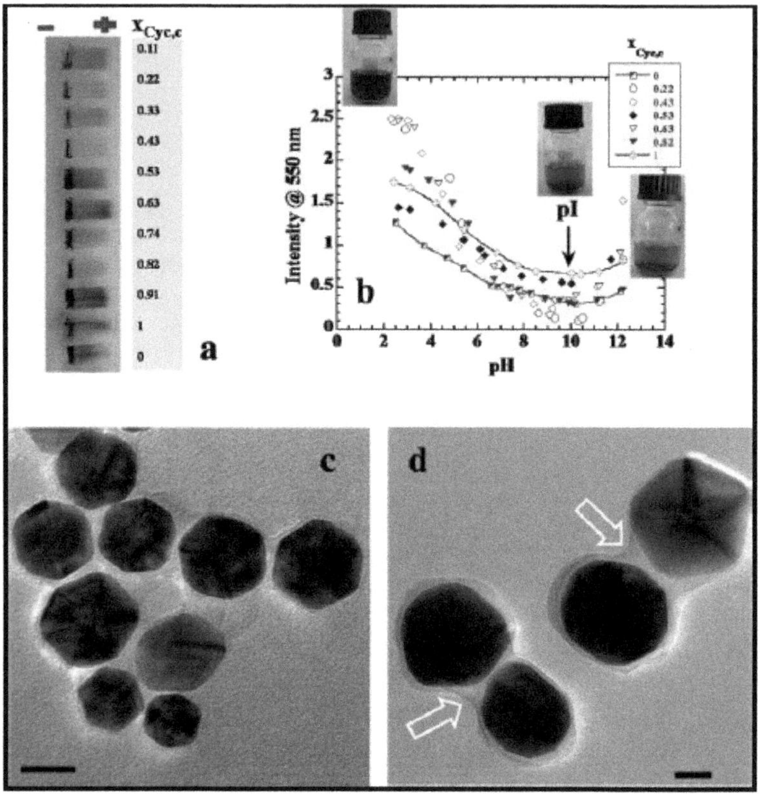

FIGURE 12.15 (a) Gel electrophoresis of protein-coated AuNPs synthesized with various mole fractions of lysozyme/cyt c mixtures. (b) Variation of the intensity of protein-coated AuNPs made with different mole fractions of lysozyme/cyt c mixtures with pH. Photographs of reaction bottles show the change in the color of the NP colloidal suspension with pH. (c and d) TEM micrographs of protein-coated AuNPs in large groups and pairs with a scale bar of 20 nm, respectively. Block arrows indicate the protein coating in the form of thin film.

Source: Reprinted with permission from American Chemical Society, ref 94.

Zein due to its water-insoluble behavior can act as a fine moisture barrier but its biodegradation cannot be avoided. Lysozyme is expected to have antimicrobial properties. Therefore, the protein mixture of Lys + zein is a good option for the industrial applications of zein. Both Lyz + cyt c and Lyz + zein complexes show remarkable adsorption of AuNPs, which is clear from their TEM images as shown below; however, the former binary combination produces pH-responsive NPs because of its amphiphilic nature, whereas the latter gives pH-sensitive NPs due to its hydrophobic nature.[94] Unfolded BSA has been used to synthesize AuNP in vitro by using different conventional surfactants. It is observed that cationic surfactants such as DTAB and 12-0-12 are more efficient than anionic SDS and zwitterionic DPS. Strong electrostatic interactions of the cationic surfactant cause unfolding of BSA even at low temperature. However, all the surfactants yield morphologically almost similar NPs, which clearly indicate the nonparticipation of these surfactants in growth process. They only serve to cause the unfolding of protein. BSA gets adsorbed on the low-energy crystal planes like [100][110] to control the crystal growth. Such protein-coated NPs do not show any hemolysis as compared with surfactant-coated NPs. Thus, they open up several possibilities of protein-coated NPs to be applied in intravenous adsorption.

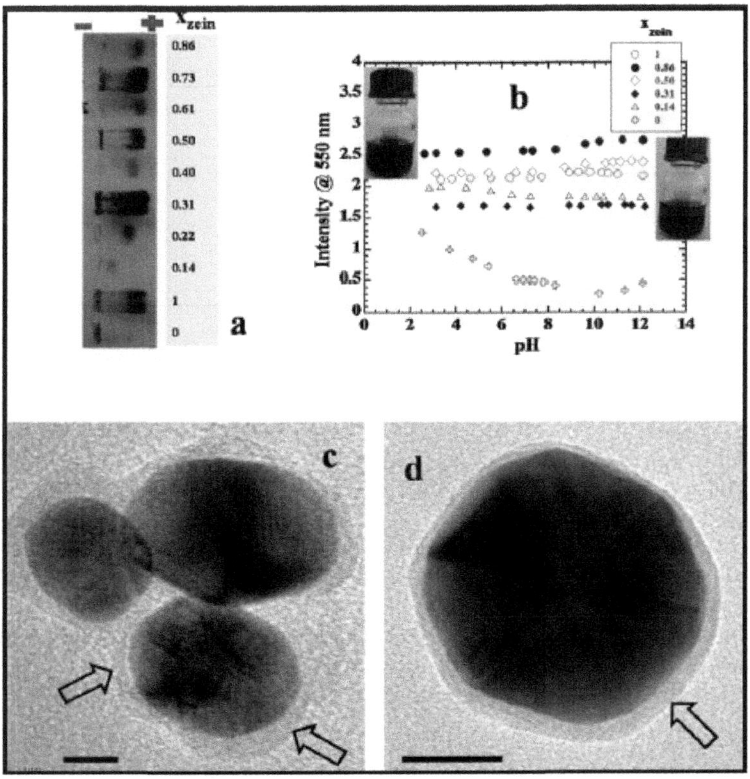

FIGURE 12.16 (a) Gel electrophoresis of protein-coated AuNPs synthesized with various mole fractions of lysozyme/zein mixtures. (b) Variation of the intensity of protein-coated AuNPs made with different mole fractions of lysozyme/zein mixtures with pH. Photographs of reaction bottles show no change in the color of the NP colloidal suspension with pH. (c and d) TEM micrographs of protein-coated AuNPs with scale bars of 5 and 20 nm, respectively. Block arrows indicate the protein coating in the form of thin film.

Source: Reprinted with permission from American Chemical Society, ref 94.

12.2 FUTURE PERSPECTIVES

This chapter introduces the applications and uses of generally nontoxic and environmentally friendly biomolecule proteins for the synthesis, characterization, and applications of nanomaterials. The protein conformation, molecular size, pH, and temperature are the important parameters that control the overall growth of nanomaterials. By the proper use of these parameters, one can easily control the shape and size of the nanomaterials. These bioconjugates act as promising candidates to be used in various biomedical applications.

KEYWORDS

- **protein template**
- **conformational changes**
- **gold nanoparticles**
- **bioconjugation**

REFERENCES

1. Wertime, T. A. The Beginnings of Metallurgy: A New Look: Arguments Over Diffusion and Independent Invention Ignore the Complex Metallurgic Crafts Leading to Iron. *Science* **1973,** *182*, 875–887.
2. Wertime, T. A. Man's First Encounter with Metallurgy. *Science* **1964,** *146*, 1257–1267.
3. Branigan, K. Lead Isotopes and the Bronze-Age Metal Trade. *Nature* **1982,** *296*, 701–702.
4. Nirmal, M.; Brus, L. Luminescence Photophysics in Semiconductor Nanocrystals. *Acc. Chem. Res.* **1999,** *32*, 407.
5. Alivisatos, A. P. Semiconductor Clusters, Nanocrystals, and Quantum Dots. *Science* **1996,** *271,* 933–937.
6. Chan, W. C. W.; Nie, S. Quantum Dot Bioconjugates for Ultrasensitive Nonisotopic Detection. *Science* **1998,** *281,* 2016–2018.
7. Dujardin, E.; Mann, S. Bio-Inspired Materials Chemistry. *Adv. Mater.* **2002,** *11,* 775.
8. Mittal, A. K.; Chisti, Y.; Banerjee, U. C. Synthesis of Metallic Nanoparticles Using Plant Extracts. *Biotech. Adv.* **2013,** *31,* 346–356.
9. Ramanavicius, A.; Kausaite, A.; Ramanaviciene, A. Biofuel Cell Based on Direct Bioelectrocatalysis. *Biosens. Bioelectron.* **2005,** *20,* 1962–1967.
10. Narayanan, K. B.; Sakthivel, N. Biological Synthesis of Metal Nanoparticles by Microbes. *Adv. Colloid Interface Sci.* **2010,** *156,* 1–13.

11. Prabhu, S.; Poulose, E. K. Silver Nanoparticles: Mechanism of Antimicrobial Action, Synthesis, Medical Applications, and Toxicity Effects. *Int. Nano Lett.* **2012,** *2,* 32–41.
12. Kim, B. M.; Hackett, M. J.; Park, J.; Hyeon, T. Synthesis, Characterization, and Application of Ultrasmall Nanoparticles. *Chem. Mater.* **2014,** *26,* 59–71.
13. Guo, S.; Wang, E. Noble Metal Nanomaterials: Controllable Synthesis and Application in Fuel Cells and Analytical Sensors. *Nano Today* **2011,** *6,* 240–264.
14. Tomar, A.; Garg, A. Short Review on Application of Gold Nanoparticles. *Global J. Pharmacol.* **2013,** *7,* 34–38.
15. Liu, X.; Atwater, M.; Wang, J.; Huo, Q. Extinction Coefficient of Gold Nanoparticles with Different Sizes and Different Capping Ligands. *Colloids Surf. A* **2007,** *58,* 3–7.
16. Biosselier, E.; Astruc, D. Gold Nanoparticles in Nanomedicine: Preparations, Imaging, Diagnostics, Therapies and Toxicity. *Chem. Soc. Rev.* **2009,** *38,* 1759–1782.
17. Sperling, R. A.; Gil, P. R.; Zhang, F.; Zanella, F.; Parak, W. J. Biological Applications of Gold Nanoparticles. *Chem. Soc. Rev.* **2008,** *37,* 1896–1908.
18. Dreaden, E. C.; Alkilany, A. M.; Huang, X.; Murphy, C. J.; El-sayed, M. A. The Golden Age: Gold Nanoparticles for Biomedicine. *Chem. Soc. Rev.* **2012,** *41,* 2740–2779.
19. Jain, P. K.; Lee, K. S.; El-Sayed, I. H.; El-Sayed, M. A. Calculated Absorption and Scattering Properties of Gold Nanoparticles of Different Size, Shape, and Composition: Applications in Biological Imaging and Biomedicine. *J. Phys. Chem. B* **2006,** *110,* 7238–7248.
20. Kong, T.; Zeng, J.; Wang, X.; Yang, X.; Yang, J.; McQuarrie, S.; McEwan, A.; Roa, W.; Chen, J.; Xing, J. Z. Enhancement of Radiation Cytotoxicity in Breast-Cancer Cells by Localized Attachment of Gold Nanoparticles. *Small* **2008,** *4,* 1537–1543.
21. Lu, Y. Z.; Chen, W. Sub-Nanometre Sized Metal Clusters: from Synthetic Challenges to the Unique Property Discoveries. *Chem. Soc. Rev.* **2012,** *41,* 3594–3623.
22. Chen, M. J.; Yin, M. Z. Design and Development of Fluorescent Nanostructures for Bioimaging. *Prog. Polym. Sci.* **2014,** *39,* 365–395.
23. Lee, K.; El-Sayed, M. A. Gold and Silver Nanoparticles in Sensing and Imaging: Sensitivity of Plasmon Response to Size, Shape, and Metal Composition. *J. Phys. Chem. B* **2006,** *110,* 19220–19225.
24. Jain, P. K.; Huang, X.; El-Sayed, I. H.; El-Sayed, M. A. Review of Some Interesting Surface Plasmon Resonance-Enhanced Properties of Noble Metal Nanoparticles and Their Applications to Biosystems. *Plasmonics* **2007,** *2,* 107–118.
25. So, C. R.; Kulp, J. L.; Oren, E. E.; Zareie, H.; Tamerler, C.; Evans, J. S.; Sarikaya, M. Molecular Recognition and Supramolecular Self-Assembly of a Genetically Engineered Gold Binding Peptide on Au{111}. *ACS Nano* **2009,** *3,* 1525.
26. Wang, X.; Muller, W. E. Marine Biominerals: Perspectives and Challenges for Polymetallic Nodules and Crusts. *Trends Biotechnol.* **2009,** *27,* 375–383.
27. Kovacas, I.; Lundany, A.; Koszegi, T.; Feher, J.; Kovacs, B.; Szolcsanyi, J.; Pinter, E. Substance P Released from Sensory Nerve Endings Influences Tear Secretion and Goblet Cell Function in the Rat. *Neuropeptides* **2005,** *39,* 395–402.
28. Schenkels, L. C. P. M.; Veerman, E. C. I.; Amerongen, A. V. N. Biochemical Composition of Human Saliva in Relation to Other Mucosal Fluids. *Crit. Rev. Oral. Biol. Med.* **1995,** *6,* 161–175. bib>
29. Yasui, T.; Fukui, K.; Nara, T.; Habata, I.; Meyer, W.; Tsukise, A. Immunocytochemical Localization of Lysozyme and Beta-Defensin in the Apocrine Glands of the Equine Scrotum. *Arch. Dermatol. Res.* **2007,** *299,* 393–397.

30. Thakur, G.; Wang, C.; Leblanc, R. M. Surface Chemistry and in Situ Spectroscopy of a Lysozyme Langmuir Monolayer. *Langmuir* **2008,** *24,* 4888–4893.
31. Bakshi, M. S.; Thakur, P.; Kaur, G.; Kaur, H.; Banipal, T. S.; Possmayer, F.; Petersen, N. O. Stabilization of PbS Nanocrystals by Bovine Serum Albumin in its Native and Denatured States. *Adv. Funct. Mater.* **2009,** *19,* 1451.
32. Lu, J. R.; Su, T. J.; Thomas, R. K.; Penfold, J.; Webster, J. Structural Conformation of Lysozyme Layers at the Air/Water Interface Studied by Neutron Reflection. *J. Chem. Soc. Faraday Trans.* **1998,** *94,* 3279–3287.
33. Yang, T.; Li, Z.; Wang, L.; Guo, C.; Sun, Y. Synthesis, Characterization, and Self-Assembly of Protein Lysozyme Monolayer-Stabilized Gold Nanoparticles. *Langmuir* **2007,** *23,* 10533–10538.
34. Jiang, X.; Jiang, J.; Jin, Y.; Wang, E.; Dong, S. Effect of Colloidal Gold Size on the Conformational Changes of Adsorbed Cytochrome C: Probing by Circular Dichroism, Uv-Visible, and Infrared Spectroscopy. *Biomacromolecules* **2005,** *6,* 46–53.
35. Lin, Y. H.; Tseng, W. L. Ultrasensitive Sensing of Hg^{2+} and CH_3Hg^+ Based on the Fluorescence Quenching of Lysozyme Type Vi-Stabilized Gold Nanoclusters. *Anal. Chem.* **2010,** *82,* 9194–9200.
36. Zhang, D.; Neumann, O.; Wang, H.; Yuwono, V. M.; Barhoumi, A.; Perham, M.; Hartgerink, J. D.; Wittung-Stafshede, P.; Halas, N. J. Gold Nanoparticles Can Induce the Formation of Protein-Based Aggregates at Physiological pH. *Nano Lett.* **2009,** *9,* 666–671.
37. de Groot, N. S. Ventura, S. Amyloid Fibril Formation by Bovine Cytochromec. *Spectroscopy* **2005,** *19,* 199–205.
38. Pertinhez, T. A.; Bouchard, M.; Tomlinson, E. J.; Wain, R.; Ferguson, S. J.; Dobson, C. M.; Smith, L. J. Amyloid Fibril Formation by a Helical Cytochrome. *FEBS Lett.* **2001,** *495,* 184–186.
39. Bakshi, M. S.; Kaur, H.; Banipal, T. S.; Singh, N.; Kaur, G. Biomineralization of Gold Nanoparticles by Lysozyme and Cytochrome c and their Applications in Protein Film Formation. *Langmuir* **2010,** *26,* 13535–13544.
40. Kluck, R. M.; BossyWetzel, E.; Green, D. R.; Newmeyer, D. D. The Release of Cytochrome c from Mitochondria: a Primary Site for Bcl-2 Regulation of Apoptosis. *Science* **1997,** *275,* 1132–1136.
41. Kluck, R. M.; Martin, S. J.; Hoffman, B. M.; Zhou, J. S.; Green, D. R.; Newmeyer, D. D. Cytochrome c Activation of CPP32-Like Proteolysis Plays a Critical Role in a Xenopus Cell-Free Apoptosis System. *EMBO J.* **1997,** *16,* 4639–4649.
42. Theorell, H.; Akesson, A.; Studies on Cytochrome c II. The Optical Properties of Pure Cytochrome c and Some of its Derivatives. *J. Am. Chem. Soc.* **1941,** *63,* 1812–1818.
43. Makarava, N.; Parfenov, A.; Baskakov, I. V. Water-Soluble Hybrid Nanoclusters with Extra Bright and Photostable Emissions: A New Tool for Biological Imaging. *Biophys. J.* **2005,** *89,* 572–580.
44. Triulzi, R. C.; Micic, M.; Giordani, S.; Serry, M.; Chiou, W. A.; Leblanc, R. M. Immunoasssay Based on the Antibody-Conjugated Pamam-Dendrimer–Goldquantum Dotcomplex. *Chem. Commun.* **2006,** 5068–5070.
45. Slocik, J. M.; Moore, J. T.; Wright, D. W. Monoclonal Antibody Recognition of Histidine-Rich Peptide Encapsulated Nanoclusters. *Nano Lett.* **2002,** *2,* 169–173.
46. Zheng, J.; Nicovich, P. R.; Dickson, R. M. Highly Fluorescent Noble-Metal Quantum Dots. *Annu. Rev. Phys. Chem.* **2007,** *58,* 409–431.

47. An, W.; Wintzinger, L.; Turner, C. H.; Bao, Y. A Combined Computational/Experimental Study of Fluorescent Gold Nanocluster Complexes. *Nano Life* **2010,** *1,* 133–143.
48. Liu, G.; Shao, Y.; Wu, F.; Xu, S.; Peng, J.; Liu, L.; DNA-hosted Fluorescent Gold Nanoclusters: Sequence-Dependent Formation. *Nanotechnology* **2013,** *24,* 015503.
49. Kawasaki, H.; Yoshimura, K.; Hamaguchi, K.; Arakawa, R. Trypsin-Stabilized Fluorescent Gold Nanocluster for Sensitive and Selective Hg^{2+} Detection. *Anal. Sci.* **2011,** *27,* 591–596.
50. Kawasaki, H.; Hamaguchi, K.; Osaka, I.; Arakawa, R. ph-Dependent Synthesis of Pepsin-Mediated Gold Nanoclusters with Blue Green and Red Fluorescent Emission. *Adv. Funct. Mater.* **2011,** *21,* 3508–3515.
51. Xu, Y.; Sherwood, J.; Qin, Y.; Crowley, D.; Bonizzonic, M.; Bao, Y. The Role of Protein Characteristics in the Formation and Fluorescence of Au Nanoclusters. *Nanoscale* **2014,** *6,* 1515–1524.
52. Wen, X.; Yu, P.; Toh, Y. R.; Hsu, A. C.; Lee, Y. C.; Tang, J. Fluorescence Dynamics in BSA-Protected Au_{25} Nanoclusters. *J. Phys. Chem. C* **2012,** *116,* 19032–19038.
53. Das, R.; Jagannathan, R.; Sharan, C.; Kumar, U.; Poddar, P. Mechanistic Study of Surface Functionalization of Enzyme Lysozyme Synthesized Ag and Au Nanoparticles Using Surface Enhanced Raman Spectroscopy. *J. Phys. Chem. C* **2009,** *113,* 21493–21500.
54. Chevrier, D. M.; Chatt, A.; Zhang, P. Properties and Applications of Protein-Stabilized Fluorescent Gold Nanoclusters: Short Review. *J. Nanophotonics* **2012,** *6,* 064504.
55. Xie, J.; Zheng, Y.; Ying, J. Protein-Directed Synthesis of Highly Fluorescent Gold Nanoclusters. *J. Am. Chem. Soc.* **2009,** *131,* 888–889.
56. Guevel, X. L.; Hotzer, B.; Jung, G.; Hollemeyer, K.; Trouillet, K.; Schneider, M. Formation of Fluorescent Metal (Au, Ag) Nanoclusters Capped in Bovine Serum Albumin Followed by Fluorescence and Spectroscopy. *J. Phys. Chem. C* **2011,** *115,* 10955–10963.
57. Yue, Y.; Liu, T.; Li, H.; Liu, Z.; Wu, Y. Microwave-Assisted Synthesis of BSA-Protected Small Gold Nanoclusters and their Fluorescence-Enhanced Sensing of Silver(I) Ions. *Nanoscale* **2012,** *4,* 2251–2254.
58. Wei, H.; Wang, Z.; Yang, L.; Tian, S.; Hou, C.; Lu, Y. Lysozyme-Stabilized Gold Fluorescent Cluster: Synthesis and Application as Hg^{2+} Sensor. *Analyst* **2010,** *135,* 1406–1410.
59. Gomes, I.; Santos, N. C.; Oliveira, L. M. A.; Quintas, A.; Eaton, P.; Pereira, E.; Franco, R. Probing Surface Properties of Cytochrome C at Au Bionanoconjugates. *J. Phys. Chem. C* **2008,** *112,* 16340–16347.
60. Ding, Y.; Shi, L.; Wei, H. Protein-Directed Approaches to Functional Nanomaterials: A Case Study of Lysozyme. *J. Mater. Chem. B* **2014,** *2,* 8268.
61. Nelson, D. L.; Cox, M. M. *Lehninger Principles of Biochemistry,* 5th ed.; W.H Freeman and Company: USA, 2008.
62. Peng, Z. G.; Hidajat, K.; Uddin, M. S. Adsorption of Bovine Serum Albumin on Nanosized Magnetic Particles. *J. Colloid Interface Sci.* **2004,** *271*(2), 277–283.
63. McDonagh, B. H.; Singh, G.; Bandyopadhyay, S.; Lystvet, S. M.; Ryan, J. A.; Volden, S.; Kim, E.; Sandvig, I.; Axel Sandvig, A.; Glomm, R. W. Controlling the Self-Assembly and Optical Properties of Gold Nanoclusters and Gold Nanoparticles Biomineralized with Bovine Serum Albumin. *RSC Adv.* **2015,** *5,* 101101–101109.
64. Xavier, P. L.; Chaudhari, K.; Baksi, A.; Pradeep, T. Protein-Protected Luminescent Noble Metal Quantum Clusters: An Emerging Trend in Atomic Cluster Nanoscience. *Nano Rev.* **2012,** *3,* 14767.

65. Muhammed, M. A. H.; Verma, P. K.; Pal, S. K.; Retnakumari, A.; Koyakutty, M.; Nair, S.; Pradeep, T. Luminescent Quantum Clusters of Gold in Bulk by Albumin-Induced Core Etching of Nanoparticles: Metal Ion Sensing, Metal-Enhanced Luminescence, and Biolabeling. *Chem.–Eur. J.* **2010,** *16*, 10103–10112.
66. Liu, C. L.; Wu, H. T.; Hsiao, Y. H.; Lai, C. W.; Shih, C.-W.; Peng, Y.-K.; Tang, K.-C.; Chang, H. W.; Chien, Y. C.; Hsiao, J. K.; Cheng, J. T.; Chou, P. T. Insulin-Directed Synthesis of Fluorescent Gold Nanoclusters: Preservation of Insulin Bioactivity and Versatility on Cell-Imaging. *Angew. Chem. Int. Ed.* **2011,** *50*, 7056.
67. Muhammed, M. A. H.; Verma, P. K.; Pal, S. K.; Kumar, R. C. A.; Paul, S.; Omkumar, R. V.; Pradeep, T. Bright, Nir-Emitting Au_{23} from Au_{25}: Characterization and Applications Including Biolabeling. *Chem.Eur. J.* **2009,** *15,* 10110–10120.
68. Lin, C. A.; Yang, T. Y.; Lee, C. H.; Huang, S. H.; Sperling, R. A.; Zanella, M.; Li, J. K.; Shen, J. L.; Wang, H. H.; Yeh, H. I.; Parak, W. J.; Chang, W. H. Synthesis, Characterization, and Bioconjugation of Fluorescent Gold Nanoclusters Toward Biological Labeling Applications. *ACS Nano* **2009,** *3,* 395–401.
69. Thompson, D.; Hermes, J. P.; Quinn, A. J.; Mayor, M. Scanning the Potential Energy Surface for Synthesis of Dendrimer-Wrapped Gold Clusters: Design Rules for True Single-Molecule Nanostructures. *ACS Nano* **2012,** *6,* 3007–3017.
70. Mathew, A.; Sajanlal, P. R.; Pradeep, T. A Fifteen Atom Silver Cluster Confined in Bovine Serum Albumin. *J. Mater. Chem.* **2011,** *21,* 11205–11212.
71. Mohanty, J. S.; Xavier, P. L.; Chaudhari, K.; Bootharaju, M. S.; Goswami, N.; Pal, S. K.; Pradeep, T. Luminescent Bimetallic Au Ag Alloy Quantum Clusters in Protein Templates. *Nanoscale* **2012,** *4,* 4255–4262.
72. Xavier, P. L.; Chaudhari, K.; Verma, P. K.; Pal, S. K.; Pradeep, T. Luminescent Quantum Clusters of Gold in Transferrin Family Protein Lactoferrin Exhibiting Fret. *Nanoscale* **2010,** *2,* 2769–2776.
73. Chaudhari, K.; Xavier, P. L.; Pradeep, T. Understanding the Evolution of Luminescent Gold Quantum Clusters in Protein Templates. *ACS Nano* **2011,** *5,* 8816–8882.
74. Baksi, A.; Xavier, P. L.; Chaudhari, K.; Goswami, N.; Pal, S. K.; Pradeep, T. Protein-Encapsulated Gold Cluster Aggregates: The Case of Lysozyme. *Nanoscale* **2013,** *5,* 2009–2016.
75. Muskens, O. L.; England, M. W.; Danos, L.; Li, M.; Mann, S. Plasmonic Response of Ag- and Au-Infiltrated Cross-Linked Lysozyme Crystals. *Adv. Funct. Mater.* **2013,** *23,* 281–290.
76. Goswami, N.; Giri, A.; Kar, S.; Bootharaju, M. S.; John, R.; Xavier, P. L.; Pradeep, T.; Pal, S. K. Protein-Directed Synthesis of NIR-Emitting, Tunable HgS Quantum Dots and Their Applications in Metal-Ion Sensing. *Small* **2012,** *8,* 3175–3184.
77. Argos, P.; Pedersen, K.; Marks, M. D.; Larkins, B. A. A Structural Model for Maize Zein Proteins. *J. Biol. Chem.* **1982,** *257,* 9984–9990.
78. Ibrahimkutty, S.; Kim, J.; Cammarata, M.; Ewald, F.; Choi, J.; Ihee, H.; Plech, A. Ultrafast Structural Dynamics of the Photo-Cleavage of Protein Hybrid Nanoparticles. *ACS Nano* **2011,** *5,* 3788–3794.
79. Bakshi, M. S.; Jaswal, V. S.; Kaur, G.; Simpson, T. W.; Banipal, P. K.; Banipal, T. S.; Possmayer, F.; Petersen, N. O. Biomineralization of BSA-Chalcogenide Bioconjugate Nano- and Micro-Crystals. *J. Phys. Chem. C* **2009,** *113,* 9121–9127.
80. Kaur, G.; Iqbal, M.; Bakshi, M. S. Biomineralization of Fine Selenium Crystalline Rods and Amorphous Spheres. *J. Phys. Chem. C* **2009,** *113,* 13670–13676.

81. Bakshi, M. S.; Thakur, P.; Kaur, G.; Kaur, H.; Banipal, T. S.; Possmayer, F.; Petersen, N. O. Stabilization of PbS Nanocrystals by Bovine Serum Albumin in its Native and Denatured States. *Adv. Funct. Mater.* **2009,** *19,* 1451–1458.
82. Deo, N.; Jockusch, S.; Turro, N. J.; Somasundaran, P. Surfactant Interactions with Zein Protein. *Langmuir* **2003,** *19,* 5083–5088.
83. Bakshi, M. S.; Kaur, H.; Khullar, P.; Banipal, T. S.; Kaur, G.; Singh, N. Protein Films of Bovine Serum Albumen Conjugated Gold Nanoparticles: a Synthetic Route from Bioconjugated Nanoparticles to Biodegradable Protein Films. *J. Phys. Chem. C* **2011,** *115,* 2982–2992.
84. Barthelemy, P.; Tomao, V.; Selb, J.; Chaudier, Y.; Pucci, B. Fluorocarbon–Hydrocarbon Non-Ionic Surfactant Mixtures: A Study of Their Miscibility. *Langmuir* **2002,** *18,* 2557–2563.
85. Asakawa, T.; Hisamatsu, H.; Miyagishi, S. Experimental Verification of Demixing Micelles Composed of Fluorocarbon and Hydrocarbon Surfactants via Fluorescence-Quenching Method. *Langmuir* **1996,** *12,* 1204–1207.
86. Mukerjee, P.; Handa, T. Adsorption of Fluorocarbon and Hydrocarbon Surfactants to Air–Water, Hexane–Water, and Perfluorohexane– Water Interfaces: Relative Affinities and Fluorocarbon–Hydrocarbon Nonideality Effects. *J. Phys. Chem.* **1981,** *85,* 2298–2303.
87. Mahal, A.; Khullar, P.; Kumar, H.; Kaur, G.; Singh, N.; Niaraki, M. J.; Bakshi, M. S. Green Chemistry of Zein Protein Toward the Synthesis of Bioconjugated Nanoparticles: Understanding Unfolding, Fusogenic Behavior, and Hemolysis. *ACS Sustainable Chem. Eng.* **2013,** *1,* 627–639.
88. Mahal, A.; Goshisht, M. K.; Kumar, H.; Kaur, G.; Singh, N.; Bakshi, M. S. Protein Mixtures of Environmentally Friendly Zein to Understand Protein–Protein Interactions Through Biomaterials Synthesis, Hemolysis, and their Antimicrobial Activities. *Phys. Chem. Chem. Phys.* **2014,** *16,* 14257.

PART IV
Polymer Composites

CHAPTER 13

PREPARATION, CHARACTERIZATION, AND APPLICATION OF SUSTAINABLE POLYMERS COMPOSITES

RAGHVENDRA KUMAR MISHRA[1,*], PRERNA[2], DINESH GOYAL[2], and SABU THOMAS[1]

[1]*International and Inter University Centre for Nanoscience and Nanotechnology, Mahatma Gandhi University, Kottayam, Kerala 686560, India*

[2]*Department of Biotechnology, Thapar University, Patiala, Punjab 147004, India*

[*]*Corresponding author. E-mail: raghvendramishra4489@gmail.com*

CONTENTS

ABSTRACT

Many countries in the world desired to initiate "green chemistry" in order to produce eco-friendly materials, which lead to the fabrication of "green products" derived from nature. Rising environmental awareness has created an immediate essentiality for mankind to strengthen the area of biodegradable materials. Sincere efforts have been made endlessly to develop commercially viable process for maneuvering polymer composites. These polymer composites are no exception to the new paradigm, thus the progression in manufacturing green composites took place. Conventionally, composites are composed of matrix and the reinforcing material. Fibers can be utilized as reinforcement in biodegradable polymers and as a source of raw material for bioenergy and biochemical production. An exhaustive classification of composites on the basis of material and geometry is described. The main focus is on the latest research and development in the field of biofiber-based composites including the modes of synthesis, polymer modification, characterization techniques involved such as light microscopy, scanning electron microscopy, and so forth. In addition, the mechanical strengths such as tensile testing and density measurements are also discussed.

13.1 INTRODUCTION

Over the past three decades, composite materials of plastics and ceramics have been predominant as an emerging material. The research in composite materials has bloomed steadily, covering and capturing new markets. Hence, composite materials are engineered materials made from two or more constituent materials with significantly different physical or chemical properties than individual materials and remain separate and distinct on a microscopic level within the finished structure. The two constituent moieties are matrix and fiber (or reinforcement). The matrix material surrounds and supports the fiber materials by maintaining their relative positions. The fibers impart their special mechanical and physical properties to enhance the matrix properties. It is a well-known fact about the composites that they are an egg in one's beer in the present scenario because of being weight saving, but on the contrary, the only hurdle is the cost of production. In order to make them more economically viable, the scientific community has been endeavoring to lead to the innovation and thus the development of manufacturing techniques. However, the improvement in the manufacturing techniques solely is insufficient to meet the overall cost hike. There is a dire necessity to look into other factors such as the design, material process, tooling, quality assurance, manufacturing, and even program management of composites to strive against metals. The main aim of this chapter is to cover the sustainable polymer composites, which are very useful in the eco-friendly and performance perspective.

13.1.1 CLASSIFICATION OF COMPOSITES

Composites are the cohesive structures of multifunctional material systems with different characteristics as that of composing discrete material. The constituent materials can be mixed in variable proportions in order to achieve desired qualities such as strength, heat resistance, or any other. They are generally better than the components alone or radically different from either of them. Composites constitute of matrix component and dispersed (reinforcing) component. Thus, they are classified in two distinct ways:

1. On the basis of matrix material (major component): This component is also called as a primary constituent and has a continuous character as it provides the bulk form holding, the embedded component in place. These include the composites in the following classes (Fig. 13.1).
2. On the basis of dispersed (reinforcing) component: This component is embedded in the matrix in a discontinuous fashion. It is the secondary constituent that is stronger than the matrix and is called as a reinforcing agent.

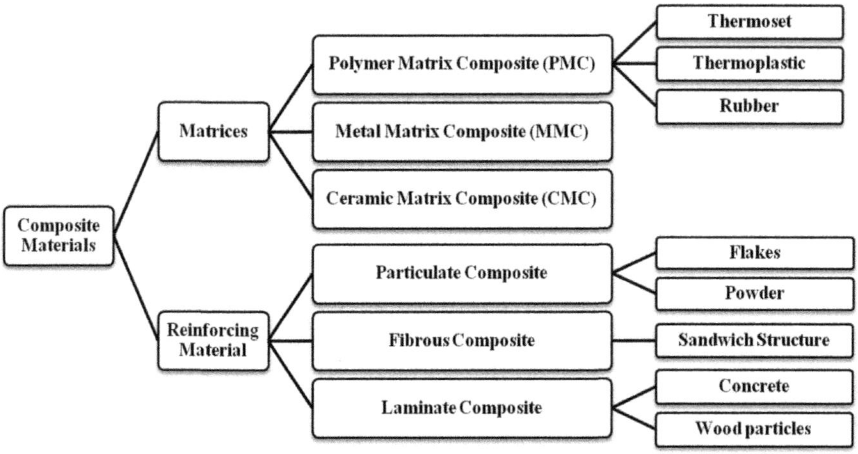

FIGURE 13.1 Classification of composites.

As discussed above, strengthening mechanism of composites depends strongly on the geometry, dispersion, distribution of the reinforcement, and compatibility of major component (matrix) and reinforcement component (dispersed). However, the geometry of the reinforcement material plays a significant role as it is the geometry of reinforcement which is responsible for the mechanical properties and high performance of the composites. A typical classification on the basis of reinforcement geometry is presented in the following Figure 13.2.

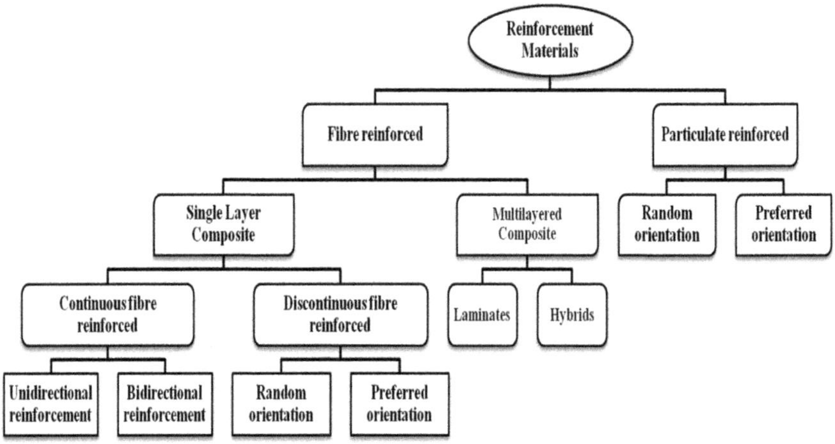

FIGURE 13.2 Classification of composites based on the reinforcement materials.

The polymer composites have been used in the various fields because they have high-performance capability; however, the present question is to recycle this composites system. The efforts to provide recycling and reduce the waste stream have appeared in the last few decades. It is noticeable for polymer composites that using the biodegradable composites and recycling of polymer composites can overcome the environmental issue because they can influence the product performance. Table 13.1 shows the typical advantages of various types of polymer composites system.

13.2 BIOFIBERS-BASED POLYMER COMPOSITES

The consistent sustainability, environmental friendliness, and eco-efficiency of green polymers have the potential to substitute the petroleum-based products flooded in the market.[1] From recent literature, the inclination of the

TABLE 13.1 Type of Composites and Their Advantage.

Type of composites	Remarks
Particulate-reinforced composites	To improve strength, toughness, resistance to ultraviolet radiation, dimension stability, lubrication, and friction wear properties
	Example: concrete, and so forth
Fiber-reinforced composites	To improve the tensile and comprehensive strength, glass transition, and thermal stability in the case of polymer composites
	Example: bulletproof jacket, automotive parts, and so forth
Laminated composites	To improve moisture, heat, and mechanical properties
	Example: space shuttle components, and so forth

scientific community is to focus on plant-derived composite materials, coined as "green composites."[2–3] In this research on "green composites," natural fibers derived from plants are added to various biodegradable matrices such as polylactic acid, poly (ε-caprolactone), and so forth, and have resulted in the development of materials with enhanced mechanical strength. To emphasize the development and applications of green composites, natural fibers such as flax, hemp, pine needles, coir, jute, sisal, and so forth are exploited exhaustively. Depending on the part of the plant from which the natural fibers are derived, they are further classified into bast or stem fibers, leaf fibers, seed hair fibers, core, pith, or stick fiber. Bast fibers include flax (linen), hemp, jute, kenaf, kudzu, nettle, okra, paper mulberry, ramie, and Roselle hemp. Examples of seed fibers include coir, cotton, kapok, and milkweed floss. Core fibers are generally present in the center part (pith) of the plant and it represents over 85% of the dry weight. The remaining fibers include roots, leaf segments, flower heads, seed hulls, and short.[4] However, the grouping of plant biomass for non-wood and wood fibers is elaborated in Figure 13.3. The present-day research on biodegradable nanomaterials is expected to yield eco-friendly, high-performance materials that can alternate the existing synthetic materials.

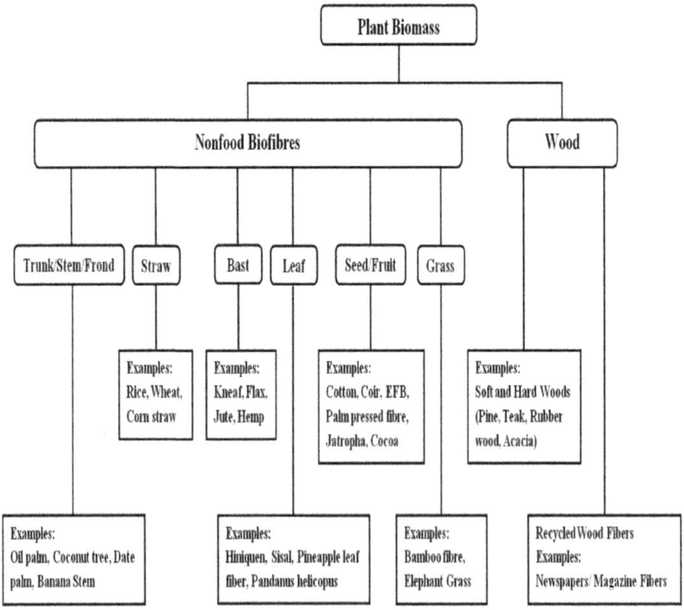

FIGURE 13.3 Classification of biofibers.

Source: Adapted with permission from ref 5. © 2012 Elsevier.

Among the naturally existing fibers, cellulose is the most abundant organic compound on the earth having a linear homopolymer consisting of D-glucopyranose units linked by β-1,4-glycosidic bonds {$C_6nH_{10}n_{+2}O_5n_{+1}$ (n=degree of polymerization of glucose)}. Cellulose has widespread application in the paper industry; it also serves as an alternate form of biofuel. With the advent of green nanotechnology, nanocrystalline cellulose is also fetching a striking interest of the researchers because of easy accessibility with comparatively better mechanical strength and surface tunable characteristics than synthetic analogs, thus, finding applications in the membrane technology. There is a strong network of hydrogen bonding in the cellulose structure dissolving the native cellulose fibers into aqueous dispersions of individual fibers without decreasing the fiber length. The natural sources of fibers contain cellulose and may also contain hemicellulose, lignin, and other extractives. Commercial cellulose production concentrates on harvested sources such as wood or on naturally highly pure sources such as cotton Table 13.2.

TABLE 13.2 The Chemical Composition of Materials Based on the Cellulose. (Source: Adapted permission from ref 5. © 2012 Elsevier.)

Type of biofiber	Composition (%)				
	Source	Cellulose	Hemicellulose	Lignin	Extract
Wood	Hardwood	43–47	25–35	16–24	2–8
	Softwood	40–44	25–29	25–31	1–5
Non-wood	Bagasse	40	30	20	10
	Coir	32–43	10–20	43–49	4
	Corn cobs	45	35	15	5
	Corn stalks	35	25	35	5
	Cotton	95	2	1	0.4
	Empty fruit bunch	50	30	17	3
	Flax (retted)	71	21	2	6
	Flax (unretted)	63	12	3	13
	Hemp	70	22	6	2
	Henequen	78	4–8	13	4
	Istle	73	4–8	17	2
	Jute	71	14	13	2
	Kenaf	36	21	18	2
	Ramie	76	17	1	6
	Sisal	73	14	11	2
	Sunn	80	10	6	3
	Wheat straw	30	50	15	5

Various studies have reported the morphology and orientation of these cellulose fibers in straws.[6] The cellulose present in vascular bundles consists of a high level of orientation in the framework and found that the thickening of the wall is because of cellulose crystals. The crystalline lamellae are found to be perpendicular to the annual rings and inclined 30–40° in the direction of the spiral line and thus proposed a model of the arrangement of cellulose chains in the vascular bundles as shown in the Figure 13.4. Cellulose also consists of sets of nanofibers assembled with a diameter range of 2–20 nm and the length in few micrometers. Following are some literatures which reviewed the fibers derived from cellulose. The modern age scientists have given different descriptions for cellulose nanofibers such as nanowhiskers, nanocrystals, and sometimes monocrystals. Cellulose fibers have certain characteristic advantages as low density, low cost and they can be recyclable and biodegradable. Along with the advantages mentioned, they possess high tensile strength and stiffness making them suitable as an efficient reinforcing material. The reduction in the amount of amorphous material in the final fiber enhances the mechanical properties. The fiber dimension of the material helps in the stress transfer through the length of the fibers.

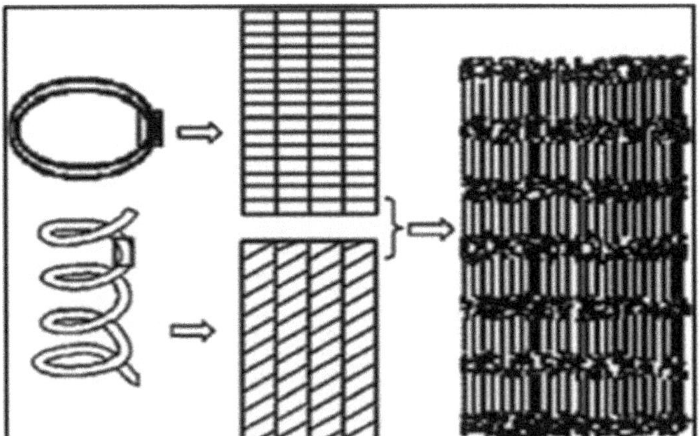

FIGURE 13.4 A schematic model of cellulose molecules in the annular and spiral vessel.
Source: Adapted with permission from ref 5. © 2012 Elsevier.

Several approaches have been tried for the production of nanocellulose composites. All of them point to diverse kinds of nanomaterials, the characteristics and properties of nanocellulose and their composites depend on the cellulose raw material composition, pretreatment techniques, and the disintegration process. Acid hydrolysis methods are extensively utilized for the extraction of amorphous cellulose.[7] Nanocellulose can be obtained from numerous sources such as wood, cotton, ramie, bacterial cellulose, wheat straw, bleached softwood pulps, tunicate cellulose, and microcrystalline cellulose through the acid hydrolysis method.[8–12] However, the deal with acid has a plenty of well-known drawbacks, such as inherent degradation of the cellulose, corrosivity, and environmental incompatibility. In subordinate parts, it undoubtedly reduces the thermal stability, mostly in the case of cellulose whiskers. The isolation of nanofibrils cellulose from cellulose fibers is a simple, low-cost, and environment-friendly method, but still has a lot of challenges.[13] Several mechanical processes have been used to extract nanofibers from cellulosic materials. These methods include treatments such as mechanical treatments (cryocrushing, grinding), high pressure homogenizing, chemical treatments (acid hydrolysis), biological treatments (enzyme-assisted hydrolysis), (2,2,6,6-tetramethylpiperidine-1-yl)oxidanyl) (TEMPO)-mediated oxidation on the surface of microfibrils and a subsequent mild mechanical treatment, electrospinning methods, ultrasonic technique, and so forth.[14–15] For example, sound energy is used in the case of ultrasonication, the chemical result of ultrasonication is received originally from hot spots through acoustic cavitations which are the creation, growth, and collapse of bubbles in a liquid,[16] as shown in Figure 13.5. However, the effect of ultrasonication in deteriorating, the polysaccharide linkages has been properly explained.[17] But these bio-micro/nanofibrils have significantly higher physical, chemical, and thermal properties than the conventional fibers. Thus, much attention has been paid in the past two decades to study how to make micro/nanofibrils and how to combine them with polymers to make nanocomposites. Furthermore, the focus on industrial application led to the choice of cheap source materials such as wood pulp and waste products from the food industry. If the source is wood pulp, then there have been reports that cellulose can be derived just through homogenization. Solvent casting is the common procedure followed by different researchers to produce a nanocomposite. The focus is on processing conditions, such as low temperature and pressure, to make it more eco-friendly. The agglomeration of fibrils during mixing with the polymer matrix is a major concern. In this situation, the surface modification of fibrils has to be done. High microfibrillated cellulose (MFC) concentration on composites and films exhibited higher mechanical strength. The high performance in a nanocellulose-reinforced starch-based matrix in mechanical and water sensitivity is because of the formation of 3-dimensional hydrogen bonds.

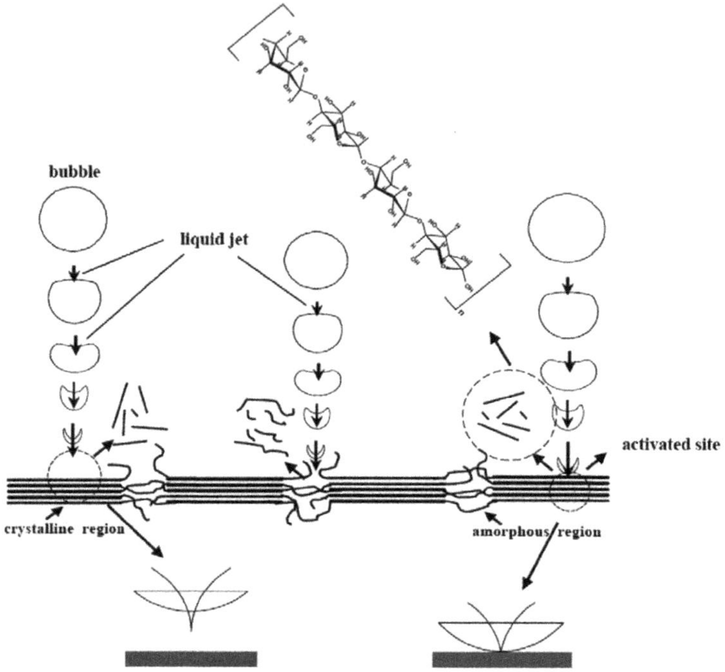

FIGURE 13.5 Schematic diagram of the microcrystalline cellulose ultrasonication process.
Source: Adapted with permission from ref 18. © 2012 Elsevier.

13.3 RECYCLED MICRO/NANOFIBRILS-REINFORCED IN SITU COMPOSITES FROM POLYMER BLENDS

A common polymer composite represents glass- or other fiber-reinforced polymer matrix, in the case of micro/nanofibril-reinforced in situ composites. The micro/nanofibrils act as a reinforcement material and the latter have significant ability to enhance the physical, chemical, and thermal properties of the polymer matrix. In the case of in situ composites, the microfibrils are not used as a separate material to reinforce the polymer matrix. Micro/nanofibrils are produced during the processing of polymer blends, subsequently, they are converted into micro/nanofibrils in situ composites by the heat treatment (isotropization). The production of MFC consists of four main steps: melt blending (blending of two polymers, they have different melting points), extrusion (blends are extruded through the extruder), cold or hot drawing (drawing is used to provide the orientation of both the blended components, this step is also called as fibrillation step), cold drawn blends consist of the micro/nanofibrils of both the blended components, finally, heat treatment (isotropization through compression molding or injection molding) is used to produce the isotropic matrix of lower melting point polymer blend component, however micro/nanofibrils of higher melting point blend polymer's component are preserved. The reinforcing nanofibrils are produced after the blending and drawing of the two polymers and after isotropization (heat treatment), playing the role of a matrix and reinforcement. In this way, the most general problem in nanocomposite technology and recycling of polymer composites is overcome.[19–20] Figures 13.6 and 13.7 shows the schematic setup for in situ composites and scanning electron microscopy (SEM) microstructure of microfibrils-reinforced in situ composites.

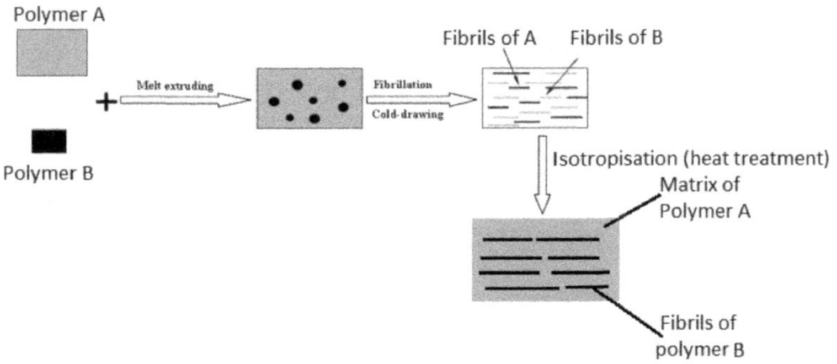

FIGURE 13.6 Schematic setup for in situ composites (A = lower melting point polymer, B = higher melting point polymer).

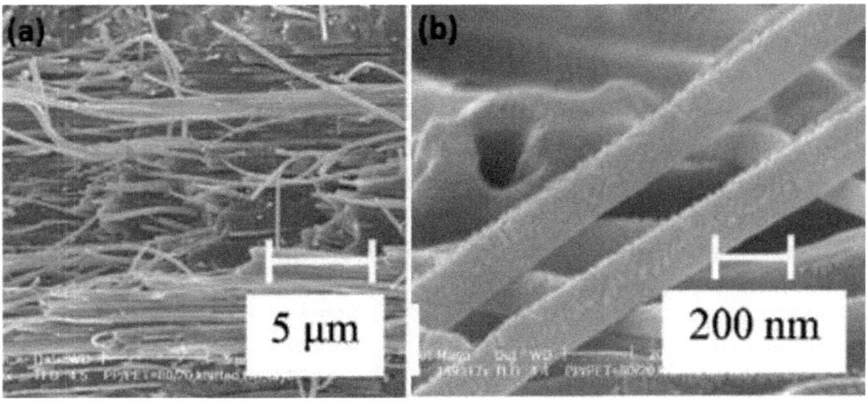

FIGURE 13.7 Scanning electron microscopy (SEM) (a and b) images of PP/PET (80/20, w/w) after: blending—drawing—compression molded (isotropisation) cryofractured surface of in situ composite.

Source: Adapted with permission from ref 19. © 2008 Elsevier.

13.4 CHARACTERIZATION

The morphology of the fibers can be analyzed by light microscopy, SEM, and atomic force microscopy. In addition to the in-depth framework of fiber, the effects of fibrillation can be accessed by mechanical testing and density measurements. Mechanical testing is used for estimating the impact of fibrillation on the mechanical properties up to the equilibrium condition.[21]

13.4.1 LIGHT MICROSCOPY

A light microscope can be used for analyzing the changes in morphology during the mechanical treatment of the fibers. Fibers are placed on a glass slide covered with a coverslip. The location of the selected images is recorded.

13.4.2 SCANNING ELECTRON MICROSCOPY (SEM)

The cellulose fiber morphology can be analyzed by the SEM. Samples are made by solvent exchange from water to the less polar solvent, ethanol, which is used to reduce the amount of hydrogen bonding between the fibers and the tendency of the aggregate formation during drying. Samples are obtained after each processing step and are extracted from water by centrifugation. The water is poured out and the fibers are dipped in ethanol using a high shear mixer. These steps are performed many times to replace most of the water with ethanol. The collected suspensions of cellulose in ethanol are dried by the freeze dryer. The dried samples are sputtered with a thin layer of gold. In the case of in situ composites, the cryo-fractured surface of samples is gold sputtered to analyze the SEM morphology.[22,23]

Microstructures are analyzed using a SEM with varying voltage in kV and magnification of 100 –1 μm.

13.4.3 TENSILE TESTING

The tensile tester is used to measure the mechanical properties of cellulose fiber films after various processing stages. In order to make the sample, the samples are dried and conditioned at low relative humidity for a couple of days and carried away from the climate chamber quickly before testing. The tests are performed at a surrounding temperature. Tensile strength and tensile modulus are determined by the stress–strain curves for at least five measurements per sample. The feed rate is fixed at 5 mm/min.

13.4.4 DENSITY MEASUREMENTS

An average of at least five measurements of cellulose fiber films is used for the weight, area, thickness, and density measurements. The weight of the dried films is measured through an analytic balance, and a micrometer screw is used to calculate the thickness. Calculation of the average of at least three independent measurements is carried out at various locations on the film such as the center, periphery, and between the former location.

13.5 CONCLUSION

As discussed in this report, research across the world has clearly validated the technical viability of synthesizing green composites exploiting the natural resources as the raw material. Generally, as per the thumb rule of demand and supply, if there is a higher acceptance, the production rate rises and the cost is expected to lessen. It is a matter of fact that the overall cost of production can be slashed down by incorporation of natural fibers but there are certain constraints to using natural fibers derived from plants such as the plant source, plant age, processing techniques, geographic origin, and climate. Thus, inexpensive biopolymers with advanced processing methods may end up helping in this rising trend. Several pretreatment processing techniques such as cryocrushing, grinding, high pressure homogenizing, chemical treatments (acid hydrolysis), biological treatments (enzyme-assisted hydrolysis), TEMPO-mediated oxidation on the surface of microfibrils, and a subsequent mild mechanical treatment, electrospinning methods, ultrasonic technique, and so forth are employed to extract the nanofibers from the plant sources. The nanofibers so derived are not only cost-effective but also equally competent to the conventional synthetic composites in terms of physical, chemical, and thermal properties.

KEYWORDS

- **polymer composite**
- **cellulose**
- **green composite**
- **matrix**
- **scanning electron microscopy**

REFERENCES

1. La Mantia, F. P.; Morreale, M. Green Composites: A Brief Review. *Compos. Part A Appl. Sci. Manuf.* **2011,** *42*(6), 579–588.
2. Jiang, L.; Hinrichsen, G. Flax and Cotton Fiber Reinforced Biodegradable Polyester Amide Composites, 2. Characterization of Biodegradation. *Angew. Makromol. Chem.* **1999,** *268*(1), 18–21.
3. Takagi, H.; Asano, A. Effects of Processing Conditions on Flexural Properties of Cellulose Nanofiber Reinforced "Green" Composites. *Compos. Part A Appl. Sci. Manuf.* **2008,** *39*(4), 685–689.
4. Khalil H. P. S. A. et al. Agro-Hybrid Composite: The Effects on Mechanical and Physical Properties of Oil Palm Fiber (EFB)/Glass Hybrid Reinforced Polyester Composites. *J. Reinf. Plast. Compos.* **2007,** *26*(2), 203–218.
5. Khalil, H. P. S. A.; Bhat, A. H.; Ireana Yusra, A. F. Green Composites from Sustainable Cellulose Nanofibrils: A Review. *Carbohydr. Polym.* **2012,** *87*(20), 963–979.
6. Yu, H.; et al. Study on Morphology and Orientation of Cellulose in the Vascular Bundle of Wheat Straw. *Polymer* **2005,** *46*(15), 5689–5694.
7. Moon, R. J.; et al. Cellulose Nanomaterials Review: Structure, Properties and Nanocomposites. *Chem. Soc. Rev.* **2011,** *40*(7), 3941–3994.
8. Revol, J.-F.; et al. Helicoidal Self-Ordering of Cellulose Microfibrils in Aqueous Suspension. *Int. J. Biol. Macromol.* **1992,** *14*(3), 170–172.
9. Revol, J. -F.; et al. Chiral Nematic Suspensions of Cellulose Crystallites; Phase Separation and Magnetic Field Orientation. *Liq. Cryst.* **1994,** *16*(10), 127–134.
10. Dong, X. M.; et al. Effects of Ionic Strength on the Isotropic–Chiral Nematic Phase Transition of Suspensions of Cellulose Crystallites. *Langmuir* **1996,** *12*(8), 2076–2082.
11. Dong, X. M.; Revol, J. -F.; Gray, D. G. Effect of Microcrystallite Preparation Conditions on the Formation of Colloid Crystals of Cellulose. *Cellulose* **1998,** *5*(1), 19–32.
12. Araki, J.; et al. Flow Properties of Microcrystalline Cellulose Suspension Prepared by Acid Treatment of Native Cellulose. *Colloids Surf. A* **1998,** *142*(1), 75–82.
13. Wang, S.; Cheng, Q. A Novel Process to Isolate Fibrils from Cellulose Fibers by HighrIntensity Ultrasonication, Part 1: Process optimization. *J. Appl. Polym. Sci.* **2009,** *113*(2), 1270–1275.
14. Leitner, J.; et al. Sugar Beet Cellulose Nanofibril-Reinforced Composites. *Cellulose* **2007,** *14*(5), 419–425.

15. Wang, B.; Sain, M. Dispersion of Soybean Stock-Based Nanofiber in Plastic Matrix. *Polym. Int.* **2006,** *56*(4), 187–208.
16. Suslick, K. S.; Choe, S-B. Sonochemical Synthesis of Amorphous Iron. *Nature* **1991,** *353*(6343), 414.
17. Tischer, P. C. S. F. et al. Nanostructural Reorganization of Bacterial Cellulose by Ultrasonic Treatment. *Biomacromolecules* **2010,** *11*(5), 1217–1224.
18. Li, W.; Yue, J.; Liu, S. Preparation of Nanocrystalline Cellulose Via Ultrasound and its Reinforcement Capability for Poly (Vinyl Alcohol) Composites. *Ultrason. Sonochem.* **2012,** *19*(3), 479–485.
19. Fakirov, S.; Bhattacharyya, D.; Shields, R. J. Nanofibril Reinforced Composites from Polymer Blends. *Colloids. Surf. A* **2008,** *313*, 2–8.
20. Fakirov, S.; et al. Contribution of Coalescence to Microfibril Formation in Polymer Blends during Cold Drawing. *J. Macromol. Sci., Part B Phys.* **2007,** *46*(1), 183–194.
21. Stelte, W.; Sanadi, A. R. Preparation and Characterization of Cellulose Nanofibers from Two Commercial Hardwood and Softwood Pulps. *Ind. Eng. Chem. Res.* **2009,** *48*(24), 11211–11219.
22. Akhtar, N.; Goyal, D.; Goyal, A. Characterization of Microwave-Alkali-Acid Pre-Treated Rice Straw for Optimization of Ethanol Production Via Simultaneous Saccharification and Fermentation (SSF). *Energy Convers. Manage.* **2016,** *141,* 133–144.
23. Akhtar, N.; Goyal, D.; Goyal, A. Physico-Chemical Characteristics of Leaf Litter Biomass to Delineate the Chemistries Involved in Biofuel Production. *J. Taiwan Inst. Chem. Eng.* **2016,** *62*, 239–246.

CHAPTER 14

DESIGN, FABRICATION, AND CHARACTERIZATION OF ELECTRICALLY ACTIVE METHACRYLATE-BASED POLYMER: ZNO NANOCOMPOSITES FOR DIELECTRICS

ILANGOVAN PUGAZHENTHI, SAKVAI MOHAMMED SAFIULLAH, and KOTTUR ANVER BASHA*

P.G. and Research Department of Chemistry, C. Abdul Hakeem College, Melvisharam, Vellore District, Tamil Nadu 632509, India
Tel.: +914172266187, Fax: +914172269487

**Corresponding author. E-mail: kanverbasha@gmail.com*

CONTENTS

ABSTRACT

In this chapter, a new poly(pyridine-4-yl-methyl) methacrylate ZnO nanocomposite (PPyMMA/ZnO) was prepared by in situ solution polymerization. The oleic acid (OA)-modified ZnONPs (OA-ZnO) were incorporated during the solution polymerization of pyridine-4-yl-methyl methacrylate (PPyMMA). The Fourier-transform infrared spectroscopy confirmed the formation of PPyMMA/ZnO with good compatibility. The X-ray diffraction studies reveal that the incorporation of ZnO into the PPyMMA leads to biphase structures of the resultant nanocomposite. The morphological changes arising due to the addition of ZnO were observed using field-emission scanning electron microscope and transmission electron microscope. The thermal behavior of PPyMMA and its ZnO nanocomposites were analyzed by thermogravimetric analysis and differential scanning calorimetry. The dielectric properties and alternating current conductivity of the polymer and its ZnO nanocomposites were studied over a wide range of frequency (100 Hz–1 MHz) and at different temperatures (30–300°C). PPyMMA/ZnO show a significantly higher dielectric constant (k = 889) and low dielectric loss (tanδ = 0.390) at 100 kHz, 30°C. However, dielectric loss decreases exponentially with an increase in frequency and becomes less at high frequency.

14.1 INTRODUCTION

High-end engineering applications of polymer demand excellent properties which are only met by incorporating fillers or reinforcers. Usually, the mixing of fillers in polymers is a tacky problem due to the difference in polarities of the polymer matrix and the constituent fillers. Improper mixing of these two results in very poor or no adhesion (compatibility), or in other words, yields products with very poor properties. In order to promote the compatibility, functionalized polymers and surface-treated nanoparticles can be used.[1]

The electronic industry is escalating nowadays, and wide research has been done on the fabrication of electronic devices with enhanced dielectric constant, excellent mechanical strength, and easy processability.[2] The polymer nanocomposites can satisfy the requirement in producing materials with high dielectric constant, low dielectric loss, and easy processability. Their properties verily rely on the morphological feature and the interfacial characteristic of individual constituents.[1] Due to these properties, the polymer nanocomposites can find applications in modern nanoelectronics, including high-speed integrated circuits and high charge storage devices.[3,4]

It is well known that the molecular motion/dynamics of nanocomposites in response to applied fields has a significant effect (ionic, interfacial, and dipole polarization) on the macroscopic properties. Several polymers have been utilized to develop polymer nanocomposites like epoxy,[5] poly(methyl methacrylate) (PMMA),[6] polypyrrole/polyvinyl alcohol (PPy-PVA),[7] low-density polyethylene (LDPE),[8] and so forth, with various nanofillers to study the dielectric behavior.[5–8] Zhou et al. reported that the combination of well-dispersed and surface-treated nanoparticles would yield excellent opportunities to engineer the dielectric behavior of a wide range of nanocomposites.[1]

It is obvious that the addition of nanoparticles into the polymers would bring a change in the electrical properties of the polymer's matrix.[9,10] So far, literatures were enlightening about the influence of nanofillers in the dielectric properties, but it is also necessary to know the stimulus effect of the polymer support on the electrical properties. It is assumed that the design of methacrylate polymer with a dipole change moiety may enhance the polymeric dipole of the resultant nanocomposites upon electrical exposure. In this present work, an effort has been taken to understand the effects of electron-rich functional group (pyridinium) in the methacrylate backbone with ZnO nanofiller on the dielectric properties. Hence, an attempt has been made to design, synthesize, and characterize a new electrically active methacrylate polymer (PPyMMA) and its nanocomposite (PPyMMA/ZnO). In this study, the in situ free-radical polymerization of a novel monomer PyMMA with various volume fractions of OA-ZnO (2, 5, and 10 wt.%) were reported. The dielectric constant, dielectric loss, and alternating current (AC) conductance of PPyMMA and PPyMMA/ZnO (2%) were studied over a wide range of frequency (100 Hz–1 MHz) at various temperatures (30–300°C).

14.2 EXPERIMENTAL SECTION

14.2.1 MATERIALS

Zinc acetate and sodium hydroxide were purchased from Sigma Aldrich, India. Pyridine-4-methanol and methacrylic acid were obtained from Merck, India, and distilled under reduced pressure before use. Azobisisobutyronitrile

(AIBN, Sigma Aldrich, India) was crystallized from ethanol at 50°C. Tetrahydrofuran (THF) acquired from Merck was dried by sodium metal before use. All other chemicals were procured from Merck and purified by standard methods.

14.2.2 PREPARATION OF NANO ZNO

About 2.2 g (25 mmol) of ZnOAc was dissolved in 200 ml of double distilled water and stirred well for about 20 min. To this, 8 g NaOH in 300 ml double-distilled water was added dropwise under vigorous stirring until the solution became homogeneous. The solution was digested at 70°C for 2 h, filtered and dried for 1 h at room temperature. The obtained precipitate was calcined for 4 h at 400°C.[11]

14.2.3 SURFACE MODIFICATION OF ZNO NANOPARTICLES BY OLEIC ACID

The surface modification of ZnONPs was carried out with oleic acid.[12] Initially, 50 ml of 2% solution of oleic acid in ethanol was prepared. To this solution, 0.5 g of ZnONPs was added and stirred at 50°C. After 4 h, the contents were centrifuged (1×106 rpm) and the grafted ZnONPs (OA-ZnO) were collected, washed with ethanol followed by acetone (5×30 ml) and dried under reduced pressure.

14.2.4 SYNTHESIS OF PYMMA

To synthesize PyMMA, esterification of pyridine-4-methanol with methacrylic acid was carried out by utilizing P_2O_5/SiO_2 as a dried solid support without using any solvent.[13] Approximately 500 mg of P_2O_5 was stirred with 1 g of SiO_2 under dry atmosphere at room temperature. To this, a 1:1 ratio of pyridine-4-methanol and methacrylic acid was added. The reaction mixture was stirred for about 2 h with moisture protection at room temperature. The organic mixture was washed with 2% sodium bicarbonate solution followed by distilled water and dried over sodium sulfate. On evaporating the solvent under reduced pressure, a pure yellow colored viscous PyMMA liquid was obtained.

14.2.5 PREPARATION OF PPYMMA/ZNO

PPyMMA/ZnO was prepared by the addition of 10 mmol of pyridine-4-yl-methyl methacrylate and 2, 5, and 10 wt.% of OA-ZnO in THF solvent. After sonication for 30 min, polymerization was carried out at 60°C using AIBN as radical initiator under N_2 atmosphere (2 h). After cooling, a white colored amorphous powder was obtained by the addition of hexane; it was further reprecipitated from chloroform to get the pure product (Scheme 14.1a). Same experimental procedure was followed (Scheme 14.1b) to synthesize pure PPyMMA polymer without ZnONPs.

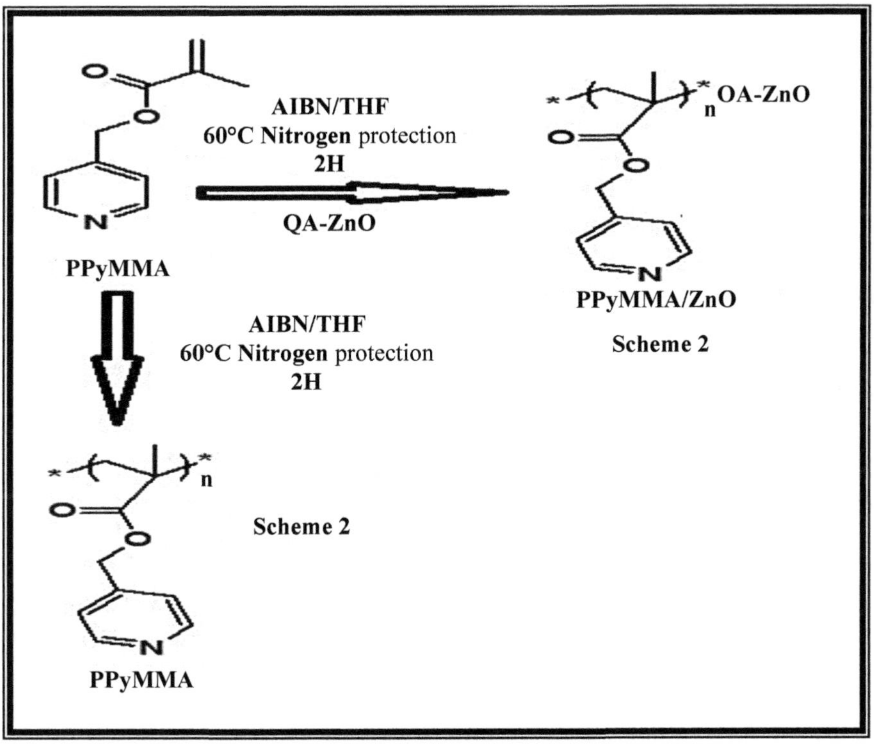

SCHEME 14.1 Preparation of (a) PPyMMA/ZnO and (b) PPyMMA.

14.2.6 INSTRUMENTATION AND CHARACTERIZATION

The ZnONPs, OA-ZnO, PPyMMA, and PPyMMA/ZnO were characterized by FTIR spectroscopy (Shimadzu IR Affinity-1S Spectrometer). The X-ray diffraction (XRD) was taken in a Bruker-D8 Advanced X-ray diffractometer with Cu-Kα (1.5418 Å). The morphology of the samples was studied using field-emission scanning electron microscope (FESEM; HITACHI SU6600). Transmission electron microscope (TEM; FEI-TECNAI G2–20 TWIN 200 kV) images of PPyMMA/ZnO were taken to know the OA-ZnO distribution. The Bruker energy-dispersive X-ray spectroscopy (EDX) with LN2 detector was used to identify the chemical compositions of the polymeric materials. The thermogravimetric analysis was recorded with TA instruments, SBC Q 600, Horizontal furnace at N_2 atmosphere. Approximately 10 mg of the sample was subjected to heating at the rate of 20°C/min from ambient to 800°C in nitrogen atmosphere. Differential scanning calorimetry (DSC) thermograms were recorded over a temperature range of 25–350°C at the scanning rate of 10°C/min. N4L impedance analyzer was used to measure the dielectric properties of the material. The sample has been prepared as pellets with 10-mm diameter and 1-mm thickness. A silver paste was coated across the sides of the sample to act as a conducting medium. Coated sample was placed across the sample holder and the measurements were carried out as a function of frequencies (100 Hz–1 MHz) at different temperatures. The temperature was changed from 30 to 300°C in 10°C steps by a thermostat operating with a liquid nitrogen flux.

14.3 RESULTS AND DISCUSSION

14.3.1 SURFACE MODIFICATION OF PRISTINE ZNO NANORODS

The surface of the ZnONPs was modified to inhibit the agglomeration and to facilitate the effective incorporation of it. The hydroxyl groups of ZnONPs were reacted with carboxylic acid group of OA to give OA-ZnO (Scheme 14.2).[14]

The chloroform/water partitioning experiments were carried out to explore the hydrophobic and hydrophilic nature of the ZnONPs before and after treatment with OA. Figure 14.1 shows the dispersion of the ZnONPs and OA-ZnO in the layers of chloroform and water at room temperature. The ZnONPs were well dispersed in the water phase, demonstrating its hydrophilic nature. The OA-ZnO exhibit good dispersion in the organic phase with a tendency toward the interface (Fig. 14.1b). This indicates that the organophilic nature of OA-ZnO promotes the compatibility of it in the polymer during the course of in situ radical polymerization.[15]

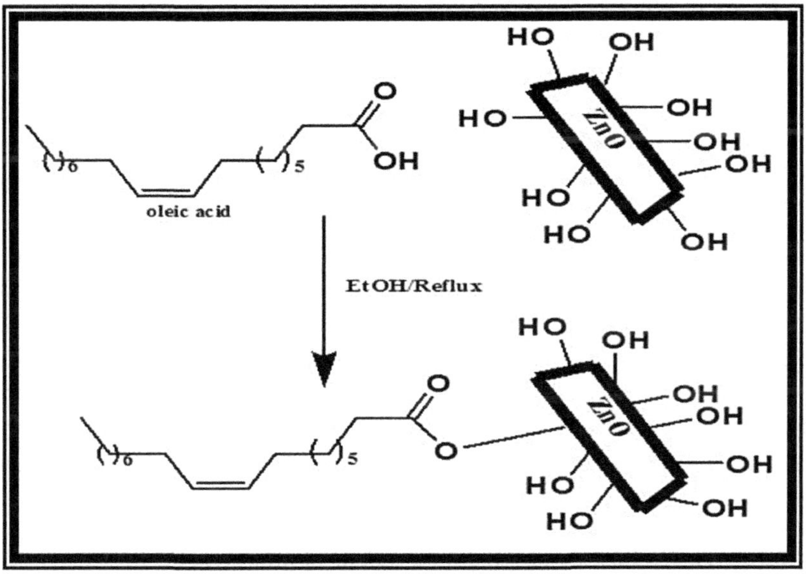

SCHEME 14.2 Surface modification of ZnO with oleic acid.

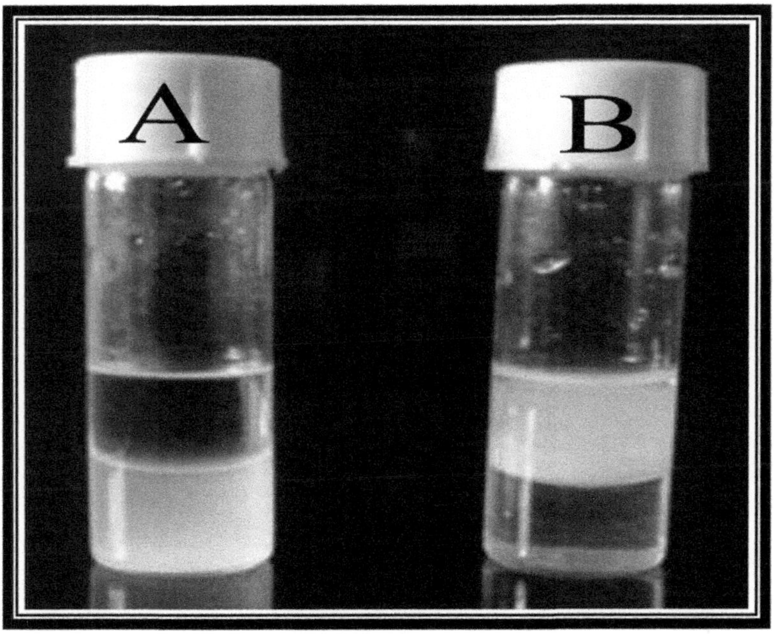

FIGURE 14.1 Dispersion of (a) bare ZnONPs and (b) OA-ZnO in chloroform/water mixture (1:1).

14.3.2 FOURIER-TRANSFORMINFRAREDCHARACTERIZATION

The FTIR spectra (Fig. 14.2a) of ZnONPs show a broad band with a low intensity at 3232.55 cm^{-1} corresponding to the vibration mode of –OH that indicates the presence of moisture on the ZnONPs. The band at 420 cm^{-1} corresponds to the Zn-O bond. Figure 14.2b illustrates the FTIR spectra of OA-grafted ZnONPs, the peaks at 1568 and 1454 cm^{-1} were assigned for symmetric and asymmetric C=O stretching of Zn-oleate which were not present in Figure 14.1a. Absence of the band at 1710 cm^{-1} corresponding to C=O stretching of oleic acid (Fig. 14.2b) confirms the formation of Zn oleate, concluding the formation of monomolecular layer on ZnONPs as zinc oleate. The FTIR spectra of pure PyMMA (Fig. 14.2d) displaying a band at 1724.3 cm^{-1} was assigned to –C=O stretching vibration of ester group. Aromatic C=C stretching is found at 1444 cm^{-1}. The C–N stretching is sensed at 1388 cm^{-1}. The peaks at 1249 and 1143 cm^{-1} are attributed to the C–O stretching. The C–H out-of-plane bending vibration of the aromatic ring is observed at 796 cm^{-1}. The 590 cm^{-1} peak is due to the C–C out-of-plane bending vibration of the aromatic ring. The FTIR spectra of PPyMMA/ZnO (Fig. 14.2c) show the same PPyMMA characteristic peaks with an additional

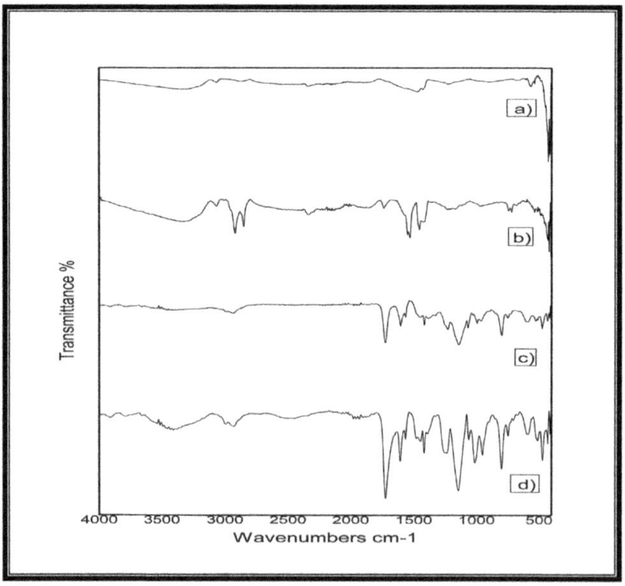

FIGURE 14.2 Fourier-transform infrared spectra of (a) ZnONPs (b) OA-ZnO, (c) PPyMMA/ZnO, and (d) PPyMMA field-emission scanning electron microscope (FESEM) and transmission electron microscope (TEM) analysis.

band at 498 cm^{-1}, corresponding to Zn-O. However, there is an evidence for the decrease in peak intensity due to the addition of OA-ZnO in the PPyMMA. This observation clearly explains about the good compatibility arising between the counterparts (OA-ZnO and PPyMMA).

14.3.3 X-RAY DIFFRACTION STUDIES

The crystallinity of pristine ZnONPs was analyzed by X-ray diffractograms (Fig. 14.3a). The sharp diffraction peaks at $2\theta=32$ (100), 34 (002), 36 (101), 47 (102), 56 (110), and 62 (103) demonstrate the crystallinity and the hexagonal wurtzite structure of ZnONPs, which is in accordance with the standard Joint Committee on Powder Diffraction Standards file no. 36–1451. To know the effect of oleic acid treatment on the crystalline size of nanoparticles, the XRD pattern of OA-ZnO was taken (Fig. 14.3b). The diffraction peaks of these two profiles are consistent with typical wurtzite structure of ZnO. The XRD patterns of the pure PPyMMA and PPyMMA/ZnO were recorded to study the effect of OA-ZnO incorporation in the PPyMMA.

The XRD pattern in Figure 14.3d shows a broad noncrystalline peak at 10–30°, confirming the amorphous nature of PPyMMA. The presence of all the characteristic diffraction peaks of ZnONPs in Figure 14.3c confirms the incorporation of it. The XRD patterns of PPyMMA/ZnO show the coexistence of broad amorphous peaks (10–30°) of the PPyMMA and crystalline peaks (30–80°) of ZnONPs.

FESEM micrographs of freshly synthesized ZnONPs and OA-ZnO provide the rod-shaped morphology. It is interesting to note that the morphology of OA-ZnO (Fig. 14.4b) is similar to the pristine ZnO nanorods (Fig. 14.4a). There is no appreciable change observed in the morphology of ZnO nanorods before and after grafting with oleic acid. The FESEM study reveals that the surface of PPyMMA is homogeneous and smooth (Fig. 14.4c). The addition of OA-ZnO in PPyMMA creates a rough and heterogeneous surface (Fig. 14.4d–f), which indicates that the incorporation of surface-modified nanoparticle has a significant influence on the morphology of the polymer. The good compatibility between the polymer and OA-ZnO may be the reason for the significant morphological changes, which was supported by TEM micrographs (Fig. 14.4g,h). The uniform dispersion of the well-separated OA-ZnO nanorods with an average size of 25 nm was observed in the PPyMMA/ZnO.

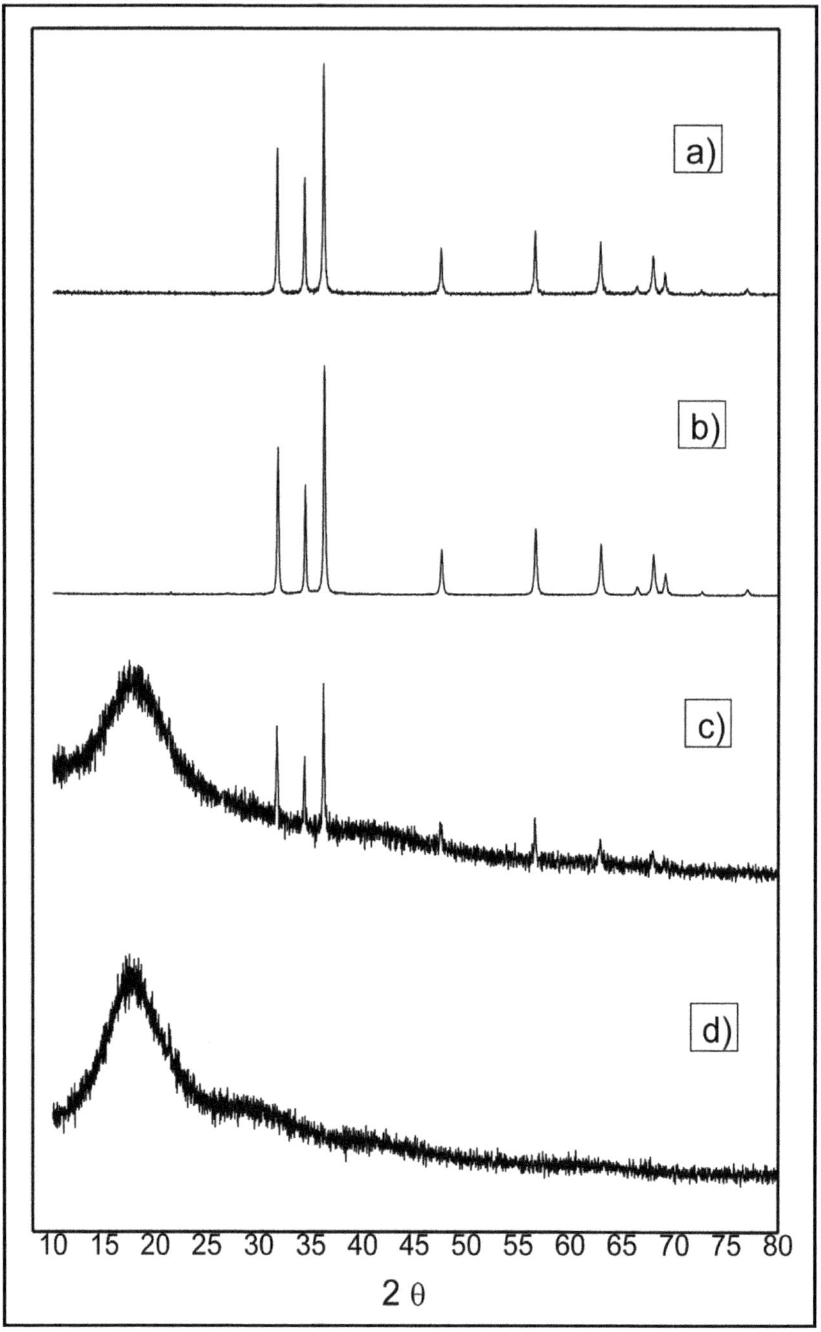

FIGURE 14.3 X-ray diffraction pattern of (a) ZnONPs, (b) OA-ZnO, (c) PPyMMA/ZnO (2%), and (d) PPyMMA.

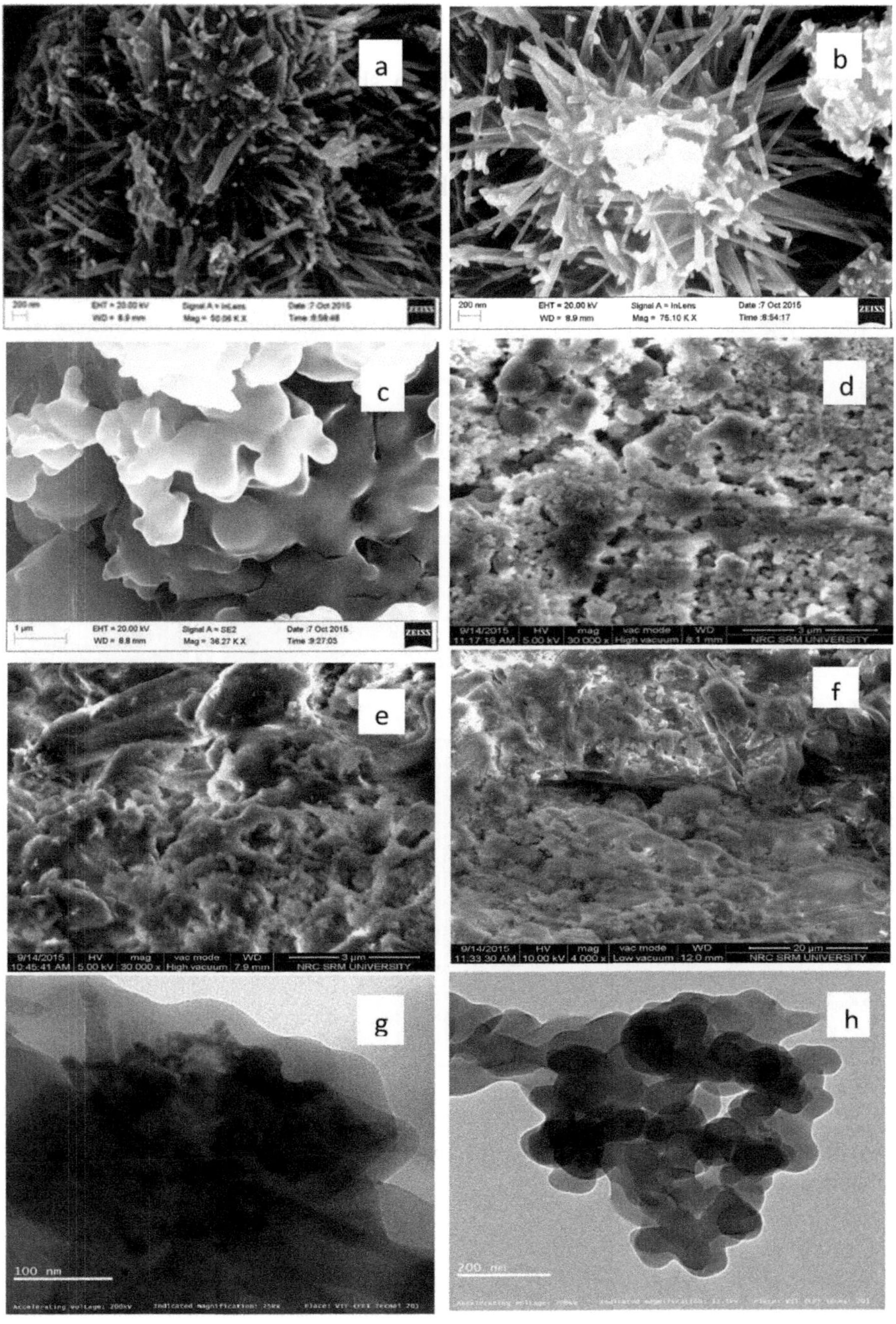

FIGURE 14.4 FESEM images of (a) ZnONPs, (b) OA-ZnO, (c) PPyMMA and PPyMMA/ZnO (d) 2%, (e) 5%, and (f) 10%, TEM Images of PPyMMA/ZnO (g) 100 nm and (h) 200 nm.

14.3.4 ENERGY-DISPERSIVE X-RAY SPECTROSCOPY ANALYSIS

The EDX spectra of OA-ZnO have been shown (Fig. 14.5b). The peaks corresponding to C, Zn, and O indicate the anchoring of OA on the surface of ZnO nanorods. The comparison of EDX spectra (Fig. 14.5c,d) of the polymer and its nanocomposite reveals the peaks corresponding to Zn and O appearing in PPyMMA/ZnO. These results clearly express the successive incorporation of ZnONPs into the PPyMMA matrix.

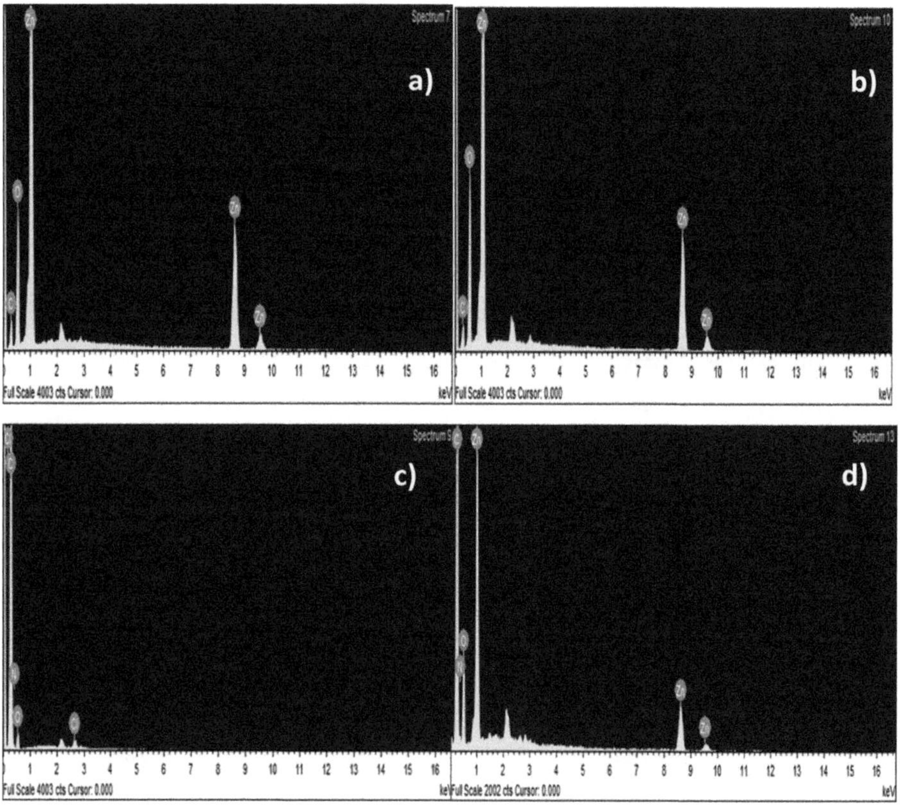

FIGURE 14.5 EDX Spectra of (a) ZnONPs, (b) OA-ZnO, (c) PPyMMA, and (d) PPyMMA/ZnO (2%).

14.3.5 THERMAL STUDIES

Thermogram of PPyMMA and its ZnO nanocomposites shows two stage of decomposition (Fig. 14.6). The initial weight loss up to 13% of polymer and its nanocomposite is observed at 150°C due to the evaporation of water and the solvent. The first-stage degradation of PPyMMA and its nanocomposite starts from 224 to 300°C which is due to the decomposition of pendent bulky pyridine moiety. The residue of 52.73% of the remaining neat PPyMMA is observed after the completion of first-stage degradation, whereas 63.63, 70.75, and 71.88% of the remaining residues were left for the corresponding PPyMMA/ZnO (2, 5, and 10 wt.%). The second-stage decomposition occurs at 363–425°C, which is due to the successive cleavage of the backbone chain of the polymer. After the final decomposition (500°C), the residue left for the remaining neat PPyMMA is only 12.94%, whereas the residues of PPyMMA/ZnO with various percentages loading of ZnO are 29.03, 34.96, and 36.54%, respectively. The thermal stability in terms of the percentage of weight loss is shown in Table 14.1. Percentage of residue increases with the increase in loading of OA-ZnO portion due to the strong interaction at the interface of PPyMMA and OA-ZnO, which in turn improves the thermal stability of the polymer.

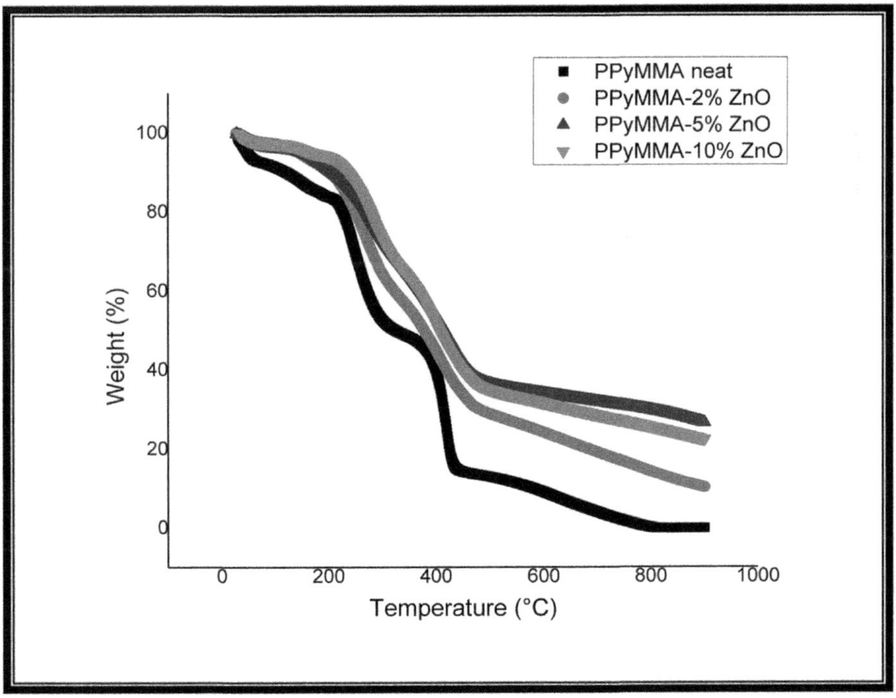

FIGURE 14.6 (See color insert.) Thermogravimetric analysis of PPyMMA and PPyMMA/ZnO with different loadings of OA-ZnO.

The DSC characteristics of pristine and nano-filled PPyMMA are shown in Figure 14.7. The addition of ZnO nanofillers improves the glass transition temperature (*Tg*) of the polymer composite. Incorporation of ZnO fillers acts as plasticizers and they can easily hinder the molecular motions which cause sub-glass activity and finally improve the *Tg* of PPyMMA polymer.[16] In Figure 14.7, an endotherm at 56.77°C corresponds to the *Tg* of PPyMMA. The composite material PPyMMA/ZnO (2, 5, and10 wt.%) shows an increased *Tg* (57.57–76.63°C) compared to the pure PPyMMA. Figure 14.8 shows the variation in *Tg* as a function of the ZnO contents, which reveals that the incorporation of 10 wt.% ZnO leads to a maximum shift in *Tg* toward higher temperature, which is due to the coordination of the nitrogen atom of pyridine ring present in PPyMMA polymer chain with ZnO. The interactions slow down the segmental mobility in the vicinity at polymer–ZnO interface.[17] Due to the high surface area of the nanoparticles, the polymer chains have strong interaction with the particle surfaces and prevent the segmental motion of the polymeric chains.[18] The improvement in *Tg* suggests that the ZnONPs stiffen the PPyMMA polymer matrix at a higher temperature (Fig. 14.8).

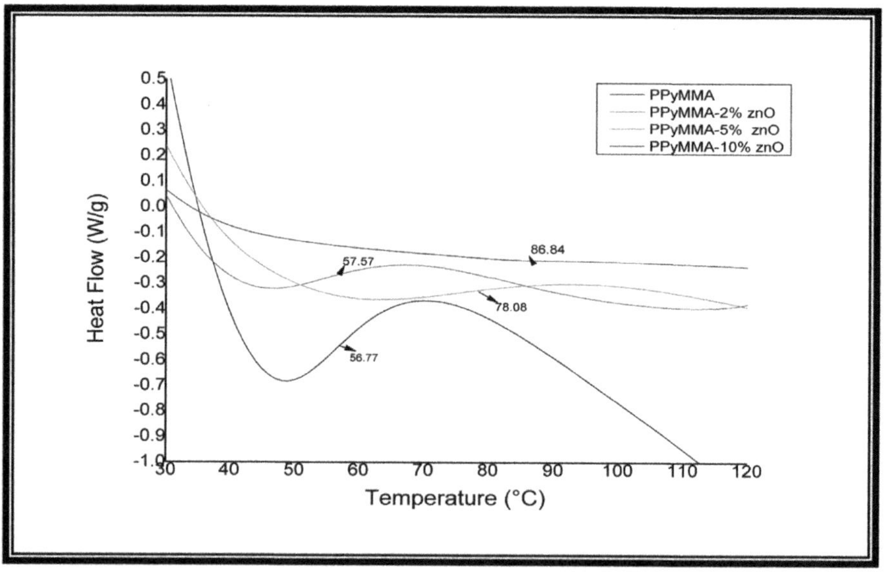

FIGURE 14.7 **(See color insert.)** Differential scanning calorimetry spectra of PPyMMA and PPyMMA/ZnO with different loadings of OA-ZnO.

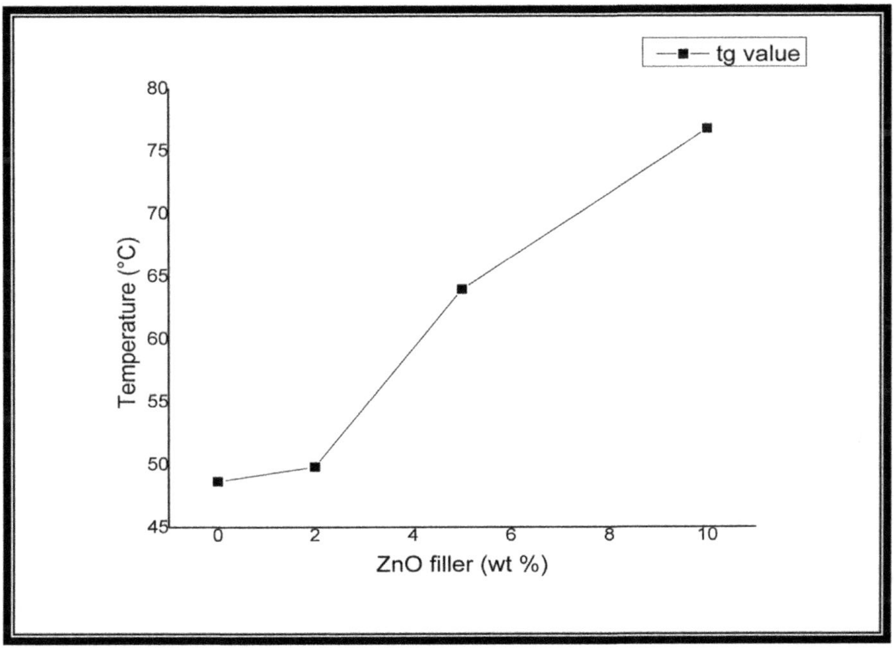

FIGURE 14.8 Variation of *Tg* values of PPyMMA with increasing loading content of OA-ZnO.

TABLE 14.1 Thermal Properties of PPyMMA and Its ZnO Nanocomposites.

Polymer nanocomposite	Percentage of residue remaining at different temperatures (%)							
	100°C	150°C	200°C	250°C	300°C	400°C	450°C	500°C
PPyMMA	91.45	87.44	83.72	69.62	52.73	39.54	14.06	12.48
PPyMMA/ZnO (2%)	95.98	94.56	90.12	80.06	63.63	45.25	32.96	29.03
PPyMMA/ZnO (5%)	97.22	95.02	90.66	84.48	70.75	52.08	39.49	34.93
PPyMMA/ZnO (10%)	97.42	95.88	93.93	86.98	71.88	53.67	42.20	36.30

14.3.6 DIELECTRIC PROPERTIES

To illustrate the effect of electron-rich functional group (pyridinium) in the methacrylate backbone, the dielectric properties of PPyMMA and its ZnO (2 wt.%) nanocomposite were studied. Figures 14.9a,b display the frequency (100 Hz–1 MHz)-dependent dielectric constant of the PPyMMA and PPyMMA/ZnO (2%) at different temperature ranges (30–300°C). The dielectric constant of both PPyMMA and its ZnO nanocomposites gradually decreases with the increase in frequency. The polarization follows the change of the electric charge at lower frequencies (100 Hz–1 kHz). The dielectric constant of the nanocomposites, which is a function of its capacitance, is proportional to the quantity of charge stored on either surface of the sample under an applied electric field.[19] The polarization of PPyMMA and OA-ZnO makes an additional contribution to the charge quantity. From this point of view, PPyMMA shows a significantly higher dielectric constant (k = 512 at 100 kHz, 30°C) compared to PMMA.[20] Further, the incorporation of OA-ZnO in the PPyMMA significantly improve the dielectric constant (k = 889 at 100 kHz, 30°C) to a higher value. This improvement in the dielectric constant is attributed to the interfacial and orientation polarization of dipoles rising due to the presence of a polymeric dipole moiety in the polymer matrix and the good compatibility between the OA-ZnO with polymer in the PPyMMA/ZnO.[1,8]

Figure 14.9c,d shows the dielectric loss of PPyMMA and PPyMMA/ZnO at various frequencies ranging from (100 Hz–1 MHz). The dielectric loss of PPyMMA and its ZnO nanocomposites decreases as the frequency increases from 100 Hz to 1 MHz. At lower frequency regions, the dielectric losses are high; these higher values are attributed to the free charge motion within the materials. A slight decrease in the dielectric loss in the low frequency regions (up to 1.5 kHz) is observed for PPyMMA/ZnO, followed by a constant frequency-independent behavior from 2 kHz to 1 MHz. This is due to the induced charges which gradually failed to follow the reversing field causing a reduction in the electronic oscillations as the frequency increased.[21] The PPyMMA/ZnO ($\tan\delta = 0.390$) exhibit more dielectric loss when compared to the PPyMMA ($\tan\delta = 0.1621$) films at the frequency of 100 kHz at room temperature. The interfacial polarization mechanism of the heterogeneous system between the polymer and OA-ZnO may be the contributing factor in this upsurge in the dielectric loss of the polymer with OA-ZnO.[22]

14.3.7 VARIATIONS OF DIELECTRIC CONSTANT AND LOSS WITH TEMPERATURE

Figure 14.9e emphasizes the plot between dielectric constant of PPyMMA and PMMA/ZnO thin films at different temperatures from 30 to 300°C. The observed dielectric constant of PPyMMA (k = 512) and PPyMMA/ZnO (k = 889) at 100 kHz is very low in the lower temperature (30°C) region due to frozen molecular dipoles. The increase in dielectric constant with

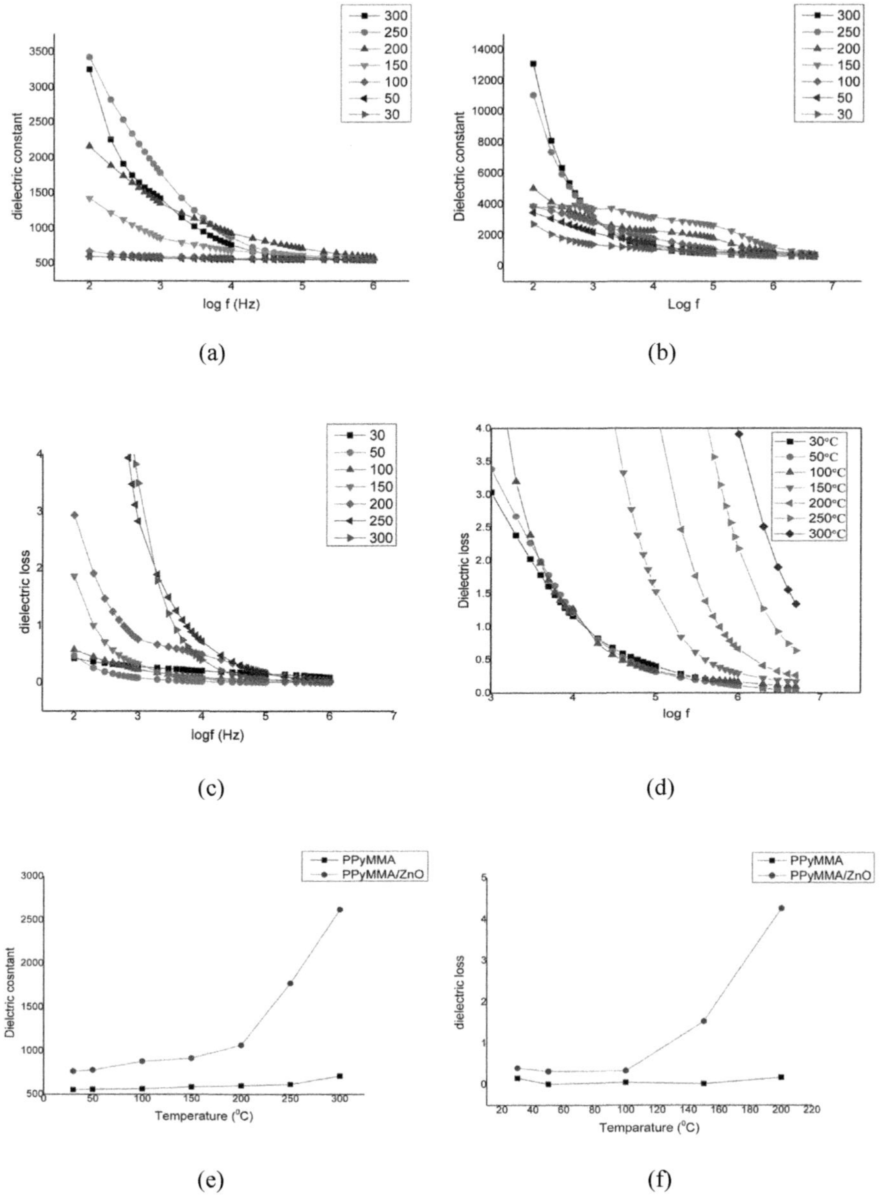

FIGURE 14.9 (See color insert.) (a) Dielectric constant of PPyMMA, (b) dielectric constant for PPyMMA/ZnO (2 wt.%), (c) dielectric loss for PPyMMA, (d) dielectric loss for PPyMMA/ZnO (2 wt.%), (e) variations of the dielectric constant with the temperature at 100 kHz, and (f) variations of dielectric loss with the temperature at 100 kHz.

increasing temperature is due to more rotational freedom of thermally activated dipoles, which leads to an increase in the dielectric constant.[23] There is a tremendous increase in the dielectric constant value (k=2621.45 at 100 kHz, 300°C) for the PPyMMA/ZnO thin film when compared to polymer PPyMMA films (k=710.35 at 100 kHz, 300°C). Figure 14.9f shows the plot between the dielectric loss and the temperature of PPyMMA and PMMA/ZnO. The rise of temperature facilitates segmental motion and dipole orientation which also leads to the mobility of ionic charges.[24] Thus, it is seen that the dielectric loss increases with the increase of temperature due to an interfacial polarization.[5]

14.3.8 ALTERNATING CURRENT CONDUCTIVITY

Figure 14.10a,b show the variation of AC electrical conductivity (σ_{ac}) of PPyMMA and PPyMMA/ZnO as a function of frequency at various temperature ranges from 30–300°C. The AC conductivity was calculated from the dielectric properties using the relation:

$$\sigma_{ac} = 2\ \pi \cdot \varepsilon_0 \varepsilon_r \omega\ (\tan\delta)$$

The AC conductivity (σ_{ac}) (100 Hz–1 MHz) of the PPyMMA/ZnO was recorded in the range of 1.12×10^{-5} to 4×10^{-4} Scm^{-1}, which is significantly higher than that of the pure PPyMMA (10^{-7} to 1.9×10^{-6} Scm^{-1}). The improvement of the AC conductivity for PPyMMA/ZnO arises from the effective dispersion of OA-ZnO in the PPyMMA matrix. It is assumed that an electrically conducting path and network of connections could be formed in the composites with the incorporation of OA-ZnO. It was observed that the conductivity (σ_{ac}) increases in both the PPyMMA and PPyMMA-ZnO with an increase in the frequency (>10 kHz). The frequency dependence of the AC conductivity is considered to be the result of interface charge polarization (Maxwell–Wagner–Sillars effect) and intrinsic electric dipole polarization, which is observed in the ZnO–polymer composites owing to the gathering of mobile charges at the interfaces and the formation of larger dipoles on OA-ZnO.[5,22]

Additionally, as the temperature increases at a higher frequency, AC conductivity also increases (Fig. 14.10c,d). This may be either due to an increase in the charge mobility with temperature or an increase in the charge concentration. Since there is no change in the charge concentration, the reason may preferably be attributed to the increase in charge mobility.

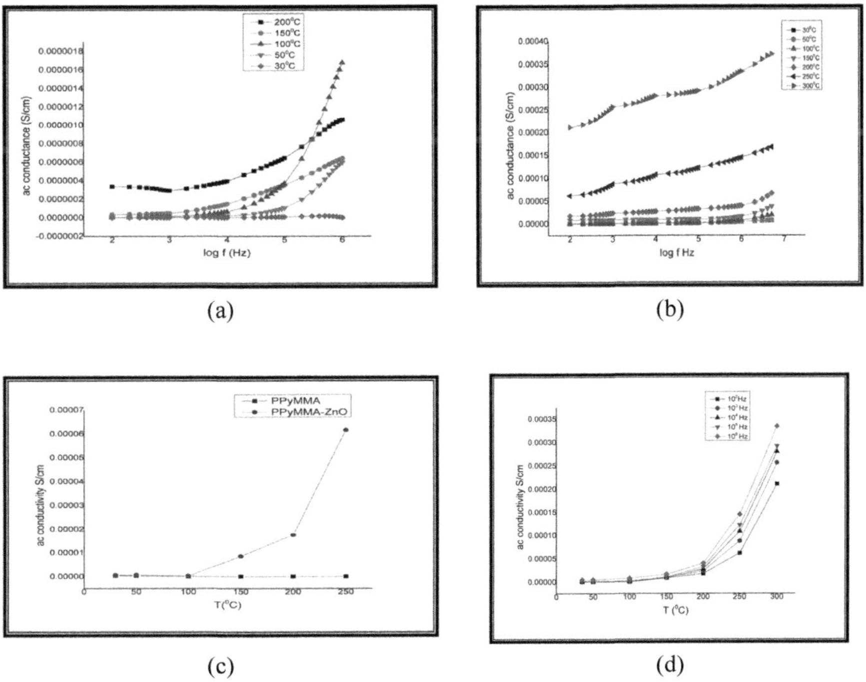

FIGURE 14.10 (See color insert.) (a) Alternating current (AC) conductance for PPyMMA, (b) AC conductance for PPyMMA/ZnO (2 wt.%), (c) AC conductivity of PPyMMA and PPyMMA/ZnO at 105 Hz, and (d) AC conductivity of PPyMMA/ZnO at different temperatures.

14.4 CONCLUSION

In summary, an attempt has been made to prepare poly(pyridine-4-yl-methyl) methacrylate ZnO nanocomposite by in situ solution polymerization. The FTIR and XRD studies confirmed the formation of PPyMMA/ZnO with good compatibility. The morphological changes arising due to the addition of OA-ZnO were observed using electron microscopic studies. The thermal analysis depicted the percentage of residues left after the gradual increase in sample decomposition with an increase in volume fraction of ZnONPs. The PPyMMA/ZnO shows a high dielectric constant (k=889), low dielectric loss (tanδ=0.390) at 100 kHz, 30°C, and a high AC conductivity. Hence, the designing of a polymeric dipole in the polymer matrix also contributes to yield excellent opportunities to engineer the dielectric behavior of the polymer nanocomposites. Thus, the PPyMMA and its ZnO nanocomposite with high dielectric constant and low dielectric loss is a promising candidate for embedded capacitors applications.

ACKNOWLEDGMENT AND FUNDING

The authors would like to thank the Defence Research and Development Organization (DRDO) India, Ref. No. ERIP/ER/1204672/M/01/1525 for funding.

KEYWORDS

- **polymer**
- **zinc oxide**
- **nanocomposites**
- **thermal properties**
- **dielectrics**
- **AC conductivity**

REFERENCES

1. Zhou, W.; Yu, D. Effect of Coupling Agents on the Dielectric Properties of Aluminium Particles Reinforced Epoxy Resin Composites. *J. Compos. Mater.* **2011,** *45*(19), 1981–1989.
2. Matchawet, S.; Kaesaman, A.; Bomlai, P.; Nakason, C. Electrical, Dielectric, and Dynamic Mechanical Properties of Conductive Carbon Black/Epoxidized Natural Rubber Composites. *J. Compos. Mater.* **2015.** DOI: 10.1177/0021998315602941.
3. Zhang, Y.; Wang, Y.; Deng, Y.; Li, M.; Bai, J. Enhanced Dielectric Properties of Ferroelectric Polymer Composites Induced by Metal-Semiconductor Zn-ZnO Core–Shell Structure. *ACS Appl. Mater. Interfaces* **2012,** *4,* 65–68.
4. Shen, Y.; Lin, Y.; Li, M.; Nan, C. W. High Dielectric Performance of Polymer Composite Films Induced by a Percolating Interparticle Barrier Layer. *Adv. Mater.* **2007,** *19,* 1418–1422.
5. Elimat, Z. M. Ac-Impedance and Dielectric Properties of Hybrid Polymer Composites. *J. Compos. Mater.* **2015,** *49*(1), 3–15.
6. Maji, P.; Choudhary, R.; Majhi, M. Structural, Optical and Dielectric Properties of ZrO_2 Reinforced Polymeric Nanocomposite Films of Polymethyl Methacrylate (PMMA). *Optik Int. J. Light Electron Optics* **2016,** *127*, 4848–4853.

7. Deshmukh, K.; Ahamed, M. B.; Pasha, S. K.; Deshmukh, R. R.; Bhagat, P. R. Highly Dispersible Graphene Oxide Reinforced Polypyrrole/Polyvinyl Alcohol Blend Nanocomposites with High Dielectric Constant and Low Dielectric Loss. *RSC Adv.* **2015,** *5*, 61933–61945.
8. Zhang, Y.; Wang, Y.; Deng, Y.; Li, M.; Bai, J. High Dielectric Constant and Low Loss in Polymer Composites Filled by Self-Passivated Zinc Particles. *Mater. Lett.* **2012,** *72*, 9–11.
9. Godovsky, D. Y. *Biopolymers·PVA Hydrogels, Anionic Polymerisation Nanocomposites;* Springer-Verlag Berlin Heidelberg: New York, 2000; pp 163–205.
10. Gangopadhyay, R.; De, A. Conducting Polymer Nanocomposites: A Brief Overview. *Chem. Mater.* **2000,** *12*, 608–622.
11. Cao, H. L.; Qian, X. F.; Gong, Q.; Du, W. M.; Ma, X. D.; Zhu, Z. K. Shape- and Size-Controlled Synthesis of Nanometre ZnO from a Simple Solution Route at Room Temperature. *Nanotechnology* **2006,** *17*, 3632–3636.
12. Hong, R.; Pan, T.; Li, H. Microwave Synthesis of Magnetic Fe_3O_4 Nanoparticles Used as a Precursor of Nanocomposites and Ferrofluids. *J. Magnetism Magnetic Mater.* **2006,** *303*, 60–68.
13. Eshghi, H.; Mirzaie, N.; Asoodeh, A. Synthesis of Fluorescein Aromatic Esters in the Presence of P_2O_5/SiO_2 as Catalyst and Their Hydrolysis Studies in the Presence of Lipase. *Dyes Pigm.* **2011,** *89*, 120–126.
14. Hong, R.; Pan, T.; Qian, J.; Li, H. Synthesis and Surface Modification of ZnO Nanoparticles. *Chem. Eng. J.* **2006,** *119*, 71–81.
15. Augustine, M. S.; Jeeju, P.; Sreevalsa, V.; Jayalekshmi, S.; Excellent, U. V. Absorption in Spin-Coated Thin Films of Oleic Acid Modified Zinc Oxide Nanorods Embedded in Polyvinyl Alcohol. *J. Phys. Chem. Solids* **2012,** *73*, 396–401.
16. Gaur, M. S.; Rathore, B. S.; Singh, P. K.; Indolia, A.; Awasthi, A. M.; Bhardwaj, S. Thermally Stimulated Current and Differential Scanning Calorimetry Spectroscopy for the Study of Polymer Nanocomposites. *J. Therm. Anal. Calorim.* **2010,** *101*, 315–321.
17. Rittigstein, P.; Torkelson, J. M. Polymer–Nanoparticle Interfacial Interactions in Polymer Nanocomposites: Confinement Effects on Glass Transition Temperature and Suppression of Physical Aging. *J. Polym. Sci. Part B: Polym. Phys.* **2006,** *44*, 2935–2943.
18. Chang, J.-H.; Seo, B.-S.; Hwang, D.-H. An Exfoliation of Organoclay in Thermotropic Liquid Crystalline Polyester Nanocomposites. *Polymer* **2002,** *43*, 2969–2974.
19. Wang, M.; Lian, Y.; Wang, X. PPV/PVA/ZnO Nanocomposite Prepared by Complex Precursor Method and its Photovoltaic Application. *Curr. Appl. Phys.* **2009,** *9*, 189–194.
20. Hayashida, K.; Takatani, Y. Poly(Methyl Methacrylate)-Grafted ZnO Nanocomposites with Variable Dielectric Constants by UV Light Irradiation. *J. Mater. Chem. C* **2016,** *4*, 3640–3645.
21. Demei Yu; Jingshen Wu; Zhou, L.; Xie, D.; Wu, S. The Dielectric and Mechanical Properties of a Potassium-Titanate-Whisker-Reinforced PP/PA Blend. *Comp. Sci. Technol.* **2000,** *60*, 499–508.
22. Park, J.; Lee, J. W.; Kim, D. W.; Park, B. J.; Choi, H. J.; Choi, J. S. Pentacene Thin-Film Transistor with Poly(methyl Methacrylate-co-methacrylic Acid)/TiO_2 Nanocomposite Gate Insulator. *Thin Solid Films* **2009,** *518*, 588–590.
23. Dang, Z.-M.; Yi-He, Z.; Tjong, S.-C. Dependence of Dielectric Behavior on the Physical Property of Fillers in the Polymer-Matrix Composites. *Synth. Met.* **2004,** *146*, 79–84.
24. Chen, F. C.; Chu, C.-W.; He, J.; Yang, Y.; Lin, J.-L. Organic Thin-Film Transistors with Nanocomposite Dielectric Gate Insulator. *Appl. Phys. Lett.* **2004,** *85*, 3295–3297.

PART V
Advanced Case Studies

CHAPTER 15

STUDIES ON PATHOGENICITY OF *VIBRIO PARAHAEMOLYTICUS* AND ITS CONTROL MEASURES

S. THANIGAIVEL[1], NATARAJAN CHANDRASEKARAN[1], AMITAVA MUKHERJEE[1], and JOHN THOMAS[1,*]

[1]*Centre for Nanobiotechnology, VIT University, Vellore, Tamil Nadu, India, Tel.: +91 416 2202876, Fax: +91 416 2243092*

[*]*Corresponding author. E-mail: john.thomas@vit.ac.in, th_john28@yahoo.co.in*

CONTENTS

ABSTRACT

Vibrio parahaemolyticus can cause severe mortality to *Penaeus monodon* in aquaculture farms. An attempt was made to isolate the bacteria from grow-out ponds in which the shrimp were reared. *V. parahaemolyticus* was isolated from the water and tested for its pathogenicity. The inoculum was injected into healthy monodon through intramuscular injection. The bacterial infection was treated using natural products as it has a wide range of biological activities including plants, seaweeds, essential oils, and probiotics were implemented against bacterial infections in freshwater aquatic animals. Hence, this study revealed the effectiveness of the neem nanoemulsion against pathogenic *Vibrio* by in vitro and in vivo analysis and result was found to be effective.

15.1 INTRODUCTION

Vibriosis is one of the major disease problems in shellfish and finfish aquaculture.[7,12,13,14] *Vibrio* spp. are widely distributed in culture facilities throughout the world. *Vibrio*-related infections frequently occur in hatcheries but epizootics also commonly occur in pond-reared shrimp species. The bacterial diseases are a main threat to *Penaeus monodon* (Asian tiger shrimp) culture. The types of bacteria which cause infection in *P. monodon* are *Vibrio, Aeromonas*, and *Pseudomonas*. Out of these three, the *Vibrio* spp. were *found* to cause mortalities on a large scale in *P. monodon* from grow-out ponds. Various *Vibrio* spp. have been documented to cause large die-off in cultured peneids in Taiwan. Vibriosis is one of the most prevalent fish diseases caused by bacteria belonging to the genus *Vibrio. Vibrio parahaemolyticus, Vibrio anguillarum, Vibrio alginolyticus*, and *Vibrio vulnificus* are the main fish pathogens which cause vibriosis in shrimp and in shellfish. *Vibrio* pathogens frequently cause diseases in fish and other aquaculture-system-reared organisms. It is mainly caused by the *Vibrio* spp.[2,4] It is responsible for the considerable mortality and economic losses in the aquaculture system.[7,14] Vibriosis mainly increases mortality rate in postlarvae and in adult of *P. monodon*, and also causes many diseases in *P. monodon* such as reddening of the body, loosening up of the body shell, black patches on the body, and necrosis in telson, rostrum region, and so forth.[16]

Preventive measures for vibriosis mainly include antibiotic treatment. However, it has long-term effects on the organism and environment.

Bacteria are known to develop resistance to most of the antibiotics used in aquaculture. Moreover, the use of antibiotics is prohibited by the Food and Drug Administration (FDA) because shrimp are mainly produced for human consumption and antibiotics contain harmful chemicals such as "nitrofuran and its metabolites," chlorine, and metals like lead, which may have adverse effects on the human body.

15.2 MATERIALS AND METHODS

15.2.1 COLLECTION AND MAINTENANCE OF ANIMALS

Shrimp (*P. monodon*, 10–15 g body weight) were collected from a grow-out pond and transported to the laboratory with continuous aeration. In the laboratory, they were maintained in 1000-l fiberglass tanks with airlift biological filters at room temperature (27–30°C) with salinity between 20 and 25 ppt. Natural seawater was used in all the experiments. It was pumped from the Bay of Bengal, near Chennai, and allowed to sediment and remove sand and other suspended particles. The seawater was first chlorinated by treating it with sodium hypochlorite at a concentration of 25 ppm and then dechlorinated by vigorous aeration, before being passed through a sand filter and then used for the experiments. The animals were fed with artificial pellet feed (CP Feed, Thailand).

15.2.2 COLLECTION OF WATER SAMPLES

Water samples from the grow-out ponds suspected to be bacterially contaminated in which the shrimp were reared, collected, and transported to the laboratory in a sterile container. Water temperature, salinity, pH, and dissolved oxygen measurements were made on the site with portable meters.

15.2.3 ISOLATION OF BACTERIA

The method of Oliver et al. (1980)[16] and Buchanan et al. (1974) was followed with slight modifications. Water samples after making tenfold dilutions were taken and plated onto thiosulfate-citrate-bile salts-sucrose (TCBS) agar and on to nutrient agar by spread plate method.[8] Plates were incubated at 37°C for 24–48 h. Bacterial colonies were carefully examined. Bacterial isolates were identified according to the taxonomic schemes of Buchanan and Gibbons[5] and West and Colwell.[23] The identification of *Vibrio* has been done by the genomic and phenotypic characteristics of the particular bacterial sequence. This was being employed for more than two decades.[19,22]

The use of 16S rRNA gene sequences to study bacterial phylogeny and taxonomy has been by far as a genetic marker for a number of reasons. These reasons include (i) its presence in almost all bacteria, often existing as a multigene family or operons; (ii) the function of the 16S rRNA gene over time has not changed, suggesting that random sequence changes are a more accurate measure of time (evolution) and (iii) the 16S rRNA gene (1500 bp) is large enough for informatics purposes[20] (Patel et al., 2001).

15.2.4 EXPERIMENTAL PATHOGENICITY OF THE BACTERIAL ISOLATE IN HEALTHY SHRIMP

15.2.4.1 INFECTION THROUGH INTRAMUSCULAR INJECTION

The adult shrimp at the rate of five per tank were maintained in 100-l capacity fiberglass tanks. The water was provided with good aeration and it was changed daily. The animals were fed with artificial pelleted feed (CP Feed, Thailand). Different concentrations (10^3, 10^4, 10^5, 10^6, 10^7, and 10^8) of bacterial suspension were used for this experiment. For adults prawn, 20 μl of bacterial suspension from each of these six concentrations was injected intramuscularly using 1-ml insulin syringes in the third abdominal segment. Control shrimp received only sterile saline. The experiment was conducted in triplicates in each bacterial concentration. Animals were checked twice daily for clinical signs of disease and mortality. Dead animals were removed.

15.2.4.2 CONFIRMATION OF PATHOGENICITY

The specific action of *V. parahaemolyticus* as a pathogen was confirmed by re-isolating the bacterium from moribund shrimp to satisfy Koch's postulates. The appendages of the shrimp were cut with a sterile blade for re-isolation of bacterial pathogen. They were homogenized separately in a sterile homogenizer. The samples were inoculated on nutrient agar and TCBS agar plates by spread plate technique for the isolation of bacterial pathogen. The isolated bacteria were identified using the procedure explained in Section 15.2.3

15.2.5 TREATMENT OF BACTERIAL INFECTION

Bacterial infection in *P. monodon* was treated using neem nanoemulsion and seaweeds. In vitro and in vivo methods were followed for treating the bacterial infection. This method of treatment is used as a natural source for treating the bacterial infection.

15.2.5.1 PREPARATION OF NEEM NANOEMULSION

Neem nanoemulsion was prepared and formulated according to the method of Anjali et al.[1] Neem oil, Tween 20 and Milli-Q water were used in the preparation of the emulsions. The emulsions, thus formed, were sonicated for 1–3 h using high-energy sonication in a sonicator (Sonics Vibra-cell Ultrasonicator, 130 W and 20 kHz). The prepared neem nanoemulsion was used for the treatment.

15.2.5.2 PREPARATION OF SEAWEED EXTRACT

Enteromorpha intestinalis was collected from Rameswaram and taken to the laboratory for further analysis. The seaweed was shade-dried and powdered and used for study. About 5 g of the dried powder was mixed with 100 ml of ethanol. It was kept in a shaker for 24 h. It was then filtered and the filtrate was taken and evaporated in a rotary evaporator. The extract was used for further analysis.

15.2.6 WELL DIFFUSION METHOD

The agar well diffusion method of Perez and Lewis[18] was followed. A sterile cotton swab was dipped into the broth culture. The cotton swab was then rotated, above the fluid level to remove excess inoculum. The agar surface of the plate was inoculated by swabbing three times, turning the plate between swabbing. The lid of the Petri dish was replaced and the plate was kept at room temperature for 5–10 min to dry the inoculum. Two wells of 8 mm diameter were punched into the agar in each plate using a sterile well cutter. Into each well, 30 μl of the neem nanoemulsion and seaweed extract was added into separate wells. Into the other two wells, Tween 20 and ethanol were added. The solutions were allowed to diffuse for 2 h. The plates were incubated at 30°C for 24–48 h. The antibacterial activity was evaluated by measuring the zone of inhibition around the well.[13,20]

15.2.7 PREPARATION OF DISC

Prepared emulsion (40 µl) and seaweed extract were added to two separate the sterile discs (HiMedia) and dried under sterile condition at room temperature. The prepared discs were used in the study.

15.2.8 COMPARATIVE STUDY

The method of Thomas et al.[20] was followed for the comparative study. The antibiotic sensitivity test was carried out by the method of Bauer et al.[3] The antibiotic disc was obtained from HiMedia. The antibiotic disc used was tetracycline. After making a lawn culture of the bacterium, the antimicrobial susceptibility disc was carefully dropped on to the surface of the Mueller-Hinton agar in the plate using aseptic techniques. The extract-loaded discs were also applied to the lawn culture of the test organism. The discs were placed 24 mm apart from each other to avoid the overlapping of the zone of inhibition. Each disc was pressed down firmly on to the surface of the agar plate using sterile forceps. The plates were incubated at 30°C for 24 h for circular clear area (zone of inhibition) in the bacterial lawn around the antibiotic disc. The diameter of the zone of inhibition was measured. The test was carried out in triplicates.

15.2.9 OPTIMUM EXPOSURE TIME OF EMULSION AND EXTRACTS

The method of Thomas et al.[20] was followed for determination of optimum exposure time of emulsion and extracts. Shrimp were experimentally infected by the bacterium isolated from water using intramuscular method as described in Section 15.2.1. The infected shrimp with clinical signs were collected and reared separately in glass tanks to which 100 mg/l of emulsion, and the extract was added to the water. The shrimp were collected at 12, 24, and 48 h of exposure. They were reared further for 40 days in aquarium tanks containing sterile seawater. The shrimp surviving at the end of 10, 20, and 30 days were counted to determine the optimum exposure time to emulsion required for the shrimp to recover from infection caused by the bacterial isolate. The toxicity of the emulsion was reported in earlier paper.[20]

15.2.10 GROWTH KINETICS

A 25-ml sterile nutrient broth was prepared in conical flasks. The isolate was inoculated into sterile broth. In total, 10 μl, 50 μl, and 100 μl of neem oil and seaweed extracts were added to separate conical flasks and were labeled. Flask without emulsion and seaweed extract served as positive control. All the flasks were incubated at 37°C at 120 rpm for 24 h. During the incubation, the absorbance at 600 nm was read every 2 h to measure the growth of the bacteria. The absorbance obtained at 600 nm was then plotted against time in hours.

15.2.11 IDENTIFICATION OF PHYTOCOMPONENTS IN NEEM OIL NANOEMULSION BY GAS CHROMATOGRAPHY–MASS SPECTROMETRY (GC-MS)

The identification of the chemical composition of neem oil nanoemulsion (*Azadirachta indica*) was performed using a gas chromatography–mass spectrometry (GC-MS) spectrograph (Agilent 6890/Hewlett–Packard 5975) fitted with electron impact mode. The emulsion (2.0 μl) of neem oil was injected with a Hamilton syringe to the GC-MS manually for total ion 125 chromatographic analysis in split mode. In quantitative analysis, selected ion monitoring (SIM) mode was employed during the GC-MS analysis. SIM plot of the ion current resulting from very small mass range with only compounds of the selected mass were detected and plotted.

15.3 RESULTS

The bacterial isolates isolated from water sample were selected and identified based on the colony morphology and biochemical and physiological characteristics. Bluish-green color colonies were observed in TCBS agar. Colonies developed on nutrient agar were Gram-negative straight rods, motile, oxidase and catalase positive, sensitive to the vibriostatic 0/129 at 150 pM, and fermentative. This bacterium was identified as *Vibrio* based on the characteristics analyzed, namely, production of urease; indole fermentation of glucose, mannitol, and cellobiose; and production of acid with gas from glucose. It was further identified that it belongs to the species parahaemolyticus based on the biological, biochemical, morphological, and physiological characters. The sequencing of *V. parahaemolyticus* revealed a homology of 100%. Therefore, it is a *Vibrio* sp. with characteristics of *V. parahaemolyticus*.

The susceptibility of *P. monodon* to *V. parahaemolyticus was* tested by intramuscular injection. The highest concentrations, 45×10^4 and 45×10^5 viable cells of *V. parahaemolyticus* per animal caused 92 and 98% mortality within 120 and 96 h post inoculation, respectively, when the animals were injected intramuscularly. The LD_{50} value of the *V. parahaemolyticus* for intramuscular route was determined at different time intervals. It was found to be 1.25×10^3 and 1.24×10^4 CFU per animal after 72 and 96 h post injection, respectively. The clinical signs included loosening up of the body shell, reddening of the body, black patches on the body surface, and necrotic telson and rostrum.

Well diffusion method was followed to evaluate the antibacterial activity against *V. parahaemolyticus.* The results revealed that the neem nanoemulsion had a zone of inhibition of 30 mm, and seaweed extract showed a zone of inhibition of 32 mm diameter. The extract-loaded disc was compared with the commercial antibiotics. The results revealed that the emulsion-loaded disc had a zone of inhibition of 20 mm and extract-loaded disc showed 23 mm diameter zone of inhibition when compared to the commercial antibiotic ciprofloxacin, which had a zone of inhibition of 30 mm. The infected shrimp showed 95 and 98% survival when injected intramuscularly with neem nanoemulsion and seaweed extract, respectively.

The growth kinetics revealed that the growth of the bacterium was inhibited with the increase of time intervals. The growth of bacterial culture treated with emulsion was found to be inhibited at 24 h and the culture which was treated with seaweed extract was found to be inhibited at 22 h. No inhibition of bacterial growth was observed in the flask, which was not treated with the extracts.

The GC-MS data revealed presence of various active compounds such as (dodecanoic acid methyl ester acetic acid, trifluoro-decyl ester, hexadecanoic acid methyl ester, heptafluorobutyric acid, n-tridecyl ester, 9-octadecenoic acid (Z-, methyl ester), hexadecanoic acid, 15-methyl-, methyl ester and 1,2-benzenedicarboxylic acid, diisooctyl ester, which have antibacterial activity. These compounds might have been responsible for curing the bacterial infection in shrimp.

15.4 DISCUSSION

Vibriosis is one of the most frequent diseases affecting crustaceans, fish, and mollusks. Vibriosis mainly increases mortality rate in postlarvae and in adult of *P. monodon*, and also causes many diseases in *P. monodon* such as reddening of the body, loosening up of the body shell, black patches on the body, necrosis in telson and rostrum region, and so forth; [17] our result agrees with this. An infection may be either localized or septicemic affecting all organs and tissues. Among the most important bacterial species affecting larval cultures are the luminescent *Vibrio harveyi* and *Vibrio splendidus* (Lavilla Pitogo et al., 1995), whereas the nursery and grow-out cultures are infected by *V. parahaemolyticus* and the recently discovered *Vibrio penaeicida*.[11] *Vibrio* is Gram-negative, ubiquitous in marine and estuarine ecosystems as well as in aquaculture farms, and comprise one of the major microbiota of these ecosystems. Many *Vibrios* are serious pathogens for animals reared in aquaculture. Vibriosis, caused by infection by *Vibrio* spp., is one of the most prevalent diseases in fish and other aquaculture-system reared organisms and is widely responsible for mortality in cultured aquaculture systems worldwide.[6] Mass mortalities of tiger shrimp (*P. monodon*) associated with *Vibrio* spp. infections have been observed in hatcheries owing to antibiotic-resistant *V. harveyi* strains in India.

Vibrio, which is a member of the normal bacterial flora of shrimp, induces mass mortalities in affected populations of shrimp. Bacterial diseases caused by members of the genus *Vibrio* such as *V. parahaemolyticus, V. alginolyticus, V. anguillarum, and V. vulnificus* have often been reported among cultured peneid shrimp (Thakur et al., 2003). Our results revealed that the pathogen isolated was confirmed as *V. parahaemolyticus*. This agrees with the report of Thakur et al. (2003).

In most of the commercial shrimp farms, *Vibrio* proliferation has been controlled by prophylaxis and chemotherapy.[9] Recently, the emergence of antibiotic-resistant *Vibrio* strains of shrimp triggered the withdrawal of synthetic antibiotics from aquaculture.[10,13] The lack of effective disease control has turned out to be the cardinal limiting factor against the realization of highly stable shrimp production. As a result, nanoemulsions have been tested.

Nanoemulsions consist of very fine emulsions, with a droplet diameter, smaller than 100 nm. Unlike microemulsions, nanoemulsions are metastable systems, whose structures depend on the process used to prepare them. The decreases in morbidity and mortality from infectious diseases over the last century were attributed mainly to the introduction of antimicrobial agents.

Nowadays, however, the resistance to antibiotics has been reaching a critical level, invalidating major antimicrobial drugs that are currently used. Neem oil, also named *margosa* oil, extracted from the seeds of the neem (*A. indica* A. Juss.), is well known for the wide application of its biological activity. Neem oil has been demonstrated to have anti-parasitic activity, antimicrobial activity, antipyretic activity, anti-inflammatory activity, antibacterial and immunostimulant activity.[20,22] Our results revealed that the neem nanoemulsion had an antibacterial activity, which inhibited the growth of *V. parahaemolyticus*, and that, in addition, the infection in *P. monodon* was completely eliminated. In conclusion, neem nanoemulsion can be used as an alternative for the treatment of *Vibrio* infection in aquaculture. Further studies will be carried out to identify the active compound identified in the GC-MS responsible for treating the infection.

ACKNOWLEDGMENT

The authors acknowledge the financial support of management of VIT University, Vellore for this research work.

KEYWORDS

- ***Penaeus monodon***
- ***Vibrio parahaemolyticus***
- **pathogenicity**
- **growth curve**
- **neem nanoemulsion**

REFERENCES

1. Anjali, C. H.; Sharma, Y.; Mukherjee, A.; Chandrasekaran, N. Neem Oil (*Azadirachta indica*) Nanoemulsion—A Potent Larvicidal Agent Against *Culex quinquefasciatus*. *Pest Manage. Sci.* **2012,** *68*, 158–163.
2. Austin, B.; Austin, D. A. Vibriosis. In *Bacterial Fish Pathogens and Disease in Farmed Wild Fish;* Ellis Harwood Ltd.: England, 1993; pp 263–287.
3. Bauer, A. W.; Kirby, W. M.; Sherris, J. C.; Turck, M. Antibiotic Susceptibility Testing by a Standardized Single Disk Method. *Am. J. Clin. Pathol.* **1966,** *45*(4), 493–496.

4. Bergh, O.; Nilsen, F.; Samuelsen, O. B. Diseases, Prophylaxis and Treatment of the Atlantic Halibut *Hippoglossus hippoglossus*: A Review. *Dis. Aquat. Org.* **2001,** *48*, 57–74.
5. Buchanan, R. E.; Gibbons, N. E. *Bergey's Manual of Determinative Bacteriology*, 8th ed, Williams & Wilkins Co.: Baltimore, 1974.
6. Chaterjee, S.; Haldar, S. Vibrio Related Diseases in Aquaculture and Development of Rapid and Accurate Identification Methods. *J. Marine Sci. Res. Dev.* **2012.** DOI: 10.4172/2155–9910.S1–002.
7. Chen, F. R.; Liu, P. C.; Lee, K. K. Lethal Attribute of Serine Protease Secreted by *Vibrio alginolyticus* Strains in Kurama Prawn *Penaeus japonicus*. *Z. Naturforsch. C* **2000,** *55*, 94–99.
8. Collins, C. H.; Lyne, P. M. *Microbiological Methods*, 4th ed; Butterworths: London, 1976.
9. Gatesoupe, F. J. E.; Arakawa, T.; Watanabe, T. The Effect of Bacterial Additives on the Production Rate and Dietary Value of Rotifers as Food for Japanese Flounder, *Paralichthys olivaceus*. *Aquaculture* **1989,** *83*(2), 39–44.
10. Hameed, A. S. S.; Rahaman, K. H.; Alagan, A.; Yoganandhan, K. Antibiotic Resistance in Bacteria Isolated from Hatchery Reared Larvae and Post-Larvae of *Macrobrachium rosenbergii*. *Aquaculture* **2003,** *217*(3), 39–48.
11. Ishimaru, K.; Akagawa-Matsushita, M.; Muroga, K. *Vibrio penaeicida* sp. nov., a Pathogen of Kuruma Prawns (*Penaeus japonicus*). *Int. J. Syst. Bacteriol.* **1995,** *45*,134–138.
12. Lightner, D. V.; Lewis, D. H. A septicemic bacterial disease syndrome of penaeid shrimp. *Mar. Fish. Rev.* **1975,** *37*(5–6), 25–28.
13. Lightner, D. V.; Redman, R. M.; Poulos, B. T.; Mari, J. L.; Bonami, J. R.; Shariff, M. Distinction of HPV-type viruses in *Penaeus chinensis* and *Macrobrachium rosenbergii* using a DNA probe. *Asian Fish. Sci* **1994,** *7,* 267–272.
14. Lavilla-Pitogo, C. R. Shrimp health research in the Asia-Pacific: present status and future directives, 1996. FAO FISHERIES TECHNICAL PAPER, 41–50.
15. Lavilla-Pitogo, C. R.; Leano, E. M.; Paner, M. G. Mortalities of Pond-Cultured Juvenile Shrimp *Penaeus monodon* Associated with Dominance of Luminescent Vibrios in the Rearing Environment. *Aquaculture* **1998,***164,* 337–349.
16. Oliver, J. D.; Warner, R. A.; Cleland, D. R. Distribution and Ecology of *Vibrio vulnificus* and Other Lactose-Fermenting Marine Vibrios in Coastal Waters of the South Eastern United States. *Appl. Environ. Microbiol.* **1980,** *44*, 1404–1414.
17. Karunasagar, I.; Pai, R.; Malathi, G. R. Mass Mortality of *Penaeus monodon* Larvae due to Antibiotic Resistant *Vibrio harveyi* Infection. *Aquaculture* **1994,** *128*, 203–209.
18. Perez, E. E.; Lewis, E. E. Use of Entomopathogenic Nematodes and Thyme Oil to Suppress Plant-Parasitic Nematodes on English Boxwood. *Plant Disease* **2006,** *90*(4), 471–475.
19. Rahaman, S.; Khan S. N.; Naser, M. N.; Karim, M. M. Isolation of Vibrio spp. from Penaeid Shrimp Hatcheries and Costal water of Cox's Bazar, Bangladesh. *Asian J. Exp. Biol. Sci.* **2010,** *1*, 288–293.
20. Thomas, J.; Jerobin, J.; Jeba Seelan, T. S.; Thanigaivel, S.; Vijayakumar, S.; Mukherjee, A.; Chandrasekaran, N. Studies on Pathogenicity of *Aeromonas salmonicida* in Catfish *Clarias batrachus* and Control Measures by Neem Nanoemulsion. *Aquaculture* **2013,** *71–75*, 396–399.

21. Vandamme, P.; Pot, B.; Gillis, M.; de Vos, P.; Kersters, K. Polyphasic Taxonomy, a Consensus Approach to Bacterial Systematics. *Microbiol. Rev.* **1996,** *60*, 407–438.
22. Van Belkum, A., Struelens, M.; de Visser, A.; Verbrugh, H.; Tibayrenc, M. Role of Genomic Typing in Taxonomy, Evolutionary Genetics, and Microbial Epidemiology. *Clin. Microbiol. Rev.* **2001,** *14*, 547–560.
23. West, P. A.; Colwell, R. R. Identification and Classification of Vibrionaceae Overview. In *Vibrios in the Environment;* Colwell, R. R. Ed.; John Wiley & sons: New York, 1984; pp 285–363.
24. Zhang, Y.-Q.; Xu, J.; Yin, Z.-Q.; Jia, R.-Y.; Lu, Y.; Yang, F.; Du, Y.-H.; Zou, P.; Lv, C.; Hu, T.-X., Liu, S.-L.; Shu, G.; Yi, G. Isolation and Identification of the Antibacterial Active Compound from Petroleum Ether Extract of Neem Oil. *Fitoterapia* **2010,** *81*, 747–750.

CHAPTER 16

REVIEW OF ANTI-INFECTIVE ACTIVITY OF BORIC ACID: A PROMISING THERAPEUTIC APPROACH

SUKHWINDER K. BHULLAR[1,*], MEHTAP OZEKMEKCI[2], and MEHMET COPUR2

[1]*Bursa Technical University, Mechanical Engineering Department, 16310, Bursa, Turkey*

[2]*Bursa Technical University, Chemical Engineering Department, 16310, Bursa, Turkey*

[*]*Corresponding author. E-mail: kaur.bhullar@btu.edu.tr*

CONTENTS

ABSTRACT

Naturally occurring in seawater, boric acid (BA) has been used for food preservation in plants and fruits and for cleaning purposes since ancient Greek times. BA can be found in antiseptics and astringents, enamels and glazes, glass fiber manufacturing, medicated powders, skin lotions, some paints, some rodent and ant pesticides, photography chemicals, powders to kill roaches, and some eyewash products. BA has antiseptic, antifungal, antibacterial, and astringent properties. For these reasons, it finds its way in many home remedies and medicinal prescriptions. The applications of this weak acid are stretched by modern research in human health, pharmaceuticals, and the treatment of many serious diseases due to its antioxidant, antifungal, antiseptic, and anticancer effects. The main focus of our review is to present a literature on the biomedical applications of BA which can be helpful for current researchers to exploit benefits of BA to further promote their candidature for the biomedical arena.

16.1 INTRODUCTION

Acidum boricum, boracic acid, hydrogen borate, and orthoboric acid are some other names known for boric acid (BA) that is a weak monobasic lewis acid of boron. Normal growth, healthy body, and serious situations such as arthritis and osteoporosis are naturally accomplished by boron. Potentials of boron reported in the literature[1–2] include treatment of arthritis by improving efficacy to absorb calcium into the cartilage and bone, reduction in inflammation and allergy, enhance the testosterone levels in male athletes and bodybuilders, minimization of osteoporosis, production of estrogen, embryonic development, cancer therapy, maintenance of cellular and organ membrane functions, prevent blood clotting, reduction in lipid accumulation, and provide protection to the body to counter strokes and heart attacks. However, more studies and research are needed to substantiate and promote candidature of boron and boron-related components for some life-threatening diseases. In addition to medical applications, boron is used in many other areas such as nourishment, pesticides, deterioration constraints, nuclear reactors, buffers, and dye substances[3–4]. Also, boron compounds are vital fertilizers for plants and many other organisms but excess use of them can be toxic[5–6]. Furthermore, boron is implicated in quorum sensing, which is an essential mechanism for antimicrobial activity.[7] Currently, ongoing research on the utilization of boron or boron compounds-containing products for medical applications includes oral drug delivery, cancer therapeutics, anticoagulants, anti-infective, antiviral, antifungal, antituberculous, and antibacterial agents.[8] Boron is an essential nutrient for animals, plants, and fungi if it is used in an appropriate amount, whereas at high concentrations BA has a toxic effect that is widely used to destroy various organisms ranging from bacteria to rodents. Also, BA has a broad antimicrobial activity that makes it a popular treatment agent for the infection in many diseases. Some of the applications of BA are illustrated in Figure 16.1.

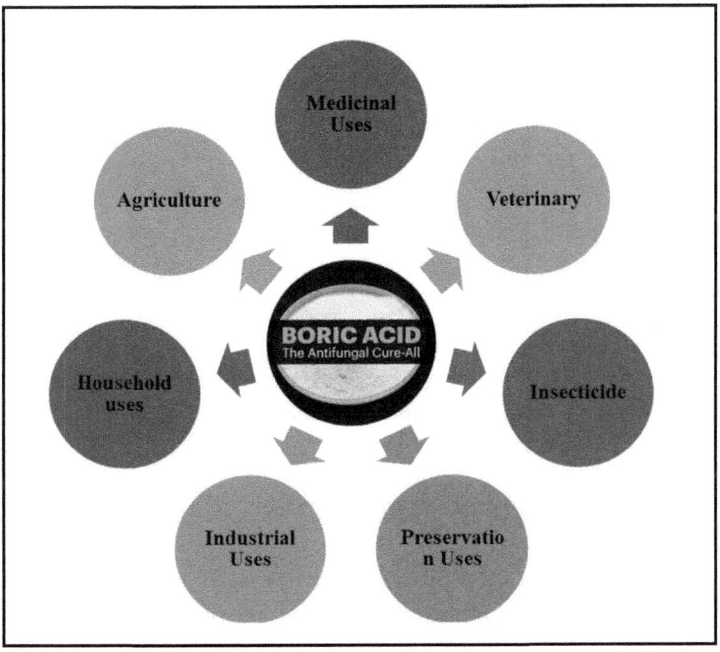

FIGURE 16.1 Applications of boric acid.

16.2 BIOMEDICAL APPLICATIONS

In this review study, our main focus is on one of the compounds of boron—BA, and its applications in the treatment of serious diseases, given in Figure 16.2 to improve human health and related literature, and the details are in next sections.

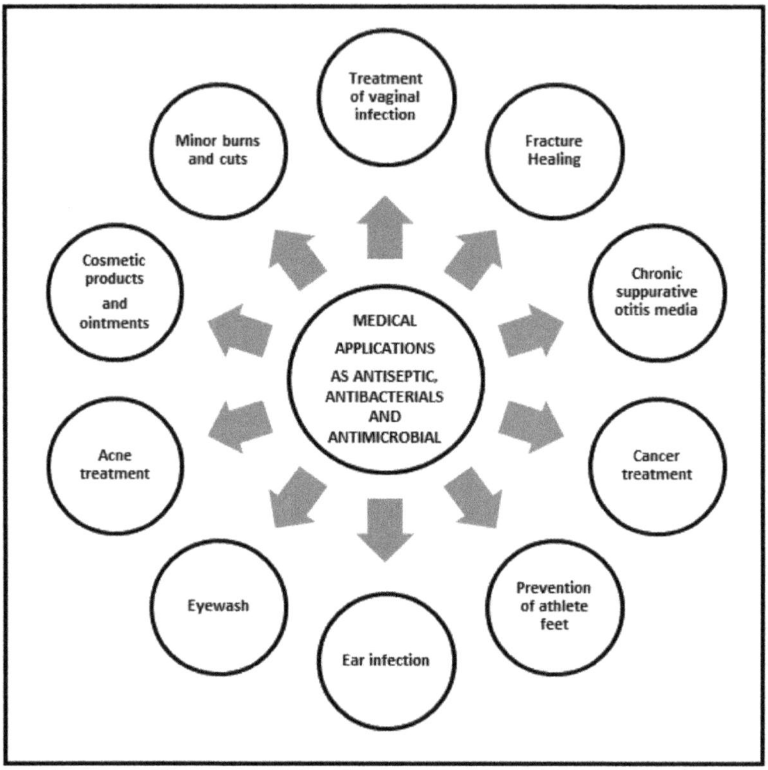

FIGURE 16.2 Applications in the treatment of some diseases.

16.2.1 YEAST/CANDIDA INFECTION

Sometimes the skin, genitals, throat, mouth, and blood can be affected by a fungal infection called candidiasis caused by the overgrowth of a type of yeast called *Candida* which is found in small amounts in the body. However, due to certain health conditions and medications more yeast can grow particularly in the moist and warm areas of the body which can cause discomfort. It can be serious sometimes. An overgrowth of normal yeast cells in the vagina is called a vaginal yeast infection; although it is not usually serious it can sometimes be annoying and itchy. Nowadays, for its treatment vaginal BA capsules are recommended by some experts, except for pregnant females.[9] With a broad antimicrobial activity, BA is used as a substitute and complementary medication for the treatment of vaginal yeast infections. A number of studies reported that compared to the current oral antifungal drugs and drug resistance, BA offers higher efficiency.[10–16]

BA has a broad antimicrobial activity, is cheap, safe, easy to use and is more effective than the over-the-counter or prescribed drugs that make it a popular treatment for vaginal infections. Moreover, BA in the form of vaginal suppositories is a natural remedy for vaginal yeast infections. Through a number of studies, superior effectiveness of BA is proved compared to the topical and oral antifungal drugs, including drug-resistant cases.[17–19] The effect of BA on morphogenesis and cell wall synthesis in yeast, using the well-established model organism *Saccharomyces cerevisiae* was studied by Schmidt et al. 2010.[17] It is discussed by authors that BA disturbs the cytoskeleton at the bud neck and impairs the assembly of the septation apparatus. Due to BA treatment cells form irregular septa and lead to the synthesis of irregular cell wall protuberances that extend far into the cytoplasm.[17] Furthermore, authors investigated that during BA exposure, thick, chitin-rich septa are formed. These septa prevent the separation of cells after abscission and cause the formation of cell chains and clumps. As a response to the BA insult, cells signal cell wall stress through the Slt2p pathway and increase chitin synthesis, apparently to repair cell wall damage. It was also found that BA is toxic if taken internally, but safe when used vaginally, and BA capsules help to relieve external itching when inserted into the vagina.[18] In another clinical study of 100 women with chronic yeast vaginitis, it was demonstrated that BA capsules 98% successfully treated the infection over a period of 2–4 weeks.[19] In some cases, if recurrent yeast infections are a problem, they can be prevented by using one capsule of BA on the vaginal area at bedtime, twice per week, beginning 1 week after menstruation.

16.2.2 CANCER TREATMENT

Boron compounds offer their potential against lung cancer cells, leukemia cells, breast cancer cells, prostate cancer cells, and ovarian cancer cells.[20–21] The tremendous importance of boronic acid promotes its candidature in the synthesis of biologically active compounds and the use of boronic acid itself as a pharmaceutical agent.[22] The incorporation of boron in some boron compounds imparts antitumor properties to different cancer cell lines. Studies and experiments demonstrated that boron compounds work as a proteasome inhibitor in cancer cells in[23] which the proteasome plays a major role in cellular pathways for the breakdown and processing of proteins to peptides and amino acids. Boron is known to be important for animal cell replication and development.[24] BA is one of the most studied boron-containing chemicals and it has been demonstrated to control the proliferation of some cancer cell types. The inhibition of cancer cells by BA involves a diversity of cellular targets such as direct enzymatic inhibition, apoptosis, receptor binding, and mRNA splicing.[25–28] Barranco and Eckhert [26] investigated BA inhibitory effects found in androgen-independent cell lines (DU-145 and PC-3), signifying that other (serine protease-independent) mechanisms could also exist. The authors have studied that BA is an inhibitor of peptidases, proteases, proteasomes, arginase, nitric oxide synthase and transpeptidases.[29–32] Based on the prostate-specific antigen (PSA), the use of BA in breast cancer as an alternative therapeutic modalities chemical therapy of prostate carcinoma has also been studied.[31,33] In another study, researchers discussed that BA inhibits the cell cycle control and proliferation of DU-145, acting against the agonist-stimulated release of Ca^{2+} from ryanodine receptor-sensitive cell stores.[34] It was reviewed[35] that in the case of melanoma cells, BA slows down the proliferation, possibly by inhibiting the second step of pre-mRNA splicing. The authors further explained that a high dose of BA (12.5–50 mM) slows cell replication and induces apoptosis in both melanoma cells and MDA231 breast cancer cells.[27–28] Also, it was investigated that the inhibition of cancer cells by BA involves a diversity of cellular targets, for example, direct enzymatic inhibition, apoptosis, receptor binding, and mRNA splicing. Furthermore, it was recently experimentally studied and demonstrated [36–37] that 1 mM of BA inhibits the ZR-75-1 breast cancer cell line, but not the MCF-7 cell line. In their study, they concluded that BA could become an anticancer agent for breast cancer, and the rise in boron intake could reduce the chance of developing this disease. It has been investigated through research study[36] that BA inhibits (with 64% less likelihood of developing prostate cancer compared to men who consumed the least amount of boron) the growth of LNCaP prostate tumors in nude mice. Two groups (10 animals/group) were dosed with BA solutions (1.7, 9.0 mgB/kg/day) orally. Control group received only water. Tumor sizes were measured weekly for 8 weeks. The size of tumors was decreased in mice exposed to the low and high dose of BA by 38% and 25%, respectively. Prostate-specific antigen (PSA) levels decreased by 88.6% and 86.4%, respectively, as compared to the control group.[38] The effects of BA and sodium tetraborate on an acute leukemia cell line and healthy human lymphocytes have been evaluated in another study.[39] In their study, they observed that BA at a concentration of 500 μm caused double nucleus and micronucleus formation in both HL-60 cells and lymphocytes.[39]

16.2.3 ANTIOXIDATIVE EFFECT

Literature has constantly demonstrated that boron is an important constituent for plants and also advantageous for humans but only in certain applications with limited quantity.[40–41] For example, supplementation of boron has a beneficial effect on the bone mineral density, brain function, cognitive performance, regulation of the normal inflammatory response, and lipid levels in serum. Boron can also be protective against lipid peroxidation, oxidative stress, DNA damage, and prostate cancer by inhibiting the prostate-specific antigen.[42–43] Boron components, particularly BA is the most widely used in the industrial, agricultural sectors and also in medicinal products as a component of pharmaceuticals.[44–45] It was studied that antimicrobial activity of BA helps to strengthen the tissue antioxidant defenses via a yet unidentified mechanism that may involve changes in oxidative metabolism.[46–48] In another study, authors report that BA also displays minimal potential for genotoxicity in bacteria and cultured mammalian cells.[49] The protective effect of BA against oxidative DNA damage has been demonstrated within these common and extreme exposure conditions.[50] It was reported by some authors [51] that the protective roles of boron compounds occurring with the effectiveness of their antioxidant capacity could be useful in the development of functional food and raw materials of medicine. Furthermore, boron's effect on myogenic differentiation and the development of scaffolds for skeletal tissue engineering, and supplements for embryonic muscle growth are discussed by authors.[52] In this study, authors explained that fine dose tuning and treatment period arrangement are highly warranted as boron treatment over the required concentrations and time might result in detrimental outcomes to myogenesis and myo-regeneration. Furthermore, vanadium compounds due to their potential uses in pharmacological and biomedical areas such as an antineoplastic, cholesterol and glucose level blood, diuretic, oxygen-hemoglobin affinity and toxic effects are of interest of several researchers.[53–56] The interaction of boron with vanadium for therapeutic uses is reported by authors[57] through the assessment of blood culture by endpoints and total antioxidant capacity (TAC) results of vanadium (5, 10 and 20 mg/L) and boron compounds (5 and 10 mg/L). The medical and therapeutic usefulness of BA and borax for increasing vanadium dosage was suggested from the obtained results. In another study, the properties of BA with high dietary supplementation were studied for antioxidant activity, lipid peroxidation, vitamin levels, and DNA damages.[58] The enhancement in the antioxidant defense mechanism and vitamin status was demonstrated from the obtained results.[59] Also, the effect of BA administration on fetal alcohol syndrome (FAS) was studied to investigate prenatal alcohol-induced oxidative stress in the cerebral cortex of newborn rat pups.[60] It was concluded from the results that BA could be influential in antioxidant mechanisms against oxidative stress resulting from prenatal alcohol exposure. Furthermore, according to research studies, boron supplementation in animal and human nutrition may have important effects on various metabolic and physiological systems of organisms.[60] The researchers also studied that BA had an antioxidant effect against bleomycin-induced pulmonary fibrosis.[61] It has been recently demonstrated through a number of research studies that BA has protective and antioxidant effects. It could be used as an antioxidant agent against hepatic ischemia/reperfusion injury which is a serious problem, especially in elective and emergency surgery of the liver such as liver transplantations, trauma, or resection.[62–63] It was also found by authors[64] that at low doses boron compounds were useful in supporting antioxidant enzyme activities in human blood cultures. In another study, the effects of BA and 2-aminoethoxydiphenyl borate (2-APB) on oxidative stress and inflammation were studied in an experimental necrotizing enterocolitis (NEC) rat model.[65] It was suggested that boron may be beneficial in preventing NEC. Moreover, the protective effect of BA against liver damage was evaluated by its attenuation of carbon tetrachloride (CCl_4)-induced hepatotoxicity by researchers in.[66] It was demonstrated by the authors that BA exhibits both the increase in antioxidant defense system activity and the inhibition of lipid peroxidation.

16.2.4 OTHER BIOMEDICAL APPLICATIONS

16.2.4.1 CHRONIC SUPPURATIVE OTITIS MEDIA

Chronic suppurative otitis media (CSOM) is a perforated tympanic membrane with persistent drainage from the middle ear (i.e., lasting >6–12 week).[67–68] Nontuberculous mycobacteria (NTM) are an increasingly recognized cause of (CSOM) in children with tympanostomy tubes and treatment of it is difficult and typically requires a combination of systemic antibiotics and surgical debridement. It was reported by authors that CSOM due to NTM can be successfully managed with topical BA powder. The challenges involved in treating this infection and BA as a potentially valuable component of therapy are discussed by Lefebcre et al.[69]

16.2.4.2 FRACTURE HEALING

According to research studies, BA has positive effects on bone tissue and the effects of BA on fracture healing were evaluated through an animal model.[70] It was suggested that BA might be useful in fracture healing. It was recommended by authors that further research is required to demonstrate the most effective local dosage and possible use of coated-BA, as well as whether BA affects fracture healing in the human body.

16.2.4.3 ATHLETE'S FOOT

BA powder can also treat fungal infections such as athlete's foot and toenail fungus. Just a few sprinkles of the BA powder in your socks or stockings or mix BA and rubbing alcohol in the ratio of 2 tsp BA to 1 cup of rubbing alcohol (a drying agent) or water and application of it with cotton swabs (or "Q-tips") can help clear mild infections and ease the itching associated with athlete's foot.[71] It can also neutralize the foot odor from athlete's foot, providing relief from stinky feet. BA is an extremely effective fungicide and often cures athlete's foot in cases where creams have failed.[72]

16.2.4.4 ADULT ACNE

BA's antibacterial properties make it useful in acne treatment simply by diluting with water.[73]

16.3 LIMITATIONS AND SIDE EFFECTS

In the right amounts, boron is an essential nutrient for animals, plants, and fungi. However, at high concentrations BA becomes an effective poison that is widely used for killing diverse organisms ranging from bacteria to rodents.[18] Also, vaginal BA treatment is not recommended for pregnant females.[17] In some cases where recurrent yeast infections have been a problem, they can be prevented by using one capsule of BA on the vaginal area at bedtime twice per week, beginning 1 week after menstruation.[74] Also, BA ointment has been historically used in the treatment of minor skin irritation and inflammation including acne, scratches, and small burns. Concerns about BA ointment mostly surround its toxicity. While a small dose is not usually dangerous, some people are allergic to BA and will experience skin reactions like swelling and redness when ointments containing BA are applied.

16.4 CONCLUDING REMARKS

BA is used in a lot of things on a daily basis. It has applications such as disinfectants, pesticides, wood preservatives, antiseptics, flame retardant, neutron absorbing precursor to other chemical compounds, eyewash, eye and eardrops to destroy any fungus. Also, it helps to promote bone and joint health as it has anti-arthritic properties and is also used for the treatment of acne, athlete's foot, toenail fungus, and yeast infections. Furthermore, BA-containing products are used for medical applications including oral drug delivery, cancer therapeutics, anticoagulants, anti-infective, antiviral, antifungal, antituberculous, and antibacterial agents. Other than the medical and daily basis applications, it is used as a lubricant on ceramic or metal surfaces, pH buffer in pools, and as thermal neutron absorbents to reduce the probability of thermal fission in nuclear power plants. In addition, BA has application in industries including fiberglass manufacturing, glass production for glass in LCD displays, and in the jewelry industry for soldering and annealing operations as well. Furthermore, in the agriculture industry it is extensively used to treat boron deficiencies in plants. Consumption of BA in the right quantity is not harmful to human beings but in high amounts it could lead to some bad side effects.

KEYWORDS

- **boric acid**
- **mRNA splicing**
- **chronic suppurative otitis media**
- **fracture healing**
- **athlete's foot**

REFERENCES

1. Yılmaz, M. T. Minimum Inhibitory and Minimum Bactericidal Concentrations of Boron Compounds Against Several Bacterial Strains. *Turk. J. Med. Sci.* **2012,** *42,* 1423–1429.
2. Proven Benefits of Boron: www.organicfacts.net/health-benefits/minerals/boron (accessed March 27, 2017).
3. Barth, S. Application of Boron Isotopes for Tracing Sources of Anthropogenic Contamination in Ground Water. *Water Res.* **1998,** *32,* 685–690.
4. Hanay, A.; Boncukcuoglu, R.; Kocakerim, M. M; Yilmaz, A. E. Boron Removal from Geothermal Waters by Ion Exchange in a Batch Reactor. *Fresenius Environ. Bull.* **2003,** *12,* 1190–1194.
5. Brown, P. H.; Bellaloui, N.; Wimmer, M. A.; Bassil, E. S.; Ruiz, J.; Hu, H.; Pfeffer, H; Dannel, F.; Römheld, V. Boron in Plant Biology. *Plant Biol.* **2002,** *4,* 205–223.
6. Reid, R. J.; Hayes, J. E.; Post, A.; Stangoulis, J. C. R.; Graham, D. A. Critical Analysis of Boron Toxicity in Plants. *Plant Cell Environ.* **2004,** *25,* 1405.
7. Camgöz, B.; Saç, M. M.; Bolca, M.; Özen, F.; Oruç, Ö. G. Demirel, N. Investigation of Radioactive and Chemical Contents of Thermal Waters Izmir Seferihisar Region Representative (in Turkish with English abstract).. *Ekoloji* **2010,** *19,* 78–87.
8. Chen, X.; Schauder, S.; Potier, N.; Van Dorsselaer, A.; Pelczer, I. Structural Identification of a Bacterial Quorum-Sensing Signal Containing Boron. *Nature* **2002,** *415,* 545–549.
9. Department of Health and Human Services Centers for Disease Control and Prevention. Vulvovaginal Candidiasis Section of Sexually Transmitted Diseases Treatment Guidelines. *MMWR* **2010,** *59,* 61–63. http://www.cdc.gov/std/treatment/2010/default.htm (accessed March 27, 2017).
10. Prutting, S. M.; Cerveny, J. D. Boric Acid Vaginal International Journal of Microbiology Suppositories: A Brief Review. *Infect. Dis. Obstet. Gynecol.* **1998,** *6,* 191–194.
11. Brittingham, A.; Wilson, W. A. The Antimicrobial Effect of Boric Acid on *Trichomonas vaginalis*. *Sex. Transm. Dis.* **2014,** *41,* 718–722.
12. Petrin, D.; Delgaty, K.; Bhatt, R.; Garber, G. Clinical and Microbiological Aspects of *Trichomonas vaginalis*. *Clin. Microbiol. Rev.* **1998,** *11,* 300–317.
13. Schwebke, J. R.; Barrientes, F. J. Prevalence of *Trichomonas vaginalis* Isolates with Resistance to Metronidazole and Tinidazole. *Antimicrob. Agents Chemother.* **2006,** *50,* 4209–4210.
14. Kirkcaldy, R. D.; Augostini, P.; Asbel, L. E.; Bernstein, K. T.; Kerani, R. P.; Mettenbrink, C. J.; Pathela, P.; Schwebke, J. R.; Secor, W. E.; Workowski, K. A.; Davis, D.; Braxton, J.; Weinstock, H. S. *Trichomonas vaginalis* Antimicrobial Drug Resistance in 6 Us Cities, Std Surveillance Network, 2009–2010. *Emerging. Infect. Dis.* **2012,** *18,* 939–943.
15. Helms, D. J.; Mosure, D. J.; Secor, W. E.; Workowski, K. A. Management of *Trichomonas vaginalis* in Women with Suspected Metronidazole Hypersensitivity. *Am. J. Obstet. Gynecol.* **2008,** *198*(4), 370.e1–370.e7.
16. Muzny, C. A. Schwebke, J. R. The Clinical Spectrum of *Trichomonas vaginalis* Infection and Challenges to Management. *Sex. Transm. Infect.* **2013,** *89,* 423–425.
17. Schmidt, M.; Schaumberg, J. Z.; Steen, C. M.; Boyer, M. P. Boric Acid Disturbs Cell Wall Synthesis in *Saccharomyces cerevisiae*. *Int. J. Microbiol.* **2010,** *2010,* 1–10.
18. Vaginitis/Vaginal Infection: Overview. http://www.diagnose-me.com/symptoms-of/vaginitis-vaginal-infection.php (accessed March 27, 2017).

19. Jovanovic, R.; Congema, E.; Nguyen, H. T. Antifungal Agents Vs. Boric Acid for Treating Chronic Mycotic Vulvovaginitis. *J. Reprod. Med.* **1991,** *36,* 593–600.
20. Estey, E. Döhner, H. Acute Myeloid Leukaemia. *Lancet* **2006,** *368,* 1894–1907.
21. Yang, W.; Gao, X.; Wang, B. Boronic Acid Compounds As Potential Pharmaceutical Agents. *Med. Res. Rev.* **2003,** *23,* 346–336.
22. Chapin, R. E. Ku, W. W. The Productive Toxicity of Boric Acid. *Environ. Health Perspect.* **1994,** *102,* 87–91.
23. Bone, R.; Shenvi, A. B.; Kettner, C. A.; Agard, D. A. Serine Protease Mechanism: Structure of an Inhibitory Complex of Alfa-Lytic Protease and a Tightly bound Peptide Boronic Acid. *Biochemistry* **1987,** *26,* 7609–7614.
24. Eckhert, C. D. Boron Stimulates Embryonic Trout Growth. *J. Nutr.* **1998,** *128,* 2488–2493.
25. Barranco, W. T.; Kim, H. T.; Stella, S. L., Jr.; Eckhert, C. D. Boric Acid Inhibits Stored Ca^{2+} Release in Du-145 Prostate Cancer Cells. *Cell Bio. Toxicol.* **2009,** *25,* 309–320.
26. Barranco, W. T.; Eckhert, C. D. Cellular Changes in Boric Acid Treated Du-145 Prostate Cancer Cells. *Br. J. Cancer* **2006,** *94,* 884–890.
27. Acerbo, A. S.; Miller, L. Assessment of the Chemical Changes Induced in Human Melanoma Cells by Boric Acid Treatment Using Infrared Imaging. *Analyst* **2009,** *134,* 1669–1674.
28. Scorei, R.; Ciubar, R.; Ciofrangeanu, C. M.; Mitran, V.; Cimpean, A.; Iordachescu, D. Comparative Effects of Boric Acid and Calcium Fructoborate on Breast Cancer Cells. *Biol. Trace Elem. Res.* **2008,** *122,* 197–205.
29. Bradke, T.; Hall, C.; Stephen, W.; Carper, S. W.; Plopper, G. E. Phenyl Boronic Acid Selectively Inhibits Human Prostate and Breast Cancer Cell Migration and Decreases Viability. *Cell Adhes. Migr.* **2008,** *2,* 153–160.
30. Hunt, C. D. Regulation of Enzymatic Activity: One Possible Role of Dietary Boron in Higher Animals and Humans. *Biol. Trace Elem. Res.* **2008,** *66,* 205–225.
31. Gallardo-Williams, M. T.; Maronpot, R. R.; Wine, R. N.; Brunssen, S. H.; Chapin, R. E. Inhibition of the Enzymatic Activity of Prostate Specific Antigen by Boric Acid and 3-Nitrophenylboronic Acid. *Prostate* **2003,** *54,* 44–49.
32. Scorei, R.; Popa, R. Boron-Containing Compounds as Preventive and Chemotherapeutic Agents for Cancer. *Anticancer Agents Med. Chem.* **2010,** *10,* 346–351.
33. Breast Cancer—Current and Alternative Therapeutic Modalities Chemical Therapy of Prostate Carcinoma. www.intechopen.com (accessed March 27, 2017).
34. Henderson, K.; Stella, S. L., Jr.; Kobylewski, S.; Eckhert, C. D. Receptor Activated Ca^{2+} Release is Inhibited by Boric Acid in Prostate Cancer Cells. *Plos One* **2009,** *4,* 1–10.
35. Shomron, N.; Ast, G. Boric Acid Reversibly Inhibits the Second Step of Pre-mRNA Splicing. *FEBS Lett.* **2003,** *552,* 219–224.
36. Meacham, S.; Karakas S; Walace A; Altun F Boron in Human Health Evidence for Dietary Recommendations and Public Policies. *Open Miner. Process. J.* **2010,** *3,* 36–53.
37. Elegbede, A. F. Mechanism of Boric Acid Analog Cytotoxicity in Breast Cancer Cells. M.S. Thesis, University of Nevada Las Vegas United States, 2007.
38. Gallardo-Williams, M. T.; Chapin, R. E.; King, P. E. Boron Supplementation Inhibits the Growth and Local Expression of Igf-1 in Human Prostate Adenocarcinoma (LNCaP) Tumors in Nude Mice. *Toxicol. Pathol.* **2004,** *32,* 73–78.
39. Canturk, Z.; Tunali, Y.; Korkmaz, S.; Gulbaş, Z. Cytotoxic and Apoptotic Effects of Boron Compounds on Leukemia Cell Line. *CytoTechnology* **2016,** *68,* 87–93.

40. Hunt, C. D. The Biochemical Effects of Physiologic Amounts of Dietary Boron in Animal Nutrition Models. *Environ. Health Perspect.* **1994,** *102,* 35–43.
41. Dinca, L.; Scorei, R. Boron in Human Nutrition and Its Regulations Use. *J. Nutr. Ther.* **2013,** *2,* 22–29.
42. Meacham, S. L.; Taper, L. J.; Volpe, S. L. Effects of Boron Supplementation on Bone Mineral Density and Dietary, Blood, and Urinary Calcium, Phosphorus, Magnesium and Boron in Female Athletes. *Environ. Health Perspect.* **1994,** *102,* 79–82.
43. Singh, N. P.; McCoy, M. T.; Tice, R. R.; Schneder, E. L. A Simple Technique for Quantitation of Low Levels of DNA Damage in Individual Cells. *Exp. Cell Res.* **1988,** *175,* 184–191.
44. Becking, G. C.; Chen, B. H. International Programme on Chemical Safety (Ipcs) Environmental Health Criteria on Boron Human Health Risk Assessment. *Biol. Trace Elem. Res.* **1998,** *66,* 439–452.
45. In Support of Summary Information on the Integrated Risk Information System (IRIS) National Center for Environmental Assessment. http:// www. Epa.gov/iris/ toxreviews/0410tr.pdf (accessed March 27, 2007).
46. De Seta, F.; Schmidt, M.; Vu, B.; Essmann, M.; Larsen, B. Antifungal Mechanisms Supporting Boric Acid Therapy of Candida Vaginitis. *J. Antimicrob. Chemother.* **2009,** *63,* 325–336.
47. Hunt, C. D.; Idso, J. P. Dietary Boron as a Physiological Regulator of the Normal Inflammatory Response: A Review and Current Research Progress. *J. Trace Elem. Exp. Med.* **1999,** *12,* 221–233.
48. Pawa, S.; Ali, S. Boron Ameliorates Fulminant Hepatic Failure by Counteracting the Changes Associated with the Oxidative Stress. *Chem-Biol. Inter.* **2006,** *160,* 89–98.
49. Moore, J. A. Expert Scientific Committee An Assessment of Boric Acid and Borax Using the Iehr Evaluative Process for Assessing Human Developmental and Reproductive Toxicity of Agents. *Reprod. Toxicol.* **1997,** *11,* 123–160.
50. Yılmaz, S.; Ustundag, A.; Ulker, O. C.; Duydu, Y. Protective Effect of Boric Acid on Oxidative DNA Damage in Chinese Hamster Lung Fibroblast V79 Cell Lines. *Cell J.* **2016,** *17,* 748–754.
51. Turkez, H.; Geyikoglu, F.; Tatar, A.; Keles, M. S.; Kaplan, I. The Effects of Some Boron Compounds Against Heavy Metal Toxicity in Human Blood. *Exp. Toxicol. Pathol.* **2011,** *64,* 93–101.
52. Apdik, H.; Doğan, A.; Demirci, S.; Aydın, S.; Şahin, F. Dose-Dependent Effect of Boric Acid on Myogenic Differentiation of Human Adipose-Derived Stem Cells (hADSCs). *Biol. Trace Elem. Res.* **2015,** *165,* 123–130.
53. Aragon, M. A.; Ayala, M. E.; Fortoul, T. I.; Bizzaro, P.; Altamirano-Lozano, M. Vanadium Induced Ultrastructural Changes and Apoptosis in Male Germ Cells. *Reprod. Toxicol.* **2005,** *20,* 127–134.
54. Rodriguez-Mercado, J.; Roldan-Reyes, E.; Altamirano-Lozano, M. Genotoxic Effects of Vanadium(Iv) in Human Peripheral Blood Cells. *Toxicol. Lett.* **2003,** *144,* 359–369.
55. Sakurai, H. Therapeutic Potential of Vanadium in Treating Diabetes Mellitus. *Clin. Calcium* **2005,** *15,* 49–57.
56. Barreto, R.; Kawakita, S.; Tsuchiya, J.; Mineli, E.; Pavasuthipaisit, K.; Helmy, A.; Marotta, F. Metal-Induced Oxidative Damage in Cultured Hepatocytes and Hepatic Lysosomal Fraction: Beneficial Effect of a Curcumin/Absinthium Compound. *Chin. J. Dig. Dis.* **2005,** *6,* 31–36.

57. Geyikoğlu, F.; Turkez, H. Boron Compounds Reduce Vanadium Tetraoxide Genotoxicity in Human Lymphocytes. *Environ. Toxicol. Pharm.* **2008,** *26,* 342–347.
58. Ince, S.; Kucukkurt, I.; Cigerci, I. H.; Fidan, A. F.; Eryavuz, A. The Effects of Dietary Boric Acid and Borax Supplementation on Lipid Peroxidation, Antioxidant Activity and DNA Damage in Rats. *J. Trace Elem. Med. Bio.* **2010,** *24,* 161–164.
59. Sogut, I.; Oglakcı, A.; Kartkaya, K.; Ol, K. K.; Sogut, M. S.; Kanbak, G.; Inal, M. E. Effect of Boric Acid an Oxidative Stress in Rats with Fetal Alcohol Syndrome. *Exp. Ther. Med.* **2015,** *9,* 1023–1027.
60. Ernster, L.; Dallner, G. Biochemical, Physiological and Medical Aspects of Ubiquinone Function. *Biochim. Biophys. Acta* **1995,** *1271,* 95–204.
61. Çelikezen, F. Ç.; Oto, G.; Özdemir, H.; Kömüroglu, U.; Yörük, I.; Demir, H.; Yeltekin, A. Ç. The Antioxidant Effect of Boric Acid and CoQ10 on Pulmonary Fibrosis in Bleomycin Induced Rats. *J. Sci. Tech.* **2012,** *2,* 27–31.
62. Gao, Z.; Li, Y. H. Antioxidant Stress and Anti-Inflammation of PPARα on Warm Hepatic Ischemia-Reperfusion Injury. *PPAR Res.* **2012,** *2012,* 1–8.
63. Basbuğ, M.; Yıldar, M.; Yaman, I.; Özkan, Ö. F.; Aksit, H.; Cavdar, F.; Sunay, F. B.; Ozyigit, M. O.; Derici, H. Effects of Boric Acid in an Experimental Rat Model of Hepatic Ischemia Reperfusion Injury. *Acta Med. Mediterr.* **2015,** *31,* 1067–1073.
64. Türkez, H.; Geyikoglu, F.; Tatar, A.; Keleş, S.; Özkan, A. Effects of Some Boron Compounds on Peripheral Human Blood. *Z. Naturforsch. C. Biosci.* **2007,** *62,* 889–896.
65. Yazıcı, S.; Akşit, H.; Korkut, O.; Sunay, B.; Çelik, T. Effects of Boric Acid and 2-Aminoethoxydiphenyl Borate on Necrotizing Enterocolitis. *Gastroenterol* **2014,** *58,* 61–67.
66. Ince, S.; Keles, H.; Erdogan, M.; Hazman, O.; Kucukkurt, I. Protective Effect of Boric Acid Against Carbon Tetrachloride-Induced Hepatotoxicity in Mice. *Drug Chem. Toxicol.* **2012,** *35,* 285–292.
67. Matsuda, Y.; Kurita, T.; Ueda, Y.; Ito, S.; Nakashima, T. Effect of Tympanic Membrane Perforation on Middle-Ear Sound Transmission. *J. Laryngol. Otol. Suppl.* **2009,** *31,* 81–90.
68. Wright, D.; Safranek, S. Treatment of Otitis Media with Perforated Tympanic Membrane. *Am. Fam. Physician* **2009,** *79,* 650–654.
69. Lefebvre, M. A.; Quach, C.; Daniel, S. J. Chronic Suppurative Otitis Media Due to Nontuberculous Mycobacteria: a Case of Successful Treatment with Topical Boric Acid. *Int. J. Pediatr. Otorhinolaryngol.* **2015,** *79,* 1158–1160.
70. Gölge, U. H.; Kaymaz, B.; Arpacı, R.; Kömürcü, E.; Göksel, F.; Güven, M.; Güzel, Y.; Cevizci, S. Effects of Boric Acid on Fracture Healing: An Experimental Study. *Biol. Trace Elem. Res.* **2015,** *167,* 264–271.
71. Athlete' Foot. http://www.diagnose-me.com/symptoms-of/athletes-foot.php.
72. Boric Acid: The Antifungal Cure-All. https://draxe.com/boric-acid.
73. Sayin, Z.; Ucan, U. S.; Sakmanoglu, A. Antibacterial and Antibiofilm Effects of Boron on Different Bacteria. *Biol. Trace Elem. Res.* **2016,** *173,* 241–247.
74. What is Boric Acid Ointment? http://www.wisegeekhealth.com/what-is-boric-acid-ointment.htm.

CHAPTER 17

CONTROL OF MAGNETISM BY VOLTAGE IN MULTIFERROICS: THEORY AND PROSPECTS

ANN ROSE ABRAHAM[1], SABU THOMAS[2], and NANDAKUMAR KALARIKKAL[1,2,*]

[1]*School of Pure and Applied Physics, Mahatma Gandhi University, Kottayam, Kerala 686560, India*

[2]*International and Inter University Centre for Nanoscience and Nanotechnology, Mahatma Gandhi University, Kottayam, Kerala 686560, India*

[*]*Corresponding author. E-mail: nkkalarikkal@mgu.ac.in*

CONTENTS

ABSTRACT

The renaissance of multiferroics has fashioned a revolution in the *next-generation emerging* memory market. This chapter gives a concise introduction to the fundamentals of multiferroics and artificial multiferroics heterostructures and also illustrates the recent advances in multiferroic heterostructures. It provides new insights into the magnetoelectric coupling in ferromagnetic–ferroelectric composites, multiferroic core–shell nanostructures, and so forth. The novel methodologies followed for the synthesis of multiferroic core–shell structures and the characterization techniques adopted are illustrated with the help of some relevant examples. Finally, the perspectives and possible breakthroughs for future multiferroics-based materials are also briefly discussed.

17.1 INTRODUCTION

The state-of-the-art technologies for the development of materials that enable low-power-consuming high-performance chips *have contributed to the wide research interest on multiferroics.* Multiferroics have been subject of intensified curiosity due to its remarkable functionality of voltage control of magnetism. Multiferroics are the materials in which several ferroic order parameters that reflect the ordering of spin, charge, orbital, and lattice degree of freedom, such as ferromagnetism (FM), ferroelectricity (FE), ferroelasticity, or ferrotoroidicity, simultaneously emerge and are coupled. Among multiferroics in which FE and magnetism coexist is of wide interest due to its promising applications in novel low-power-consuming spintronic devices where spin of electron can be controlled electrically. The cross-coupling effects between electric and magnetic order parameters in multiferroics contribute to the electric field control of magnetism that is crucial for the development of emerging low-power magnetoelectric (ME) devices.[16]

The electric field control of magnetism has turned out to be a very vibrant focus of research in the field of solid-state physics[14] and driven by the production and synthesis of novel materials and artificial hybrid multiferroic heterostructures. The prospect of manipulating the magnetic or electric structure by means of either electric or magnetic fields is made possible by the renaissance of multiferroics.[5] Great success has been made in this field by the continued efforts of the researchers to develop hybrid multiferroic structures that exhibit a significant value of ME coupling at room temperatures (RTs). Manipulated hybrid structures with desired material properties are obtained through various strategies such as functionalization, interfacing layers, epitaxial growth, and other techniques. The proximity effects and interfacial effects between magnetic and ferroelectric materials are key factors to be well thought-out while designing and developing multiferroics with strong ME coupling between phases.[14]

17.2 BRIEF THEORY

17.2.1 PRIMARY FERROIC ORDERS

Ferroic materials display a spontaneous magnetization (M_i), electric polarization (P_i), or strain (ε_{ij}) that can be switched from one orientation to another by the conjugate electric (E_i), magnetic (H_i), and stress (σ_{ij}) fields, respectively.[21]

17.2.1.1 FERROMAGNETISM (FM)

Charge and spin are two attributes of an electron. Magnetism originates from the spin of electron. The quantum mechanical phenomenon of exchange[32] gives rise to a stable and switchable magnetization, exhibited by a ferromagnetic crystal. The difference in energy of two electrons in a system with antiparallel and parallel spins is called the exchange energy, E_{ex}.

$S_i{\cdot}S_j$ is the spin angular momentum and J is a numerical quantity called exchange integral. If there is an exchange coupling between the magnetic moments of neighboring atoms, a magnetic order on a macroscopic scale may be formed at low temperatures. If the sign of the coupling is positive (J_{ex} is positive), E_{ex} is a minimum when the magnetic moments are aligned parallel to each other (i.e., FM) and if J_{ex} is negative, E_{ex} is minimum when the spins are antiparallel, leading to antiferromagnetism. The critical temperature of a ferromagnetic material is called the Curie temperature (T_c) and that of an AFM material, the Néel temperature (T_N). The critical temperature at which this magnetic order is lost is higher if the coupling is stronger.

17.2.1.2 FERROELECTRICITY (FE)

A ferroelectric crystal exhibits reversible spontaneous polarization. Barium titanate (BTO, $BaTiO_3$), lead titanate (PTO, $PbTiO_3$), and other common ferroelectrics such as potassium niobate ($KNbO_3$), lead lanthanum zirconate titanate (PLZT, $PbLaZrTiO_3$), lead zirconate titanate (PZT, $PbZrTiO_3$) enjoy the perovskite structure. The perovskite structure can be considered as a cubic close-packed arrangement of large A and O ions, and the smaller B ions at the octahedral interstitial positions. Barium titanate (BTO) is one of the most extensively studied ferroelectric materials as it exhibits ferroelectric properties at and above RT (~ 125°C) and its chemical and mechanical stability. In the ferroelectric phase, modifications to cation and anion positions give rise to relative displacements of ions inside the unit cell, resulting in reversible spontaneous dipole moments. The moment which develops polarization, $P = q\frac{d}{v}$, where q is the electric charge on the displaced ion, d is the relative displacement, and V is the volume of the unit cell. This moment is related to the electric displacement as,

$$D = \varepsilon_0 E + P$$

where ε_0 and ε are the free space and relative susceptibilities or permittivities, respectively.

17.2.2 CHALLENGE OF COUPLING MAGNETISM AND FE

There exists a contradiction between the mechanism for FM which requires d-electrons while FE occurs only in the absence of d-electrons.[40] This is the underlying reason for the scarcity of ferromagnetic ferroelectrics. This is referred to as the "d^0 versus d^n problem." Multiferroism is mainly achieved in magnetic perovskite-structure oxides by the stereochemical activity of the lone pair on the large (A-site) cation that provides the FE, while the small (B-site) cation remains magnetic. This is the mechanism for FE in the Bi-based magnetic ferroelectrics such as bismuth ferrite, $BiFeO_3$.

17.3 MAGNETOELECTRIC (ME) EFFECT

Magnetism and FE associated with local spins and off-center structural distortions, respectively, coexist in rare materials, termed multiferroics.[25] In a solid, electric field (E) induces electric polarization (P) and magnetic field (H) induces magnetization (M). While the dynamic relations between E and H can be described by Maxwell equations, those between P and M are a highly nontrivial issue. The coupling between P and M should be mediated by electrons on the crystalline lattice, namely electron spin, orbital, and charge degrees of freedom in a solid are all relevant. The P–M coupling, if any, enables the ME effect, that is alternation of M (or in more general magnetic structure) by application of E and conversely change of P by H. The nontrivial spin–lattice coupling[24] in multiferroics has been manifested by the ME effects, polarization change through field-induced phase transition, magneto-dielectric effect and dielectric anomalies that arise at magnetic transition temperatures. Arguably, ME effects have been studied the most in multiferroics such as perovskite-type $BiFeO_3$ or $BiMnO_3$, the boracites family and the families of $BaMF_4$ (M, divalent transition metal ion), hexagonal $RMnO_3$ (R, rare earths), and the rare-earth molybdates.[24]

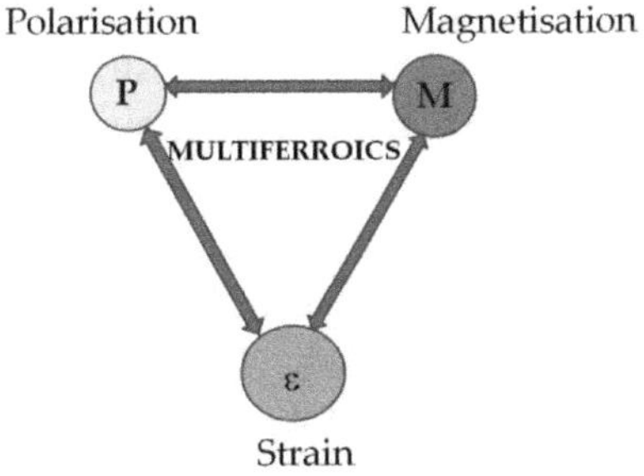

FIGURE 17.1 *P–M* coupling that enables the magnetoelectric (ME) effect.

17.3.1 THEORY OF ME EFFECT

According to Landau theory, the ME effect in a single-phase multiferroic is described by the free energy F of the system in terms of an applied magnetic field H whose ith component is denoted as H_i, and an applied electric field E whose ith component is denoted as E_i. Based on general thermodynamic formalism, the Landau free energy function F, under the Einstein summation convention in SI units, is expressed in the following way:

$$F = U - TS - E.P - M.H$$

where U is the internal energy per unit volume, T the temperature, S the entropy per unit volume, E the electric field, H the magnetic field, P the polarization, M the magnetization.

$$F(E,H) = F_0 - P_i^s E_i - M_i^s H_i - \frac{1}{2}\varepsilon_0\, \varepsilon_{ij} E_i E_j - \frac{1}{2}\mu_0 \mu_{ij} H_i H_j$$

$$- \alpha_{ij} E_i H_j + \frac{\beta_{ijk}}{2} E_i H_j H_k + \frac{\Upsilon_{ijk}}{2} H_i E_j E_k$$

with E and H as the electric field and magnetic field, respectively. Differentiating the free energy equation with respect to electric and magnetic fields, respectively, leads to polarization (μC/cm^2) and magnetization (μ_B, where μ_B is the Bohr magneton),[19] which are as follows:

$$P_i = -\frac{\partial F(E,H)}{\partial E_i} = P_i^s + \varepsilon_0\, \varepsilon_{ij} E_j + \alpha_{ij} H_j + \frac{\beta_{ijk}}{2} H_j H_k + \ldots$$

$$M_i = -\frac{\partial F(E,H)}{\partial H_i} = M_i^s + \mu_0\, \mu_{ij} H_j + \alpha_{ij} E_j + \frac{\beta_{ijk}}{2} E_i H_j + \ldots$$

where P^S and M^S denote the spontaneous polarization and magnetization, whereas ε and μ are the electric and magnetic susceptibilities. The tensor α corresponds to induction of polarization by a magnetic field or magnetization by an electric field, which is designated as the linear ME effect. It is supplemented by higher-order ME effects such as those parameterized by the tensors β and γ. Normally, the prefix "linear" is omitted and the linear manifestation is referred to as the "ME effect." Magnetic field control of electric polarization can be described by the dynamic ME coefficient: $\alpha = \partial P/H$ or $\alpha_E = nE/{=}H$, where P, E, and H are the electric polarization, electric field, and magnetic field, respectively, and α_E is most often used.

17.3.2 TYPES OF MULTIFERROICS

Depending on the origin of the polarization of the material, multiferroics are classified into two families: Type-I and Type-II multiferroics[26]. In Type-I multiferroics, magnetism and FE exist independently, while in Type-II multiferroics, FE is due to certain type of magnetic ordering.[26] Type-II multiferroics are a group of magnetically induced ferroelectrics.

17.3.2.1 TYPE-I MULTIFERROICS

Type-I multiferroics are those in which FE arise from various probable sources. In Type-I multiferroics, FE and FM have different sources, and hence, FE and FM occur at different temperatures. The FE ordering temperature is usually higher than the magnetic ordering temperature in these Type-I materials. Type-1 multiferroics, where the magnetism and the FE have different origins, often present high polarizations and high critical temperatures.[51] They are regular ferroelectrics which can also be ferromagnetic. FE and magnetism exist independent of each other still a small degree of coupling exists and coupling between them leads to development of single-phase ME multiferroic materials. Bismuth manganite ($BiMnO_3$), hexagonal manganites, strong ferroelectric frustrated $LuFe_2O_4$ are all examples, bismuth ferrite ($BiFeO_3$) being the leading Type-I multiferroic.[9] Bismuth ferrite (BFO) nanoparticles exhibit attractive magnetic properties and nonlinear optical response which opens an avenue of applications including bio-applications.[49]

17.3.2.1.1 Lone Pair Multiferroics (Bismuth-Based Compounds)

In lone pair multiferroics, the magnetic and ferroelectric orderings are associated with two chemically different cations, with not essentially strong coupling.[3] The stereochemical activity of the lone pair on the large (A-site) cation provides the FE, while keeping the small (B-site) cation magnetic.[39] BFO is perchance the single material that enjoys both magnetic and a strong ferroelectric ordering at RT. It takes up the perovskite, but not the ferrite structure. AFM and ferroelectric ordering coexists in BFO.[3] The ordering of lone pairs of electrons that do not participate in chemical bonds contributes to the polarization in the lone pair ferroelectrics. For example, the ordering of Bi^{3+} ions causes FE in bismuth ferrite ($BiFeO_3$/BFO) and $BiMnO_3$. The magnetism in BFO is associated with the 3d electrons of the Fe^{3+} cations. Bismuth manganite ($BiMnO_3$) is an interesting multiferroic material that exhibits both ferromagnetic and ferroelectric properties making it attractive for various technological applications.[8] $BiMnO_3$ exhibits a ferromagnetic transition temperature (T_C) at around $-168.15°C$ and a ferroelectric transition temperature (T_E) at 476.85°C. $BiFeO_3$ is the principal Type-I multiferroic.[9,35]

Bi^{3+} has the foremost role in the origin of FE, in $BiFeO_3$ and $BiMnO_3$. Two outer 6s electrons in these ions, called as lone pairs, or dangling bonds, do not participate in chemical bonds. They meet the requirement for high polarizability of FE. FE in these systems originates from the ordering of the lone pairs. The ordering of lone pairs (of Bi^{3+} ions) contributes to the polarization. In BFO, the magnetism is due to the 3d electrons of the Fe^{3+} cations. The ME coupling in BFO is related to the presence of a magnetic cycloid structure that allows a coupling of M and P on an atomic level even in AFM structures, where on average the linear ME effect is forbidden.

17.3.2.1.2 Charge-Ordered Multiferroics

The simultaneous existence of inequivalent sites with different charges, and inequivalent (short and long) bonds cause FE, in charge-ordered systems.[9] The most capable class of correlations-induced ferroelectrics is the charge-order-induced ferroelectrics. The competing charge interactions or charge which leads to a charge instability that carries ferroelectric polarization, in these materials. Perovskite manganites of the Type $Pr_{1-x}Ca_xMnO_3$, and $LuFe_2O_4$ of double-layered structure, with a triangular iron lattice within each layer (entirely different structure from that of spinel structure) are examples.

The structure of $LuFe_2O_4$ is entirely different from that of spinel structure, even though its chemical formula resembles that of a spinel. It possesses a double-layered structure, with a triangular iron lattice within each layer. The competing charge interactions arise from an intermediate valence iron ion which is arranged in bilayers of triangular lattices. The lowest energy state of this strongly correlated charge lattice is a charge-ordered state which carries ferroelectric polarization. The charge–charge interactions drive the charge order, and hence, the ordering temperature and ferroelectric polarization are high. At low temperatures, magnetic ordering is achieved due to the magnetic degrees of freedom linked with the charge-ordered iron ions, and it increases the ferroelectric polarization.

17.3.2.1.3 Geometric Multiferroics

Geometric multiferroics are those in which different sub-lattices or parts of the lattice are accountable for the ferroelectric and magnetic orders. Spin glasses and geometrically frustrated magnets are the two classes where frustration effects are observed. The first material identified as a geometric ferroelectric is the AFM ferroelectric hexagonal manganite, $YMnO_3$. FE in $YMnO_3$ arises from the rotation of MnO_5 bipyramid which separates two-dimensional triangular planes of Y ions. In $YMnO_3$, the tilting or rotation of rigid MnO_5 block causes FE (and spinning of Mn at the center at lower temperatures causes multiferroicity). The Y–O bonds form dipoles because of the tilting, and there come into view two "down" dipoles per one "up" dipole, and hence, the system becomes ferroelectric.

17.3.2.2 TYPE-II MULTIFERROICS (OR MAGNETICALLY DRIVEN MULTIFERROICS)

Type-II multiferroics are remarkable because of the interesting physics involved in them. Mato et al. have defined "Type-II multiferroics" as those magnetically ordered phases in which FE arises as a magnet-induced effect.[37] In Type-II multiferroics, the electric polarization is induced by the magnetic long-range order. In Type-II multiferroics or magnetic multiferroics,[51] magnetism induces FE, rather than a non-centrosymmetric crystal structure. Both orders are strongly coupled in these, and hence, they are more interesting and important. The drawbacks associated with Type-II multiferroics are that they have low critical temperatures (T_c) and low polarization (P) values (~ 10^{-2} μC/cm^2). The ME coupling is very strong because FE is driven by magnetic order and do not exist without the latter. Hence, to overcome the limitations, significant effort is focused on the design of new magnetic multiferroics. The complex manganites, RMn_2O_5 are the foremost illustration of Type-II multiferroics. The collinear magnetic structured manganite Ca_3CoMnO_6 that has a relatively high polarization (~400 μC/m^2) is another example.[9]

Type-II multiferroics are subdivided into two classes. One in which FE is induced by magnetic spiral structures and the second one in which FE is induced by collinear magnetic structures. FE and magnetism are strongly coupled in Type-II multiferroics. In these, one of the orders exists as a consequence of the other. FE is completely due to magnetism in Type-II multiferroics. FE arises entirely due to magnetism and occurs in the magnetically ordered state only. FE occurs at the same temperature as that of magnetic ordering and is induced by it. Spiral magnetic ordering, for example, can give rise to Type-II multiferroicity.

17.3.2.2.1 Spiral Type-II Multiferroics

In these Type-II multiferroics, the FE is due to noncollinear (spiral) magnetic structure. The mechanism of ME coupling in spiral multiferroics is spin–orbit coupling. The driving mechanism of the polarization in these is an inverse Dzyaloshinskii effect, which operates in systems with noncollinear, usually spiral magnetic structures. It requires the direct action of the relativistic spin–orbit interaction. Type-II multiferroicity works in systems such as $TbMnO_3$, $Ni_3V_2O_8$, and $MnWO_4$, in multiferroic pyroxenes and in some other systems. Below T_N = 28 K, in $TbMnO_3$, the Mn spins order such that a cycloid is swept out by the tip of the spins. FE in these materials arises from a magnetic spiral structure that breaks inversion symmetry and creates a polar axis. Magnetic order and FE are thus directly coupled, since there occurs no ferroelectric polarization without magnetic order. Altering the symmetry of the magnetic order can lead to switching or suppression of FE. The maximum ferroelectric polarization observed in spin–spiral ferroelectrics is very small, of the order of 0.2 μC/cm^2, probably due to the small size of the ME coupling that is mediated by spin–orbit interactions.[28]

According to Katsura, Balatsky, Nagaosa, and Mostovoy, in a cycloidal spiral, the polarization, P[27] is given by:

$$P \sim r_{ij} \times [S_i \times S_j]$$

$$P \sim [Q \times e]$$

where, r_{ij} is the vector linking the adjacent spins S_i and S_j, Q is the wave vector that describes the spiral, and $e \sim [S_i \times S_j]$ is the spin rotation axis.

The mechanism of this polarization is linked with the spin–orbit interaction. Some of the electric polarizations arise from the so-called spin-current effect also, where the exchange of electrons from a noncollinear structure leads to electric polarization. In insulators, magnetic frustration leads to spiral magnetic ordering, and hence, these Type-II multiferroics are generally observed in frustrated systems.[27] The maximum ferroelectric polarization observed in spin–spiral ferroelectrics is very small, of the order of 0.2 μC/cm^2, which can be due to the small ME coupling that is mediated by spin–orbit interactions.

17.3.2.2.2 Collinear Type-II Multiferroics

The FE appears in collinear magnetic structures, in collinear Type-II multiferroics. All magnetic moments are aligned along a particular axis. In these collinear spin ferroelectrics, FE is mediated by symmetric exchange strain, which does not depend on the presence of the spin–orbit interaction and is of the order of the dominant exchange interactions in a material. The magnetic coupling changes with the atomic positions and hence polarization appears in these materials as a result of exchange striction. These materials display ferroelectric polarization, larger than any previously observed ferroelectric polarization, in a magnetically induced ferroelectric. The ferroelectric polarization of the order of 2–6 μC/cm^2 is predicted for these magnetically induced ferroelectrics with the commensurate collinear magnetic order.[28] Ca_3CoMnO_6, orthorhombic $HoMnO_3$, $TmMnO_3$, and rare-earth-nickelates $ReNiO_3$ in their charge-ordered state are examples.

The mechanism of creation of polarization is based on magnetostriction (MS). For the MS to show multiferroic behavior, the presence of inequivalent magnetic ions with different charges is required. These, in turn, may be either just different transition metal (TM) ions or the same element in different valence states.

17.3.3 SIZE-DEPENDENT MULTIFERROIC PROPERTIES

Multiferroic properties of nanocrystalline $BaTiO_3$ was reported by Mangalam et al.[31] In $BaTiO_3$, a typical ferroelectric material, FE decreases as particle size reduces. However, at the same time, FM cannot occur in its bulk form. By making use of experiment and first principles simulations, C. N. R. Rao et al. have illustrated that multiferroic nature emerges in intermediate-size nanocrystalline $BaTiO_3$. FM arises from the oxygen vacancies at the surface and FE arises from the core. The observed magnetocapacitance effect at RT, formed as a consequence of robust coupling that occurs between a surface polar phonon and spin, confirms the multiferroic nature of BTO. The multiferroic nature of nanocrystalline $BaTiO_3$ reveals new prospects of novel electro-magneto-mechanical devices.

The multiferroic nature of BTO is rendered possible by the surface magnetism of the nanocrystalline BTO. C. N. R Rao et al. have reported the unforeseen surface magnetism and the coupling of ferroelectric and magnetic properties in nanocrystalline BTO. Interestingly, the ferroelectric and magnetic properties are coupled as shown by the observation of magnetocapacitance.

17.3.4 MULTIFERROIC DOMAIN WALLS

A multiferroic system[43] is also fragmented into domains,[20] just like any other ferroic material. The boundaries among adjacent domains or the regions of transition are called "domain walls" or "domain boundaries."[44] Electric and magnetic domain walls exist in multiferroic materials.[34] The multiferroic domains are characterized by an assembly of at least two order parameters. Engineering of domain walls open new prospectives for developing multiferroics with enhanced functionalities.[58] FE and FM domains can be clearly distinguished in intrinsic or artificial multiferroics FM on FE and vice versa.[33]

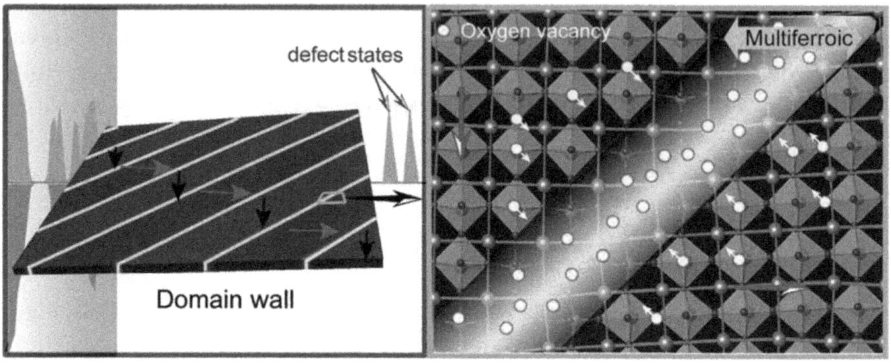

FIGURE 17.2 (See color insert.) Multiferroic domain walls.
Source: Reprinted (adapted) with permission from ref 58. © 2016 American Chemical Society.

17.4 SINGLE-PHASE MULTIFERROICS

Single-phase multiferroics such as $BiFeO_3$, $BaTiO_3$, $CdCr_2S_4$, $LuFe_2O_4$, $RMnO_3$, and RMn_2O_5 (R = rare earth, Y, and Bi) enthusiastically show signs of strong ferroelectric and ferromagnetic ordering and thus display ME-coupled MF (MECMF) property. MECMF has enormous applications in the budding field of multistate logic, nonvolatile memory storage, spintronics, spin valves, air-conditioning filters, and sensors. The problem associated with single-phase MFs is that the magnetic and electric ordering occurs at different temperatures, and hence, the coupling between ordered parameters is rather feeble/absent.[52] They display a high ferroelectric transition temperature and a low magnetic transition temperature. ME coupling interactions demonstrated by Type-I single-phase multiferroic materials are considerably weak and they also display ME coupling at low temperatures with a small value of polarization. Until now, at any RT, single-phase multiferroic possessing large, strong electric and magnetic polarizations has never been recognized.[24] These limitations confine the use of single-phase multiferroics and have encouraged the investigations and research on hybrid multiferroic structures.

$BaTiO_3$ (BTO) is acknowledged as a lead-free perovskite material with RT ferroelectric property for more than six decades. The ferroelectric property of BTO is due to the off-centering of cation at the B site (Ti^{4+}) with respect to the centrosymmetric cubic perovskite. BTO with tetragonal crystal system is strongly ferroelectric when compared to hexagonal system. $BaTiO_3$ (BTO) is an outstanding ferroelectric crystal ABO_3-type perovskite structure and the bandgap of about 3.8 eV and the ferroelectric-to-paraelectric transition temperature above 400 K.

The single-phase ME multiferroic[52] that shows the greatest potential and which is widely explored is $BiFeO_3$ (BFO). Bismuth ferrite ($BiFeO_3$, BFO)[11] is an excellent multiferroic material that enjoys high ferroelectric (Curie temperature, T_C~1103 K) and AFM (Néel temperature, T_N~643 K) transition temperatures. At high temperatures, it displays both ferroelectric and AFM ordering. However, high leakage current makes it unrealistic for practical applications.[52] $BiFeO_3$ is as a unique multiferroic material at RT.[54] The ME coupling constant of $BiFeO_3$ is greater than those of characteristic multiferroics such as $TbMnO_3$ and it indicates the strong coupling between the magnetic and dielectric properties in $BiFeO_3$ (BFO). BFO enjoys G-type AFM order related to spins of iron Fe^{3+} below T_N=643 K along with superimposed long-range incommensurate cycloidal spiral structure. BFO is subject to many transitions as observed in the magnetic properties near the temperatures 50, 140, 200, and 643 K. The transition at 50 K is interpreted in terms of spin glass and the anomaly at about 140 K is ascribed to spin reorientations, while the transition near 200 K shows magnon softening. The most distinct one is the transition between AFM–paramagnetic phase, which occurs at T_N=643 K.

Field-induced variation in cycloidal spin ordering that occurs in $BiFeO_3$ was reported. The AFM order was detected from the hysteresis loops below the Néel temperature T_N=646 K. The cycloid vanishes above the critical field H_c. The anomaly in $M(H)$, which occurred in the low magnetic field range, was attributed to the field-induced transition from circular cycloid to the anharmonic cycloid. The coincidence between the anomalies in the magnetic and electric properties is explained by the critical behavior of chemical potential (μs) that is related to the magnetic phase transition.[3]

17.5 COMPOSITE MULTIFERROICS

Owing to the shortage and limitations of single-phase multiferroics, artificial ME materials[42] with the magnetic and ferroelectric transition temperatures beyond the RT[18] are of great interest. Hence, the research on composite multiferroics that yield a giant ME coupling response even above RT started to bloom. Composite-type multiferroics, with ferroelectric and magnetic phases, yield a giant ME coupling response even above RT. In such systems, neither the ferroelectric (thus piezoelectric) nor the magnetic phase shows the ME effect, but composites of these two phases have a remarkable ME effect. Thus, the ME effect is a result of the product of the magnetostrictive effect (magnetic/mechanical effect) in the magnetic phase and the piezoelectric effect (mechanical/electrical effect) in the piezoelectric one, that is, the coupling between both components is mediated by strain; for instance, the magnetic phase changes its shape magnetostrictively when a magnetic field is applied and, in turn, the strain is passed along to the piezoelectric phase giving rise to an electric polarization.[28]

In artificial multiferroic heterostructures,[14] one of the two phases should be piezomagnetic or magnetostrictive and the other should be piezoelectric or electrostrictive.[24] Ferroelectrics in combination with ferromagnets are generally employed for this purpose. These systems can be put into contact in the form of composites, laminates, or epitaxial layers of thin films.

A mechanical strain brings about a dielectric polarization change in the ferroelectric phase due to the piezoelectric effect. The single-phase materials show low values of ME coupling coefficient at RT, making it complicated to fabricate suitable devices for practical application. This has necessitated research on the development of composite multiferroic systems. The ME coupling coefficient achieved is in the order of magnitude larger than in single-phase multiferroics. The two-phase multiferroics composed of ferromagnetic and ferroelectric materials show advantages in realizing such device due to its strong strain-mediated ME interaction. Momentous ME coupling response can be realized in multiferroic composites containing ferroelectric and ferromagnetic phases by manipulation of magnetization rotation by the electric control at RT. These hybrid multiferroic systems help to bring about tunable and nonvolatile control of magnetism by means of an electric field at RT.

The multiferroic nanocomposites are developed in different geometries, for example, (0–3, 2–2, 1–3) (i.e., 0–3-type particulate films, 2–2-type layered or laminate heterostructures, or 1–3-type fiber or rod-like vertical heterostructures). Strong ME coupling effect is expected in strain-mediated multiferroic systems.[7] At the nanoscale, even the charge- or exchange-bias (EB)-mediated ME coupling can be engineered.[23]

Several multiferroic systems have been studied so far. Multiferroic systems such as $CoFe_2O_4$–$PbTiO_3$ systems,[56] Fe_3O_4–$PbTiO_3$ core–shell multiferroic composites,[29] $x$$BiFeO_3$–$(1-x)$ $PbTiO_3$, $(1-x-y)$ $PbTiO_{3-x}$Bi-$(Ni_{1/2}Ti_{1/2})$ $O_{3-y}$$BiFeO_3$,[22] $(1-x)$ $BaTiO_3$–x$LaYbO_3$ ceramics,[15] ferroelectric and dielectric polymer (PVDF)–$BaFe_{12}O_{19}$[4] have been subject of investigation. Zhai et al. have reported the magnetic–dielectric properties of $NiFe_2O_4$/PZT composites.[60] Electrical properties and ME effect in $(x)\ Ni_{0.5}Zn_{0.5}Fe_2O_4$ + $(1-x)\ Ba_{0.8}Pb_{0.2}Zr_{0.8}Ti_{0.2}O_3$ (barium lead zirconate titanate) composites were also reported.[6] The influence of the piezomagnetic $NiFe_2O_4$ phase on the piezoelectric Pb $(Mg_{1/3}Nb_{2/3})_{0.67}Ti_{0.33}O_3$ phase is also reported.[45]

17.6 MULTIFERROIC CORE–SHELL STRUCTURES

Design and synthesis of core–shell nanostructures with multifunctional properties are of great interest.[13] Hybrid multiferroic core–shell structures with ferrite[53]–ferroelectric components have been subject of increased research interest[1] as core–shell interfaces play an important role in nanoscale FE and multiferroicity.[12] Ferromagnetic–ferroelectric composites exhibit strain-mediated coupling between the magnetic and electric phases.[47] Ferromagnetic nanostructures have wide applications in nanodevices[59] and have attracted huge interest. Ferrites have extensive applications in low-temperature ceramic technology.[61] Magnetic nanoparticles have promising applications in the field of biomedicine.[36]

Magnetically directed self-assembly of nanoparticles into superstructures of one, two, and three dimensions is of mounting demand. Magnetic-field-assisted-assembly of nickel ferrite (NFO)-BTO core–shell particles were reported by Sreenivasulu et al.[48] The core–shell nanoparticles synthesized by chemical self-assembly were assembled into linear chains and 2D and 3D arrays by magnetic-field-directed assembly technique. The magnetic dipole moment linked with the nanoparticles guides the assembly of nanocomposites into superstructures such as linear chains and arrays.[55] Fabrication of magnetic microstructures with multifarious two-dimensional geometric shapes employing the magnetically assembled iron oxide (Fe_3O_4) and cobalt ferrite ($CoFe_2O_4$) nanoparticles by the magnetic assembly method were reported by Velez et al.[55] Investigation of strain-mediated ME effects in such assembled superstructure of NFO–BTO core–shell particles is reported.

Liu et al. have introduced a procedure for the synthesis of multiferroic core–shell nanosystems with controlled core and shell thicknesses.[30] Ferrite/perovskite oxide core–shell nanostructures were synthesized by a hydrothermal method combined with annealing process. Perovskite oxide (ABO_3) layer was coated on spherical ferrite particles in two steps. Fe_3O_4 microspheres were prepared from 2 mmol of $FeCl_30{\cdot}6H_2O$ and 0.3 g of NaOH dissolved in ethylene glycol and sealed in a Teflon-lined stainless steel autoclave and heated at 200°C for 10 h. For the synthesis of ferrite/ $PbTiO_3$ core–shell particles, 0.6 g of the as-synthesized Fe_3O_4 microspheres were dispersed in 5 ml of PEG 400 with sonication and then in 2 mmol of Ti $(SO_4)_2$ to get a uniform coating of Ti hydroxide. The Fe_3O_4/Ti hydroxide particles were redispersed in 2 mmol of Pb $(NO_3)_2$ and 0.4 g of KOH and again sealed in a 50-ml Teflon-lined stainless steel autoclave, maintained at 160°C for 6 h. The product was dried in a vacuum oven at 60°C for 4 h, after the reaction, to obtain the core–shell particles. These core–shell particles could enhance the ME coupling to a large extent for practical applications and serve as supreme models for the synthesis of ferrite/perovskite oxide core–shell nanostructures and investigation of the complex physical mechanism of the ME effect.

Tesfa et al. have reported the structural, dielectric, and multiferroic properties of $CaFe_2O_4$–$BaTiO_3$ core–shell and mixed composites.[57] The core–shell $CaFe_2O_4$–$BaTiO_3$ and $(1-x)$ $CaFe_2O_{4-x}BaTiO_3$ (x = 0.3, 0.5, 0.7) mixed composites were synthesized using a combination of solution processing and solid-state reaction method, respectively, followed by high-temperature calcinations. In both core–shell and mixed nanocomposites, the occurrence of the spinel ferrite phases were confirmed by X-ray diffraction (XRD) patterns and transmission electron microscopy. High-resolution transmission electron microscope images indicate a clear view of ferrite and ferroelectric phases in each core–shell and mixed composite, where the ME coupling effect of ferrite and ferroelectric phase happened. Dependence of dielectric constant (ε_0), loss tangent (tan d), and AC conductivity (S_{AC}) with frequency (100 Hz to 2 MHz) and temperature (25–550°C) with increasing probing frequency (1 kHz to 2 MHz) of the composites have been investigated. The peak observed in mixed composites and core–shell nanostructures in the low-temperature range (130–150°C) is attributed to ferroelectric-to-paraelectric-phase transition of $BaTiO_3$. The core–shell composite did not show any magnificent improvement in dielectric properties than mixed composites. The *P–E* and *M–H* hysteresis loops reflect the presence of ordered FE and FM behavior at RT. The core–shell composite shows drastic improvement in ferroelectric, magnetic, and ME properties than the mixed composites. In addition, the calculated linear ME coupling coefficient results reveal strong ME coupling effect and the maximum recorded value is ~30.32 mV/cm·Oe for the core–shell composite. Larger saturation magnetization and remanent magnetization contribute to the higher ME coupling coefficient values of the core–shell composites.

Development of multiferroic core–shell nanocomposites using a "click" chemistry approach was reported by Sreenivasulu et al.[47] Barium titanate and nickel ferrite nanoparticles of 10–100 nm were functionalized with complementary coupling groups and allowed to self-assemble in the presence of a catalyst. The barium titanate and nickel ferrite nanoparticles were functionalized with O-propargyl citrate and alkyl azide groups, respectively. Two types of core–shell particles were synthesized through the CuAAC reaction magnetic core with ferroelectric shell—100 nm diameter NFO core−50 nm BTO shell (sample-A) and ferroelectric core with ferromagnetic shell—50 nm BTO core–10 nm NFO shell (sample-B), and 100 nm BTO core–10 nm NFO shell (sample-C). Structural and compositional studies on as-assembled particles were carried out by XRD, scanning electron microscopy (SEM), transmission electron microscopy (TEM), and scanning probe microscopy (SPM). The purity of the samples was confirmed by the XRD pattern that shows the peaks expected for NFO and BTO. The core–shell nature of clusters was observed by SEM and magnetic force microscopy images. The chemical composition of the particles and clusters in BTO and NFO phases was examined by the energy-dispersive X-ray spectroscopy.

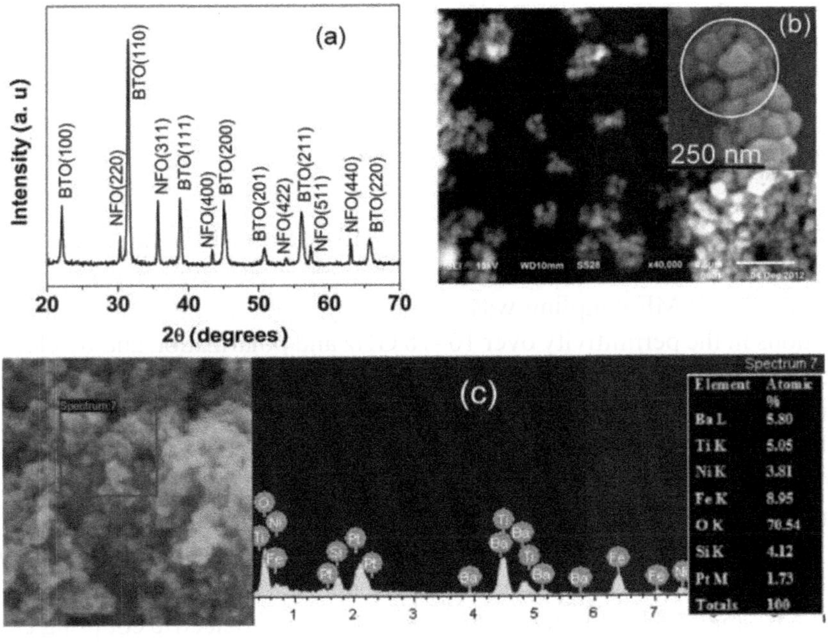

FIGURE 17.3 (See color insert.) (a) X-ray diffraction (XRD) data for as-assembled core–shell particles with 100 nm NFO core and 50 nm BTO shell (sample-A). (b) SEM micrograph of clusters of sample-A. (c) energy-dispersive X-ray spectroscopy data for sample-A.
Source: Reprinted with permission from ref 47. © 2014 AIP Publishing LLC.

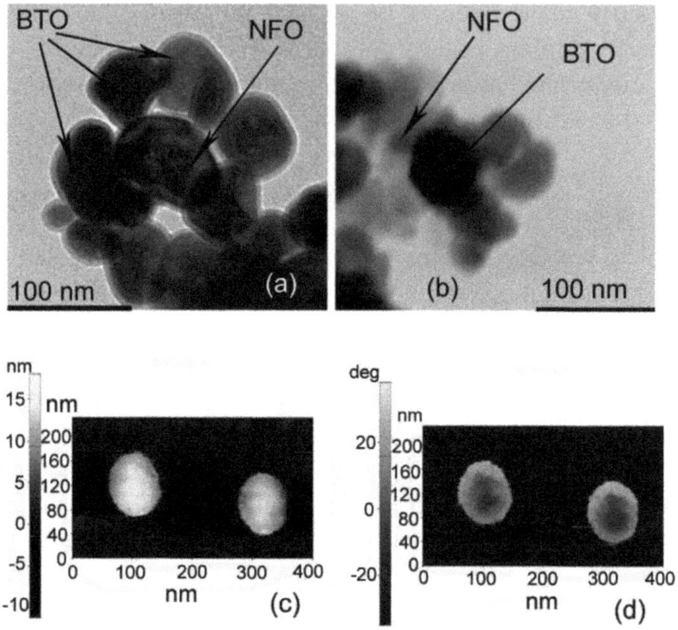

FIGURE 17.4 (See color insert.) (a) TEM micrograph showing core–shell structures for sample-A. (b) Similar TEM micrograph for clusters of 50 nm BTO core and 10-nm NFO shell (sample-B). MFM (c) amplitude and (d) phase image of sample-B.

Source: Reprinted with permission from ref 47. © 2014 AIP Publishing LLC.

The dynamics of ME interactions in self-assembled ferrite–ferroelectric core–shell nanoparticles were studied and reported. Evidence for strong strain-mediated ME coupling was obtained by static magnetic field-induced variations in the permittivity over 16–18 GHz and polarization and by electric field induced by low-frequency AC magnetic fields. The low-frequency ME response of the samples was studied. The ME voltage coefficient ($\delta V/(t \cdot \delta H)$) or the ME response of sample-A measured as a function of H is shown in Figure 17.5. The ME response displayed by the core–shell samples were compared with that reported for similar ferromagnetic–ferroelectric composites. The MEVC for the as-assembled BTO–NFO core–shell samples is found to be much smaller than reported values for bulk and similar composites. The relatively small MEVC is due to smaller MS and piezoelectric coupling coefficient and a large leakage current in the nanocomposites. It is observed that while induced polarization and MDE in single-phase multiferroics reported is of the order 1% or less, the core–shell system shows changes in ε and P in the range 1–5% for fields on the order of 4 kOe. The indication of strong ME coupling was obtained through H-induced polarization, magneto-dielectric effects over 16–18 GHz, and low-frequency ME effects.

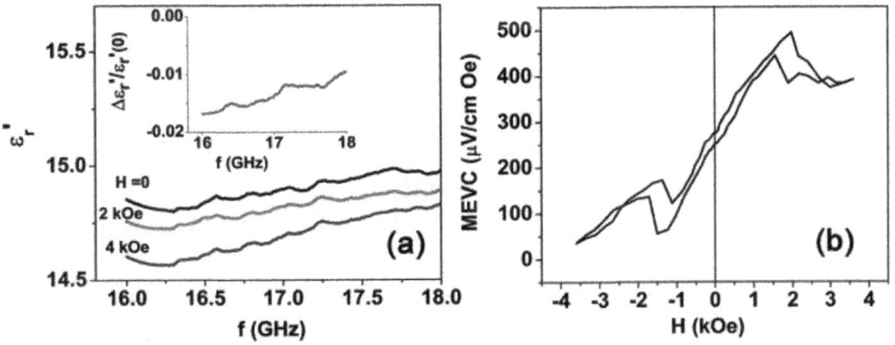

FIGURE 17.5 **(See color insert.)** (a) Real part of the relative permittivity ε_r' versus frequency for an as-assembled sample-A and the relative variation in ε_r' in H i4 kOe estimated from the data. (b) Low-frequency ME voltage coefficient versus bias field H data for a film of sample-A.

Source: Reprinted with permission from ref 47. © 2014 AIP Publishing LLC.

Corral-Flores et al. have reported a method for the preparation of magnetostrictive–piezoelectric core–shell particulate composites that show enhanced ME effect.[10] A shell of barium titanate was formed around nanoparticles of cobalt ferrite, varying the composition of the cobalt ferrite magnetostrictive phase from 20 to 60 wt.%. Cobalt ferrite nanoparticles were synthesized by coprecipitation method and then added to the precursor gel of barium titanate, allowing the in situ formation of the composite and thereby restricting the contact of the ferrite particles during sintering. The samples were sintered at a temperature ranging from 1100 to 1250°C for 12 h, followed by a plating step to be electrically poled. Additional samples were prepared by conventional mechanical milling for comparison, starting from cobalt ferrite prepared either by coprecipitation or the solgel technique and commercial barium titanate. Samples of same compositions prepared by different methods and sintered under the same conditions showed different behavior.

Multiferroics combining ferroelectric and ferromagnetic properties[2] have attracted countless attention due to their prospective applications in novel nonvolatile information storage and innovative ME sensors. Shi et al. have reported the synthesis of artificial exchange-biased two-phase core–shell nanostructures consisting of ferromagnetic (Ni) and multiferroic bismuth ferrite ($BiFeO_3$) materials, by a two-step method.[46] An exchange bias effect was observed and studied, which indicates that it is possible to fabricate ferromagnetic–multiferroic nanostructures to utilize the combined ferroelectric and AFM functionalities of bismuth ferrite.

Core–shell $CoFe_2O_4$–$BaTiO_3$ nanoparticles and nanotubes[38] with multiferroic and ME properties were prepared using a combination of solution processing and high-temperature calcination. $CoFe_2O_4$ nanoparticles of 12 nm diameter were prepared by the hydrothermal treatment using a mixture of $Co(NO_3)_2 \cdot 6H_2O$, $Fe(NO_3)_3 \cdot 9H_2O$, and polyvinylpyrrolidone. About 0.1 g of $CoFe_2O_4$ nanoparticles were dispersed in 60 ml of the $BaTiO_3$ precursor solution, containing a mixture of 30 ml aqueous solution of 0.029 g of $BaCO_3$ and 0.1 g of citric acid with 30 ml ethanol solution of 1 g of citric acid and 0.048 ml titanium isopropoxide. The mixture was dried at 60°C under stirring and calcined at 780°C for 5 h to obtain $CoFe_2O_4$–$BaTiO_3$ core–shell nanoparticles. Both the core–shell nanostructures exhibit magnetic and dielectric hysteresis at RT and ME effect. From the transmission electron microscopy images, the diameter of the core–shell nanoparticles was found to be between 40 and 60 nm. The temperature-dependent magnetization data in the 10–390 K range at 100 Oe revealed the divergence between field cooled and zero field cooling that increased with decreasing temperature. Temperature dependence of dielectric constant of $CoFe_2O_4$–$BaTiO_3$ nanoparticles was also studied and reported. The dielectric constants of both the nanocomposites were observed to decrease upon application of magnetic field. The core–shell nanoparticles were found to exhibit 1.7% change in magnetocapacitance around 134 K at 1T, while the core–shell nanotubes display an amazing 4.5% change in magnetocapacitance around 310 K at 2T.

Extensive research on multiferroics[33,41,50] has triggered novel and functional device applications incorporating ferromagnetic and ferroelectric counterparts.[17,33]

17.7 CONCLUSION

Manipulation of functionalities at the nanoscale is possible through the control of their size, composition, and morphology of the nanoparticles. Nanomaterials that integrate two or more ferroic orders such as core–shell nanoparticles have ever-increasing demand due to the multifunctionality that is achieved through the formation of multiple shells or manipulation of core–shell materials with varied functionalities. The core–shell geometric structures offer huge storage capacities, low densities, tunable optical, and magnetic properties that enable novel technological advancements for spintronics applications such as ME and nonvolatile memories.

The theory and mechanisms of multiferroics are highlighted in the chapter. Single-phase multiferroic systems and also artificial multiferroic systems exhibiting RT magneto-dielectric coupling are presented in the chapter. Multiferroic heterostructures present high potential for low-energy-consumption spintronics devices. The novel design strategies of various multifunctional core–shell nanostructures are discussed in the chapter. Hybrid multiferroic systems with RT spin–charge coupling are auspicious for developing next-generation multifunctional devices with ultralow power consumption.

KEYWORDS

- **ferroics**
- **multiferroics**
- **single-phase multiferroics**
- **core–shell structures**
- **magnetization**
- **dielectrics**
- **magnetoelectric coupling**

REFERENCES

1. Abraham, A. R.; et al. Magnetic Response of Superparamagnetic Multiferroic Core–Shell Nanostructures. 2016, May, 050151. http://scitation.aip.org/content/aip/proceeding/aipcp/10.1063/1.4947805.
2. Abraham, A. R.; et al. Realization of Enhanced Magnetoelectric Coupling and Raman Spectroscopic Signatures in 0–0 Type Hybrid Multiferroic Core–Shell Geometric Nanostructures. *J. Phys. Chem. C* **2017.** DOI: p.acs.jpcc.6b12461. http://pubs.acs.org/doi/abs/10.1021/acs.jpcc.6b12461.
3. Andrzejewski, B. Field Induced Changes in Cycloidal Spin Ordering and Coincidence Between Magnetic and Electric Anomalies in $BiFeO_3$ Multiferroic. *J. Magn. Magn. Mater.* **2013,** *342*, 17–26.
4. Anithakumari, P.; et al. Enhancement of Dielectric, Ferroelectric and Magneto-Dielectric Properties in PVDF–$BaFe_{12}O_{19}$ Composites: A Step Towards Miniaturizated Electronic Devices. *RSC Adv.* **2016,** *6*(19), 16073–16080. DOI: C5RA27023E, http://xlink.rsc.org/?.
5. Anantharaman, M. R.; Myung, G. H.; Xaiobo, Z.; Kumarasiri, A.; Tyagi, A. K.; Mandal, B. P.; Narayanan, T. N.; Lawes, G.; Ajayan, P. M. Hybrid Multiferroic Nanostructure with Magnetic-Dielectric Coupling, 2012.

6. Bammannavar, B. K.; Naik, L. R. Electrical Properties and Magnetoelectric Effect in (x) $Ni_{0.5}Zn_{0.5}Fe_2O_4$ + (1−x) BPZT Composites. *Smart Mater. Struct.* **2009,** *18*(8). <Go to ISI>://000268251300013.
7. Barone, P.; Picozzi, S. Mechanisms and Origin of Multiferroicity. *C. R. Phys.* **2015,** *16*(2), 143–152. http://dx.doi.org/10.1016/j.crhy.2015.01.009. 2015.
8. Branković, Z.; et al. Multiferroic Bismuth Manganite Prepared by Mechanochemical Synthesis. *J. Eur. Ceram. Soc.* **2010,** *30*(2), 277–281.
9. van den Brink, J.; Khomskii, D. I. Multiferroicity due to Charge Ordering. *J. Phys.: Condens. Matter* **2008,** *20*(43), 434217. http://arxiv.org/abs/0803.2964\nhttp://stacks.iop.org/0953–8984/20/i=43/a=434,217?key=crossref.3dda4587a2b2cde6c0af2466031d1e4f.
10. Corral-Flores, V.; et al. Enhanced Magnetoelectric Effect in Core–Shell Particulate Composites. *J. Appl. Phys.* **2006,** *99*(8), 1–4.
11. Das, A.; et al. Enhanced Magnetoelectric Properties of $BiFeO_3$ on Formation of $BiFeO_3$/$SrFe_{12}O_{19}$ Nanocomposites. *J. Appl. Phys.* **2016,** *119*(23), 234102.
12. Duan, C. G.; et al. Interface Effect on Ferroelectricity at the Nanoscale. *Nano Lett.* **2006,** *6*(3), 483–487.
13. El-Toni, A. M.; et al. Design, Synthesis and Applications of Core–Shell, Hollow Core, and Nanorattle Multifunctional Nanostructures. *Nanoscale* **2016.** DOI: C5NR07004J, http://xlink.rsc.org/?.
14. Fernandes Vaz, C. A.; Staub, U. Artificial Multiferroic Heterostructures. *J. Mater. Chem. C* **2013,** *1*(41), 6731. DOI: c3tc31428f, http://www.scopus.com/inward/record.url?eid=2-s2.0–84,885,113,248&partnerID=40&md5=8d493e79cfba9477c19f2560ea2713a1\nhttp://xlink.rsc.org/?.
15. Feteira, A.; Sinclair, D. C. The Influence of Nanometric Phase Separation on the Dielectric and Magnetic Properties of (1−x) $BaTiO_3$–$xLaYbO_3$ ($0 \leq x \leq 0.60$) Ceramics. *J. Mater. Chem.* **2009,** *19*(3), 356. DOI: b816039b, http://xlink.rsc.org/?.
16. Garcia, V.; Bibes, M.; Barthelemy, A.Artificial Multiferroic Heterostructures for an Electric Control of Magnetic Properties. *C. R. Phys.* **2015,** *16*(2), 168–181. http://dx.doi.org/10.1016/j.crhy.2015.01.007.
17. Guo, R.; et al. Photovoltaic Effect. *Nat. Commun.* **2013,** *4*(May), 1–5. http://dx.doi.org/10.1038/ncomms2990.
18. Guo, Z.; et al. Structural, Magnetic and Dielectric Properties of Fe-Doped $BaTiO_3$ Solids. *Mod. Phys. Lett. B* **2012,** *26*(09), p 1250056. http://www.worldscientific.com/doi/abs/10.1142/S021798491250056X.
19. Hajra, P.; Maiti, R.; Chakravorty, D. Nanostructured Multiferroics. **2011,** *70*(June), 53–64.
20. Hoffmann, T.; et al. Time-Resolved Imaging of Magnetoelectric Switching in Multiferroic $MnWO_4$. *Phys. Rev. B—Condens. Matter Mater. Phys.* **2011,** *84*(18), 1–6.
21. Hu, J. M.; Chen, L. Q.; Nan, C. W. Multiferroic Heterostructures Integrating Ferroelectric and Magnetic Materials. *Adv. Mater.* **2016,** *28*(1), 15–39.
22. Hu, P.; et al. Magnetic Enhancement and Low Thermal Expansion of (1–x–y) $PbTiO_3$-XBi(Ni1/2Ti1/2) O_3-$YBiFeO_3$. *J. Mater. Chem.* **2011,** *21*(40), 16205. DOI: c1jm12410b, http://xlink.rsc.org/?.
23. Hu, Z.; Sun, N. X. Epitaxial Multiferroic Heterostructures. *Compos. Magnetoelectr.* **2015,** 87–101. http://linkinghub.elsevier.com/retrieve/pii/B9781782422549000056.
24. Huang, W.; Yang, S.; Li, X. Multiferroic Heterostructures and Tunneling Junctions. *J. Materiomics* **2015,** *22*. http://linkinghub.elsevier.com/retrieve/pii/S2352847815000581.

25. Hur, N.; et al. Electric Polarization Reversal and Memory in a Multiferroic Material Induced by Magnetic Fields. *Nature* **2004,** *429*(May), 392–395.
26. Khomskii, D. Classifying Multiferroics: Mechanisms and Effects. *Physics* **2009a,** *2*.
27. Khomskii, D. Classifying Multiferroics: Mechanisms and Effects. *Physics* **2009b,** *2*, 20. http://link.aps.org/doi/10.1103/Physics.2.20.
28. Kreisel, J.; Kenzelmann, M. Multiferroics—The Challenge of Coupling Magnetism and Ferroelectricity. *Europhys. News* **2009,** *40*(5), 17–20.
29. Liu, R.; et al. Multiferroic Ferrite/Perovskite Oxide Core/Shell Nanostructures. *J. Mater. Chem.* **2010a,** 20(47), 10665. DOI: c0jm02602f, http://xlink.rsc.org/?.
30. Liu, R. et al. Multiferroic Ferrite/Perovskite Oxide Core/Shell Nanostructures. *J. Mater. Chem.* **2010b,** *20*(47), 10665.
31. Mangalam, R. V. K.; et al. Multiferroic Properties of Nanocrystalline $BaTiO_3$. **2009,** *149*, 1–5.
32. Martin, L. W.; Ramesh, R. Multiferroic and Magnetoelectric Heterostructures. *Acta Mater*. **2012,** *60*(August), 2449–2470.
33. Matzen, S.; Fusil, S. Domains and Domain Walls in Multiferroics. *C. R. Phys.* **2015,** *16*(2), 227–240. http://dx.doi.org/10.1016/j.crhy.2015.01.013.
34. Meier, D. Functional Domain Walls in Multiferroics. *J. Phys. Condens. Matter* **2015,** *27*(46), 463003. http://iopscience.iop.org/article/10.1088/0953–8984/27/46/463003.
35. Nayek, P.; Li, G. Superior Electro-Optic Response in Multiferroic Bismuth Ferrite Nanoparticle Doped Nematic Liquid Crystal Device. *Sci. Rep.* **2015,** *5*(May), 10845. http://www.nature.com/srep/2015/150604/srep10845/full/srep10845.html.
36. Pankhurst, Q. A.; et al. Applications of Magnetic Nanoparticles in Biomedicine. *J. Phys. D: Appl. Phys.* **2003,** *36*(13), 167–181. http://iopscience.iop.org/article/10.1088/0022–3727/36/13/201\nhttp://iopscience.iop.org/0022–3727/36/13/201.
37. Perez-Mato, J. M.; et al. Symmetry Conditions for Type II Multiferroicity in Commensurate Magnetic Structures. *J. Phys. Condens. Matter* **2016,** *28*(28), 286001. http://stacks.iop.org/0953–8984/28/i=28/a=286,001?key=crossref.f12a0378d1678724bb39a6683c276c64.
38. Raidongia, K.; et al. Multiferroic and Magnetoelectric Properties of Core–Shell $CoFe_2O_4$ @ $BaTiO_3$ Nanocomposites. *Appl. Phys. Lett.* **2010,** *97*(6), 2014–2017.
39. Ramesh, R.; Spaldin, N. A. Multiferroics: Progress and Prospects in Thin Films. *Nat. Mater.* **2007,** *6*(1), 21–29. http://www.nature.com/doifinder/10.1038/nmat1805.
40. Rao, C. N. R.; Rayan, C. New Routes to Multiferroics. *J. Mater. Chem.* **2007,** *17*, 4931–4938.
41. Schileo, G. Recent Developments in Ceramic Multiferroic Composites Based on Core/Shell and Other Heterostructures Obtained by Sol–Gel Routes. *Prog. Solid State Chem.* **2013,** *41*(4), 87–98. http://dx.doi.org/10.1016/j.progsolidstchem.2013.09.001.
42. Schileo, G.; et al. Yttrium Iron Garnet/Barium Titanate Multiferroic Composites. *J. Am. Ceram. Soc.* **2016,** *99*(5), 1609–1614.
43. Scott, J. F. Applications of Magnetoelectrics. *J. Mater. Chem.* **2012,** *22*(11), 4567.
44. Scott, J. F.; et al. Hydrodynamics of Domain Walls in Ferroelectrics and Multiferroics: Impact on Memory Devices. *Appl. Phys. Lett.* **2016,** *109*(4). http://dx.doi.org/10.1063/1.4959996.
45. Sheikh, A. D.; Mathe, V. L. Effect of the Piezomagnetic $NiFe_2O_4$ Phase on the Piezoelectric Pb $(Mg_{1/3}\ Nb_{2/3})$ 0.67 $Ti_{0.33}O_3$ Phase in Magnetoelectric Composites. *Smart Mater. Struct.* **2009,** *18*(6), 65014. http://stacks.iop.org/0964–1726/18/i=6/a=065,014.
46. Shi, D-W.; et al. Exchange-Biased Hybrid Ferromagnetic–Multiferroic Core–Shell Nanostructures. *Nanoscale* **2014,** 6(13), 7215. DOI=c4nr00393d, http://xlink.rsc.org/?.

47. Sreenivasulu, G.; et al. Controlled Self-Assembly of Multiferroic Core–Shell Nanoparticles Exhibiting Strong Magneto-Electric Effects. *Appl. Phys. Lett.* **2014,** *104*(5), 052901. http://scitation.aip.org/content/aip/journal/apl/104/5/10.1063/1.4863690.
48. Sreenivasulu G. et al. Superstructures of Self-Assembled Multiferroic Core–Shell Nanoparticles and Studies on Magneto-Electric Interactions. *Appl. Phys. Lett.* **2014,** *105*(7). http://dx.doi.org/10.1063/1.4893699.
49. Staedler, D.; et al. Cellular Uptake and Biocompatibility of Bismuth Ferrite Harmonic Advanced Nanoparticles. *Nanomed.: Nanotechnol., Biol. Med.* **2014,** *11*(4), 815–824. http://dx.doi.org/10.1016/j.nano.2014.12.018.
50. Stephanovich, V. A.; Laguta, V. V. Transversal Spin Freezing and Reentrant Spin Glass Phases in Chemically Disordered Fe-Containing Perovskite Multiferroics. *Phys. Chem. Chem. Phys.* **2016,** *18*(10), 7229–7234. http://dx.doi.org/10.1039/C6CP00054A.
51. Su et al, J.; et al. Magnetism-driven Ferroelectricity in Double Perovskite Y 2 $NiMnO_6$. *ACS Appl. Mater. Interfaces* **2015,** *7*, 13260. http://pubs.acs.org/doi/abs/10.1021/acsami.5b00911.
52. Sundararaj, A.; et al. Room Temperature Magnetoelectric Coupling in $BaTi_1$-$xCrxO_3$ Multiferroic Thin Films. *J. Appl. Phys.* **2016,** *119*(2), 1–7. http://dx.doi.org/10.1063/1.4939068.
53. Thankachan, R. M.; et al. Cr^{3+}-Substitution Induced Structural Reconfigurations in the Nanocrystalline Spinel Compound $ZnFe_2O_4$ as Revealed from X-ray Diffraction, Positron Annihilation and Mössbauer Spectroscopic Studies. *RSC Adv.* **2015,** *5*(80), 64966–64975.
54. Tokunaga, M.; Azuma, M.; Shimakawa, Y. High-field Study of Multiferroic $BiFeO_3$. *J. Phys.: Conf. Ser.* **2010,** *200*(1), 012206. http://stacks.iop.org/1742–6596/200/i=1/a=012206?key=crossref.ad2ed54b4e0cc40c6c9b25fe044c5859.
55. Velez, C.; et al. Magnetic Assembly and Cross-Linking of Nanoparticles for Releasable Magnetic Microstructures. *ACS Nano* **2015,** *9*(10), 10165–10172.
56. Wang, B. Y.; et al. Effect of Geometry on the Magnetic Properties of $CoFe_2O_4$–$PbTiO_3$ Multiferroic Composites. *RSC Adv.* **2013,** *3*(21), 7884. DOI=c3ra00104k, http://xlink.rsc.org/?.
57. Woldu, T.; et al. A Comparative Study on Structural, Dielectric and Multiferroic Properties of $CaFe_2O_4$/$BaTO_3$ Core–Shell and Mixed Composites. *J. Alloys Compd.* **2017,** *691*(Nov 2016), 644–652. http://dx.doi.org/10.1016/j.jallcom.2016.08.277.
58. Xu, T.; et al. Multiferroic Domain Walls in Ferroelectric $PbTiO_3$ with Oxygen Deficiency. *Nano Lett.* **2016,** *16*(1), 454–458.
59. Yao, Z. N.; et al. Detection of Domain Wall Distribution Nucleation in Ferromagnetic Nanocontact Structures by Magnetic Force Microscopy. *J. Magn. Magn. Mater.* **2013,** *342*, 1–3.
60. Zhai, J.; et al. Magnetic-dielectric Properties of $NiFe_2O_4$/PZT Particulate Composites. *J. Phys. D: Appl. Phys.* **2004,** *37*(6), 823–827. http://iopscience.iop.org/0022–3727/37/6/002.
61. Zhang, H-W.; et al. Development and Application of Ferrite Materials for Low Temperature Co-fired Ceramic Technology. *Chin. Phys. B* **2013,** *22*(11), 117504. http://stacks.iop.org/1674–1056/22/i=11/a=117,504?key=crossref.884925d1cef000106d4e6c43ebd90579.

CHAPTER 18

LOW-COST MATERIALS FOR THE REMOVAL OF CONTAMINANTS FROM WASTEWATER

THERESA O. EGBUCHUNAM[1,*], GRACE OBI[1,2], FELIX E. OKIEIMEN[3], and SENEM YETGIN[4]

[1]*Department of Chemistry, Federal University of Petroleum Resources, Effurun, Delta State, Nigeria*

[2]*obi.grace@fupre.edu.ng*

[3]*Centre for Biomaterials Research, University of Benin, Benin City, Edo State, Nigeria, E-mail: felix.okieimen@uniben.edu*

[4]*Department of Food Engineering, Kastamonu University, Kastamonu, Turkey, E-mail: syetgin@kastamonu.edu.tr*

[*]*Corresponding author. E-mail: egbuchunam.theresa@fupre.edu.ng*

CONTENTS

ABSTRACT

This article presents a review on the use of low-cost adsorbents from locally available materials in the removal of organic and inorganic contaminants such as heavy metals, phenolic compounds, etc. from wastewater. Many studies on adsorption properties of various low cost adsorbents, such as activated carbons based on agricultural waste, have been reported in recent years. This review provides recent literature demonstrating the usefulness of low-cost adsorbents from various sources in the adsorption of various contaminants in wastewater.

18.1 INTRODUCTION

The use of low-cost adsorbents obtained from local materials as a replacement for adsorbents that are expensive for the effective removal of organic and inorganic contaminants from aqueous solutions has been of great interest in recent times. Natural materials or waste products can be modified using various techniques with little cost to improve their surface properties and adsorption capacity for common organic and inorganic contaminants in the aquatic environment such as heavy metals, phenolic compounds, volatile organic substances (VOCs) among others. Organic and inorganic contamination exists in aqueous wastewaters of many industries such as petrochemical, mining, textile, metal plating, tannery, etc. Some contaminants associated with these activities tend to accumulate in living organisms causing various diseases and disorders. Several technologies have been developed over the years by different researchers to remove organic and inorganic contaminants from wastewaters: precipitation and co-precipitation, ion exchange, membrane separation, reverse osmosis, and adsorption, among others. Adsorption process has been found to be one of the most effective methods for removing contaminants in wastewater because it is inexpensive. Carbonaceous adsorbents such as activated carbon have been shown to be one of the most efficient products in the elimination of contaminants from wastewater. In spite of being the most popular and widely used adsorbent in wastewater applications, activated carbon remains an expensive material making it less attractive in small-scale industries. In the light of this, researches geared toward alternative low-cost adsorbents which are readily available have intensified in recent years. Attention has been focused on exploring alternatives from low-cost available materials and agricultural by-products which have been used to prepare activated carbon and are considered important sources of adsorbents as they have been reported to contain certain functional groups such as hydroxyl, carboxyl, carbonyl, ester, amino, and phosphor groups constituting the source of surface acidity responsible for their effectiveness in the removal of contaminants.[1,2] These adsorbents having high carbonaceous content have been prepared from wood, charcoal, saw dust, coconut shell, and cow bones, among others,[3] and reports from a literature survey[4] affirm that these low-cost adsorbents have demonstrated outstanding removal capabilities for certain metal ions when compared with activated carbon. These adsorbent materials made from locally available raw materials are modified to improve their surface properties as the adsorption process on carbons have been shown to depend on the metallic species to be removed, adsorbent surface chemistry, and adsorption conditions. Agricultural by-products and wastes present highly recommendable sources because their use provides a two-fold environment and economic advantage: a recycling path is devised and new adsorbents are produced from a low-cost material for the use in wastewater treatment plans.

A wide range of potentially low-cost adsorbents has been studied over the years: for example, natural clay minerals in raw and modified forms (such as montmorillonite, kaolinite, and illite) for the removal of inorganic and organic contaminants from aqueous solutions,[5–13] peanut husks have been used to remove metal ions from wastewater,[14] corn cob fragments have been used to remove cadmium,[15] chicken eggshell have been used for the removal of ionic pollutants from aqueous solutions,[16] and the ability of bone char produced from the carbonization of animal bones has been shown to adsorb considerable quantities of organic compounds and metal ions from aqueous medium.[17–22] The widespread use of low-cost adsorbents in industries for wastewater treatment applications today are strongly recommended due to their local availability, technical feasibility, engineering applicability, and cost-effectiveness.

18.2 CLAY MINERALS

Clay minerals are common weathering and low-temperature hydrothermal alteration products. They are hydrous aluminium phyllosilicates which can be divided into three main groups: the Kaolin group which includes kaolinite, dickite, halloysite, and nacrite; the montmorillonite—smectite group and the illite group which includes the clay micas. Clay minerals can also be seen as phyllosilicate or sheet-like structures consisting of layers of hydrous oxides of aluminium, silicon, or magnesium. The clay mineral structure is based on the combination of tetrahedral and octahedral sheets[23,24] and can be categorized depending on the way their tetrahedral and octahedral sheets are packaged into layers.

Clay minerals are layer-lattice silicates made up of combinations of two structural units: a silicon–oxygen tetrahedron and an aluminium oxygen–hydroxyl octahedron. For example, montmorillonite is known as a 2:1-type aluminosilicate as its crystalline structure presents an alumina octahedral between two layers of silica while kaolinite is termed the 1:1 clay mineral because there is a layer of a single tetrahedral sheet and a single octahedral sheet. The most common clays used as adsorbents are montmorillonite/smectite groups which refer to nonmetallic clays composed of hydrated sodium aluminum silica with a formula of $(Ca, Na, H)(Al, Mg, Fe, Zn)_2(Si, Al)_4O_{10}(OH)_2nH_2O$ and kaolinite group clays with the general formula of $Al_2Si_2O_5(OH)_4$.[25]

Clays find wide range of applications in various areas of science due to their natural abundance and the propensity with which they can be chemically and physically modified to suit practical technological needs. The properties of clay minerals such as high cation exchange capacity; swelling and high surface areas make clays such as bentonite (clays rich in smectite) favorable materials to be widely used as rheology control reagents, membranes in under storage tanks (USTs), sorbents in pollution prevention and environmental remediation. The ion-exchange capability of clay minerals, in particular smectites, influences their unique physical properties such as the cation retention and diffusion processes of charged and uncharged molecules. The adsorption capability of clay results from a net negative charge on the structure of fine-grain silicate materials. Natural clays contain inorganic exchange cations such as sodium, Na+ and calcium, Ca2+, that are strongly hydrated in the presence of water producing a hydrophilic environment at the clay surface making them ineffective as sorbents.[26] However, ion exchange of the inorganic cations present in the clay mineral with large quaternary ammonium cations (QACs) of the form $[(CH_3)_3NR]^+$ and $[(C_2H_5)_3NR]^+$ where R is an alkyl, phenyl, or benzyl functional group may be used in replacing the natural inorganic cations thereby changing the clay surfaces from hydrophilic to hydrophobic. Clay minerals and their modified derivatives have composed a large family of adsorbents which can be used for the adsorption of most of the chemical contaminants from aqueous solution. Among this family of adsorbents, those based on montmorillonite (Mt), a typical 2:1 type clay mineral, have been most extensively studied.

18.3 ORGANOCLAYS

Naturally occurring clay minerals can be transformed into useful materials by modifying the surface properties so as to enhance their capacity to remediate environmental contaminants thereby responding to the related recycling and environmental challenges. Chemically modified clay minerals represent an innovative and promising class of low cost sorbent materials. Organic modification of the clay minerals surface by the replacement of inorganic exchange cations with large organic cations of the form $[(CH_3)_3NR]^+$ or $[(CH_3)_2NR_2]^+$ where R is a large ($>C_{12}$) alkyl hydrocarbon, yields organoclays with organophilic properties which has been shown to significantly increase the attenuation of some organic compounds.[26–29] When hydrophobic modification of a clay mineral's surface is undertaken, a variety of organoclays can be formed and it is an important consequence of replacing inorganic cations with organic cations as the clay surface takes on a hydrophobic character instead of hydrophilic. A relatively easy method of modifying the clay surface, making it more compatible with an organic matrix is ion-exchanging, that is, introducing organic molecules into the clay mineral structure. The cations present in the clay material are not strongly bound to the clay surface, so small molecule cations can replace the cations present in the clay. The intercalation of cationic surfactants not only changes the surface properties from hydrophilic to hydrophobic but the basal spacing of the layers increase as well; thereby, enhancing the capacity of the materials to remove organic contaminants.[11] QACs containing long-chain alkyl group are generally characterized by linear isotherms over a wide range of solute concentrations,[25] due to the favorable interlayer microenvironment created by the long-chain alkyl ammonium ions for the partitioning of organic molecules. More recently, the use of organic modified clays for the sorptive removal of organic contaminants from water has been widely studied in the last few years.[6,9,10,12,13] The interest in the use of natural clay minerals in raw and modified forms (such as montmorillonite, kaolinite, and illite) for the removal of inorganic and organic contaminants from aqueous solutions has been on the increase in recent years on account of their relative low cost in comparison with activated carbons. For example, organoclays prepared using montmorillonite, a typical 2:1 type clay mineral with one O sheet being sandwiched between two T sheets, have been shown to be suitable for use in the uptake of large organic cations due to its expandable interlayer spaces and its adsorption capacity found to be higher than commercial activated carbon.[30–34] More recently, kaolinite, a 1:1 clay material, have been modified with the organic cation, cetyl trimethylammonium (CTA^+) and the results showed improved interlayer microenvironment occasioned by CTA^+ modification resulting in surface increase in the capacity of the clay material to remove petroleum hydrocarbon fractions from aqueous medium.[8,13,27] Organic modified clay minerals are promising low-cost and high-efficient adsorbents suitable for use in the removal of cationic contaminants in aqueous medium and they could be used as precursor materials to synthesize potential adsorbents in commercial scale for treating organic compounds in wastewater.

18.4 BONE CHAR

Carbonaceous adsorbents such as activated carbon have been shown to be one of the most efficient products in the elimination of volatile organic compounds from wastewater. However, activated carbons are expensive materials which require complex activation processes and therefore, it is necessary from an economic point of view to explore alternatives from low cost available materials to regenerate and reuse them. To decrease treatment costs, attempts are made to find inexpensive alternatives from waste materials of industrial, domestic, and agricultural wastes. The processing of waste materials is arising from vast natural resources by transforming them into useful materials; thereby, responding to the related recycling and environmental challenges is an important factor when conducting a research.

The use of adsorbents of biological origin has emerged in the last decade as one of the most promising alternatives to conventional treatment technologies. A large number of apatite-based materials (mineral phosphates, synthetic apatite, bone meal, and bone char) have been considered as matrixes for remediation of metal-contaminated water and soil.[35] Apatites are often identified by the general formula $M_{10}(XO_4)Y_2$ where M^{2+} is a divalent cation, $(XO_4)^3$ is a trivalent anion, and Y is a monovalent anion.[36] Apatites of different origins (mineral, synthetic and derived from animal and fish bones) have been used as sorbents for heavy metals such as Pb, Zn, Cu, Cd, Co, and Sb.[37–40] Bone char is a naturally occurring biological hydroxyapatite derived from the carbonization of crushed animal bones under high temperatures in the range of 400–500°C in an oxygen-depleted atmosphere. Bone char consists mainly of calcium phosphate and a small amount of carbon. It is a mixed compound adsorbent in which carbon is distributed throughout the porous structure of hydroxyapatite, $(Ca_{10}(PO_4)_6(OH)_2)$, which is a good adsorbent and successfully used as a permeable reactive barrier for immobilizing various heavy metals and treating acid mine drainage[41] and its synthetic form has been shown to have high removal capacity for divalent heavy metal ions from aqueous media.[37,42–45] These reports indicate that the possible reaction mechanisms for metal immobilization include ion-exchange processes,[37,46] surface complexation,[43,47] dissolution of hydroxyapatite and precipitation of new metal phosphates, and substitution of Ca in hydroxyapatite by other metals during recrystallization (co-precipitation).[37]

Bone char has been extensively used in the removal of various toxic metals and pollutants from water and its quality is usually controlled by the amount of oxygen present in the charring atmosphere. Recent studies[22] have shown that bone char prepared by pyrolyzing cow bones collected from an abattoir in Agbarho, Delta State, Nigeria, having high surface area and fairly well-developed pore structure (Fig. 18.1) has the potential of removing p-nitrophenol from aqueous media. The processing and transformation of animal bones into bone char with good adsorption properties would alleviate problems of disposal and management of these waste by-products while producing value-added products from animal bones for water and wastewater treatment.

FIGURE 18.1 Micrograph of bone char obtained at 10 μm.[22]

18.5 COMMON CONTAMINANTS FOUND IN WASTEWATER

Contamination of water bodies, soils, and sediments can seriously affect plants, animals, and human beings due to their bioaccumulation, nonbiodegradability, and toxicity even at low concentrations. Industrial wastewaters arising from coal gasification, steel production, petroleum and petrochemical processing, rubber proofing, mining, pharmaceuticals, etc., contain these toxins and can be classified into organic and inorganic contaminants. Generally, pollution caused by organic and inorganic contaminants pose adverse effects on the environment all over the world, the examples including but not limited to, a variety of toxic effects on aquatic organisms; exposure leading to adverse health effects which include cancer, irritation of mucosal membranes, hematological changes, impairment of the central nervous system, respiratory problems, and disruption of liver and kidneys, effect on human health directly or indirectly through drinking water, touching with skin and food chain; in agroecological environments especially in soils causing great harm to crop growth, yield, and quality.[6,48–50] Common organic contaminants encountered in the environment especially in aquatic systems include phenolic compounds and benzene, toluene, ethylbenzene, and xylene (BTEX), while inorganic contaminants include heavy metals such as copper, zinc, cadmium, lead, nickel, among others. The removal of these organic pollutants from contaminated wastewater and water bodies is critical to ensuring the safety of water supplies all over the world. Adsorption on solid substrate materials has been found to be one of the most frequently used and suitable process for the removal of these contaminants from solution. Solids prepared from low-cost materials such as biomass, fish bones, organoclays, animal bones, etc., have been used over the years for pollution attenuation.

18.6 APPLICATION OF LOW-COST MATERIALS FOR CONTAMINANTS' REMOVAL

The adsorption process is widely used for treatment of industrial wastewater from organic and inorganic pollutants. Different types of adsorbents are classified into natural adsorbents and synthetic adsorbents. Natural adsorbents include charcoal, clays, clay minerals, zeolites, and ores. These natural materials, in many instances, are relatively cheap, abundant in supply, and have significant potential for modification and ultimately enhancement of their adsorption capabilities. Each adsorbent has its own characteristics such as porosity, pore structure, and nature of its adsorbing surfaces. Efforts have been geared toward exploiting new low-cost materials from biological origins that are inexpensive, readily available, and environmentally friendly for use as adsorbents or precursor materials to synthesize potential adsorbents for treating organic and inorganic compounds in wastewater.

Clays in their raw and modified forms have received much attention as low-cost adsorbents for the removal of organic contaminants such as phenolic compounds, BTEX, and heavy metals such as Fe, Co, Ni, and Pb.[5–10,12,13,25–30,50–55] These adsorbents have been prepared and characterized from kaolinite, illite, and montmorillonite representing the three main classes of clay minerals. Of these three, reports[54,56] show that montmorillonite has the smallest crystals, largest surface area, and largest cation-exchange capacity. Montmorillonite, being a 2:1 clay type has high surface charges resulting from the spread of isomorphous substitution in tetrahedral and octahedral sheets, while 1:1 layered kaolinite has little isomorphous substitution and this may account for the higher adsorption capacity of montmorillonite to adsorb cations.

Bone char prepared from animal bones is characterized by a mesoporous structure having large surface areas within the range 60–120 m^2/g.[18–22,35,41,57] Energy-dispersive X-ray (EDX) analysis carried out on bone char reveals that the adsorbent is mainly composed of calcium (Ca) and phosphorus (P) and other minor elements such as oxygen (O) and carbon (C).[22,58] Several studies have reported the removal of a variety of organic and inorganic toxic compounds from aqueous solutions with bone char, for example, phenol and its derivatives,[22] heavy metal ions[18–21,35,41,59] showed high affinity for the removal of these contaminants from aqueous media. The adsorption capacities of adsorbents obtained from some studies in literature varied depending on the characteristics of the individual adsorbents, the extent of chemical modification, and the concentration of the adsorbate solution. Table 18.1 shows the adsorption capacities obtained for some adsorbents by various authors in the last few years. In some studies, equilibrium sorption data obtained were represented by the Langmuir isotherm suggesting homogeneous adsorption sites on the surfaces of the adsorbents. The apparent suggestion from these data explains that both the adsorbate concentration and the reactions of the solid–aqueous layer interface may be important in determining the capacity of adsorption.

TABLE 18.1 Adsorption Capacities and Adsorption Kinetic Parameters for Adsorbents.

Adsorbent	Contaminant removed	Model/adsorption capacity (mg/g)	Model/rate constant	Reference
Organokaolinite clay	Benzene, toluene, ethylbenzene, and xylene	Langmuir/396.42	Pseudo-second order/4500 g/mg/min	[13]
Bone char	p-nitrophenol	Langmuir/365.76	Pseudo-second order/2.5×10^{4} g/mg/min	[22]
Bone char	As(V)	Langmuir/0.335	Pseudo-second order/4.5×10^{-5} g/mg/min	[19]
Bone char	Co(II)	Langmuir/108.70	Pseudo-second order/0.032 g/mg/min	[20]
Unmodified kaolinite clay	Pb(II)	Langmuir/4.73	Pseudo-second order/0.032 g/mg/min	[55]
Modified kaolinite clay	Pb(II)	Langmuir/32.2	Pseudo-second order/1.08 g/mg/min	[55]
Tripolyphosphate-treated kaolinite clay	Pb^{2+} and Cd^{2+}	Langmuir/126.58 and 113.64, respectively	–	[52]
Sodium tetraborate-modified kaolinite clay	Pb^{2+} and Cd^{2+}	Langmuir/42.92 and 44.05, respectively	–	[56]

18.7 MECHANISM FOR UPTAKE OF CONTAMINANTS ON ADSORBENTS

Among the possible techniques for water treatments, the adsorption process by solid adsorbents shows potential as one of the most efficient methods for the treatment and removal of organic and inorganic contaminants in wastewater treatment.[60] Adsorption has advantage over the other methods because of its simple design and is recognized as an effective and low-cost technique for the removal of pollutants from water and wastewater. Adsorption is a physical phenomenon which permits the removal of contaminants from gases and liquids (fluid). In adsorption, molecules dissolved in a fluid preferentially accumulate at a solid surface. The cause of the preferential accumulation is thought to be weak physical and chemical bonds between the adsorbate and the adsorbent. Although the exact nature of these bonds is not known, the end result is that adsorbate molecules are in a lower energy state on the surface than they are in the fluid (liquid or gaseous). The adsorbed molecules distribute between the liquid phase (adsorbate) and the solid phase (adsorbent) when the adsorption process reaches equilibrium.[61] Adsorption can also be said to be the physical adhesion of chemicals onto the surface of a solid. The effectiveness of this solid is directly proportional to the surface area available for the reaction. In solution, as the adsorption process proceeds, the sorbed solute tends to desorb into the solution. Equal amounts of solute eventually are being adsorbed and desorbed simultaneously. The contaminants adsorbed can be reduced to as low a level as desired. The molecules of the contaminant are trapped and held by the internal surface of the adsorbent. Consequently, the rates of adsorption and desorption will attain an equilibrium state, called adsorption equilibrium.[62]

From the literature reports,[63–71] the related mechanisms for the uptake of contaminants include surface adsorption, partition, surface precipitation, and structural incorporation. Surface adsorption refers to the concentration of contaminants in solution adhering onto or near the surface or pores of the adsorbent, which includes physical adsorption (driven by London–van der Waals forces) and chemisorption (involving the formation of chemical bonds). The adsorption of hydrophobic organic contaminants onto the activated carbon has been shown to be a typical physical adsorption process, while the adsorption of heavy metal cations and oxyanions on metal hydroxides involve chemisorption.[64,65] Partition refers to the distribution of contaminants between two phases and this implies that the contaminants will penetrate into the entire network of a bulk phase rather than concentrate onto the surface of the adsorbent. Surface precipitation involves the formation of precipitates on the surface of the adsorbent which generally needs a relatively high concentration of cations and anions.[63,65,69] Finally, structural incorporation refers to incorporating ions into the solid phase of the adsorbent, for example, sequestration of metal cations into the crystal structure into the crystal structure of minerals by isomorphous substitution.[65] In all of these mechanisms, the sorption of such contaminants onto modified adsorbents usually depends on experimental conditions such as pH, adsorbate concentration, competing ions, and particle size.[70]

In adsorption equilibrium studies, the relationship between the amount adsorbed (q_e) and its equilibrium concentration in solution (C_e) is best described by adsorption isotherms usually derived from mathematical correlations and based on a set of assumptions related to the heterogeneity or homogeneity of the adsorbents, the type of coverage and the possibility of the interaction between the adsorbates.[71] The mathematical correlation is usually depicted graphically by expressing the solid phase against its residual concentration and the isotherms best describe how contaminants in solution interact with solid adsorbent materials which is a critical tool for optimizing the adsorption mechanism pathway, analyzing the surface properties and capacities of the adsorbents and the effective design of the adsorption system. Several two or three parameter models are common and published in the literature for the application in describing experimental data obtained using adsorption isotherms. Generally, curve fitting into $R^2 = 1.000$ (correlation coefficient) is used to indicate the "goodness of fit" of the experimental data with the model, and the equation parameters from each model are used to express the surface properties and affinity of the adsorbent. Various models are listed in Table 18.2.[67,71–73]

Characteristic adsorption properties such as surface property, the adsorption affinity of the adsorbent, and maximum adsorption capacity are usually determined from the adsorption isotherm and correlative constants. The information derived from these isotherms is the most important aspect when determining the adsorption mechanism of an adsorption process.

Kinetics of the adsorption process is also applied when investigating the adsorption mechanism and rate-controlling steps such as mass transport and chemical reaction processes. The rate-limiting step in the adsorption process is the key factor to be determined in order to understand the overall mechanism. Common kinetic models in the literature are given in Table 18.3.[20,28,73,74]

TABLE 18.2 Lists of Common Adsorption Isotherm Models and Equations.

Isotherm model/type	Model equation		Graphical plot (model parameters)
	Nonlinear form	Linear form	
Langmuir/two-parameter model	$q_e = \frac{q_{max} K_L C_e}{1 + K_L C_e}$	$\frac{C_e}{q_e} = \frac{C_e}{q_{max}} + \frac{1}{q_{max} K_L}$	C_e/q_e versus C_e ($q_{max,}$ K_L)
		$\frac{1}{q_e} = \frac{1}{q_{max}} + \frac{1}{q_{max} K_L C_e}$	$1/q_e$ versus $1/C_e$
		$q_e = q_{max} - \frac{q_e}{K_L C_e}$	q_e *versus* $\frac{q_e}{K_L C_e}$
		$\frac{q_e}{C_e} = K_L q_{max} - K_L q_e$	$\frac{q_e}{C_e}$ versus q_e
Freundlich/two-parameter model	$q_e = K_F C_e^{\frac{1}{n}}$	$\log q_e = \log K_F + \frac{1}{n} \log C_e$	$\log q_e$ versus $\log C_e$ (K_F, n)
Sips/three-parameter model	$q_e = \frac{q_{max}\ b C_e^{\frac{1}{n}}}{1 + b C_e^{\frac{1}{n}}}$	$\frac{1}{n} \ln C_e = -\ln\left(\frac{q_{max}}{q_e}\right) + \ln b$	$\ln \frac{q_{max}}{q_e}$ versus $\ln C_e$ (q_{max}, $b\ n$)
Temkin/two-parameter model	$q = \frac{RT}{b_T} \ln(a_T C_e)$	$q_e = +\frac{RT}{b} \ln C_e$	q_e versus $\ln C_e$ (a_T, b_T)
Redlich–Peterson/three-parameter model	$q_e = \frac{K_R\ C_e}{1 + a_R C_e^{\hat{a}}}$	$\ln\left(K_R \frac{C_e}{q_e} - 1\right) = \beta \ln(C_e) + \ln(a_R)$	$\ln\left(K_R \frac{C_e}{q_e} - 1\right)$ versus $In\ (C_e)$ (K_R, α_R, β)

TABLE 18.2 *(Continued)*

Isotherm model/type	Model equation		Graphical plot (model parameters)
	Nonlinear form	Linear form	
Dubinin–Radushkevich/two-parameter model	$q_e = q_D\, exp(-B_D\left[RT\, In\left(1+\frac{1}{C_e}\right)\right]^2$	$In\, q_e = \ln q_D - B_D\left[RT\, ln\left(1+\frac{1}{C_e}\right)\right]^2$	$In\, q_e$ versus $\left[\ln\left(1+\frac{1}{C_e}\right)\right]^2$ (q_D, B_D)
Toth/three-parameter model	$q_e = \frac{K_T\, C_e}{(a_T + C_e)^{\frac{1}{t}}}$	$In\frac{q_e}{K_T} = \ln(C_e) - \frac{1}{t}\ln(a_T + C_e)$	$\ln\frac{q_e}{K_T}$ versus $\ln(C_e)$ (K_T, a_T)

TABLE 18.3 Common Kinetic Models.

Kinetic model	Model equation		Linear plot	Kinetic constants
	General form	Integrated form		
Lagergren's pseudo first order	$\frac{d_q}{d_t}=k_1(q_e-q_t)$	$log(q_e-q_t)=log\ q_e-\left(\frac{k_1}{2.303}\right)t$	$\log(q_e-q_t)$ versus t	k_1
Ho and McKay's pseudo second order	$\frac{d_q}{d_t}=k_2\ (q_e-q_t)^2$	$\frac{t}{q_t}=\frac{1}{k_2 q_e^2}+\frac{1}{q_e}t$	$\frac{t}{q_t}$ versus t	k_2
Intraparticle diffusion	$q_t\ k_{id}\ t^{\frac{1}{2}}+c$	–	q_t versus $t^{\frac{1}{2}}$	k_{id}
Elovich	$\frac{dq_t}{dt}=\alpha e^{\beta}\ q_t$	$q_t=\frac{1}{\beta}ln(\alpha\beta)+\frac{1}{\beta}\ln t$	q_t versus $\ln t$	α, β

18.8 CONCLUSION

The use of low-cost materials as adsorbents has emerged in the last decade as one of the most promising alternatives to conventional treatment technologies for the removal of organic and inorganic pollutants from wastewater. The widespread uses of low-cost adsorbents in industries for wastewater treatment applications today are strongly recommended due to their local availability, technical feasibility, engineering applicability, and cost-effectiveness. Industrial wastewaters contain many contaminating organic and inorganic materials such as phenolic compounds, heavy metals, and dyes. These common substances are being generated from various industrial processes and should be reduced to the maximum permissible concentration. The removal of these organic pollutants from contaminated wastewater and water bodies is critical to ensuring the safety of water supplies all over the world. A large number of studies in the literature show that a large number of alternative adsorbents from low-cost materials have been studied to replace activated carbons; however, efforts are still geared toward producing these adsorbents on a larger scale for as they show potential in removing contaminants from wastewater. A growing exploitation to evaluate the feasibility and suitability of natural, renewable, and low-cost materials as alternative adsorbents in pollution control studies should be intensified by regulatory bodies.

KEYWORDS

- **low cost adsorbents**
- **contaminants**
- **adsorption**
- **adsorption capacities**

REFERENCES

1. Park, S.-J.; Jang Y.-S. Pore Structure and Surface Properties of Chemically Modified Activated Carbons for Adsorption Mechanism and Rate of Cr (IV). *J. Colloid Interface Sci.* **2002,** *249*, 458–463.

2. Banat, F.; Al-Asheh, S.; Mohai, F. Batch Zinc Removal from Aqueous Solution Using Dried Animal Bones. *Sep. Purif. Technol.* **2000,** *21*, 155–164.
3. Alvarez-Merino, M. A.; Lopez-Ramon, V.; Moreno-Castilla, C. A Study of the Static and Dynamic Adsorption of Zn(II) Ions on Carbon Materials from Aqueous Solutions. *J. Colloid Interface Sci.* **2005,** *288*, 335–341.
4. Babel, S.; Kurniawan, T. A. Low-cost Adsorbents for Heavy Metals Uptake from Contaminated Water: A Review. *J. Hazard Mater.* **2003,** *B97*, 219–243.
5. Sharmasarkar, S.; Jaynes, W. F.; Vance, G. F. BTEX Sorption by Montmorillonite Organoclays: TMPA, ADAM, HDTMA. *Water, Air Soil Pollut.* **2000,** *119*, 257–273.
6. Nourmoradi, H.; Nikaeen, M.; Khiadani, M. Removal of Benzene, Toluene, Ethylbenzene and Xylene (BTEX) from Aqueous Solutions by Montmorillonite Modified with Non-ionic Surfactant: Equilibrium, Kinetic and Thermodynamic Study. *Chem. Eng. J.* **2012,** *191*, 341–348.
7. Park, Y.; Ayoko, G. A.; Frost, R. L. Characterization of Organoclays and Adsorption of p-nitrophenol: Environmental Application. *J. Colloid Interface Sci.* **2011,** *360*, 440–456.
8. Fafard, J.; LyubiMova, O.; Stoyanov, S. R.; Dedzo, G. K.; Gusarov, S.; Kovalenko, A.; Detellier, C. Adsorption of Indole on Kaolinite in Nonaqueous Media: Organoclay Preparation and Characterization, and 3D-RISM-KH Molecular Theory of Solvation Investigation. *J. Phys. Chem. C* **2013,** *117*, 18556–18566.
9. Barreto, E. P.; Lemos, M. S.; Aranha, I. B.; Buchler, P. M.; Dweck, J. Partially Exchanged Organophilic Bentonites Part II: Phenol Adsorption. *J. Therm. Anal. Calorim.* **2011,** *105*, 915–920.
10. Koh, S. M.; Dixon, J. B. Preparation and Application of Organominerals as Sorbents of Phenol, Benzene and Toluene. *Appl. Clay Sci.* **2001,** *18*(3), 111–122.
11. Xi, Y.; Zhou, Q.; Frost, R. L.; He, H. Thermal Stability of Octadecyltrimethylammonium Bromide Modified Montmorillonite Organoclay. *J. Colloid Interface Sci.* **2007,** *311*, 347–353.
12. Park, Y.; Ayoko, G. A.; Horvath, E.; Kurdi, R.; Kristof, J.; Frost, R. L. Structural Characterization and Environmental Application of Organoclays for the Removal of Phenolic Compounds. *J. Colloid Interface Sci.* **2013,** *393*, 319–330.
13. Egbuchunam, T. O.; Obi, G.; Okieimen, F. E.; Tihminliogluc, F. Removal of BTEX from Aqueous Solution Using Organokaolinite. *Int. J. Appl. Environ. Sci.* **2016,** *11*(2), 505–513.
14. Li, Q.; Zhai, J.; Zhang, W.; Wang, M.; Zhou, J. Kinetic Studies of Adsorption of Pb(II), Cr(III) and Cu(II) from Aqueous Solution by Sawdust and Modified Peanut Husk. *Hazard. Mater.* **2007,** *141*(1), 163–167.
15. Garg, U. K.; Kaur, M. P.; Garg, V. K.; Sud, D. Removal of Hexavalent Chromium from Aqueous Solution by Agricultural Waste Biomass. *J. Hazard Mater.* **2007,** *140*(1–2), 60–68.
16. Tsai, W. T.; Hsein, K. J.; Hsu, H. C.; Lin, K. Y.; Chiu, C. H. Utilization of Ground Eggshell Waste as an Adsorbent for the Removal of Dyes from Aqueous Solution. *Bioresour. Technol.* **2008,** *99*(6), 1623–1629.
17. Chen, S.-B.; Zhu, Y.-C.; Ma, Y.-B.; McKay, G. Effect of Bone Char Application on Pb Bioavailability on a Pb-contaminated Soil. *Environ. Pollut.* **2006,** *139*(3), 433–439.
18. Choy, K. K. H.; McKay, G. Sorption of Cadmium, Copper and Zinc Ions onto Bone Char Using Crank Diffusion Model. *Chemosphere* **2005,** *60*(8), 1141–1150.

19. Liu, J.; Huang, X.; Liu, J.; Wang, W.; Zhang, W.; Dong, F. Adsorption of Arsenic (V) on Bone Char: Batch, Column and Modelling Studies. *Environ. Earth Sci.* **2014,** *72*, 2081–2090.
20. Pan, X.; Wang, J.; Zhang, D. Sorption of Cobalt to Bone Char: Kinetics, Competitive Sorption and Mechanism. *Desalination* 2009, 249, 609–614.
21. Cheng, C.; Porter, J.; McKay, G. Sorption Kinetic Analysis for the Removal of Cadmium Ions from Effluents Using Bone Char. *Water Res.* **2001,** *35*, 605–612.
22. Egbuchunam, T. O.; Obi, G.; Okieimen, F. E.; Yetgin, S. Adsorptive Removal of p-nitrophenol from Aqueous Solutions by Bone Char: Equilibrium and Kinetic Studies. *J. Chem. Chem. Eng.* **2016,** *10*, 325–335.
23. McLauchlin, A. R.; Thomas, N. L. Preparation and Characterization of Organoclays Based on an Amphoteric Surfactant. *J. Colloid Interface Sci.* **2008,** *321*, 39–43.
24. Kooli, F.; Yan, L.; Tan, S. X.; Zheng, J. Organoclays from Alkaline-Treated Acid-Activated Clays: Properties and Thermal Stability. *J. Therm. Anal. Calorim.* **2014,** *115*, 1465–1475.
25. Park, Y.; Ayoko, G. A.; Frost, R. L. Application of Organoclays for the Adsorption of Recalcitrant Organic Molecules from Aqueous Media. *J. Colloid Interface Sci.* **2011,** *354*, 292–305.
26. Park Y; Ayoko GA; Frost RL; Characterization of Organoclays and Adsorption of p-nitrophenol: Environmental Application. *J. Colloid Interface Sci.* **2011,** *360*, 440–456.
27. Egbuchunam, T. O.; Obi, G.; Okieimen, F. E.; Tihminliogluc, F. Effect of Exchanged Surfactant Cation on the Structure of Kaolinitic Clay. *J. Mater. Sci. Appl.* **2015,** *2015*, 1–14.
28. Alkaram, U. F.; Mukhlis, A. A.; Al-Dujaili, A. H. The Removal of Phenol from Aqueous Solutions by Adsorption Using Surfactant Modified Bentonite and Kaolinite. *J. Hazard Mater.* **2009,** *169*, 324–332.
29. Smith, J. A.; Bartlett-Hunt, S. L.; Burns, S. E. Sorption and Permeability of Gasoline Hydrocarbons in Organobentonites Porous Media. *J. Hazard Mater.* **2003,** *B96*, 91–97.
30. Zhu, R.; Chen, Q.; Zhou, Q.; Xi, Y.; Zhu, J.; He, H. Adsorbents Based on Montmorillonite for Contaminant Removal from Water: A Review. *Appl. Clay Sci.* **2016,** *123*, 239–258.
31. Wei, J.; Zhu, R.; Zhu, J.; Ge, F.; Yuan, P.; He, H.; Ming, C. Simultaneous Sorption of Crystal Violet and 2-Naphthol to Bentonite with Different CECs. *J. Hazard Mater.* **2009,** *166*, 195–199.
32. Xi, Y.; Ding, Z.; Frost, R. L. Structure of Organoclays—An X-ray Diffraction and Thermogravimetric Analysis Study. *J. Colloid Interface Sci.* **2004,** *277*(1), 116–120.
33. Moraru, V. N. M. Structure Formation of Alkylammonium Montmorillonites in Organic Media. *Appl. Clay Sci.* **2001,** *19*, 11–26.
34. Zhu, J.; He, H.; Zhu, I.; Wen, X.; Deng, F. Characterization of Organic Phases in the Interlayer of Montmorillonite Using FTIR and 13C NMR. *J. Colloid Interface Sci.* **2005,** *286*, 239–244.
35. Ko, D. C. K.; Cheung, C. W.; Choy, K. K. H.; Porter, J. F.; McKay, G. Sorption Equilibria of Metal Ions on Bone Char. *Chemosphere* **2004,** *54*, 273–281.
36. Deydier, E.; Guilet, R. Sharrock, P. Beneficial Use of Meat and Bone Meal Combustion Residue: An Efficient Low Cost Material to Remove Lead from Aqueous Effluent. *J. Hazard Mater.* **2003,** *101*, 55–64.
37. Gomez del Rio, J. A.; Morando, P. J.; Cicerone, D. S. Natural Materials for Treatment of Industrial Effluents: Comparative Study of the Retention of Cd, Zn and Co

by Calcite and Hydroxyapatite Part 1: Batch Experiments. *J. Environ. Manage.* **2004,** *71*, 169–177.
38. Young, J. L.; Evert, J. E.; Richard, J. R. Sorption Mechanisms of Zinc on Hydroxyapatite: Systematic Uptake Studies on EXAFS Spectroscopy Analysis. *Environ. Sci. Technol.* **2005,** *39*, 4042–4048.
39. Sheha, R. R. Sorption Behaviour of Zn(II) Ions on Synthesized Hydroxyapatite. *J. Colloid Interface Sci.* **2007,** *310*, 18–26.
40. Chaturvedi, P. K.; Seth, C. S.; Misra, V. Sorption Kinetics and Leachability of Heavy Metal from Contaminated Soil Amended with Immobilizing Agent (Humus Soil and Hydroxyapatite). *Chemosphere* **2006,** *64*, 1109–1114.
41. Liu, J.; Huang, X.; Liu, J.; Wang, W.; Zhang, W.; Dong, F. Adsorption of Arsenic (V) on Bone Char: Batch, Column and Modelling Studies. *Environ. Earth Sci.* **2014,** *72*, 2081–2090.
42. Chen, S. B.; Ma, Y. B.; Chen, L.; Xian, K. Adsorption of Aqueous Cd^{2+}, Pb^{2+} and Cu^{2+} Ions by Nano-Hydroxyapatite: Single- and Multi-Metal Competitive Adsorption Study. *Geochem. J.* **2010,** *44*, 233–239.
43. Leyva, A. G.; Marrero, J.; Smichowski, P.; Cicerone, D. Sorption of Antimony onto Hydroxyapatite. *Environ. Sci. Technol.* **2001,** *35*, 3669–3675.
44. Peld, M.; Tonsuaadu, K.; Bender, V. Sorption and Desorption of Cd^{2+} and Zn^{2+} Ions in Apatite-Aqueous Systems. *Environ. Sci. Technol.* **2004,** *38*, 5626–5631.
45. Corami, A.; Ferrini, V.; Mignardi, S.; Synthetic Phosphates as Binding Agent of Pb, Zn, Cu and Cd in the Environment. *World Res. Rev.* **2005,** *17*, 121–135.
46. Chen, Z. S. Hseu, Z. Y. In Situ Immobilization of Cadmium and Lead by Different Amendments in Two Contaminated Soils. *Water Air Soil Pollut.* **2002,** *140*, 73–84.
47. Vega, E. D.; Pedregosa, J. C.; Narda, G. E.; Morando, P. J. Removal of Oxovanadium (IV) from Aqueous Solutions by Using Commercial Crystalline Calcium Hydroxyapatite. *Water Res.* **2003,** *37*, 1776–1782.
48. Yantesee, W.; Warner, C. I.; Sangvanich, T.; Addleman, R. S.; Carter, T. G.; Wiacek, R. J.; Fryxell, G. E.; Timchalk, C.; Warner, M. G. Removal of Heavy Metals from Aqueous Systems with Thiol Functionalized Superparamagnetic Nanoparticles. *Environ. Sci. Technol.* **2007,** *41*, 5114–5119.
49. Feng, V.; Gong, J.-L.; Zeng, G.-M.; Niu, Q.-Y.; Zhang, H.-Y.; Niu, C.-G.; Deng, J.-H.; Yan, M. Adsorption of Cd(II) and Zn(II) from Aqueous Solutions Using Magnetic Hydroxyapatite Nanoparticles as Adsorbents. *Chem. Eng. J.* **2010,** *162*, 487–494.
50. Aivalioti, M.; Pothoulaki, D.; Papoulias, P.; Gidarakos, E. Removal of BTEX, MTBE, and TAME from Aqueous Solutions by Adsorption onto Raw and Thermally Treated Lignite. *J. Hazard Mater.* **2012,** 207–208, 136–142.
51. Bhattacharyya, K. G. Gupta, S. S. Adsorption of a Few Heavy Metals on Natural and Modified Kaolinite and Montmorillonite: A Review. *Adv. Colloid Interface Sci.* **2008,** *140*, 114–131.
52. Unuabonah, E. I.; Olu-Owolabi, B. I.; Adebowale, K. O.; Ofomaja, A. E. Adsorption of Lead and Cadmium Ions from Aqueous Solutions by Tripolyphosphate-Impregnated Kaolinite Clay. *Colloids Surf. A: Physicochem. Eng. Aspects* **2007,** *292*, 202–211.
53. Adebowale, K. O.; Olu-Owolabi, B. I.; Unuabonah, E. I. The Effect of Some Operating Variables on the Adsorption of Lead and Cadmium Ions on Kaolinite Clay. *J. Hazard Mater.* **2006,** *B134*, 130–139.

54. Gupta, S. S.; Bhattacharyya, K. G. Immobilization of Pb (II), Cd (II) and Ni (II) ions on Kaolinite and Montmorillonite Surfaces from Aqueous Medium. *J. Environ. Manage.* **2008,** *87*, 46–58.
55. Jiang, M.-Q.; Wang, Q.-P.; Jin, X.-Y.; Chen, Z.-L. Removal of Pb(II) from Aqueous Solution Using Modified and Unmodified Kaolinite Clay. *J. Hazard Mater.* **2009,** *170*, 332–339.
56. Unuabonah, E. I.; Adebowale, K. O.; Olu-Owolabi, B. I.; Yang, L. Z.; Kong, L. X. Adsorption of Pb(II) and Cd(II) from Aqueous Solutions onto Sodium Tetrabromate-modified Kaolinite Clay: Equilibrium and Thermogravimetric Studies. *Hydrometallurgy* **2008,** *93*, 1–9.
57. Leyva-Ramos, R.; Rivera-Utrilla, J.; Medellin-Castillo, N. A. Kinetic Modeling of fluoride Adsorption from Aqueous Solution onto Bone Char. *Chem. Eng. J.* **2010,** *158*, 458–467.
58. Rojas-Mayorga, C. K.; Mendoza-Castillo, D. I.; Bonilla-Petriciolet, A.; Silvestre-Albero, J. Tailoring the Adsorption Behaviour of Bone Char for Heavy Metal Removal from Aqueous Solution. *Adsorpt. Sci. Technol.* **2016,** *4*(6), 368–387.
59. Cheng, C.; Porter, J.; McKay, G. Sorption Kinetic Analysis for the Removal of Copper and Zinc from Effluents Using Bone Char. *Sep. Purif. Technol.* **2000,** *19*, 55–64.
60. Carvalho, M. N.; Da Motta, M.; Benachour, M.; Sales, D. C. S. Abreu, C. A. M. Evaluation of BTEX and Phenol Removal from Aqueous Solution by Multi-solute Adsorption onto Smectite Organoclay. J. *Hazard Mater.* **2012,** *239–240*, 95–101.
61. Sawyer, C. N.; McCarty, P. L.; *Chemistry for Environmental Engineering,* 3rd ed.; McGraw-Hill: Singapore, 1978.
62. Faust, S. D.; Aly, O. M. *Adsorption Processes for Water Treatment;* Butterworth Publishers: Heinemann, 1987.
63. Waychunas, G. A.; Fuller, C. C.; Davis, J. A. Surface Complexation and Precipitate Geometry for Aqueous Zn(II) Sorption on Ferrihydrite I: X-ray Absorption Extended Fine Structure Analysis. *Geochim. Cosmochim. Acta* **2002,** *66*, 1119–1137.
64. Li, W.; Feng, J.; Kwon, K. D.; Kubicki, J. D.; Phillips, B. L. Surface Speciation of Phosphate on Boehmite (γ–AlOOH) Determined from NMR Spectroscopy. *Langmuir* **2010,** *26*, 4753–4761.
65. O'Day, P. A. Vlassopoulos, D. Mineral-based Amendments for Remediation. *Element* **2010,** *6*, 375–381.
66. Liu, J.; Zhu, R.; Xu, T.; Xu, Y.; Ge, F.; Xi, Y.; Zhu, J.; He, H. Co-adsorption of Phosphate and Zinc(II) on the Surface of Ferrihydrite. *Chemosphere* **2016,** *144*, 1148–1155.
67. Malamis, S.; Katsou, E. A Review on Zinc and Nickel Adsorption on Natural and Modified Zeolite, Bentonite and Vermiculite: Examination of Process Parameters, Kinetics and Isotherms. *J. Hazard Mater.* 2013, 252–253, 428–461.
68. Ler, A.; Stanforth, R. Evidence for Surface Precipitation of Phosphate on Goethite. *Environ. Sci. Technol.* **2003,** *37*, 2644–2700
69. Xie, J.; Wang, Z.; Lu, S. Y.; Wu, D. Y.; Zhang, Z. J.; Kong, H. N. Removal and Recovery of Phosphate from Water by Lanthanum Hydroxide Materials. *Chem. Eng. J.* **2014,** *254*, 163–170.
70. Schlegel, M. L. Manceau, A. Zn incorporation in hydroxyl-Al- and Keggin Al13-Intercalated Montmorillonite: A Powder and Polarized EXAFS Study. *Environ. Sci. Technol.* **2007,** *41*, 1942–1948.
71. Foo, K. Y. Hameed, B. H. Insights into the Modeling of Adsorption Isotherm Systems. *Chem. Eng. J.* **2010,** *156*, 2–10.

72. Hamdaoui, O.; Naffrechoux, E. Modeling of Adsorption Isotherms of Phenol and Chlorophenols onto Granular Activated Carbon Part II: Models with More than Two Parameters. *J. Hazard Mater.* **2007,** *147*, 401–411.
73. Varank, G.; Demir, A.; Yetilmezso, K.; Top, S.; Sekman, E.; Bilgili, M. S. Removal of 4-nitrophenol from Aqueous Solution by Natural Low-Cost Adsorbents. *Ind. J. Chem. Technol.* **2012,** *19*, 7–25.
74. Ho, Y. S.; McKay, G. Pseudo-second Order Model for Sorption Processes. *Process Biochem.* **1999,** *34*, 451–465.

INDEX

D

F

Y

Z